T0318984

COST EFFECTIVE TECHNOLOGIES FOR SOLID WASTE AND WASTEWATER TREATMENT

Advances in Environmental Pollution Research Series

COST EFFECTIVE TECHNOLOGIES FOR SOLID WASTE AND WASTEWATER TREATMENT

Edited by

Srujana Kathi
ICSSR Postdoctoral Research Fellow, Department of Applied Psychology, Pondicherry University, Puducherry, India

Suja Devipriya
Associate Professor, School of Environmental Studies, Cochin University of Science and Technology, Cochin, India

Kaliannan Thamaraiselvi
Professor of Environmental Biotechnology at Bharathidasan University, Tiruchirappalli, India

Elsevier
Radarweg 29, PO Box 211, 1000 AE Amsterdam, Netherlands
The Boulevard, Langford Lane, Kidlington, Oxford OX5 1GB, United Kingdom
50 Hampshire Street, 5th Floor, Cambridge, MA 02139, United States

Library of Congress Cataloging-in-Publication Data
A catalog record for this book is available from the Library of Congress

British Library Cataloguing-in-Publication Data
A catalogue record for this book is available from the British Library

ISBN: 978-0-12-822933-0

For information on all Elsevier publications
visit our website at https://www.elsevier.com/books-and-journals

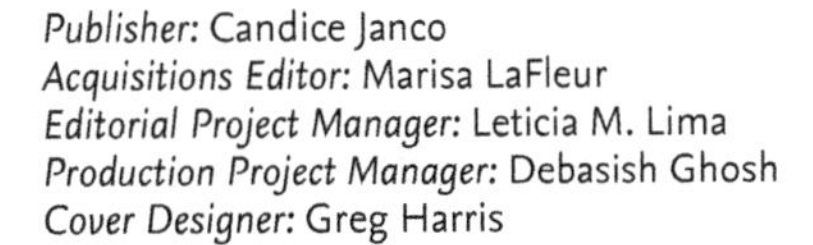

Publisher: Candice Janco
Acquisitions Editor: Marisa LaFleur
Editorial Project Manager: Leticia M. Lima
Production Project Manager: Debasish Ghosh
Cover Designer: Greg Harris

Typeset by STRAIVE, India

Contents

7. Assessment of biochar application in decontamination of water and wastewater

ALAA EL DIN MAHMOUD AND SRUJANA KATHI

8. In situ chemical oxidation (ISCO) remediation: A focus on activated persulfate oxidation of pesticide-contaminated soil and groundwater

RAMA MOHAN KURAKALVA

9. Composting of food waste: A novel approach

M. ANAND, T.B. ANJALI, K.B. AKHILESH, AND JOHN K. SATHEESH

10. Biological pretreatment for enhancement of biogas production

AISHIKI BANERJEE, BINOY KUMAR SHOW, SHIBANI CHAUDHURY, AND S. BALACHANDRAN

11. Recent trends in bioremediation of pollutants by enzymatic approaches

M. SRINIVASULU, M. SUBHOSH CHANDRA, NAGA RAJU MADDELA, NARASIMHA GOLLA, AND BELLAMKONDA RAMESH

12. Phytoremediation of heavy metals and petroleum hydrocarbons using *Cynodon dactylon (L.) Pers.*

SRUJANA KATHI

13. Proteobacteria response to heavy metal pollution stress and their bioremediation potential

VERONICA FABIAN NYOYOKO

14. Treatment of harvested rainwater and reuse: Practices, prospects, and challenges

15. Phytoremediation: A wonderful cost-effective tool

16. Vermicomposting: An efficient technology for the stabilization and bioremediation of pulp and paper mill sludge

17. Potential of solid waste prevention and minimization strategies

18. Nanoremediation of pollutants: A conspectus of heavy metals degradation by nanomaterials

19. Excess fluoride issues and mitigation using low-cost techniques from groundwater: A review

20. Cost-effective biogenic-production of inorganic nanoparticles, characterizations, and their antimicrobial properties

Contributors

K.B. Akhilesh School of Environmental Studies, Cochin University of Science and Technology, Cochin, Kerala, India

Esayas Alemayehu African Centre of Excellence for Water Management, Addis Ababa University, Addis Ababa; Faculty of Civil & Environmental Engineering, Jimma Institute of Technology, Jimma University, Jimma, Ethiopia

M. Anand School of Environmental Studies, Cochin University of Science and Technology, Cochin, Kerala, India

Munisamy Anbarashan Department of Ecology, French Institute of Pondicherry, Puducherry, India

T.B. Anjali School of Environmental Studies, Cochin University of Science and Technology, Cochin, Kerala, India

Michael Attia Irrigation Engineering and Hydraulics Department, Faculty of Engineering, Alexandria University, Alexandria, Egypt

S. Balachandran Department of Environmental Studies, Siksha-Bhavana, Visva-Bharati, Santiniketan, India

Aishiki Banerjee Department of Environmental Studies, Siksha-Bhavana, Visva-Bharati, Santiniketan, India

Shaileshkumar Baskaran School of Bio Sciences and Technology, Vellore Institute of Technology (VIT), Vellore, India

M. Subhosh Chandra Department of Microbiology, Yogi Vemana University, Kadapa, Andhra Pradesh, India

Shibani Chaudhury Department of Environmental Studies, Siksha-Bhavana, Visva-Bharati, Santiniketan, India

Suja Purushothaman Devipriya School of Environmental Studies, Cochin University of Science and Technology, Cochin, Kerala, India

Hani Ezz Environmental Engineering Department, Egypt-Japan University of Science and Technology (E-JUST), Alexandria, Egypt

A.J. Francis Division of Advanced Nuclear Engineering, Pohang University of Science and Technology (POSTECH), Pohang, South Korea; Environmental and Climate Sciences Department, Brookhaven National Laboratory, Upton, New York, United States

Narasimha Golla Applied Microbiology Laboratory, Department of Virology, Sri Venkateswara University, Tirupati, Andhra Pradesh, India

Mona G. Ibrahim Environmental Engineering Department, Egypt-Japan University of Science and Technology (E-JUST); Environmental Health Department, High Institute of Public Health, Alexandria University, Alexandria, Egypt

Suh Jeong-Min Department of Bioenvironmental Energy, College of Natural Resources and Life Science, Pusan National University, Miryang, Gyeongsangnam-do, Republic of Korea

Kishore Kumar Kadimpati Department of Pharmaceutical Biotechnology, Narayana Pharmacy College, Nellore, Andhra Pradesh, India

Adane Woldemedhin Kalsido African Centre of Excellence for Water Management, Addis Ababa University, Addis Ababa; Department of Hydraulic & Water Resources Engineering, College of Engineering and Technology, Wachemo University, Hossana, Ethiopia

Madhuraj Palat Kannankai School of Environmental Studies, Cochin University of Science and Technology, Cochin, Kerala, India

Srujana Kathi ICSSR Postdoctoral Research Fellow, Department of Applied Psychology, Pondicherry University, Puducherry, India

Abhishek Kumar Soil Ecosystem and Restoration Ecology Lab, Department of Botany, Panjab University, Chandigarh, India

Rama Mohan Kurakalva Hydrogeochemistry Group, CSIR-National Geophysical Research Institute (CSIR-NGRI), Hyderabad, Telangana; Faculty of Physical Sciences, Academy of Scientific & Innovative Research (AcSIR), Ghaziabad, India

Thyagarajan Lakshmi Priya Department of Civil Engineering, Environmental Engineering Division, Government College of Technology, Coimbatore, India

Naga Raju Maddela Departamento de Ciencias Biológicas, Facultad de Ciencias de la Salud, Universidad Técnica de Manabí, Portoviejo, Manabí, Ecuador

Alaa El Din Mahmoud Environmental Sciences Department, Faculty of Science; Green Technology Group, Faculty of Science, Alexandria University, Alexandria, Egypt

Beshah Mogessie African Centre of Excellence for Water Management, Addis Ababa University; Water Development Commission, Ministry of Water, Irrigation, and Energy, Addis Ababa, Ethiopia

Azhaguchamy Muthukumaran Department of Biotechnology, Kalasalingam Academy of Research and Education, Krishnankoil, Tamil Nadu, India

NT Nandhini School of Bio Sciences and Technology, Vellore Institute of Technology (VIT), Vellore, India

G. Narasimha Applied Microbiology Laboratory, Department of Virology, Sri Venkateswara University, Tirupati, Andhra Pradesh, India

Mahmoud Nasr Sanitary Engineering Department, Faculty of Engineering, Alexandria University; Environmental Engineering Department, Egypt-Japan University of Science and Technology (E-JUST), Alexandria, Egypt

Veronica Fabian Nyoyoko Department of Microbiology, University of Nigeria, Nsukka, Nigeria

Anbarashan Padmavathy Department of Ecology and Environmental Sciences, Pondicherry University, Puducherry, India

Arunkumar Patchaiyappan Department of Ecology and Environmental Sciences, School of Life Sciences, Pondicherry University; Department of Social Sciences, French Institute of Pondicherry, Pondicherry, India

Duraisamy Prabha Department of Environmental Sciences, Bharathiar University, Coimbatore, Tamil Nadu, India

Yi Pyoung-In Department of Bioenvironmental Energy, College of Natural Resources and Life Science, Pusan National University, Miryang, Gyeongsangnam-do, Republic of Korea

Bellamkonda Ramesh Department of Food Technology, Vikrama Simhapuri University, Nellore, Andhra Pradesh, India

John K. Satheesh School of Environmental Studies, Cochin University of Science and Technology, Cochin, Kerala, India

Mythili Sathiavelu School of Bio Sciences and Technology, Vellore Institute of Technology (VIT), Vellore, India

Jang Seong-Ho Department of Bioenvironmental Energy, College of Natural Resources and Life Science, Pusan National University, Miryang, Gyeongsangnam-do, Republic of Korea

Binoy Kumar Show Department of Environmental Studies, Siksha-Bhavana, Visva-Bharati, Santiniketan, India

Anand Narain Singh Soil Ecosystem and Restoration Ecology Lab, Department of Botany, Panjab University, Chandigarh, India

Siril Singh Department of Environment Studies; Soil Ecosystem and Restoration Ecology Lab, Department of Botany, Panjab University, Chandigarh, India

Subpiramaniyam Sivakumar Department of Bioenvironmental Energy, College of Natural Resources and Life Science, Pusan National University, Miryang, Gyeongsangnam-do, Republic of Korea

M. Srinivasulu Division of Advanced Nuclear Engineering, Pohang University of Science and Technology (POSTECH), Pohang, South Korea; Department of Microbiology, Sri Krishnadevaraya University, Anantapuramu; Department of Biotechnology, Yogi Vemana University, Kadapa, Andhra Pradesh, India

Hong Sung-Chul Department of Bioenvironmental Energy, College of Natural Resources and Life Science, Pusan National University, Miryang, Gyeongsangnam-do, Republic of Korea

Beteley Tekola African Centre of Excellence for Water Management; School of Chemical and Bio Engineering, Addis Ababa University, Addis Ababa, Ethiopia

Kaliannan Thamaraiselvi Professor of Environmental Biotechnology at Bharathidasan University, Tiruchirappalli, India

Rajeswari Uppala Department of Biotechnology, Kalasalingam Academy of Research and Education, Krishnankoil, Tamil Nadu, India

Rajni Yadav Soil Ecosystem and Restoration Ecology Lab, Department of Botany, Panjab University, Chandigarh, India

Perspectives and Foreword

It is a great pleasure for me to write a foreword for the book *Cost-Effective Technologies for Solid Waste and Wastewater Treatment*, which deals with problems associated with the environment and their restoration using low-cost treatment techniques and methods.

The factors disturbing the environment are assorted, and methods to find solutions are frequently well associated with current or traditional approaches to improve the system. In recent years, environmental issues with respect to liquid and solid wastes are of major concern to the society, public authorities, researchers, and industries. Most of the domestic and industrial activities release solid and liquid wastes into the ecosystem, which contains detrimental levels of toxic pollutants. The traditional methods of treatment are inadequate because of the identification of new combinations of pollutants, population growth, and increasing industrial activities to increase economic growth. The conventional methods provide limited solution due to new challenges in technological advancement. Development and designing of new low-cost, efficient, and eco-friendly approaches of decontamination/treatment is an emerging research field. Protecting the environment from the consequences of solid and liquid wastes has become a main concern for the society, and it is a very difficult task to maintain the sustainability of a proper environment. In this context, the authors have made an excellent attempt to bring out the issues that can solve the problems associated with solid and liquid waste treatment. The important idea of this book is to carry forward the issues of environment and their treatment methods using proper approach/procedures.

The book comprises different chapters related to topics in various techniques, processes, and methodologies on waste treatment, which will be very much useful to the readers.

I congratulate the editors and contributors for having conceived this idea and bringing out a book that is the need of the hour to all the concerned.

Dr. K. Kadirvelu
Scientist F
Defence Bio-Engineering & Electromedical Laboratory (DEBEL)
Defence Research and Development Organisation (DRDO)
Ministry of Defence, Government of India
OiC DRDO—Bharathiar University Centre for Life Sciences, Coimbatore, Tamil Nadu, India

Foreword

Today, nearly eight billion of us inhabit this planet, producing waste every day, more than half of which is not collected, treated, or safely disposed of, resulting in a monumental problem for humankind. The goal of sustainable development, as envisaged in the Sustainable Development Goals (SDGs), cannot be achieved without recognizing the vital importance of this issue, learning to manage the waste we create responsibly, and endeavoring to create new value out of it, thereby attempting to outweigh the costs.

This book on waste management is therefore very topical, as it provides a comprehensive coverage of eco-friendly, low cost, and state-of-the-art technologies for removing a range of contaminants of emerging concern including organic matter, chemicals, and micropollutants, with minimal effects on the environment. The focus is on innovative techniques and appropriately modified conventional techniques to convert or reuse the wastes—solid wastes, sewage, and wastewater—generated by anthropogenic activities.

This book is a practical guide for researchers, policymakers, and environmental managers in waste management, environmental science and engineering, chemical engineering, and biotechnology, providing information not only on the designs of systems for treatment and reuse of wastes but also on their scale, efficiency, and effectiveness. The case studies illustrating specific techniques and methodologies for treatment and environmental cleanup, such as bioaugmentation and biostimulation, phytoremediation, bioremediation, biosorption, anaerobic digestion, and in situ chemical oxidation (ISCO), for contaminants such as plastics, radioactive pollutants, heavy metals, petroleum hydrocarbons, and paper and pulp sludge, are expected to be of interest to the waste management community.

Of particular interest are the chapters covering nanoremediation of heavy metals, biogenic production and evaluation of inorganic nanoparticles, downflow hanging sponge technology, application of biochar for wastewater treatment, ISCO for pesticide-contaminated soil, groundwater fluoride mitigation using low-cost techniques, constructed wetlands, and rainwater harvesting.

The sections related to solid waste management such as composting of food waste, single-cell protein production from solid waste, and biological enhancement for biogas production provide interesting and informative material for students.

I compliment the editors for conceiving this book and the contributors for sharing their experience in this topical area, through well-researched, in-depth, and eminent practical chapters. All in all, this is a book that is expected to be a single reference source for students and professionals alike to address existing needs as well as meet upcoming challenges in the waste management.

Dr. Chitra Rajagopal
Former DS and DG (R&M)
Defence Research and Development Organisation
(DRDO)

Preface

Due to the rapid population exploitation, increasing consumerism, swelling use of plastics, etc., management of waste—both liquid and solid—is emerging as a problem of concern. Appropriate management of solid and liquid wastes is very much essential to improve the habitat and sanitation of any community. If these wastes are not given adequate attention, the health and living environment of the community will be miserably affected. Numerous methods for solid and liquid waste management are available and are commonly summarized as 4Rs: Reduction, Reuse, Recycling, and Recovery. *Cost-Effective Technologies for Solid Waste and Wastewater Treatment* provides complete information on sustainable technologies for management and remediation of liquid and solid wastes.

The book includes a review of different recent techniques developed for economic management of waste water and solid wastes. The book focusses on case studies and efficient methods for removing contaminants released from industrial activities and reuse of wastes. It elaborates on cost-effective technologies applicable for the remediation of the ecosystem, such as bioaugmentation and biostimulation for plastic degradation, bioremediation of radiocesium, plant-based coagulants for water treatment, downflow hanging sponge technology for wastewater treatment, biochar for decontamination of water treatment, in situ chemical oxidation of pesticides, biogas production, enzymatic approach for remediation of pollutants, phytoremediation of heavy metals and hydrocarbons, microbial remediation of heavy metals, nanoremediation of pollutants, and fluoride mitigation. In addition, the book deals with utilization of waste for the production of single-cell protein, composting of food waste, vermicomposting of pulp and paper mill sludge, solid waste prevention and minimization strategies, and rainwater harvesting and reuse. The book will be an enlightening practical guide for researchers, students, and managers in the fields of environmental science and engineering, wastewater management and life sciences.

Kaliannan Thamaraiselvi
Editor

Abbreviations

AD	anaerobic digestion
AFM	atomic force microscopy
AgNPs	silver nanoparticles
AOPs	advanced oxidation process
API	analytical profile index
APO	activated persulfate oxidation
ASTM	American Society for Testing and Materials
ATR-FTIR	attenuated total reflection-Fourier transform infrared
AuNPs	gold nanoparticles
BATH	bacterial adherence to hydrocarbons
BHB	Bushnell-Haas Broth
BOD	biological oxygen demand
CE	circular economy
CEC	cation exchange capacity
COD	chemical oxygen demand
CTA	cerium-based adsorbent
CW	constructed wetland
DCIP	2,6-dichlorophenolindophenol
DD	donnan dialysis
DDT	dichloro diphenyl trichloroethane
DHS	downflow hanging sponge
DLS	dynamic light scattering
DMF	*N,N*-dimethylformamide
DNA	deoxyribo nucleic acid
DNAPL	dense nonaqueous phase liquid
DO	dissolved oxygen
DPPH	1-diphenyl-1-2-picrylhydrazyl
DSC	differential scanning calorimetry
ECs	emerging contaminants
ED	electro-dialysis
EPR	electron paramagnetic resonance
EPs	emerging pollutants
ESI-MS	electron spray ionization-mass spectroscopy
FADH	flavin adenine dinucleotide
FSCW	free surface constructed wetlands
FTIR	Fourier transform infrared spectroscopy
FW	food waste
GC-MS	gas chromatography
HDPE	high-density poly ethylene
HFCW	horizontal flow constructed wetlands
HT-GPC	high-temperature gel permeation chromatography
IC_{50}	inhibitory concentration
ISCO	in situ chemical oxidation
ISWM	integrated solid waste management
LDPE	low-density poly ethylene
LiP	lignin peroxidase
LLDPE	linear low-density poly ethylene
MALDI-TOF	matrix associated laser desorption ionization-time of flight
MATH	microbial adhesion to hydrocarbons assay
MBT	mechanical biological treatment
MIC	minimum inhibitory concentration
MNP	manganese-dependent peroxidase
MNPs	metal nanoparticles
MSW	municipal solid waste
NADH	nicotinamide adenine dinucleotide hydrogen
NADPH	nicotinamide adenine dinucleotide phosphate hydrogen
NAPL	nonaqueous phase liquid
NEERI	National Environmental Engineering Research Institute
NMR	nuclear magnetic resonance spectrometer
OECD	Organisation for Economic Co-operation and Development
PAH	polycyclic aromatic hydrocarbons
PCBs	polychlorinated biphenyls
PEG	polyethylene glycol
PET	poly ethylene terephthalate
PGPs	plant growth promoters
PHC	petroleum hydrocarbons
PPCPs	pharmaceutical and personal care products
Pt NPs	platinum nanoparticles
RDF	refuse-derived fuel
RO	reverse osmosis
SDG	Sustainable Development Goals
SEM	scanning electron microscopy
SERS	surface-enhanced Raman spectroscopy
TEM	transmission electron microscopy
TGA	thermo gravimetric analysis
TOC	total organic carbon
TOG	total organic grease
TPH	total petroleum hydrocarbon
UASB	upflow anaerobic sludge blanket reactor
USDA-ARS	United States Department of Agriculture-Agricultural Research Service
USEPA	United States Environmental Protection Agency
UTM	universal tensile meter

VFCW vertical flow constructed wetlands
VP versatile peroxidase
WTE waste to energy
XRD X-ray diffraction
ZnO NPs zinc nanoparticles
ZOI zone of inhibition

CHAPTER

1

An introduction to cost-effective technologies for solid waste and wastewater treatment

Madhuraj Palat Kannankai and Suja Purushothaman Devipriya

School of Environmental Studies, Cochin University of Science and Technology, Cochin, Kerala, India

OUTLINE

1.1 Introduction

The first two decades of the 21st century witnessed unprecedented technological advancements in every sphere of life. As a result of the advent of cost-effective and sustainable technologies in the waste management sector, the efficiency and effectiveness of solid waste and water treatment systems have increased manifold. However, even this promising rate of growth in the waste management sector is still insufficient to deal with an estimated annual generation of 2 billion tonnes of municipal solid waste globally (The World Bank, 2018). Further, as predicted by the World Bank, the amount of municipal solid waste generated is set to cross 3 billion tonnes by 2050. However, this, too, seems underestimated as it did not take into account the extreme surge in the volume of wastes generated while tackling the COVID-19 pandemic, which has rendered our waste management systems overwhelmed (Klemeš et al., 2020) and led to the large-scale discharge of untreated solid and liquid wastes into the open environment. Presently, 46% of the world population lacks access to basic waste management facilities (The World Bank, 2018). This leads to the disposal of about 70% of total municipal solid wastes in landfills or in the open. Also, unhygienic and improper waste management practices often cause environmental damages and public health hazards. These detrimental effects are mostly reflected in developing rather than developed countries due to high population density, unregulated waste dumping, and the subsequent spread of pathogens and contaminants (Nanda and Berruti, 2020). Even though developed countries generate significantly higher quantities of per capita solid waste than third-world countries, their implementation of sustainable waste management practices accompanied by strict rules and regulations keep many of the solid waste-related issues in check.

The demand for water in domestic, industrial, agricultural, and municipal sectors is exponentially increasing as a result of rapid population growth and economic prosperity. Since 1990, approximately 2.6 billion people have gained access to potable water; however, 663 million still struggle for it (Tortajada and Biswas, 2018). The current

Cost Effective Technologies for Solid Waste and Wastewater Treatment
https://doi.org/10.1016/B978-0-12-822933-0.00013-9

rate of technological growth in the wastewater treatment sector is satisfactory but slow-paced because the climate change-induced water scarcity and decline in water quality will further exacerbate the existing global water crisis in the coming decades, and hence, the affordable and cutting-edge solutions to wastewater treatment will become indispensable for ensuring drinking water to the masses.

In this critical juncture, to manage the ever-increasing solid and liquid waste volumes, there is a necessity to invent new technologies having a significant edge over traditional approaches in terms of reliability and cost-effectiveness without failing to meet regulatory requirements. For example, the conventional filtration/chlorination water treatment techniques are now being substituted with advanced technologies, such as membrane processes, advanced oxidation processes (AOPs), adsorption on activated carbon and other materials, etc. Similarly, landfill sites in developed countries are now being used for the installation of machineries for conducting modern solid waste treatment techniques, such as incineration, pyrolysis, plasma gasification, aerobic and anaerobic digestion, and deep slurry injections (Saleem, 2018). Moreover, these countries strictly adhere to the concept of 3Rs: reduce, reuse, and recycle. The 3R concept is closely associated with the system of circular economy (CE), which promotes reuse, repair, refurbishment, sharing, remanufacturing, and recycling to minimize the use of raw materials, and thereby, reducing waste generation and carbon emission (Geissdoerfer et al., 2017). This closed-loop system of regenerative approaches is becoming a substitute for the traditional linear economy, which does not involve regeneration or recovery of waste materials (Sugawara and Nikaido, 2014). Sophisticated waste management systems are an integral part of CE (Seadon, 2010) and such systems adopt a holistic approach to solve the problems posed by solid waste, essentially by taking control over various processes and steps involved in the life cycle of a product—from designing the product to the extraction of raw materials to consumption to recycling and recovery of materials—in order to meet the sustainability goals.

1.2 Emerging technologies in solid waste management

The uncontrolled disposal of solid wastes in landfills, widely practiced in developing countries, poses a serious threat of groundwater and surface water contamination due to the leakage of leachate through damaged landfill liners (Mishra et al., 2018). Further, landfill gases (e.g., CH_4 and CO_2) are potent greenhouse gases that contribute to increasing global temperature (Ritzkowski and Stegmann, 2007). While first-world countries are shifting toward the zero-waste concept (Zaman and Lehmann, 2013) to minimize net solid waste generation, third-world countries are trailing behind due to the lack of expertise, ethics, funds, and stringent legal frameworks. Table 1.1 shows some of the emerging technologies implemented by the solid waste management sector for maintaining environmental sustainability while effectively reducing waste and waste-related hazards.

Waste to energy (WTE) technologies are receiving increasing importance mainly due to their potential use as a renewable source of energy. Different types of solid wastes are processed through different techniques to derive energy. Food and yard wastes are generally treated using anaerobic digestion, whereas, plastic wastes are considered to be appropriate for gasification, while, incineration is suitable for the treatment of all types of waste (Kumar and Samadder, 2017). However, proper management of municipal solid waste is achieved only when there is clear knowledge regarding the composition and quantity of wastes being generated (Yadav and Samadder, 2017). Moreover, WTE plants are constructed by taking into account a set of characteristics related to wastes, including density, moisture content, size of materials, and calorific value (Aleluia and Ferrão, 2016). Recent trends in solid waste management indicate a growth in the production of high value-added products such as SCP (single-celled protein) from wastes as an alternative for agricultural-based proteins for the purpose of providing nutrition to humans and animals. Also, SCP can be cultivated from a wide variety of biodegradable industrial wastes in an eco-friendly way (Spalvins et al., 2018).

The efforts taken by researchers to develop eco-friendly approaches for reclaiming soils contaminated by chemicals released as a result of unregulated solid waste disposal have given rise to a range of bioremediation techniques, which involve the use of microorganisms to detoxify/remove contaminants from soil. The major examples for bioremediation methods are bioreactor (Benson et al., 2007), biostimulation, bioaugmentation (Omokhagbor Adams et al., 2020), bioleaching (Mishra et al., 2005), windrow composting (Manjula et al., 2013), bioventing, and biosparging (Azubuike et al., 2016). Bioremediation is carried out via two broad methods, viz.: in situ and ex situ. In the former method, the treatment is applied at the contaminated site itself whereas in the latter the contaminated medium is excavated and transported into a treatment station for detoxification (Singh and Tiwari, 2014). Phytoremediation is a technique very similar to bioremediation but uses green plants to absorb and biomagnify elements from a

TABLE 1.1 Emerging technologies implemented by the solid waste management sector.

Method	Technology	Description
Recycling	Enzymatic deinking (Saxena and Singh Chauhan, 2017)	Use of cellulolytic, hemicellulolytic, and ligninolytic enzymes to remove printing ink from paper fibers
	Anaerobic digestion and biogas production (Lora Grando et al., 2017)	Methane gas generated as a result of anaerobic digestion of organic materials and biodegradable plastics can be burned to generate heat and electricity
	Cullet remanufacturing (Saleem, 2018)	The glass cullets are ground and melted to remanufacture glass containers. The energy consumed by this process is significantly lower than that of manufacturing new glasses
Processing	Fluffing (Torbert et al., 2011)	Fluff is a pulp-like material formed following the separation and sterilization of the organic fraction of recyclable municipal garbage
	Melting (Nishino et al., 2009)	Waste melting involves gasification and burning of combustible biomass and melting of incombustible materials, such as metals
	Vermicomposting (Ali et al., 2015)	A method for decomposing organic wastes using worms (e.g., red wigglers, white worms, and earthworms) to yield nutrient-rich compost
	Plasma pyrolysis/gasification (Tang et al., 2013)	This technique utilizes thermochemical properties of plasma for conducting pyrolysis and gasification of carbonaceous solid wastes at a rate much faster than conventional thermal treatment technologies
	Refuse-derived fuel (RDF) (Gendebien et al., 2003)	RDF consists of high-calorific combustible fractions of solid waste. Hence, they are used as fuel for electricity generation
	Fluidized-bed technology (Arena, 2012)	This combustion technique is used for the processing of finely divided homogeneous solid waste materials including coal and sewage sludge. The finely divided solids are converted into a dynamic fluid-like stage
	Dry anaerobic composting (DAC) (Rocamora et al., 2020)	DAC is appropriate for treating wastes with high solid content (20%–40%). Here, the decomposition of biomass by anaerobic bacteria takes place in the absence of oxygen
Bioconversion	Bioreactor (Kumar et al., 2011)	Rapid degradation and stabilization of organic wastes take place with the help of microbes. The degradation process is enhanced by the addition of liquid and air into the reactor
Disposal	Microturbine (Medeiros et al., 2017)	The turbine operates using landfill gases and produces energy, especially in those landfills where the gas output is not sufficient for larger engines to run
Landfill gas recovery	Microbial fuel cell (MFC) (Khan et al., 2017)	The electrons and protons derived from the microbial metabolism of organic/inorganic substances are received by the electrodes and converted into electricity

contaminated site and convert them into biomolecules through metabolization. In addition, this will also enhance the esthetics of a contaminated site by spreading greenery (Nagendran et al., 2006). Following are some of the important plants generally used for phytoremediation purposes: *Brassica juncea* L. (Lim et al., 2004), *Tithonia diversifolia* and *Helianthus annuus* (Adesodun et al., 2010), *Salix alba* and *Salix viminalis* (Mleczek et al., 2010), *Thlaspi caerulescens* (Zhao et al., 2003), and *Solanum lycopersicum* L. (Almasi et al., 2019). For further enhancing the efficiency of bioremediation, particularly in harsh environments where microbial populations fail to thrive due to factors like extreme contaminant concentration, pH, temperature, or salinity, the enzyme-mediated processes are preferred as they are rapid, easy to control, highly specific, and cost-effective than plant or microbe-mediated bioremediation methods (Kumar and Bharadvaja, 2019).

The proper functioning of waste management systems and the attainment of sustainability goals are highly dependent on the proper collection of waste from the source and its segregation, sorting, and storage. In most cases, the responsibility of waste collection is entitled to the municipality/local body to implement, whereas segregation/sorting is carried out through the joint efforts of individual citizens and staff working at segregation centers. In Guangzhou (China), sorting of solid wastes, such as plastic, paper, and metal, was done with equal work contribution from sanitary workers and individuals. This approach, in addition to recycling a significant amount of solid wastes, also ensured economic and environmental benefits (Tang et al., 2018). However, there are certain factors that influence the process of waste sorting: socioeconomic background of individuals, distance to collection centers, and public awareness and engagement.

1.3 Emerging technologies in wastewater management

The wastewater treatment sector is developing much faster than the solid waste management sector mainly to fix the increasing imbalance between global per capita water consumption and freshwater availability. The conventional approaches followed in wastewater treatment involve four basic stages wherein a combination of physical, chemical, and biological processes takes place, resulting in the removal of solids, carbonaceous materials, and occasionally, nutrients from wastewater (Fig. 1.1).

Rapid population growth and industrial development rendered our freshwater resources scarcer than ever (Seow et al., 2016). Besides, climate change-induced water stress and the threat of emerging contaminants (EC), such as pharmaceutical and personal care products (PPCPs), pesticides, and hormones, have further worsened the existing water crisis. The removal of potentially harmful chemicals from polluted water could be achieved only with advanced treatment technologies, such as membrane filtration systems (Fig. 1.2) (Hube et al., 2020), advanced oxidation processes (AOP) (Babu et al., 2019), UV irradiation (Hora et al., 2019), etc. (Table 1.2). The selection of an appropriate membrane depends on the purposes of filtration, which include pathogen removal, organic matter removal, heavy metal removal, and softening of water. Among the pressure-driven membrane filtration processes, RO (reverse osmosis) is highly efficient than others because of its ability to separate up to 99.5% of small particles and monovalent ions from wastewater (Ezugbe and Rathilal, 2020). Advanced oxidation processes (AOP), such as photocatalysis (Devipriya and Yesodharan, 2005; Saran et al., 2016), Fenton processes, and ozonation, are generally applied in combination with ozone (O_3), catalyst, or ultraviolet (UV) irradiation techniques to maximize treatment efficiency (Ghime and Ghosh, 2020). Among these processes, exhaustive researches were carried out particularly on

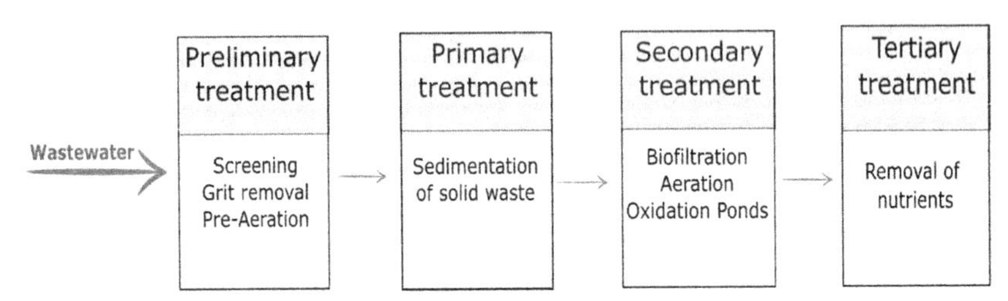

FIG. 1.1 Schematic representation of four basic stages involved in the conventional wastewater treatment process.

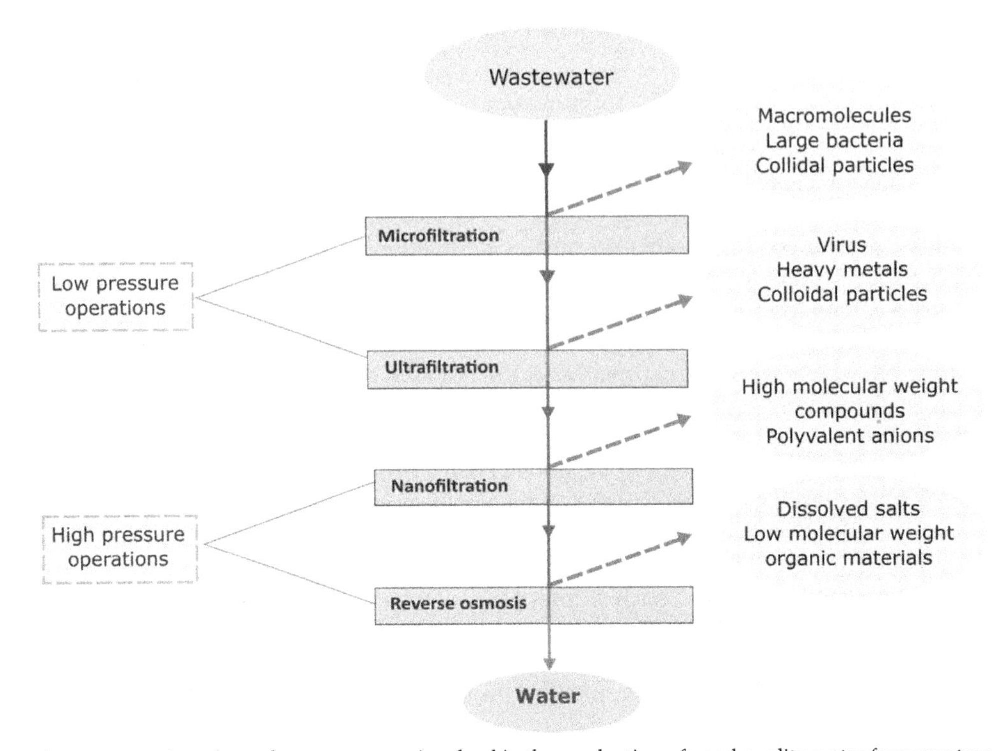

FIG. 1.2 Schematic representation of membrane processes involved in the production of good quality water from wastewater.

TABLE 1.2 Emerging technologies in wastewater treatment.

Method	Description
Microfiltration (MF)	This is a pressure-driven technique involving the separation of colloidal particles, large bacteria, and other macromolecules from a solution while allowing the solutes that are smaller than the membrane pores (0.1–10 μm) to pass through (Charcosset, 2012)
Ultrafiltration (UF)	The pore size of UF membranes ranges between 0.001 and 0.05 μm. Hence, they can efficiently separate heavy metals, suspended solids, viruses, and proteins from fluids (Singh and Hankins, 2016)
Nanofiltration (NF)	This process prevents high-molecular-weight compounds, oligosaccharides, and polyvalent anions from passing through the semipermeable membrane of a pore size < 0.002 μm (Abdel-Fatah, 2018)
Reverse osmosis (RO)	RO takes place when there is a greater differential pressure than the osmotic pressure. It allows water to pass through but rejects dissolved salts and low-molecular-weight organic materials (Mikhak et al., 2019)
Advanced oxidation processes (AOP)	This process is widely used for the degradation of recalcitrant organic components in wastewater through the in situ generation of highly reactive hydroxyl radicals, which can be achieved via a set of technologies, such as ozonation, sonochemical degradation, electrochemical remediation, heterogeneous photocatalysis, and Fenton-based reactions (Gautam and Chattopadhyaya, 2016)
Nanosorbents	Nanomaterials effectively adsorb dissolved organic pollutants

various homogeneous and heterogeneous photocatalytic processes because of their ability to degrade anthropogenic and natural organic contaminants, including carcinogenic dyes. For example, Ag-deposited titanium dioxide (TiO_2) was found to be enhancing the rate of dye decolorization in an acidic environment (Saran et al., 2017). Also, a methodology for the recovery and reuse of sedimented TiO_2 using plant-based coagulant was proposed and it will ensure efficient utilization of the photocatalyst (Patchaiyappan et al., 2016).

The emergence of nanotechnology has significantly increased treatment cost efficiency by improving a range of properties of water treatment systems through the introduction of techniques such as nanophotocatalysis and materials such as nanosorbents (Qu et al., 2013). Nanosorbents are being increasingly popular compared to traditional sorbents (e.g., clay minerals, activated carbons, and chelating materials), owing to their unique properties like the presence of multiple sorption sites, large surface area, strong solution mobility, tunable pore size, etc. (El-sayed, 2020; Santhosh et al., 2016).

Among the numerous wastewater treatment methods, the constructed wetlands (CWs) are gaining wider attention as an alternative to the activated sludge process, mainly due to their high treatment efficiency for organics and suspended solids. CWs, such as free surface-constructed wetlands (FSCWs), horizontal flow-constructed wetlands (HFCWs), and vertical flow wetlands (VFCWs), resemble natural wetlands in their functioning. Further, hybrid-constructed wetlands developed by combining two or more CWs (usually HFCWs and VFCWs) are more recommended for treating domestic or municipal wastewater as this combination can enhance removal of total nitrogen (Vymazal, 2010).

Another affordable and promising wastewater treatment technology called downflow hanging sponge (DHS) is quite similar to trickling filters in terms of their working principle. It was primarily developed as a posttreatment method to upflow anaerobic sludge blanket (UASB) reactor in order to produce high-quality effluent (Nurmiyanto and Ohashi, 2019). This technique involves the usage of polyurethane sponges which are 90% void, and hence they provide ample surface area for the microbial processes to take place (Maharjan et al., 2020).

However, the expansion of cost-effective and mechanized treatment technologies to the rural and remote regions of many developing countries is a laborious and protracted task. In such circumstances, on-site water treatment techniques such as coagulation can be employed. But, the unavailability of chemical-based coagulants to rural communities highlights the importance of natural coagulants as an alternative to be used for drinking water treatment (Yin, 2010). Natural coagulants derived from plants such as *Strychnos potatorum* (Arunkumar et al., 2019; Deshmukh et al., 2013), *Moringa oleifera* (Camacho et al., 2017), *Azadirachta indica* (Pandey et al., 2020), *Dolichos lablab* (Asrafuzzaman et al., 2011), and *Hibiscus rosa-sinensis* (Nidheesh et al., 2017) were found to be effectively reducing the turbidity of wastewater.

Biochar is the carbon-rich solid by-product of pyrolysis and is becoming one of the novel methods for eliminating heavy metals and organic compounds from wastewater. The distinctive characteristics of biochar, including specific surface area, microporosity, ion-exchange capacity (IEC), and adsorption capacity, are determined by pyrolysis temperature and feedstock type. Moreover, these characteristics will govern the rate and type of pollutants to be removed from wastewater (Pokharel et al., 2020).

Natural zeolites include several hydrated aluminosilicate minerals that are widely available at low cost and can be used as adsorbents in wastewater purification processes, particularly for removing ammonium and heavy metal ions. They possess a high cation-exchange capacity in addition to enhanced molecular sieve properties, and these properties also help in the removal of noxious odors from the surrounding environment. Furthermore, the adsorption capacity of natural zeolites can be improved through modification processes, such as ion exchange, surfactant functionalization, and acid treatment (Wang and Peng, 2010).

1.4 Conclusion

The radical technological innovations taking place in the 21st century have provided a wide range of solutions to overcome burgeoning issues associated with waste management. However, the expansion of modern waste remediation techniques to rural regions of peripheral countries requires strenuous efforts from various stakeholders. At present, numerous waste management techniques are being rigorously developed across the major cities of the world in order to keep them sustainable and liveable for years to come. In a predictive analytical model using compositional Bayesian regression, even though the global organic waste generation showed a decreasing trend from 2015 to 2050, all other waste types exhibited an increasing trend worldwide (Chen et al., 2020). Similarly, the global water demand tends to increase exponentially over the next three decades, and by 2050, the growth in demand for water in nonagricultural sectors will surpass that in the agricultural sector (Boretti and Rosa, 2019). Further, the projected decline in groundwater availability is likely to exacerbate the predicted global water demand, and the climate change-related threats can further worsen the existing water crisis. Fortunately, the recent developments in cost-effective wastewater treatment technologies indicate a massive transformation in some key areas of waste management, including recycling rates, landfill modernization, turnaround times, and safety measures. Hence, at this crucial point in time, the government needs to come to the fore with financial support and stringent rules and regulations to strengthen the waste management sector with the active involvement of local stakeholders.

References

Abdel-Fatah, M.A., 2018. Nanofiltration systems and applications in wastewater treatment: review article. Ain Shams Eng. J. 9 (4), 3077–3092. https://doi.org/10.1016/j.asej.2018.08.001.

Adesodun, J.K., Atayese, M.O., Agbaje, T.A., Osadiaye, B.A., Mafe, O.F., Soretire, A.A., 2010. Phytoremediation potentials of sunflowers (*Tithonia diversifolia* and *Helianthus annuus*) for metals in soils contaminated with zinc and lead nitrates. Water Air Soil Pollut. 207 (1–4), 195–201. https://doi.org/10.1007/s11270-009-0128-3.

Aleluia, J., Ferrão, P., 2016. Characterization of urban waste management practices in developing Asian countries: a new analytical framework based on waste characteristics and urban dimension. Waste Manag. 58, 415–429. https://doi.org/10.1016/j.wasman.2016.05.008.

Ali, U., Sajid, N., Khalid, A., Riaz, L., Rabbani, M.M., Syed, J.H., Malik, R.N., 2015. A review on vermicomposting of organic wastes. Environ. Prog. Sustain. Energy 34 (4), 1050–1062. https://doi.org/10.1002/ep.12100.

Almasi, A., Mohammadi, M., Mosavi, S.A., Eghbali, S., 2019. Phytoremediation potential of sewage sludge using native plants: *Gossypium hirsutum* L. and *Solanum lycopersicum* L. Int. J. Environ. Sci. Technol. 16 (10), 6237–6246. https://doi.org/10.1007/s13762-018-2030-2.

Arena, U., 2012. Process and technological aspects of municipal solid waste gasification. A review. Waste Manag. 32 (4), 625–639. https://doi.org/10.1016/j.wasman.2011.09.025.

Arunkumar, P., Sadish Kumar, V., Saran, S., Bindun, H., Devipriya, S.P., 2019. Isolation of active coagulant protein from the seeds of *Strychnos potatorum*–a potential water treatment agent. Environ. Technol. 40 (12), 1624–1632. https://doi.org/10.1080/09593330.2018.1427798.

Asrafuzzaman, M., Fakhruddin, A.N.M., Hossain, M.A., 2011. Reduction of turbidity of water using locally available natural coagulants. ISRN Microbiol. 2011, 1–6. https://doi.org/10.5402/2011/632189.

Azubuike, C.C., Chikere, C.B., Okpokwasili, G.C., 2016. Bioremediation techniques–classification based on site of application: principles, advantages, limitations and prospects. World J. Microbiol. Biotechnol. 32 (11). https://doi.org/10.1007/s11274-016-2137-x.

Babu, D.S., Srivastava, V., Nidheesh, P.V., Kumar, M.S., 2019. Detoxification of water and wastewater by advanced oxidation processes. Sci. Total Environ. 696. https://doi.org/10.1016/j.scitotenv.2019.133961.

Benson, C.H., Barlaz, M.A., Lane, D.T., Rawe, J.M., 2007. Practice review of five bioreactor/recirculation landfills. Waste Manag. 27 (1), 13–29. https://doi.org/10.1016/j.wasman.2006.04.005.

Boretti, A., Rosa, L., 2019. Reassessing the projections of the world water development report. Npj Clean Water 2 (1). https://doi.org/10.1038/s41545-019-0039-9.

Camacho, F.P., Sousa, V.S., Bergamasco, R., Ribau Teixeira, M., 2017. The use of *Moringa oleifera* as a natural coagulant in surface water treatment. Chem. Eng. J. 313, 226–237. https://doi.org/10.1016/j.cej.2016.12.031.

Charcosset, C., 2012. Membrane processes in biotechnology and pharmaceutics. In: Membrane Processes in Biotechnology and Pharmaceutics., https://doi.org/10.1016/C2011-0-04261-8.

Chen, D.M.C., Bodirsky, B.L., Krueger, T., Mishra, A., Popp, A., 2020. The world's growing municipal solid waste: trends and impacts. Environ. Res. Lett. 15 (7). https://doi.org/10.1088/1748-9326/ab8659.

Deshmukh, B.S., Pimpalkar, S.N., Rakhunde, R.M., Joshi, V.a., 2013. Evaluation performance of natural strychnos potatorum over the synthetic coagulant alum, for the treatment of turbid water. Int. J. Innov. Res. Sci. 2 (11), 6183–6189.

Devipriya, S., Yesodharan, S., 2005. Photocatalytic degradation of pesticide contaminants in water. Sol. Energy Mater. Sol. Cells 86 (3), 309–348. https://doi.org/10.1016/j.solmat.2004.07.013.

El-sayed, M.E.A., 2020. Nanoadsorbents for water and wastewater remediation. Sci. Total Environ. 739. https://doi.org/10.1016/j.scitotenv.2020.139903.

Ezugbe, E.O., Rathilal, S., 2020. Membrane technologies in wastewater treatment: a review. Membranes 10 (5). https://doi.org/10.3390/membranes10050089.

Gautam, R.K., Chattopadhyaya, M.C., 2016. Advanced oxidation process–based nanomaterials for the remediation of recalcitrant pollutants. In: Nanomaterials for Wastewater Remediation, pp. 33–48, https://doi.org/10.1016/b978-0-12-804609-8.00003-0.

Geissdoerfer, M., Savaget, P., Bocken, N.M.P., Hultink, E.J., 2017. The circular economy—a new sustainability paradigm? J. Clean. Prod. 143, 757–768. https://doi.org/10.1016/j.jclepro.2016.12.048.

Gendebien, A., Leavens, A., Blackmore, K., Godley, A., Lewin, K., Whiting, K.J., Davis, R., Giegrich, J., Fehrenback, H., Gromke, U., del Bufalo, N., Hogg, D., 2003. Refuse derived fuel, current practice and perspectives. In: Current Practice, pp. 1–219.

Ghime, D., Ghosh, P., 2020. Advanced oxidation processes: a powerful treatment option for the removal of recalcitrant organic compounds. In: Advanced Oxidation Processes—Applications, Trends, and Prospects., https://doi.org/10.5772/intechopen.90192.

Hora, P.I., Novak, P.J., Arnold, W.A., 2019. Photodegradation of pharmaceutical compounds in partially nitritated wastewater during UV irradiation. Environ. Sci.: Water Res. Technol. 5 (5), 897–909. https://doi.org/10.1039/c8ew00714d.

Hube, S., Eskafi, M., Hrafnkelsdóttir, K.F., Bjarnadóttir, B., Bjarnadóttir, M.Á., Axelsdóttir, S., Wu, B., 2020. Direct membrane filtration for wastewater treatment and resource recovery: a review. Sci. Total Environ. 710. https://doi.org/10.1016/j.scitotenv.2019.136375.

Khan, M.D., Khan, N., Sultana, S., Khan, M.Z., Sabir, S., Azam, A., 2017. Microbial fuel cell: waste minimization and energy generation. In: Modern Age Environmental Problems and Their Remediation, pp. 129–146, https://doi.org/10.1007/978-3-319-64501-8_8.

Klemeš, J.J., Van Fan, Y., Tan, R.R., Jiang, P., 2020. Minimising the present and future plastic waste, energy and environmental footprints related to COVID-19. Renew. Sust. Energ. Rev. 127. https://doi.org/10.1016/j.rser.2020.109883.

Kumar, L., Bharadvaja, N., 2019. Enzymatic bioremediation: a smart tool to fight environmental pollutants. In: Smart Bioremediation Technologies: Microbial Enzymes, pp. 99–118, https://doi.org/10.1016/B978-0-12-818307-6.00006-8.

Kumar, A., Samadder, S.R., 2017. A review on technological options of waste to energy for effective management of municipal solid waste. Waste Manag. 69, 407–422. https://doi.org/10.1016/j.wasman.2017.08.046.

Kumar, S., Chiemchaisri, C., Mudhoo, A., 2011. Bioreactor landfill technology in municipal solid waste treatment: an overview. Crit. Rev. Biotechnol. 31 (1), 77–97. https://doi.org/10.3109/07388551.2010.492206.

Lim, J.M., Salido, A.L., Butcher, D.J., 2004. Phytoremediation of lead using Indian mustard (*Brassica juncea*) with EDTA and electrodics. Microchem. J. 76 (1–2), 3–9. https://doi.org/10.1016/j.microc.2003.10.002.

Lora Grando, R., de Souza Antune, A.M., da Fonseca, F.V., Sánchez, A., Barrena, R., Font, X., 2017. Technology overview of biogas production in anaerobic digestion plants: a European evaluation of research and development. Renew. Sust. Energ. Rev. 80, 44–53. https://doi.org/10.1016/j.rser.2017.05.079.

Maharjan, N., Hewawasam, C., Hatamoto, M., Yamaguchi, T., Harada, H., Araki, N., 2020. Downflow hanging sponge system: a self-sustaining option for wastewater treatment. In: Wastewater Treatment [Working Title]., https://doi.org/10.5772/intechopen.94287.

Manjula, G., Ravikannan, S.P., Meenambal, T., 2013. Bioremediation of municipal solid waste by windrow composting. J. Environ. Sci. Eng. 55 (4), 466–471.

Medeiros, G.P., Pinto, L.S., Calixto, W.P., Stach, A.H.M., Domingues, E.G., Costa, A.N., Neto, D.P., Tschudin, M.H.E., 2017. Technical and economic feasibility of using microturbines for the energy utilization of landfill gas. In: 2017 CHILEAN Conference on Electrical, Electronics Engineering, Information and Communication Technologies, CHILECON 2017—Proceedings, 2017-January, 1–7., https://doi.org/10.1109/CHILECON.2017.8229736.

Mikhak, Y., Torabi, M.M.A., Fouladitajar, A., 2019. Refinery and petrochemical wastewater treatment. In: Sustainable Water and Wastewater Processing, pp. 55–91, https://doi.org/10.1016/B978-0-12-816170-8.00003-X.

Mishra, D., Kim, D.-J., Ahn, J.-G., Rhee, Y.-H., 2005. Bioleaching: a microbial process of metal recovery; A review. Met. Mater. Int. 11 (3), 249–256. https://doi.org/10.1007/bf03027450.

Mishra, H., Karmakar, S., Kumar, R., Kadambala, P., 2018. A long-term comparative assessment of human health risk to leachate-contaminated groundwater from heavy metal with different liner systems. Environ. Sci. Pollut. Res. 25 (3), 2911–2923. https://doi.org/10.1007/s11356-017-0717-4.

Mleczek, M., Rutkowski, P., Rissmann, I., Kaczmarek, Z., Golinski, P., Szentner, K., Strazyńska, K., Stachowiak, A., 2010. Biomass productivity and phytoremediation potential of *Salix alba* and *Salix viminalis*. Biomass Bioenergy 34 (9), 1410–1418. https://doi.org/10.1016/j.biombioe.2010.04.012.

Nagendran, R., Selvam, A., Joseph, K., Chiemchaisri, C., 2006. Phytoremediation and rehabilitation of municipal solid waste landfills and dumpsites: a brief review. Waste Manag. 26 (12), 1357–1369. https://doi.org/10.1016/j.wasman.2006.05.003.

Nanda, S., Berruti, F., 2020. Municipal solid waste management and landfilling technologies: a review. Environ. Chem. Lett. https://doi.org/10.1007/s10311-020-01100-y.

Nidheesh, P.V., Thomas, P., Nair, K.A., Joju, J., Aswathy, P., Jinisha, R., Varghese, G.K., Gandhimathi, R., 2017. Potential use of hibiscus rosa-sinensis leaf extract for the destabilization of turbid water. Water Air Soil Pollut. 228 (1). https://doi.org/10.1007/s11270-016-3232-1.

Nishino, M., Nishimura, S., Katafuchi, M., 2009. Waste melting systems offered by JFE engineering. JFE Tech. Rep. 13, 63–66.

Nurmiyanto, A., Ohashi, A., 2019. Downflow hanging sponge (DHS) reactor for wastewater treatment—a short review. MATEC Web Conf. 280. https://doi.org/10.1051/matecconf/201928005004, 05004.

Omokhagbor Adams, G., Tawari Fufeyin, P., Eruke Okoro, S., Ehinomen, I., 2020. Bioremediation, biostimulation and bioaugmention: a review. Int. J. Environ. Bioremed. Biodegrad. 3 (1), 28–39. https://doi.org/10.12691/ijebb-3-1-5.

Pandey, P., Khan, F., Ahmad, V., Singh, A., Shamshad, T., Mishra, R., 2020. Combined efficacy of *Azadirachta indica* and *Moringa oleifera* leaves extract as a potential coagulant in ground water treatment. SN Appl. Sci. 2 (7). https://doi.org/10.1007/s42452-020-3124-2.

Patchaiyappan, A., Saran, S., Devipriya, S.P., 2016. Recovery and reuse of TiO_2 photocatalyst from aqueous suspension using plant based coagulant - A green approach. Korean J. Chem. Eng. 33 (7), 2107–2113. https://doi.org/10.1007/s11814-016-0059-9.

Pokharel, A., Acharya, B., Farooque, A., 2020. Biochar-assisted wastewater treatment and waste valorization. In: Applications of Biochar for Environmental Safety., https://doi.org/10.5772/intechopen.92288.

Qu, X., Alvarez, P.J.J., Li, Q., 2013. Applications of nanotechnology in water and wastewater treatment. Water Res. 47 (12), 3931–3946. https://doi.org/10.1016/j.watres.2012.09.058.

Ritzkowski, M., Stegmann, R., 2007. Controlling greenhouse gas emissions through landfill in situ aeration. Int. J. Greenhouse Gas Control 1 (3), 281–288. https://doi.org/10.1016/S1750-5836(07)00029-1.

Rocamora, I., Wagland, S.T., Villa, R., Simpson, E.W., Fernández, O., Bajón-Fernández, Y., 2020. Dry anaerobic digestion of organic waste: a review of operational parameters and their impact on process performance. Bioresour. Technol. 299. https://doi.org/10.1016/j.biortech.2019.122681.

Saleem, W., 2018. Latest technologies of municipal solid waste management in developed and developing countries: a review. Int. J. Adv. Sci. Res. 1 (October 2016), 22–29.

Santhosh, C., Velmurugan, V., Jacob, G., Jeong, S.K., Grace, A.N., Bhatnagar, A., 2016. Role of nanomaterials in water treatment applications: a review. Chem. Eng. J. 306, 1116–1137. https://doi.org/10.1016/j.cej.2016.08.053.

Saran, S., Kamalraj, G., Arunkumar, P., Devipriya, S.P., 2016. Pilot scale thin film plate reactors for the photocatalytic treatment of sugar refinery wastewater. Environ. Sci. Pollut. Res. 23 (17), 17730–17741. https://doi.org/10.1007/s11356-016-6964-y.

Saran, S., Manjari, G., Arunkumar, P., Devipriya, S.P., 2017. Solar photocatalytic decolorization of synthetic dye solution using pilot scale slurry type falling film reactor. Korean J. Chem. Eng. 34 (11), 2984–2992. https://doi.org/10.1007/s11814-017-0204-0.

Saxena, A., Singh Chauhan, P., 2017. Role of various enzymes for deinking paper: a review. Crit. Rev. Biotechnol. 37 (5), 598–612. https://doi.org/10.1080/07388551.2016.1207594.

Seadon, J.K., 2010. Sustainable waste management systems. J. Clean. Prod. 18 (16–17), 1639–1651. https://doi.org/10.1016/j.jclepro.2010.07.009.

Seow, T., Lim, C., Nor, M., Mubarak, M., Lam, C., Yahya, A., Ibrahim, Z., 2016. Review on wastewater treatment technologies. Int. J. Appl. Environ. Sci. 11 (1), 111–126.

Singh, R., Hankins, N.P., 2016. Introduction to membrane processes for water treatment. In: Emerging Membrane Technology for Sustainable Water Treatment, pp. 15–52, https://doi.org/10.1016/B978-0-444-63312-5.00002-4.

Singh, S.P., Tiwari, G., 2014. Application of bioremediation on solid waste management: a review. J. Bioremed. Biodegr. 05 (06). https://doi.org/10.4172/2155-6199.1000248.

Spalvins, K., Zihare, L., Blumberga, D., 2018. Single cell protein production from waste biomass: comparison of various industrial by-products. Energy Procedia 147, 409–418. https://doi.org/10.1016/j.egypro.2018.07.111.

Sugawara, E., Nikaido, H., 2014. Properties of AdeABC and AdeIJK efflux systems of *Acinetobacter baumannii* compared with those of the AcrAB-TolC system of *Escherichia coli*. Antimicrob. Agents Chemother. 58 (12), 7250–7257. https://doi.org/10.1128/AAC.03728-14.

Tang, L., Huang, H., Hao, H., Zhao, K., 2013. Development of plasma pyrolysis/gasification systems for energy efficient and environmentally sound waste disposal. J. Electrost. 71 (5), 839–847. https://doi.org/10.1016/j.elstat.2013.06.007.

Tang, J., Wei, L., Su, M., Zhang, H., Chang, X., Liu, Y., Wang, N., Xiao, E., Ekberg, C., Steenari, B.M., Xiao, T., 2018. Source analysis of municipal solid waste in a mega-city (Guangzhou): challenges or opportunities? Waste Manag. Res. 36 (12), 1166–1176. https://doi.org/10.1177/0734242X18790350.

The World Bank, 2018. What a Waste: An Updated Look into the Future of Solid Waste Management. September 20, 2018 https://www.worldbank.org/en/news/immersive-story/2018/09/20/what-a-waste-an-updated-look-into-the-future-of-solid-waste-management.

Torbert, H.A., Gebhart, D.L., Busby, R.R., 2011. New municipal solid waste processing technology reduces volume and provides beneficial reuse applications for soil improvement and dust control. In: Integrated Waste Management—Volume I., https://doi.org/10.5772/20433.

Tortajada, C., Biswas, A.K., 2018. Achieving universal access to clean water and sanitation in an era of water scarcity: strengthening contributions from academia. Curr. Opin. Environ. Sustain. 34, 21–25. https://doi.org/10.1016/j.cosust.2018.08.001.

Vymazal, J., 2010. Constructed wetlands for wastewater treatment. Water 2 (3), 530–549. https://doi.org/10.3390/w2030530.

Wang, S., Peng, Y., 2010. Natural zeolites as effective adsorbents in water and wastewater treatment. Chem. Eng. J. 156 (1), 11–24. https://doi.org/10.1016/j.cej.2009.10.029.

Yadav, P., Samadder, S.R., 2017. A global prospective of income distribution and its effect on life cycle assessment of municipal solid waste management: a review. Environ. Sci. Pollut. Res. 24 (10), 9123–9141. https://doi.org/10.1007/s11356-017-8441-7.

Yin, C.Y., 2010. Emerging usage of plant-based coagulants for water and wastewater treatment. Process Biochem. 45 (9), 1437–1444. https://doi.org/10.1016/j.procbio.2010.05.030.

Zaman, A.U., Lehmann, S., 2013. The zero waste index: a performance measurement tool for waste management systems in a "zero waste city". J. Clean. Prod. 50, 123–132. https://doi.org/10.1016/j.jclepro.2012.11.041.

Zhao, F.J., Lombi, E., McGrath, S.P., 2003. Assessing the potential for zinc and cadmium phytoremediation with the hyperaccumulator *Thlaspi caerulescens*. Plant Soil 249 (1), 37–43. https://doi.org/10.1023/A:1022530217289.

CHAPTER

2

Bioaugmentation and biostimulation of dumpsites for plastic degradation

Shaileshkumar Baskaran and Mythili Sathiavelu

School of Bio Sciences and Technology, Vellore Institute of Technology (VIT), Vellore, India

OUTLINE

2.1 Introduction

Plastics are one of the anthropogenic solid wastes. Globally, 6300 million metric tons (MT) of plastic waste was generated from 1950 to 2015. It may increase up to 12,000 MT by 2050 (Geyer et al., 2017). The hike in littering and improper solid waste management leads to landfill dumping, which is the origin of microplastics. Around 5 trillion plastic chunks of the micro (< 4.75 mm) and macro (> 4.75 mm) size were floating in the sea from 2007 to 2013. Among microplastics, the size ranges of 0.33–1.00 mm and 1.01–4.75 mm influence the sea compared with macroplastics (Eriksen et al., 2014). The mismanaged plastic waste about 4.8 to 12.7 million MT intruded the ocean in 2010 from 190 coastal countries globally, in which China contributes the highest by more than 5 million MT (Jambeck et al., 2015). The microplastics and nanoplastics enter the ocean due to fragmentation in the environment and the breakdown of polymer microbeads applied in various sectors. The microplastics (MP) get into our diet through sea salt. 1 to 10 MPs/kg of salt with 149 μm in size was identified from 17 salt brands across eight countries (Karami et al., 2017). A single cup of tea contains 11.6 billion microplastics and 3.1 billion nanoplastics due to migration from the tea plastic bags (Hernandez et al., 2019). Similarly, styrene, the monomer of polystyrene widely used as single-use food container, was migrated into food (Khaksar and Ghazi-Khansari, 2009). A recent study stated that 27 new compounds were migrated from plastic packaging to foods that were not included in any regulations (Ibarra et al., 2019). Plastic pollution poses a threat to human health.

Cost Effective Technologies for Solid Waste and Wastewater Treatment
https://doi.org/10.1016/B978-0-12-822933-0.00015-2

The current management practices of plastic waste are recycling, dumping in landfills, and energy recovery, which includes mechanical recycling, incineration, pyrolysis, and gasification. Pyrolysis is a promising energy recovery process nevertheless with limitations that includes heavy metal impurities in processed waste and lower yield while using mixed plastic waste in landfills (Canopoli et al., 2018). Mechanical recycling involves shredding, washing, drying, and re-granulating waste plastics to generate recycled pellets. The quality of recycled plastics is bothered due to the plastic additives of the waste stream, which hinders the quality while reprocessing (Horodytska et al., 2018). Globally, conventional recycling is lesser than dumping in landfills. The dumpsites include open dumps and controlled dumps that are nonengineered waste disposal practices where municipal solid wastes are dumped in a barren land without any baseliners. However, sanitary landfill is a well-engineered solid waste management system, which has geomembrane baseliners, leachate management system, gas ventilation circuits, and top cover (Swati et al., 2018). Biodegradation of synthetic polymers was widely reported in laboratory scale and in open environment conditions, but due to recalcitrant nature, hydrophobicity, and low bioaccessibility, the biodegradation of polymers is slow. How a plastic dumpsite or landfill can be cleaned up biologically based on the recent findings of plastic biodegradation?

Bioaugmentation and biostimulation are an emerging tool for the remediation of contaminated sites. Both bioremediation tools suit for major contaminants but more studies are required to tackle plastic waste. Bioaugmentation and biostimulation of polycontaminated sites excluding synthetic plastics were established. This chapter outlines the recent findings, limitations, and prospects on plastic biodegradation, and also proposes bioaugmentation and biostimulation approaches for plastic degradation in dumpsites and landfills.

2.2 Microorganisms degrading synthetic polymers

The plastics are categorized as thermoplastics and thermosets. The thermoplastics are recyclable, including polyethylene (PE), polyethylene terephthalate (PET), polypropylene (PP), polystyrene (PS), and polyvinyl chloride (PVC). They are widely used from packaging to engineering applications. Thermosets are nonrecyclable, which include epoxy resins, polyurethane (PUR), phenolics, etc. Geyer et al. (2017) stated that most of the plastic wastes were generated from the packaging and transporting sector with 141 and 17 MT in 2015, respectively; of them, thermosets were majorly littered in the environment as shown in Fig. 2.1. It depicts that the PE and PP account for higher waste generation among all other polymers.

2.2.1 Polyethylene biodegradation

Polyethylene and polypropylene are represented as polyolefins. Polyethylene and polypropylene polymers account for 97 and 55 MT of waste littering among other polymers in 2015, respectively (Geyer et al., 2017). Polyethylene is the largest littered plastic, which includes high-density polyethylene (HDPE), linear low-density polyethylene (LLDPE), and low-density polyethylene (LDPE). It is widely applied in single-use plastics and food packaging. HDPE accounts for 40 million MT, while LDPE and LLDPE cause 57 million MT plastic waste generation in 2015 alone (Geyer et al., 2017). Polyolefin is hard to degrade due to the absence of functional groups that facilitate biodegradation, its high

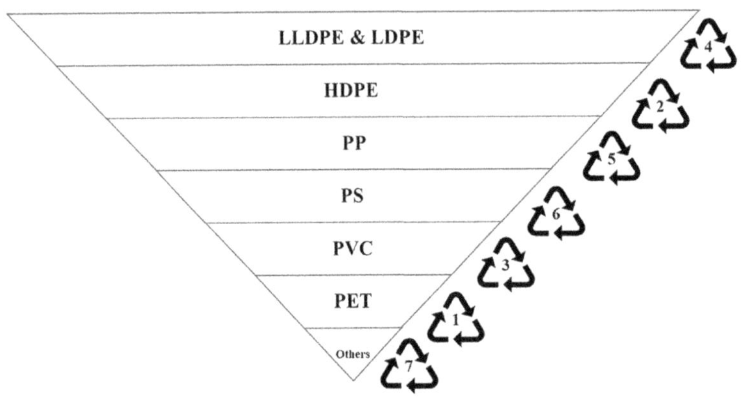

FIG. 2.1 Global polymer waste generation in 2015. *HDPE*, high-density polyethylene; *LDPE*, low-density polyethylene; *LLDPE*, linear low-density polyethylene; *PET*, polyethylene terephthalate; *PS*, polystyrene; *PVC*, polyvinyl chloride. The codes represent the particular polymer resin for recycling. "Others" classification includes polyester polyurethane, polycarbonate, and flexible packaging. *Data source: Geyer, R., Jambeck, J.R., Law, K.L., 2017. Production, use and fate of all plastics ever made. Sci. Adv. 3, 1–5. https://doi.org/10.1126/sciadv.1700782.*

molecular weight, and stable C–C and C–H bonds (Yang et al., 2014). Attempts were made to biodegrade polyolefin with various sources of microbial resources like soil, plastic contaminated soil, mulched film, endophytes, and insect gut microbiota, as shown in Table 2.1. Yang et al. (2014) isolated eight bacterial strains from the gut microbiota of the wax worm larvae *Plodia interpunctella*; of them, *Enterobacter asburiae* YT1 and *Bacillus sp.* YP1 can degrade LDPE up to 6.1% ± 0.3% and 10.7% ± 0.2%, respectively, which is the first report on PE degradation by bacteria from waxworm gut. Bombelli et al. (2017) proceeded by using wax moth *Galleria mellonella* that assimilated a commercial PE bag with 92% weight loss in 12 h.

Many reports focused on the effect of various pretreatments on polyolefin biodegradation. Gamma-irradiation, UV radiation, and heat treatments are the prominent methods used in polyethylene biodegradation. Novotný et al. (2018) compared the virgin and gamma-irradiated cum heat-treated PE degradability by *Bacillus amyloliquefaciens*, which is isolated from composted plastics. Preliminary screening for degradation was carried out with Azure B dye. Pretreated LLDPE was degraded better than LDPE, whereas virgin LDPE and LLDPE were intact. Awasthi et al. (2017) isolated *Klebsiella pneumoniae* CH001 from plastic dumpsite by serial dilution and screened for HDPE

TABLE 2.1 Microorganisms involved in polyolefin biodegradation.

Plastic	Source of microorganism/insect	Microorganism/insect	Characterization of degradation	Time	References
Polyethylene (PE)					
Untreated pure LDPE ([a]M_n: 88200)	*Plodia interpunctella* Gut microbiota	*Enterobacter asburiae* *Bacillus* sp.	6.1% ± 0.3% weight loss 10.7 ± 0.2% weight loss	60 days	Yang et al. (2014)
Commercial PE bag	NA[b]	*Galleria mellonella*	2.2% ± 1.2 holes/worm/h and 92% weight loss	1 h	Bombelli et al. (2017)
γ-irradiated/[c]T_{90}-treated LLDPE	Plastics from composting plant	*Bacillus amyloliquefaciens*	3.2% ± 1.3% weight loss	60 days	Novotný et al. (2018)
γ-irradiated/T_{90}-treated LDPE	Plastics from composting plant	*Bacillus amyloliquefaciens*	0.5% ± 0.1% weight loss	60 days	Novotný et al. (2018)
HDPE	Soil from a plastic dump yard	*Penicillium oxalicum* *Penicillium chrysogenum*	53.34% weight loss 58.59% weight loss	90 days	Ojha et al. (2017)
LDPE	Soil from a plastic dump yard	*Penicillium oxalicum* *Penicillium chrysogenum*	36.60% weight loss 34.35% weight loss	90 days	Ojha et al. (2017)
[d]T_{70}-treated HDPE	Plastic waste dumpsite	*Klebsiella pneumoniae*	18.4% weight loss[e]	60 days	Awasthi et al. (2017)
LDPE-titania-starch blend	Soil from a plastic dump yard	*Stenotrophomonas pavanii*	Increase in the vinyl bond index, keto and ester carbonyl index of the polymer	56 days	Tahir et al. (2016)
LDPE	Soil from a municipal dump yard	*Aspergillus versicolor*	19% weight loss	60 days	Das et al. (2018)
UV-B treated LDPE with cobalt stearate pro-oxidant	NA	Bacterial consortia *Bacillus pumilus* *Bacillus halodenitrificans* *Bacillus cereus*	8.4% ± 1.37% weight loss	20–24 months	Roy et al. (2008)
Commercial LDPE bag	NA	Soil burial	Biofilm formation	22 months	Mumtaz et al. (2010)
Xylene-treated LDPE powder	Soil from a municipal dump yard	*Bacillus amyloliquefaciens* *Bacillus amyloliquefaciens*	12% mineralization[f] 14.7% mineralization	60 days	Das and Kumar (2015)
LDPE	NA	*Pseudomonas aeruginosa*	20% weight loss	120 days	Myint and Ravi (2012)

Continued

TABLE 2.1 Microorganisms involved in polyolefin biodegradation—cont'd

Plastic	Source of microorganism/insect	Microorganism/ insect	Characterization of degradation	Time	References
PE microplastics (MP) 250 μm < MP > 1000 μm.	NA	*Zalerion maritimum*	Change in intensity of functional groups	28 days	Paço et al. (2017)
LDPE	NA	*Rhodococcus ruber*	0.86% weight loss/ week	60 days	Sivan et al. (2006)
PE	NA	*Lysinibacillus fusiformis* *Bacillus cereus*	Based on the ecotoxicological effect of degraded broth on plants	60 days	Shahnawaz et al. (2016a)
LDPE + Bran	*Tenebrio molitor* Mealworm	*Citrobacter sp.* *Kosakonia sp.*[g]	40.1% ± 8.5% molecular mass loss	32 days	Brandon et al. (2018)
Commercial PE bag	Rhizosphere soil of *Avicennia marina*	*Lysinibacillus fusiformis* *Bacillus cereus*	21.87% ± 6.37% weight loss 13.87% ± 3.6% weight loss	60 days	Shahnawaz et al. (2016b)
Commercial PE bag	Marine soil sediment from Eckernförde Bay, Baltic Sea	Biostimulated marine sediments	No sign of biodegradation	98 days	Nauendorf et al. (2016)
HDPE	Marine soil sediment	*Aspergillus tubingensis* *Aspergillus flavus*	6.02% ± 0.2% weight loss 8.51% ± 0.1% weight loss	30 days	Sangeetha Devi et al. (2015)
Polypropylene (PP)					
γ (60 KGy)-irradiated PP	Leaf and stem of *Psychotria flavida*	*Lasiodiplodia theobromae*	0.3% weight loss	90 days	Sheik et al. (2015)
LMWPP-1[h] LMWPP-2[i] HMWPP[j]	Soil from a municipal dump yard	*Stenotrophomonas panacihumi*	20.3% ± 1.39% weight loss 16.6% ± 1.70% weight loss 12.7% ± 0.97% weight loss	90 days	Jeon and Kim (2016)
TT-PP[k] UT-PP[l]	Soil from a plastic dump yard	*Bacillus flexus*	10.70% weight loss 0.40% weight loss	12 months	Arkatkar et al. (2009)
UV-treated MI-PP[m]	NA	*Phanerochaete chrysosporium* *Engyodontium album*	18.8% weight loss 9.42% weight loss	12 months	Jeyakumar et al. (2013)
UV-treated PP	Soil from a municipal dump yard	*Bacillus flexus*	2.5% weight loss	12 months	Arkatkar et al. (2010)
γ-irradiated isotactic PP/wood biocomposite (7:3)	Natural substrate	*Bjerkandera adusta*	Cracks and scrap in the film surface	49 days	Butnaru et al. (2016)
PP with prooxidants	NA	*Rhodococcus Rhodochrous*	Low ADP/ATP ratio	180 days	Fontanella et al. (2013)
Isotactic PP microplastics (M_n: 67000)	Mangrove sediments	*Bacillus sp.* *Rhodococcus sp.*	4.0% weight loss 6.4% weight loss	40 days	Auta et al. (2018)

a M_n : Number average molecular weight.
b NA : Not applicable.
c T_{90}: High-temperature treatment at 90°C.
d T_{70}: High-temperature treatment at 70°C.
e Based on universal tensile machine analysis.
f Based on CO_2 estimation.
g Based on the next-generation sequencing, direct ingestion of LDPE by mealworm.
h LMWPP-1: Low molecular weight polypropylene (M_n: 790).
i LMWPP-2: Low molecular weight polypropylene (M_n: 5200).
j HMWPP: High molecular weight polypropylene.
k TT-PP: Thermally treated polypropylene (80°C for 10 days).
l UT-PP: Untreated polypropylene.
m MI-PP: Metal ion-blended polypropylene.

degradation in nutrient broth by varying polymer concentrations. LDPE-degrading organisms were isolated from the mulched plastic from a dumpsite, and enrichment was done with LDPE powder as a sole carbon source supplemented with the mineral salt medium for a week and repeated for 2 weeks (Tahir et al., 2016).

2.2.2 Polypropylene biodegradation

Polypropylene (PP) is highly recalcitrant to degrade than polyethylene due to repetitive pendant methyl groups (Jeon and Kim, 2016). The degradation studies on polypropylene are scarce when compared with other polymers. The pretreatment of these polymers favors biodegradation, as shown in Table 2.1. The pretreatment includes thermal treatment, chemical treatment, UV, and gamma irradiation. The application of pretreatment in the laboratory scale should mimic the dumpsite or environment littered plastics, which helps in understanding the effect of such external factors. The gamma-irradiated PP can be degraded by *Bjerkandera adusta* and *Lasiodiplodia theobromae* (Sheik et al., 2015; Butnaru et al., 2016). The natural weathering of PP and PE (HDPE and LDPE) in seawater for 12 months was reported by Artham et al. (2009). Among the various polymers, LDPE resulted in the maximum weight loss (1.9%) after 12 months, whereas PP showed 0.65% weight loss.

The prooxidants like iron stearate (Fe^{3+})-, magnesium stearate (Mn^{2+})-, calcium stearate (Ca^{2+})-, and cobalt stearate (Co^{2+})-filled polyolefins favor photodegradation. Iron directly acts as photooxidant, which results in oxidation and chain scission, and turns the hydrophobic nature of the polymer into hydrophilic when exposed into the atmosphere. The PP with calcium stearate provides maximum degradation with 8.73% (Mandal et al., 2017). The polypropylene/poly(ethylene oxide)/TiO_2/octacalcium phosphate with succinic acid (OCPC) was 20% mineralized by microorganisms after photodegradation (24 h) followed by soil microorganisms (80 days) (Miyazaki et al., 2012). Carvalho et al. (2013) stated a thermoplastic polymer, polyacetal, as prooxidant with PP. The addition of prooxidants initiates photodegradation that can be approached in production plants; meanwhile, it cannot be a solution for prolonged dumped plastics in the landfill and dumpsites. Jeon and Kim (2016) stated the biodegradation of untreated PP by *Stenotrophomonas panacihumi* within 90 days, which results in a higher degradation rate than other studies in a short time. The biodegradation of UV-treated PP by *Phanerochaete chrysosporium* after 12 months was observed (Jeyakumar et al., 2013).

2.2.3 Polyethylene terephthalate biodegradation

Polyethylene terephthalate (PET) polymer is widely applied in the packaging sector followed by polyolefins. A research outbreak discovered that a novel bacterium *Ideonella sakaiensis* can completely assimilate PET within 6 weeks, which was isolated from a consortium No. 46 among the 250 samples. PETase and MHETase were the two extracellular enzymes responsible for PET assimilation that results in terephthalic acid (TPA) and ethylene glycol (EG) as its end products. The enzyme PETase acts on polyethylene terephthalate and results in mono(2-hydroxyethyl) terephthalic acid (MHET) and TPA; further, the enzyme MHETase hydrolyzes MHET to TPA and EG (Yoshida et al., 2016). The characterization of PETase revealed that it has properties of lipases and cutinases. Protein engineering studies stated that PETase is an aromatic polyesterase that does not degrade aliphatic polyester but it degrades a bio-based semi-crystalline polymer polyethylene-2,5-furandicarboxylate, which mimics PET. The PETase action was due to its evolution in a polymer-rich environment, and further protein engineering studies are required (Austin et al., 2018). Apart from *Ideonella sakaiensis*, various microorganisms degrade PET as shown in Table 2.2, and the review (Koshti et al., 2018) highlights the same but only *Ideonella sakaiensis* can completely assimilate PET. Studies are required to isolate a potent microbiota that completely assimilates PET in a short duration.

2.2.4 Biodegradation of other polymers

The polymers that are minimal in waste include polystyrene, polyvinyl chloride, and polyurethane. Polystyrene (PS) foam was degraded by the gut microbiota of *Tenebrio molitor* worm; bacterium *Exiguobacterium sp.* of the worm gut microbiota can degrade PS up to 7.4% ± 0.4% after 60 days (Yang et al., 2015a,b, 2018). High-impact polystyrene (HIPS) is highly dumped in the landfill/dumpsites, which poses a major threat. The degradation of HIPS with flame retardants hexabromobiphenyl oxide and antimony trioxide by *Bacillus* sp. with 23% weight loss in 30 days was observed (Mohan et al., 2016). The biodegradation experiment that focuses on identifying the microbiota or microbes which degrade the mixed plastics as the sole carbon source is necessary.

TABLE 2.2 Microorganisms involved in biodegradation of PET, PS, PUR, and PVC.

Plastic	Source of microorganism	Microorganism/insect	Characterization of degradation	Time	References
Polyethylene terephthalate (PET)					
PET	Rhizosphere soil of *Salix viminali* grown in a plastic dump yard	*Arthrobacter sulfonivorans* *Serratia plymuthica* *Clitocybe sp.* *Laccaria laccata*	Increase in carbon dioxide evolution	6 months	Janczak et al. (2018)
PET (500 μm) PET (420 μm) PET (300 μm) PET (212 μm)	Soil	*Streptomyces sp.*	49.2% weight loss 57.4% weight loss 62.4% weight loss 68.8% weight loss	18 days	Farzi et al. (2019)
PET	Soil from a municipal dump yard	*Ideonella sakaiensis*	Complete degradation	42 days	Yoshida et al. (2016)
Poly(ethylene terephthalate-*co*-lactate)	Thermophilic digester	Thermophilic sludge	Reduction in molecular weight	394 days	Hermanová et al. (2015)
Polystyrene (PS)					
PS:PLA PS:PLA: OMMT	NA	*Pseudomonas aeruginosa*	10.25% degradation	28 days	Shimpi et al. (2012)
Starch: PS (80:20) Starch: PS (70:30)	NA	Compost	No evidence of PS degradation; meanwhile, complete starch degradation was observed	39 days	Pushpadass et al. (2010)
PS	*Tenebrio molitor*	*Tenebrio molitor* gut microbiota	47.7%	16 days	Yang et al. (2015a)
PS (10%) + Bran (90%)	*Tenebrio molitor*	*Tenebrio molitor* gut microbiota	16.9 ± 1.9 mg PS/ 100 mealworms per day	32 days	Yang et al. (2018)
PS	*Tenebrio molitor* gut microbiota	*Exiguobacterium sp.*	7.4 ± 0.4%	60 days	Yang et al. (2015b)
Polyester polyurethane (PUR)					
PUR	*Psidium guajava*	*Pestalotiopsis microspora*	Complete degradation	14 days	Russell et al. (2011)
PUR	Soil	*Aspergillus tubingensis*	Complete fragmentation	60 days	Khan et al. (2017)
PUR	Soil from a dump yard	*Pseudomonas aeruginosa*	Reduction in molecular weight	30 days	Shah et al. (2013)
PUR	NA	*Bacillus subtilis and Pseudomonas aeruginosa co-culture*	Increase in carbon dioxide evolution	30 days	Shah et al. (2016)
PUR	Compost	*Arthrographis kalrae (Dominant in the microbiota)*	Significant deterioration	28 days	Zafar et al. (2014)
PUR	Aircraft microbiome	*Papiliotrema laurentii*	Increase in carbon dioxide evolution	8 days	Hung et al. (2019)
Polyvinyl chloride (PVC)					
Chemical treated (PVC)	Soil from plastic recycling site	*Mucor rouxii*	Reduced thermal stability	90 days	Singh and Pant (2016)
PVC with 30% additives	NA	*Pseudomonas citronellolis* *Bacillus flexus*	10% reduction in molecular weight	45 days	Giacomucci et al. (2019)
Starch-blended PVC	PVC buried in garden soil with sewage sludge	*Phanerochaete chrysosporium*	80,275 to 78,866 Da molecular weight reduction	84 days	Ali et al. (2014)
PVC	Waste dump site	*Consortia* *Pseudomonas otitidis, Bacillus cereus, and Acanthopleurobacter pedis*	66.61 to 15.63 Da molecular weight reduction	9 months	Anwar et al. (2016)

2.3 Bioaugmentation and biostimulation

Bioaugmentation involves the introduction of potent microbial isolate, consortia, or microbiota in a contaminated site for remediation. It is preferable only when 1. biostimulation failed due to no trace of pollutant-degrading microbes, 2. the pollutants and their intermediates are toxic to indigenous microbiota, 3. polycontaminant is degraded with a chain of mechanisms, 4. microbes are required for long adaptation in polluted sites, 5. scarcity is observed in targeted xenobiotic-degrading microorganisms, and 6. the conventional recycling methods are costlier than bioremediation in small sites. Even bioaugmentation can be implemented in the remediation of closed landfills. The selection of appropriate microbes or consortia or microbiota for remediation is mandated, which should have the following properties: (a) The highest survival rate is found for the indigenous microbiota, (b) the introduced microbiota should not induce ecological succession, (c) it should be a potent degrader of targeted xenobiotics (hereafter contaminant), and (d) it should not pose a threat to human health (Mrozik and Piotrowska-seget, 2010; Lebeau, 2011).

Biostimulation involves enrichment of the indigenous microbiota of the site by providing nutrients. The addition of chemical nutrients, its absorption, soil properties, and equity in the bioavailability of nutrients to indigenous microbiota are essential in biostimulation. The added nutrients must stimulate the indigenous microbial flora. The prolonged exposure of microorganisms to contaminants makes them evolved. Several nutrients favor the remediation of contaminants with rate-limiting nutrients, which include oxygen, nitrogen, and phosphorous. The xenobiotics will act as a carbon source, and oxygen is required when the concentration of contaminants exceeds the oxygen input. The supplement of nitrogen promotes remediation in anoxic conditions. The addition of carbon source is necessary when the contaminants are low and recalcitrant in a site (Hazen, 2010).

2.4 Bioaugmentation and biostimulation approaches in the dumpsite/landfill

Bioaugmentation and biostimulation of various contaminants were reported, which include the remediation of heavy metals, polyaromatic hydrocarbons (PAH), nonaqueous-phase liquid (NAPL), and dense nonaqueous-phase liquid (DNAPL). Suja et al. (2014) demonstrated the cumulative bioaugmentation and biostimulation approach for treating total petroleum hydrocarbons both in pilot-scale and in field studies by applying native microbiota and supplementing C: N: P: K in the ratio of 100:10:1:3. The degradation rate constant of pilot-scale and field studies was 0.0390 and 0.0339 per day, respectively. Very limited studies on bioremediation of plastics in landfills and dumpsites were observed due to various factors. The discussion on those prior findings and road map for further research needs in site-specific in situ bioremediation is elucidated in Fig. 2.2. It is a cumulative bioaugmentation and biostimulation

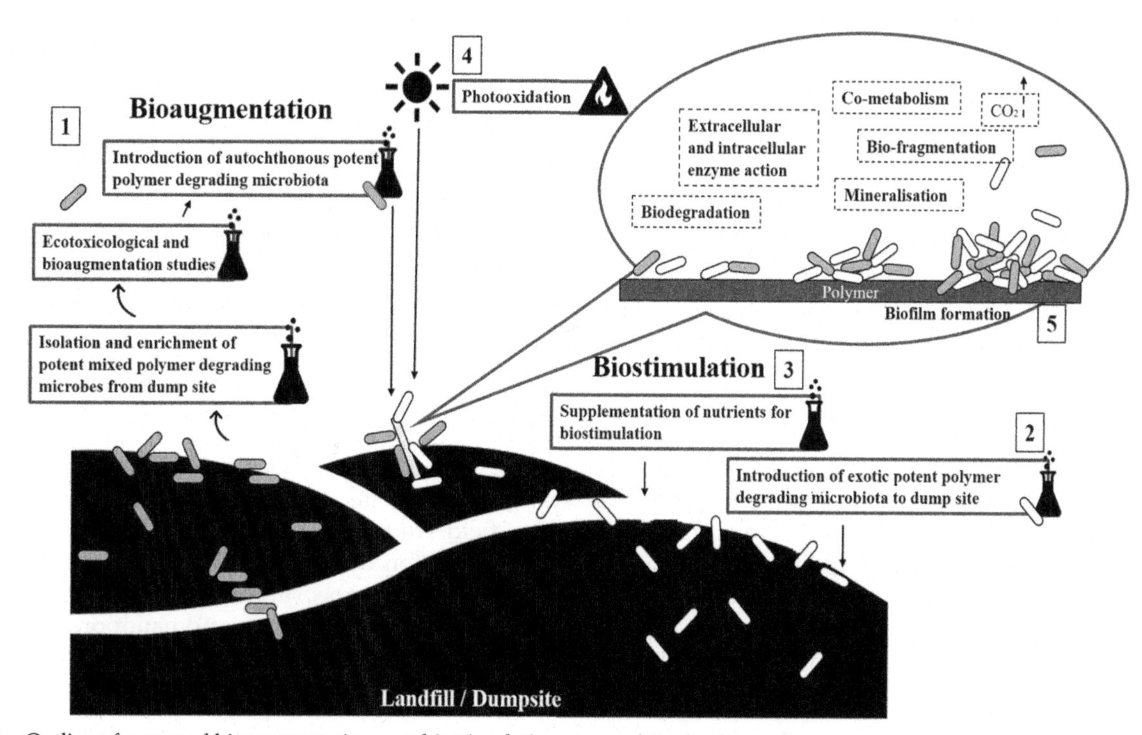

FIG. 2.2 Outline of proposed bioaugmentation cum biostimulation approach in the dumpsite.

approach focused on the selection of appropriate plastic-degrading indigenous microbiota from a particular site for remediating the same site by microbial enrichment (bioaugmentation), meanwhile supplementing the right nutrients to stimulate the indigenous microbiota for bioremediation of plastics (biostimulation), and if the prior two strategies do not work well, then the introduction of exotic microbiota that are capable of degrading the plastics into the site is appreciated.

2.4.1 Landfill or dumpsite site characterization

The modern landfill (sanitary landfill) is designed with precautionary measures to safeguard the environment; however, the atmospheric emissions and land scarcity should be noticed. The dumpsites are conventional waste disposal systems in developing countries either as an open dump or a controlled dump without any skilled manpower and prone to ecotoxicological risk (Hettiaratchi, 2012). Remediation of dumpsites with piles of leftover waste dumps after recycling and energy recovery is necessary rather than the focus on remediating a modern landfill; the former has higher risk and needs to remediate due to the non-engineered layout for gas removal and leachate management. Characterization and monitoring of the well-established sanitary landfill are possible by recurrent gas emission and leachate concentration measures. The dumpsite characterization before applying bioaugmentation and biostimulation includes the assessment of groundwater and surface water contamination due to leachates, microplastics, nanoplastics, and heavy metals like arsenic, chromium, uranium, mercury, cadmium, which is necessary. The other factors that are to be characterized include the organic carbon content, total organic carbon (TOC), biological oxygen demand (BOD), chemical oxygen demand (COD), indigenous microbiota, and antimicrobial-resistant microbial niche of dumpsites.

2.4.2 Identification of microbes in landfill plastisphere

"The life on plastic debris in the aquatic environment" defines the plastisphere (Zettler et al., 2020). The microbial niche on the plastisphere was explored from various origins on marine and freshwater ecosystems, whereas the studies on naturally weathered plastic litter-inhabitant microbiota in terrestrial environment are scarce. However, many microbial isolates from dumpsite soil were observed. The microbiota that survives on plastic debris in the landfill or dumpsite is stated hereafter as a landfill plastisphere. The nonrecyclable municipal solid waste includes plastic as a heap in the dumpsites where the microorganisms in the bottom of the waste heap are anaerobic methanogens and the mid-zone comprises acid formers and the top layer of the heap constitutes aerobic microorganisms. The next-generation sequencing and multi-omics approach can enlighten the microbiota, which naturally colonizes the plastic film in dumpsites. The microbiota of plastisphere with different origins shows the microbial diversity that survives in that recalcitrant xenobiotics, which includes diatoms, natural biofilm formers, cyanobacteria, bacteria, and fungus (Zettler et al., 2020). The plastisphere also provides space to pathogens and antimicrobial-resistant microbes; among the pathogens, *Vibrio sp.*, dominate the plastisphere, followed by *Campylobacteraceae, Aeromonas salmonicida, or Arcobacter sp.*, (Zettler et al., 2020; Zettler et al., 2013). Septicemia, nosocomial, diarrhea associated with human pathogens, *Pseudomonas* and *Acinetobacter* were found abundant in multiple plastic debris obtained from the Urumqi River with an increased antibiotic gradient (Xue et al., 2020). The various microbiota that colonize and degrade the plastisphere can be found elsewhere (Jacquin et al., 2019). A recent study revealed that the color of plastic debris matters for microbiota colonization in 30 days. Unique species of microbes were found in blue-colored plastic debris when compared with in yellow-colored and transparent plastic debris (Wen et al., 2020). So, a variety of colored plastic debris is to be collected while collecting the sample for landfill plastisphere identification. The identification of the landfill plastisphere is necessary since it provides the knowledge of inhabitant microbiota that can be used for effective bioaugmentation.

2.4.3 Prerequisite for bioaugmentation and biostimulation

The pilot-scale and field studies on bioaugmentation for the remediation of various contaminants provide us a fundamental understanding to proceed with the remediation of plastic waste in dumpsites. Bioaugmentation and biostimulation studies of various contaminants from scratch to field studies were reported. The laboratory-scale studies should enrich the landfill plastisphere toward plastic degradation by assessing its potential to assimilate plastic debris with an experimental setup that mimics field parameters like microcosm studies, which are depicted in Fig. 2.3. The next-generation sequencing includes pyrosequencing, ion torrent, Illumina, and nanopore sequencing, which provides the microbiota of the landfill plastisphere from a particular site by analyzing the microbial niche on various plastic debris of different polymers.

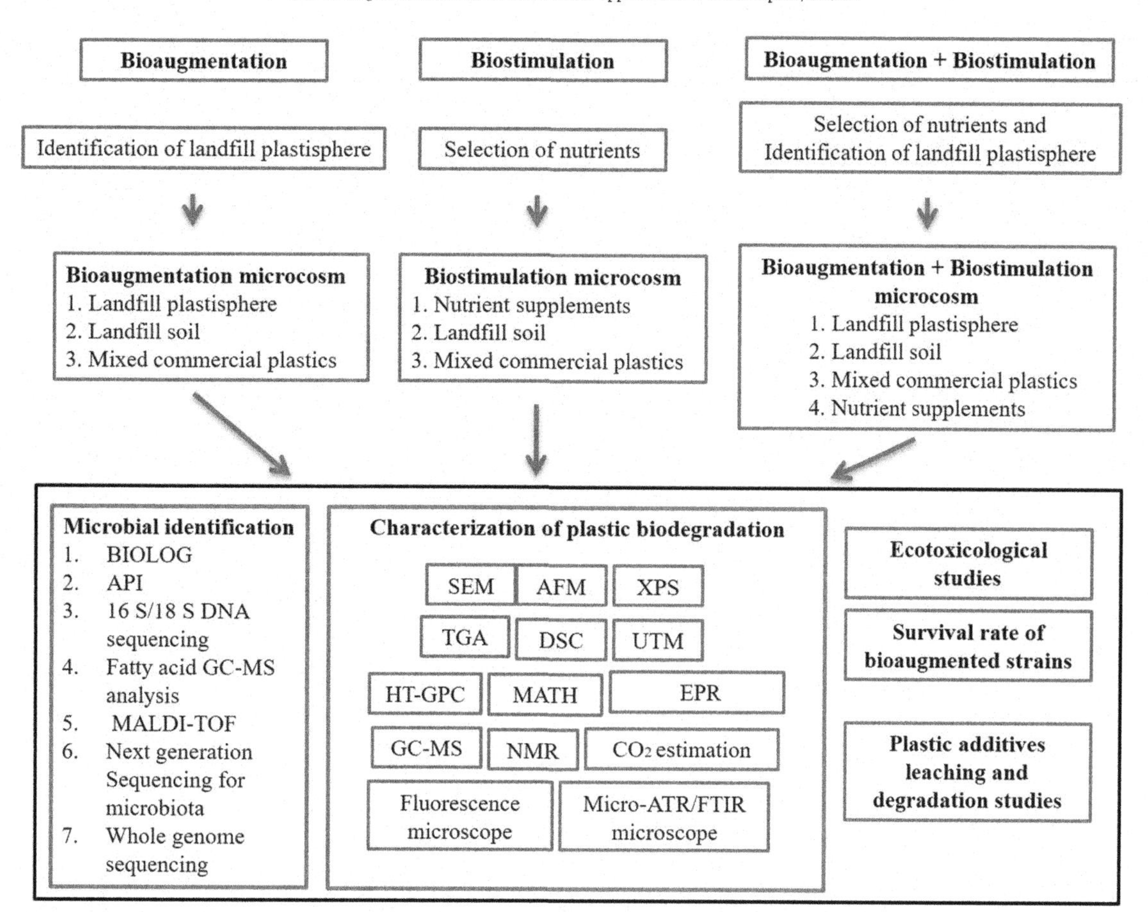

FIG. 2.3 Bioaugmentation cum biostimulation microcosm strategy. *AFM*, atomic forced microscopy; *ATR-FTIR*, attenuation total reflection Fourier transform infrared spectroscopy; *API*, analytical profile index; *DSC*, differential scanning calorimetry; *HT-GPC*, high-temperature gel permeation chromatography; *ESI-MS*, electron spray ionization–mass spectroscopy; *EPR*, electron paramagnetic resonance spectroscopy; *GC–MS*, gas chromatography; *MATH*, microbial adhesion to hydrocarbons assay; *SEM*, scanning electron microscopy; *TGA*, thermogravimetric analysis; *UTM*, universal tensile meter; *NMR*, nuclear magnetic resonance spectrometer, *MALDI-TOF*, matrix-associated laser desorption ionization-time of flight.

The bioaugmentation, biostimulation, cumulative bioaugmentation, and biostimulation microcosm studies can reveal the suitability of the promising method that can be applied in the dumpsites. Bioaugmentation microcosm constitutes landfill plastisphere as inoculum, various commercial plastics like PP, PET, PE, PVS, PS, and unsterilized landfill soil. The control for bioaugmentation microcosm includes sterile landfill soil and unsterilized landfill soil without the addition of landfill plastisphere to make sure the effect of the plastisphere on degradation. The biostimulation microcosm comprises cost-effective nutrient supplements that favor the growth of landfill soil microbiota. so that it can acclimatize the commercial plastics added. The control for biostimulation includes the unsterilized landfill soil and commercial plastics without the nutrients, which convey the effect of nutrient addition and soil microbe's efficacy on plastic degradation. The bioaugmentation cum biostimulation microcosm includes the addition of plastisphere and nutrient supplement to favor the degradation of plastics by both bioaugmented microbiota and indigenous microbiota. The characterization of polymer degradation in microcosm studies through analytical techniques is depicted in Fig. 2.3.

The commercial plastics added in each microcosm should be characterized by attenuated total reflection-Fourier transform infrared (ATR-FTIR) spectroscopy followed by thermogravimetric analysis (TGA) and differential scanning calorimetry (DSC). Skariyachan et al. (2019) directed the degradation of polyolefins in detail, who discussed various techniques; this chapter emphasizes the current findings, limitations, and other analytical instrumentation techniques for plastic biodegradation. Jung et al. (2018) stated the identification of plastic marine debris by using ATR-FTIR spectroscopy. The respective functional groups and the spectral pattern of polymers can be correlated with spectral libraries to identify the polymers. The degradation of polyolefins was characterized by ATR-FTIR spectroscopy based on the carbonyl, methyl, hydroxyl, and vinyl bond index, which states the oxidation and surface degradation. A diverse wavenumber (cm^{-1}) was used in the calculation of the carbonyl index in different research papers: ASTM WK65360, 2018 "New test method for determining the carbonyl index of a polyolefin plastic material using Infra-Red spectrometry (FT-IR)" is in progress to resolve this issue. The oxidation of polypropylene was calculated by the methyl index instead of the carbonyl index by measuring absorbance at 1456 cm^{-1} (Rouillon et al., 2016). It is necessary to detach the biofilm remains in the microbially treated plastics while preparing samples for ATR-FTIR spectroscopy.

The biofilm formation and surface deformation of polymers can be observed by using scanning electron microscopy (SEM). The optimization of glutaraldehyde or osmium tetrachloride fixation and successive ethanol dehydration of plastic samples for biofilm observation is a crucial step. The fixation time and concentration of glutaraldehyde will vary for bacteria, fungi, and algae. The chemicals used should be inert for plastic, which can be inferred by observing the control film undergoing the same procedure. For surface deformation, the film with saline wash is required. The SEM-EDAX analysis depicts the elemental composition of the plastics. The microbiota in each microcosm should also be isolated and tested for swarming and swimming tests in agar plates, and further conventional biofilm assay in 96-well microtiter plates (non-surface-treated) were used to assess the potent biofilm-forming microbes. Atomic forced microscopy (AFM) will elucidate the root mean square surface roughness; the raw data requires second-order flatness treatment (Arkatkar et al., 2009; Butnaru et al., 2016). Microbial adherence to hydrocarbons (MATH) assay or bacterial adherence to hydrocarbons (BATH) assay and salt aggregation test aims to assess the affinity of the microbial cell towards hydrocarbons or hydrophobic liquids by measuring cell surface hydrophobicity. An alternative method is appraised against classical MATH assay. The direct cell counts and aqueous suspension microscopic analysis assure the cell surface hydrophobicity abandoning the noises from hydrocarbons used (Warne et al., 2010).

Thermogravimetric analysis (TGA) of control and degraded polymers reveals the reduced thermal stability in treated polymers, which ensures the percentage of thermal weight loss. The degradation of additives, polymer chain length, and crystallinity can also be observed (Jeyakumar et al., 2013). Differential scanning calorimetry (DSC) provides the crystallinity, melting temperature, and changes in the phase transition by comparing the control and treated polymers. The melting point of naturally weathered or microbially treated polymer samples will decrease when compared with untreated films (Arkatkar et al., 2009; Xiong et al., 2017). The GC–MS analysis of plastics will state the volatile degraded end products and plastic additives degradation by comparing the chromatogram peaks with the standard library. Various methods of sample preparation strategies are there in which solid-phase microextraction (SPME) improves the quantitative analysis of degradation and additives leaching. Ultrasonication followed by GC–MS analysis of polymers and soil also provides the degraded end products (Contat-Rodrigo et al., 2001). The GC–MS analysis of plastic end products reveals the degradation of polypropylene into monomers and the formation of intermediates after the action of fungal strains *Phanerochaete chrysosporium* and *Engyodontium album* (Jeyakumar et al., 2013). Headspace solid-phase microextraction technique coupled with GC–MS was used to identify the additives leaching and in the identification of pungent plastic smell in recycled plastics by Fuller et al. (2020). Such a method can be adopted to characterize the additives leaching from the microbially treated plastics. High temperature–gel permeation chromatography is used to determine the molecular weight distribution of polymers; both number average molecular weight (Mn) and weight average molecular weight (Mw) based on the reduction of molecular weight in treated plastics can be correlated (Jeon and Kim, 2016). The optimized procedure for each plastic is necessary; the characterization of the molecular weight of a polymer differs from that of the plastics. Plastic additives added with the polymer in plastics manufacturing are taken into account during analysis, and the selection of respective solvents and temperature for polymers is necessary. MALDI TOF MS and NMR are also applied to estimate the molecular weight of polymers (Izunobi and Higginbotham, 2011). Raddadi and Fava (2019) critiqued the analytical techniques used for plastic degradation. Each technique has its pros and cons, by subsequent training, accurate sample preparation, providing international standards for the characterization of plastic degradation in the laboratory to field scale in all aspects and advancement in analytical techniques to rectify those pitfalls in the future for betterment of the scope of plastic degradation.

2.4.4 Bioremediation of plastics in landfill

A new emerging approach landfill bioreactor is appraised for the remediation of dumpsites or landfills, which is similar to sanitary landfill but differs in some perspective. A landfill bioreactor supports remediation and degradation of wastes by adding water and landfill leachate recirculation, which enriches the indigenous microflora. This bioreactor can be aerobic, anaerobic, or hybrid, which can stabilize the waste within 5 years; meanwhile, the conventional landfill took 30 years or longer. The microorganisms employed in the landfill must stabilize the waste by the sequential process that includes hydrolysis, acidogenesis, acetogenesis, methanogenesis, and stabilization (Sang et al., 2012). How it can be correlated with plastic degradation in landfills? Zettler et al. (2020) stated biodegradation as per international standards that "In aerobic conditions, the complete breakdown of plastics into carbon dioxide and water meanwhile in anaerobic conditions plastic transformed into methane, carbon dioxide and biomass where 90% of carbon must be assimilated within 180 days in an industrial composting unit," and claimed that microorganism which partially degrade the plastics should not overstate as biodegradation. In the landfill, partially degraded end products of plastics by a microorganism can act as the carbon source for another microorganism, which leads to the assimilation of plastics by a chain reaction and results in CO_2 and water or methane, CO_2, and biomass.

The alternate landfill bioreactor approach was proposed, in which the stabilized waste and municipal waste were mixed in a proportion to avoid the leachate leak. This approach provides methane increment and acetoclastic methanogenesis (Ali and Yue, 2020). Gu et al. (2020) compared the waste stabilization among the anaerobic sanitary landfill, anaerobic bioreactor landfill, and semi-aerobic bioreactor landfill—all of those include plastic waste. The semi-aerobic landfill bioreactor stabilizes the waste better than others in both laboratory-scale and pilot-scale studies. The Fukuoka method in Japan is a simultaneous aerobic and anaerobic landfill bioreactor where greenhouse gas is reduced, thus improving the quality of leachate and resulting in better stabilization of waste. Similarly, the biocell concept of landfill bioreactor is operated with three phases. Anaerobic phase with leachate recirculation is followed by aerobic phase for waste stabilization and the mining of waste for energy recovery (Hettiaratchi, 2012).

The cumulative bioaugmentation and biostimulation approach is shown in Fig. 2.2, with a semi-aerobic landfill bioreactor that is to be approached. The introduction of a potent landfill plastisphere into the dumpsite as bioaugmentation for bioremediation of plastics requires safe delivery to the site. Various microbial carriers like lignocellulosic biomass, which include sugarcane bagasse, corncob, sawdust, and plant fibers, were used. Even microbial immobilization enhances the degradation of various pollutants (Dzionek et al., 2016). But for plastic degradation, immobilization may hinder the assimilation rate. A new eggshell-based formulation has eggshell, cocopeat, tween, and sodium bicarbonate in the ratio of 8:1:0.6:0.4 (w/w), respectively. It was used to deliver the indigenous microbe *Janibacter cremeus* into the respective site for the remediation of hexahydro-1,3,5-trinitro-1,3,5-triazine (RDX)-contaminated unsaturated soil, which resulted in 62% cleanup of RDX (Kalsi et al., 2021). A notable study on the degradation of plastic mixture HDPE (56%): LDPE (29%): PP(12%): PS (3%) (W/W) by biostimulation with the stabilized waste as nutrients in simulated lysimeters, which mimic the open dumpsite, reveals that methanotrophs are abundant in that microbiota. HDPE has the highest kinetic decay of 0.128 per year (Muenmee et al., 2015). The supplementation of the right electron donor and electron acceptor is necessary. The valorization of methane generated in landfill bioreactors can be transformed into value-added products like polyhydroxyalkanoates (PHA) and liquid fuel, or used as biogas (Ge et al., 2014; Liu et al., 2020).

Characterization of in situ degradation and landfill plastisphere survival is crucial to validate the effectiveness of the landfill bioreactor performance. The microbiota that colonizes and forms an active biofilm in plastic debris after bioaugmentation and biostimulation can be identified by "combinatorial labeling and spectral imaging–fluorescence *in situ* hybridization" (CLASI-FISH) in confocal microscopy. The microbiota can be identified by using multiple probes at the same time to identify the diversity of the plastisphere (Schlundt et al., 2019). Application of Nanopore and Illumina platforms can reveal the microbiota of landfills. MiSeq high throughput sequencing was applied to identify the landfill bacterial microbiota. Firmicutes, bacteroidetes and proteobacteria were found dominant in the landfill microbiota. Various factors which affect bacterial diversity includes moisture, pH, dissolved organic carbon, nitrogen elements like ammonia-nitrogen and nitrate (Xu et al., 2017). The elemental analyzer/isotope ratio mass spectrometry (EA/IRMS) is used to characterize the degradation of polymers based on the variation in $\delta^{13}C$ values (Berto et al., 2017). The degradation of polymers in environmental conditions is favored by photodegradation or thermo-degradation, which results in chain scission and radicle formation. The analysis of those radical formations, biradical and triplets, can be identified by using electron paramagnetic resonance (EPR) spectroscopy; the spin traps observe the free radical and convert it into a stable one that can be analyzed by EPR. 5,5-Dimethyl-1-pyrroline-*N*-oxide (DMPO) is a commonly used spin trap (Naveed et al., 2018). The Strum test, or respirometry method, was used to estimate the rate of degradation based on CO_2. Two international standards ASTM D5338 and ISO 14855 were used to estimate the degradation of plastics in soil, and a new method was proposed based on ISO 14855, which can be applied to estimate the rate of degradation (Briassoulis et al., 2020; Silva et al., 2020).

2.5 Conclusion

Dumpsite composes diverse contaminants compared with plastics; the holistic approach to bioremediation of dumpsites with site-specific bioaugmented microbiota which favors the degradation and methanogenesis aided with biostimulation may enhance environmental safety. The management of landfill leachate either by on-site or by off-site treatment is necessary. The commercial plastics include additives like antioxidants, plasticizers, and flame retardants, which should be degraded completely. The microplastics generated in the dumpsites should also be completely degraded and assimilated by the introduction of microbiota. So, the intrusion of microplastics and nanoplastics into the water bodies can be minimized.

References

Ali, M., Yue, D., 2020. Population dynamics of microbial species under high and low ammonia nitrogen in the alternate layer bioreactor landfill (ALBL) approach. Bioresour. Technol. 315, 123787. https://doi.org/10.1016/j.biortech.2020.123787.

Ali, M.I., Ahmed, S., Javed, I., Ali, N., Atiq, N., Hameed, A., Robson, G., 2014. Biodegradation of starch blended polyvinyl chloride films by isolated *Phanerochaete chrysosporium* PV1. Int. J. Environ. Sci. Technol. 11, 339–348. https://doi.org/10.1007/s13762-013-0220-5.

Anwar, M.S., Kapri, A., Chaudhry, V., Mishra, A., Ansari, M.W., Souche, Y., Nautiyal, C.S., Zaidi, M.G.H., Goel, R., 2016. Response of indigenously developed bacterial consortia in progressive degradation of polyvinyl chloride. Protoplasma 253, 1023–1032. https://doi.org/10.1007/s00709-015-0855-9.

Arkatkar, A., Arutchelvi, J., Bhaduri, S., Veera, P., Doble, M., 2009. Degradation of unpretreated and thermally pretreated polypropylene by soil consortia. Int. Biodeter. Biodegr. 63, 106–111. https://doi.org/10.1016/j.ibiod.2008.06.005.

Arkatkar, A., Juwarkar, A.A., Bhaduri, S., Uppara, P.V., Doble, M., 2010. Growth of *Pseudomonas* and *Bacillus* biofilms on pretreated polypropylene surface. Int. Biodeter. Biodegr. 64, 530–536. https://doi.org/10.1016/j.ibiod.2010.06.002.

Artham, T., Sudhakar, M., Venkatesan, R., Nair, C.M., Murty, K.V.G.K., Doble, M., 2009. Biofouling and stability of synthetic polymers in sea water. Int. Biodeter. Biodegr. 63, 884–890. https://doi.org/10.1016/j.ibiod.2009.03.003.

ASTM WK65360, 2018. New test method for determining the carbonyl index of a polyolefin plastic material using Infra-Red spectrometry (FT-IR). ASTM International, West Conshohocken. https://www.astm.org/DATABASE.CART/WORKITEMS/WK65360.htm. Accessed on 01.10.2020.

Austin, H.P., Allen, M.D., Donohoe, B.S., Rorrer, N.A., Kearns, F.L., Silveira, R.L., Pollard, B.C., Dominick, G., Duman, R., Omari, K.E., Mykhaylyk, V., Wagner, A., Michener, W.E., Amore, A., Skaf, M.S., Crowley, M.F., Thorne, A.W., Johnson, C.W., Lee Woodcock, H., McGeehan, J.E., Beckham, G.T., 2018. Characterization and engineering of a plastic-degrading aromatic polyesterase. Proc. Natl. Acad. Sci. U. S. A. 115, E4350–E4357. https://doi.org/10.1073/pnas.1718804115.

Auta, H.S., Emenike, C.U., Jayanthi, B., Fauziah, S.H., 2018. Growth kinetics and biodeterioration of polypropylene microplastics by *Bacillus sp.* and *Rhodococcus sp.* isolated from mangrove sediment. Mar. Pollut. Bull. 127, 15–21. https://doi.org/10.1016/j.marpolbul.2017.11.036.

Awasthi, S., Srivastava, P., Singh, P., Tiwary, D., Mishra, P.K., 2017. Biodegradation of thermally treated high-density polyethylene (HDPE) by *Klebsiella pneumoniae* CH001. 3 Biotech 7, 1–10. https://doi.org/10.1007/s13205-017-0959-3.

Berto, D., Rampazzo, F., Gion, C., Noventa, S., Ronchi, F., Traldi, U., Giorgi, G., Maria, A., Giovanardi, O., 2017. Preliminary study to characterize plastic polymers using elemental analyser / isotope ratio mass spectrometry (EA/IRMS). Chemosphere 176, 47–56. https://doi.org/10.1016/j.chemosphere.2017.02.090.

Bombelli, P., Howe, C.J., Bertocchini, F., 2017. Polyethylene bio-degradation by caterpillars of the wax moth *Galleria mellonella*. Curr. Biol. 27, R292–R293. https://doi.org/10.1016/j.cub.2017.02.060.

Brandon, A.M., Gao, S.H., Tian, R., Ning, D., Yang, S.S., Zhou, J., Wu, W.M., Criddle, C.S., 2018. Biodegradation of polyethylene and plastic mixtures in mealworms (larvae of Tenebrio molitor) and effects on the gut microbiome. Environ. Sci. Technol. 52, 6526–6533. https://doi.org/10.1021/acs.est.8b02301.

Briassoulis, D., Mistriotis, A., Mortier, N., Tosin, M., 2020. A horizontal test method for biodegradation in soil of bio-based and conventional plastics and lubricants. J. Clean. Prod. 242, 118392. https://doi.org/10.1016/j.jclepro.2019.118392.

Butnaru, E., Darie-Nita, R.N., Zaharescu, T., Balaes, T., Tanase, C.T., Hitruc, G., Doroftei, F., Vasile, C., 2016. Gamma irradiation assisted fungal degradation of the polypropylene/biomass composites. Radiat. Phys. Chem. 125, 134–144. https://doi.org/10.1016/j.radphyschem.2016.04.003.

Canopoli, L., Fidalgo, B., Coulon, F., Wagland, S.T., 2018. Physico-chemical properties of excavated plastic from landfill mining and current recycling routes. Waste Manag. 76, 55–67. https://doi.org/10.1016/j.wasman.2018.03.043.

Carvalho, C.L.D., Silveira, A.F., Rosa, S., 2013. A study of the controlled degradation of polypropylene containing pro-oxidant agents. Springerplus 2, 1–11.

Contat-Rodrigo, L., Haider, N., Ribes-Greus, A., Karlsson, S., 2001. Ultrasonication and microwave assisted extraction of degradation products from degradable polyolefin blends aged in soil. J. Appl. Polym. Sci. 79, 1101–1112. https://doi.org/10.1002/1097-4628(20010207)79:6.

Das, M.P., Kumar, S., 2015. An approach to low-density polyethylene biodegradation by *Bacillus amyloliquefaciens*. 3 Biotech 5, 81–86. https://doi.org/10.1007/s13205-014-0205-1.

Das, M.P., Kumar, S., Das, J., 2018. Fungal-mediated deterioration and biodegradation study of low-density polyethylene (LDPE) isolated from municipal dump yard in Chennai, India. Energy, Ecol. Environ. 3, 229–236. https://doi.org/10.1007/s40974-018-0085-z.

Dzionek, A., Wojcieszyńska, D., Guzik, U., 2016. Natural carriers in bioremediation: a review. Electron. J. Biotechnol. 23, 28–36. https://doi.org/10.1016/j.ejbt.2016.07.003.

Eriksen, M., Lebreton, L.C.M., Carson, H.S., Thiel, M., Moore, C.J., Borerro, J.C., Galgani, F., Ryan, P.G., 2014. Plastic Pollution in the World ' s Oceans : more than 5 trillion plastic pieces weighing over 250,000 tons afloat at sea. PLoS One 9, 1–15. https://doi.org/10.1371/journal.pone.0111913.

Farzi, A., Dehnad, A., Fotouhi, A.F., 2019. Biodegradation of polyethylene terephthalate waste using *Streptomyces* species and kinetic modeling of the process. Biocatal. Agric. Biotechnol. 17, 25–31. https://doi.org/10.1016/j.bcab.2018.11.002.

Fontanella, S., Bonhomme, S., Brusson, J., Pitteri, S., Samuel, G., Pichon, G., Lacoste, J., Fromageot, D., Lemaire, J., Delort, A., 2013. Comparison of biodegradability of various polypropylene films containing pro-oxidant additives based on Mn, Mn / Fe or Co. Polym. Degrad. Stab., 1–10. https://doi.org/10.1016/j.polymdegradstab.2013.01.002.

Fuller, J., White, D., Yi, H., Colley, J., Vickery, Z., Liu, S., 2020. Analysis of volatile compounds causing undesirable odors in a polypropylene - high-density polyethylene recycled plastic resin with solid-phase microextraction. Chemosphere 260, 127589–127595. https://doi.org/10.1016/j.chemosphere.2020.127589.

Ge, X., Yang, L., Sheets, J.P., Yu, Z., Li, Y., 2014. Biological conversion of methane to liquid fuels: status and opportunities. Biotechnol. Adv. 32, 1460–1475. https://doi.org/10.1016/j.biotechadv.2014.09.004.

Geyer, R., Jambeck, J.R., Law, K.L., 2017. Production, use and fate of all plastics ever made. Sci. Adv. 3, 1–5. https://doi.org/10.1126/sciadv.1700782.

Giacomucci, L., Raddadi, N., Soccio, M., Lotti, N., Fava, F., 2019. Polyvinyl chloride biodegradation by *Pseudomonas citronellolis* and *Bacillus flexus*. N. Biotechnol. 52, 35–41. https://doi.org/10.1016/j.nbt.2019.04.005.

Gu, Z., Chen, W., Wang, F., Li, Q., 2020. A pilot-scale comparative study of bioreactor landfills for leachate decontamination and municipal solid waste stabilization. Waste Manag. 103, 113–121. https://doi.org/10.1016/j.wasman.2019.12.023.

Hazen, T.C., 2010. Biostimulation. In: Timmis, K.N. (Ed.), Handbook of Hydrocarbon and Lipid Microbiology. Springer, Berlin Heidelberg, Berlin, Heidelberg, pp. 4517–4530. https://doi.org/10.1007/978-3-540-77587-4_355.

Hermanová, S., Šmejkalová, P., Merna, J., Zarevúcka, M., 2015. Biodegradation of waste PET based copolyesters in thermophilic anaerobic sludge. Polym. Degrad. Stab. 111, 176–184. https://doi.org/10.1016/j.polymdegradstab.2014.11.007.

Hernandez, L.M., Xu, E.G., Larsson, H.C.E., Tahara, R., Maisuria, V.B., Tufenkji, N., 2019. Plastic teabags release billions of microparticles and nanoparticles into tea. Environ. Sci. Technol. https://doi.org/10.1021/acs.est.9b02540.

Hettiaratchi, J.P.A., 2012. Landfill bioreactors. In: Meyers, R.A. (Ed.), Encyclopedia of Sustainability Science and Technology. Springer New York, New York, NY, pp. 5720–5732. https://doi.org/10.1007/978-1-4419-0851-3_114.

Horodytska, O., Valdés, F.J., Fullana, A., 2018. Plastic flexible films waste management – a state of art review. Waste Manag. 77, 413–425. https://doi.org/10.1016/j.wasman.2018.04.023.

Hung, C.S., Barlow, D.E., Varaljay, V.A., Drake, C.A., Crouch, A.L., Russell, J.N., Nadeau, L.J., Crookes-Goodson, W.J., Biffinger, J.C., 2019. The biodegradation of polyester and polyester polyurethane coatings using *Papiliotrema laurentii*. Int. Biodeter. Biodegr. 139, 34–43. https://doi.org/10.1016/j.ibiod.2019.02.002.

Ibarra, V.G., Quirós, A.R.B.D., Losada, P.P., Sendón, R., 2019. Non-target analysis of intentionally and non intentionally added substances from plastic packaging materials and their migration into food simulants. Food Packag. Shelf Life 21, 100325. https://doi.org/10.1016/j.fpsl.2019.100325.

Izunobi, J.U., Higginbotham, C.L., 2011. Polymer molecular weight analysis by 1 H NMR spectroscopy. J. Chem. Educ. 88, 1098–1104. https://doi.org/10.1021/ed100461v.

Jacquin, J., Cheng, J., Odobel, C., Pandin, C., Conan, P., Pujo-Pay, M., Barbe, V., Meistertzheim, A.L., Ghiglione, J.F., 2019. Microbial ecotoxicology of marine plastic debris: A review on colonization and biodegradation by the "plastisphere.". Front. Microbiol. 10, 1–16. https://doi.org/10.3389/fmicb.2019.00865.

Jambeck, J.R., Andrady, A., Geyer, R., Narayan, R., Perryman, M., Siegler, T., Wilcox, C., Lavender Law, K., 2015. Plastic waste inputs from land into the ocean. Science (80) 347, 768–771. https://doi.org/10.1126/science.1260352.

Janczak, K., Hrynkiewicz, K., Znajewska, Z., Dąbrowska, G., 2018. Use of rhizosphere microorganisms in the biodegradation of PLA and PET polymers in compost soil. Int. Biodeter. Biodegr. 130, 65–75. https://doi.org/10.1016/j.ibiod.2018.03.017.

Jeon, H.J., Kim, M.N., 2016. Isolation of mesophilic bacterium for biodegradation of polypropylene. Int. Biodeter. Biodegr. 115, 244–249. https://doi.org/10.1016/j.ibiod.2016.08.025.

Jeyakumar, D., Chirsteen, J., Doble, M., 2013. Synergistic effects of pretreatment and blending on fungi mediated biodegradation of polypropylenes. Bioresour. Technol. 148, 78–85. https://doi.org/10.1016/j.biortech.2013.08.074.

Jung, M.R., Horgen, F.D., Orski, S.V., Viviana Rodriguez, C., Beers, K.L., Balazs, G.H., Jones, T.T., Work, T.M., Brignac, K.C., Royer, S., Hyrenbach, K.D., Jensen, B.A., Lynch, J.M., 2018. Validation of ATR FT-IR to identify polymers of plastic marine debris, including those ingested by marine organisms. Mar. Pollut. Bull. 127, 704–716. https://doi.org/10.1016/j.marpolbul.2017.12.061.

Kalsi, A., Celin, S.M., Bhanot, P., Sahai, S., Sharma, J.G., 2021. A novel egg shell-based bio formulation for remediation of RDX (hexahydro-1,3,5-trinitro-1,3,5-triazine) contaminated soil. J. Hazard. Mater. 401, 123346. https://doi.org/10.1016/j.jhazmat.2020.123346.

Karami, A., Golieskardi, A., Keong Choo, C., Larat, V., Galloway, T.S., Salamatinia, B., 2017. The presence of microplastics in commercial salts from different countries. Sci. Rep. 7, 1–11. https://doi.org/10.1038/srep46173.

Khaksar, M.R., Ghazi-Khansari, M., 2009. Determination of migration monomer styrene from GPPS (general purpose polystyrene) and HIPS (high impact polystyrene) cups to hot drinks. Toxicol. Mech. Methods 19, 257–261. https://doi.org/10.1080/15376510802510299.

Khan, S., Nadir, S., Ullah, Z., Ali, A., Hasan, F., 2017. Biodegradation of polyester polyurethane by *Aspergillus tubingensis*. Environ. Pollut. 225, 469–480. https://doi.org/10.1016/j.envpol.2017.03.012.

Koshti, R., Mehta, L., Samarth, N., 2018. Biological recycling of polyethylene terephthalate: a mini-review. J. Polym. Environ. 26, 3520–3529. https://doi.org/10.1007/s10924-018-1214-7.

Lebeau, T., 2011. Bioaugmentation for in situ soil remediation: How to ensure the success of such a process. In: Singh, A., Parmar, N., Kuhad, R.C. (Eds.), Bioaugmentation, Biostimulation and Biocontrol. Springer, Berlin Heidelberg, Berlin, Heidelberg, pp. 129–186. https://doi.org/10.1007/978-3-642-19769-7_7.

Liu, L.-Y., Xie, G.-J., Xing, D.-F., Liu, B.-F., Ding, J., Ren, N.-Q., 2020. Biological conversion of methane to polyhydroxyalkanoates: current advances, challenges and perspectives. Environ. Sci. Ecotechnol 2, 100029. https://doi.org/10.1016/j.ese.2020.100029.

Mandal, D.K., Bhunia, H., Bajpai, P.K., Kumar, A., Madhu, G., Nando, G.B., 2017. Biodegradation of pro-oxidant filled polypropylene films and evaluation of the Ecotoxicological impact. J. Polym. Environ. https://doi.org/10.1007/s10924-017-1016-3.

Miyazaki, K., Arai, T., Shibata, K., Terano, M., Nakatani, H., 2012. Study on biodegradation mechanism of novel oxo-biodegradable polypropylenes in an aqueous medium. Polym. Degrad. Stab. 97, 2177–2184. https://doi.org/10.1016/j.polymdegradstab.2012.08.010.

Mohan, A.J., Sekhar, V.C., Bhaskar, T., Nampoothiri, K.M., 2016. Microbial assisted high impact polystyrene (HIPS) degradation. Bioresour. Technol. 213, 204–207. https://doi.org/10.1016/j.biortech.2016.03.021.

Mrozik, A., Piotrowska-seget, Z., 2010. Bioaugmentation as a strategy for cleaning up of soils contaminated with aromatic compounds. Microbiol. Res. 165, 363–375. https://doi.org/10.1016/j.micres.2009.08.001.

Muenmee, S., Chiemchaisri, W., Chiemchaisri, C., 2015. Microbial consortium involving biological methane oxidation in relation to the biodegradation of waste plastics in a solid waste disposal open dump site. Int. Biodeter. Biodegr. 102, 172–181. https://doi.org/10.1016/j.ibiod.2015.03.015.

Mumtaz, T., Khan, M.R., Hassan, M.A., 2010. Study of environmental biodegradation of LDPE films in soil using optical and scanning electron microscopy. Micron 41, 430–438. https://doi.org/10.1016/j.micron.2010.02.008.

Myint, B., Ravi, K., 2012. Biodegradation of Low Density Polythene (LDPE) by *Pseudomonas*. Species 52, 411–419. https://doi.org/10.1007/s12088-012-0250-6.

Nauendorf, A., Krause, S., Bigalke, N.K., Gorb, E.V., Gorb, S.N., Haeckel, M., Wahl, M., Treude, T., 2016. Microbial colonization and degradation of polyethylene and biodegradable plastic bags in temperate fine-grained organic-rich marine sediments. Mar. Pollut. Bull. 103, 168–178. https://doi.org/10.1016/j.marpolbul.2015.12.024.

Naveed, K., Wang, L., Yu, H., Ullah, R.S., 2018. Recent progress in the electron paramagnetic resonance study of polymers. Polym. Chem. 9, 3306–3335. https://doi.org/10.1039/c8py00689j.

Novotný, Č., Malachová, K., Adamus, G., Kwiecień, M., Lotti, N., Soccio, M., Verney, V., Fava, F., 2018. Deterioration of irradiation/high-temperature pretreated, linear low-density polyethylene (LLDPE) by *Bacillus amyloliquefaciens*. Int. Biodeter. Biodegr. 132, 259–267. https://doi.org/10.1016/j.ibiod.2018.04.014.

Ojha, N., Pradhan, N., Singh, S., Barla, A., Shrivastava, A., Khatua, P., Rai, V., Bose, S., 2017. Evaluation of HDPE and LDPE degradation by fungus, implemented by statistical optimization. Sci. Rep. 7, 1–13. https://doi.org/10.1038/srep39515.

Paço, A., Duarte, K., João, P., Santos, P.S.M., Pereira, R., Pereira, M.E., Freitas, A.C., Duarte, A.C., Rocha-santos, T.A.P., 2017. Biodegradation of polyethylene microplastics by the marine fungus *Zalerion maritimum*. Sci. Total Environ. 586, 10–15. https://doi.org/10.1016/j.scitotenv.2017.02.017.

Pushpadass, H.A., Weber, R.W., Dumais, J.J., Hanna, M.A., 2010. Biodegradation characteristics of starch-polystyrene loose-fill foams in a composting medium. Bioresour. Technol. 101, 7258–7264. https://doi.org/10.1016/j.biortech.2010.04.039.

Raddadi, N., Fava, F., 2019. Biodegradation of oil-based plastics in the environment : existing knowledge and needs of research and innovation. Sci. Total Environ. 679, 148–158. https://doi.org/10.1016/j.scitotenv.2019.04.419.

Rouillon, C., Bussiere, P.O., Desnoux, E., Collin, S., Vial, C., Therias, S., Gardette, J.L., 2016. Is carbonyl index a quantitative probe to monitor polypropylene photodegradation? Polym. Degrad. Stab. 128, 200–208. https://doi.org/10.1016/j.polymdegradstab.2015.12.011.

Roy, P.K., Titus, S., Surekha, P., Tulsi, E., Deshmukh, C., Rajagopal, C., 2008. Degradation of abiotically aged LDPE films containing pro-oxidant by bacterial consortium. Polym. Degrad. Stab. 93, 1917–1922. https://doi.org/10.1016/j.polymdegradstab.2008.07.016.

Russell, J.R., Huang, J., Anand, P., Kucera, K., Sandoval, A.G., Dantzler, K.W., Hickman, D., Jee, J., Kimovec, F.M., Koppstein, D., Marks, D.H., Mittermiller, P.A., Nu, S.J., Santiago, M., Townes, M.A., Vishnevetsky, M., Williams, N.E., Boulanger, L., Bascom-slack, C., Strobel, S.A., 2011. Biodegradation of polyester polyurethane by endophytic Fungi. Appl. Environ. Microbiol. 77, 6076–6084. https://doi.org/10.1128/AEM.00521-11.

Sang, N.N., Soda, S., Ishigaki, T., Ike, M., 2012. Microorganisms in landfill bioreactors for accelerated stabilization of solid wastes. J. Biosci. Bioeng. 114, 243–250. https://doi.org/10.1016/j.jbiosc.2012.04.007.

Sangeetha Devi, R., Rajesh Kannan, V., Nivas, D., Kannan, K., Chandru, S., Robert Antony, A., 2015. Biodegradation of HDPE by *Aspergillus spp.* from marine ecosystem of gulf of Mannar. India. Mar. Pollut. Bull. 96, 32–40. https://doi.org/10.1016/J.MARPOLBUL.2015.05.050.

Schlundt, C., Mark Welch, J.L., Knochel, A.M., Zettler, E.R., Amaral-Zettler, L.A., 2019. Spatial structure in the "Plastisphere": molecular resources for imaging microscopic communities on plastic marine debris. Mol. Ecol. Resour. 620–634. https://doi.org/10.1111/1755-0998.13119.

Shah, Z., Hasan, F., Krumholz, L., Fulya, D., Ali, A., 2013. Degradation of polyester polyurethane by newly isolated *Pseudomonas aeruginosa* strain MZA-85 and analysis of degradation products by GC-MS. Int. Biodeter. Biodegr. 77, 114–122. https://doi.org/10.1016/j.ibiod.2012.11.009.

Shah, Z., Gulzar, M., Hasan, F., Shah, A.A., 2016. Degradation of polyester polyurethane by an indigenously developed consortium of *Pseudomonas and Bacillus* species isolated from soil. Polym. Degrad. Stab. 134, 349–356. https://doi.org/10.1016/j.polymdegradstab.2016.11.003.

Shahnawaz, M., Sangale, M.K., Ade, A.B., 2016a. Bacteria-based polythene degradation products: GC-MS analysis and toxicity testing. Environ. Sci. Pollut. Res. 23, 10733–10741. https://doi.org/10.1007/s11356-016-6246-8.

Shahnawaz, M., Sangale, M.K., Ade, A.B., 2016b. Rhizosphere of *Avicennia marina* (Forsk .) Vierh. as a landmark for polythene degrading bacteria. Environ. Sci. Pollut. Res., 14621–14635. https://doi.org/10.1007/s11356-016-6542-3.

Sheik, S., Chandrashekar, K.R., Swaroop, K., Somashekarappa, H.M., 2015. Biodegradation of gamma irradiated low density polyethylene and polypropylene by endophytic fungi. Int. Biodeter. Biodegr. 105, 21–29. https://doi.org/10.1016/j.ibiod.2015.08.006.

Shimpi, N., Borane, M., Mishra, S., Kadam, M., 2012. Biodegradation of polystyrene (PS)-poly(lactic acid) (PLA) nanocomposites using *Pseudomonas aeruginosa*. Macromol. Res. 20, 181–187. https://doi.org/10.1007/s13233-012-0026-1.

Silva, S.A., Hinkel, E.W., Lisboa, T.C., Selistre, V.V., da Silva, A.J., da Silva, L.O.F., Faccin, D.J.L., Cardozo, N.S.M., 2020. A biostimulation-based accelerated method for evaluating the biodegradability of polymers. Polym. Test. 91, 106732. https://doi.org/10.1016/j.polymertesting.2020.106732.

Singh, R., Pant, D., 2016. Polyvinyl chloride degradation by hybrid (chemical and biological) modification. Polym. Degrad. Stab. 123, 80–87. https://doi.org/10.1016/j.polymdegradstab.2015.11.012.

Sivan, A., Szanto, M., Pavlov, V., 2006. Biofilm development of the polyethylene-degrading bacterium *Rhodococcus ruber*. Appl. Microbiol. Biotechnol. 72, 346–352. https://doi.org/10.1007/s00253-005-0259-4.

Skariyachan, S., Manjunath, M., Shankar, A., Bachappanavar, N., Patil, A.A., 2019. Application of novel microbial consortia for environmental site remediation and hazardous waste management toward low- and high-density polyethylene and prioritizing the cost-effective, eco-friendly and sustainable biotechnological intervention. In: Hussain, C.M. (Ed.), Handbook of Environmental Materials Management. Springer International Publishing, Cham, pp. 431–478. https://doi.org/10.1007/978-3-319-73645-7_9.

Suja, F., Rahim, F., Taha, M.R., Hambali, N., Rizal Razali, M., Khalid, A., Hamzah, A., 2014. Effects of local microbial bioaugmentation and biostimulation on the bioremediation of total petroleum hydrocarbons (TPH) in crude oil contaminated soil based on laboratory and field observations. Int. Biodeter. Biodegr. 90, 115–122. https://doi.org/10.1016/j.ibiod.2014.03.006.

Swati, T.I.S., Vijay, V.K., Ghosh, P., 2018. Scenario of landfilling in India: Problems, challenges and recommendations. In: Hussain, C.M. (Ed.), Handbook of Environmental Materials Management. Springer International Publishing, Cham, pp. 1–16. https://doi.org/10.1007/978-3-319-58538-3_167-1.

Tahir, C., Qazi, I.A., Hashmi, I., Bhargava, S., 2016. Biodegradation of low density polyethylene (LDPE) modified with dye sensitized titania and starch blend using *Stenotrophomonas pavanii*. Int. Biodeter. Biodegr. 113, 276–286. https://doi.org/10.1016/j.ibiod.2016.01.025.

Warne, C., Tufenkji, N., Ghoshal, S., 2010. A modified microbial adhesion to hydrocarbons assay to account for the presence of hydrocarbon droplets. J. Colloid Interface Sci. 344, 492–496. https://doi.org/10.1016/j.jcis.2009.12.043.

Wen, B., Liu, J.H., Zhang, Y., Zhang, H.R., Gao, J.Z., Chen, Z.Z., 2020. Community structure and functional diversity of the plastisphere in aquaculture waters: does plastic color matter? Sci. Total Environ. 740, 140082. https://doi.org/10.1016/j.scitotenv.2020.140082.

Xiong, J., Liao, X., Zhu, J., An, Z., Yang, Q., Huang, Y., Li, G., 2017. Natural weathering mechanism of isotactic polypropylene under different outdoor climates in China. Polym. Degrad. Stab. 146, 212–222. https://doi.org/10.1016/j.polymdegradstab.2017.10.012.

Xu, S., Lu, W., Liu, Y., Ming, Z., Liu, Y., Meng, R., Wang, H., 2017. Structure and diversity of bacterial communities in two large sanitary landfills in China as revealed by high-throughput sequencing (MiSeq). Waste Manag. 63, 41–48. https://doi.org/10.1016/j.wasman.2016.07.047.

Xue, N., Wang, L., Li, W., Wang, S., Pan, X., Zhang, D., 2020. Increased inheritance of structure and function of bacterial communities and pathogen propagation in plastisphere along a river with increasing antibiotics pollution gradient. Environ. Pollut. 265, 114641. https://doi.org/10.1016/j.envpol.2020.114641.

Yang, J., Yang, Y., Wu, W.M., Zhao, J., Jiang, L., 2014. Evidence of polyethylene biodegradation by bacterial strains from the guts of plastic-eating waxworms. Environ. Sci. Technol. 48, 13776–13784. https://doi.org/10.1021/es504038a.

Yang, Y., Yang, J., Wu, W., Zhao, J., Song, Y., Gao, L., Yang, R., Jiang, L., 2015a. Biodegradation and mineralization of polystyrene by plastic-eating mealworms: part 1. Chemical and Physical Characterization and Isotopic Tests. Environ. Sci. Technol. https://doi.org/10.1021/acs.est.5b02661.

Yang, Y., Yang, J., Wu, W., Zhao, J., Song, Y., Gao, L., Yang, R., Jiang, L., 2015b. Biodegradation and mineralization of polystyrene by plastic-eating mealworms: part 2. Role of Gut Microorganisms. Environ. Sci. Technol. https://doi.org/10.1021/acs.est.5b02663.

Yang, S.S., Brandon, A.M., Andrew Flanagan, J.C., Yang, J., Ning, D., Cai, S.Y., Fan, H.Q., Wang, Z.Y., Ren, J., Benbow, E., Ren, N.Q., Waymouth, R.M., Zhou, J., Criddle, C.S., Wu, W.M., 2018. Biodegradation of polystyrene wastes in yellow mealworms (larvae of *Tenebrio molitor* Linnaeus): factors affecting biodegradation rates and the ability of polystyrene-fed larvae to complete their life cycle. Chemosphere 191, 979–989. https://doi.org/10.1016/j.chemosphere.2017.10.117.

Yoshida, S., Hiraga, K., Takehana, T., Taniguchi, I., Yamaji, H., Maeda, Y., Toyohara, K., Miyamoto, K., Kimura, Y., Oda, K., 2016. A bacterium that degrades and assimilates poly(ethylene terephthalate). Science (80-.) 351, 1196–1199. https://doi.org/10.1126/science.aad6359.

Zafar, U., Nzeram, P., Langarica-fuentes, A., Houlden, A., Heyworth, A., Saiani, A., Robson, G.D., 2014. Biodegradation of polyester polyurethane during commercial composting and analysis of associated fungal communities. Bioresour. Technol. 158, 374–377. https://doi.org/10.1016/j.biortech.2014.02.077.

Zettler, E.R., Mincer, T.J., Amaral-Zettler, L.A., 2013. Life in the "plastisphere": microbial communities on plastic marine debris. Environ. Sci. Technol. 47, 7137–7146. https://doi.org/10.1021/es401288x.

Zettler, L.A., Zettler, E.R., Mincer, T.J., 2020. Ecology of the plastisphere. Nat. Rev. Microbiol. 18, 139–151. https://doi.org/10.1038/s41579-019-0308-0.

CHAPTER

3

Bioremediation approach for treatment of soil contaminated with radiocesium

M. Srinivasulu[a,b], G. Narasimha[c], and A.J. Francis[a,d]

[a]Division of Advanced Nuclear Engineering, Pohang University of Science and Technology (POSTECH), Pohang, South Korea [b]Department of Microbiology, Sri Krishnadevaraya University, Anantapuramu, Andhra Pradesh, India [c]Applied Microbiology Laboratory, Department of Virology, Sri Venkateswara University, Tirupati, Andhra Pradesh, India [d]Environmental and Climate Sciences Department, Brookhaven National Laboratory, Upton, New York, United States

OUTLINE

3.1 Introduction

Presence of radiocesium in the environment is a great concern because of its high solubility as an alkaline metal ion, its long half-life (30.2 years for ^{137}Cs), and its easy assimilation by living organisms (Bostick et al., 2002; Tsukada et al., 2002). The solubility and retention of radiocesium are strongly dependent on the adsorption properties of the solid phase, including mineralogy, pH, ionic strength, competitive cations, and organic matter (Kim and Kirkpatrick, 1997; Dumat and Staunton, 1999; Cha et al., 2006; Giannakopoulou et al., 2007; Nakamaru et al., 2007; Bellenger and Staunton, 2008; Fan et al., 2012). The Great East Japan Earthquake occurred on March 11, 2011, and caused an explosion at the Tokyo Electric Power Company (TEPCO) Fukushima Daiichi Nuclear Power Plant, which resulted in the release of the large amount of radionuclides into the environment (Water Supply Division (WSD), Health Service Bureau (HSB), Ministry of Health, Labour, and Welfare (MHLW), 2011). The radioactive cesium (^{134}Cs and ^{137}Cs) was detected in drinking water after the explosion (WSD, HSB, MHLW, 2011; Ikemoto and Magara, 2011).

The radiation contamination after the Fukushima Daiichi Nuclear Power Plant accident drew a major concern throughout the world. Several countries, regions, and oceans were impacted. An effective remediation strategy is urgently needed. Although many studies have been carried out, very little progress has been made with radiocesium

Cost Effective Technologies for Solid Waste and Wastewater Treatment
https://doi.org/10.1016/B978-0-12-822933-0.00006-1

decontamination (Dahu et al., 2016). The Chernobyl nuclear accident has caused a great concern and an interest in the environmental fate of radiocesium because radioactive air plumes have reached several European countries and contaminated large territories, particularly forested areas (Nimis, 1996). The isotope ^{137}Cs was one of the most important radionuclides involved in that contamination. It causes considerable environmental problems because of its quite long half-life. Approximately 6% of the European territory showed a contamination level of ^{137}Cs above 20 kBq/m^2—a level that was above 40 and 1480 kBq/m^2 in 2.0% and 0.03% of the European territory, respectively (Izrael et al., 1996). The ecosystems contaminated include agricultural, natural, and semi-natural environments. In all these environments soil acts as a major sink-source compartment in ^{137}Cs fluxes through plant and food chains (Alexakhin et al., 1996; Prister et al., 1996). Actually, ^{137}Cs soil contamination generates a long-term burden on food chains. For instance, food intake in contaminated regions can contribute to 60% of the total radiation dose in humans (Strand et al., 1996). A number of in situ measurements after the Chernobyl accident showed that nutrient cycling and storage led to a much longer persistence of radiocesium in semi-natural ecosystems in comparison with agricultural cropping systems (Frissel et al., 1990; Myttenaere et al., 1992; Gerzabek et al., 1998; Strebl et al., 1999).

Phytoremediation has been suggested as a cost-effective strategy for the remediation of soils contaminated with radionuclides and heavy metals. Cesium may be particularly amenable to phytoremediation because Cs has a similar ionic potential to the plant nutrient potassium (K) and has been observed to exhibit many similarities to K with respect to plant uptake (Gommers et al., 2000; White and Broadly, 2000). Knowledge of the nature and reversibility of Cs sorption to the soil solid phase is essential to evaluate the potential of remediation for Cs removal from the soil and to develop contaminant removal technologies.

Phyllosilicate minerals comprise much of the active solid phase in many low organic matter temperate soils. Cesium interactions with phyllosilicate mineral surfaces are relatively strong due to the formation of inner sphere complexes. Cesium bound to external planar surface sites can generally be exchanged by other cations, whereas that bound in collapsed (dehydrated) interlayer sites or at frayed edges of a collapsed interlayer is more strongly sorbed and difficult to remove via the ion exchange (Comans et al., 1991). The availability of interlayer sites for cation exchange depends on interlayer spacing, which is influenced by the degree of hydration of the ions present in the interlayer. Cations with low hydration energy such as Cs, K, rubidium (Rb), and ammonium (NH_4) can shed their hydration shell and enter clay interlayers. Dehydration of these ions permits a close approach to the tetrahedral silicate layers and the formation of polar bonds with structural oxygen atoms. Cesium bound to interlayer sites is not readily exchanged by other cations and is generally considered fixed (Comans et al., 1991).

A strong surface association between Cs and the soil solid phase occurs at frayed edge sites (FES) of micaceous phyllosilicate minerals. The exudates of plant roots, microorganisms, and organic matter decomposition products may enhance the bioavailability of Cs in the rhizosphere by accelerating weathering at frayed edges of phyllosilicate minerals and thus releasing Cs sorbed to frayed edge and interlayer sites and makes the ion available for uptake. The FES of illite, a secondary clay mineral similar to muscovite mica, have a particularly high affinity for Cs (Francis and Brinkley, 1976; Staunton and Roubaud, 1997). Characterization of Cs desorption from illite will provide an insight regarding Cs dynamics in contaminated rhizosphere soils, in particular, the potential applicability of bioremediation for the removal of Cs from contaminated soils.

A selectivity coefficient describing the exchange properties of regular (planar) exchange sites or FES with respect to Cs can be determined by comparing the retention of Cs in relation to a competing ion, such as K or NH_4. Cation exchange on accessible sites of expandable phyllosilicate minerals or soil organic matter occurs easily with relatively small differences in selectivity among alkali and alkaline earth metals. Exchange on sites at collapsible interlayers such as those found in micas and vermiculite, on the other hand, shows high selectivity for large alkali metals (K, NH_4, Cs) over their smaller counterparts (Na, Li) or alkaline earth metals (Ca, Mg). The less negative hydration energy of the large alkali metals allows dehydration and collapse of the interlayer or sites near the edges of collapsed interlayers of micas . As a result, FES can exhibit very high selectivity for large alkali metals ($Cs > NH_4 > K$) relative to other cations (Comans et al., 1991; Staunton and Roubaud, 1997; Kruyts and Delvaux, 2002; Wauters et al., 1996; Rigol et al., 2002). FES will retain sorbed Cs and K against exchange with a large excess of Ca but not against exchange with a large excess of NH_4 (Wauters et al., 1996).

3.2 Deposition of radiocesium

The atmospheric wet and dry depositions of radiocesium from nuclear weapon tests in the 1950s and 1960s were estimated up to 1.5×10^{18} Bq, whereas that of the Chernobyl accident in 1986 amounted to 7.8×10^{16} Bq (Izrael et al., 1996). The former deposits represent dispersed, relatively low level but chronic inputs, whereas the latter was a pulse input involving a relatively high contamination level. As demonstrated after the Chernobyl fallout, rainfall patterns have a decisive influence on the ^{137}Cs contamination of lands (De Vries and Van der Kooy, 1986; Fowler et al., 1987). The Chernobyl ^{137}Cs far-fallout occurred in a soluble form in Western Europe, whereas the ^{137}Cs near-fallout consisted of particle forms of uranate and/or ferrite type (Auerbach, 1986; Rayyes et al., 1993). The very low solubility of these particles delays the mobilization of radiocesium by the above-standing vegetation (Konoplev et al., 1993; Sanzharova et al., 1994).

Cs radionuclides are considered among the most hazardous products of nuclear fission, capable to reach the top of food chain and thus to become a risk factor for human health. It is assumed that the fate of radionuclides in the environment follows the behavior of stable elements. According to many authors (Tsukada et al., 2002; Yoshida et al., 2004), the soil-to plant-transfer factor of radioactive Cs is well correlated with that of stable Cs, indicating that the properties of stable Cs in different ecosystems may be regarded as a useful tool in predicting the behavior of ^{137}Cs and ^{134}Cs. Hence, to facilitate experimental studies on radiocesium sorption and mobility in the environment, the utilization of stable Cs was applied.

The sorption reactions at soil-water interfaces are one of the most important phenomena, which determine the transport of Cs in aquatic and soil environments, fate, and bioavailability to biota. A key factor for Cs sorption is the content of clay and especially the type of clay minerals present in soils (Shender and Eriksson, 1993). Cesium has very small hydration energy; thus, the electrostatic attraction of Cs ions by clay particles is large and therefore are preferentially sorbed (Hakem et al., 1999). Besides the Cs sorption on phyllosilicate minerals, it also correlates strongly with the cation exchange capacity of the soil (Grutter et al., 1990).

3.3 Target ecosystems

The environmental contamination of radiocesium rapidly leads to its accumulation in soil, in agricultural areas as well as in natural and semi-natural environments. Radionuclide contamination of agricultural lands is obviously highly sensitive, because of the direct impact on the food chain. Semi-natural areas are recognized as environments showing little interference with humans (forests, upland pastures, moorland, marshland, peatland, alpine meadows, unimproved meadows, and tundras). Semi-natural and natural areas are defined as follows: In the latter, the flora and fauna are original and spontaneous, and the vegetation has not undergone structural modifications by humans. In the semi-natural countryside, the flora and the fauna are still mainly native but the vegetation is greatly modified by human intervention; the relationships between biological forms have been modified. The dominant biological form has often been replaced and the derived vegetation no longer belongs to the same formation as the original vegetation (Poldini, 1990). The semi-natural environments produce human food as well as various other materials. The use of these ecosystems is important for some population groups (Skuterud et al., 1997; Wright et al., 1997). After the Chernobyl accident, it was observed that the vegetation in natural and semi-natural environments was highly contaminated with ^{137}Cs than crops from agricultural lands (Frissel et al., 1990; Myttenaere et al., 1992; Tsukada et al., 1998). Higher ^{137}Cs biological half-lives were measured in these environments versus agricultural areas, demonstrating a longer persistence in the natural and semi-natural ecosystems (Muck and Gerzabek, 1996). Consequently, one can expect an increase of radiocesium doses to humans from the contribution of these highly complex ecosystems (Skuterud et al., 1997; Howard and Howard, 1997). Among these environments, forest ecosystems are particularly complex as they involve several soil and vegetative strata as well as diverse plant communities.

3.4 Radiocesium incorporation in the biogeochemical cycle

Once in the surface horizons of the soil, the fate of radiocesium can follow the three main routes: strong retention by soil particles, migration with percolating soil solution, and uptake by plant roots and microorganisms. As radiocesium is strongly retained in soil mineral phase, Cs migration rates are quite small and generally $< 2\,cm\,yr^{-1}$ (Beckmann and Faas, 1992; Bunzl et al., 1994). The major pathway for plant contamination is root uptake. Traditionally, the assessment of the contamination of above-standing vegetation with radiocesium is commonly based on the determination of transfer factor (TF) or aggregated transfer factor (T_{agg}) coefficients expressing the ratio of radionuclide concentration in the plant to that in the soil. The large number of *in situ* measurement studies, particularly after Chernobyl, show that TFs exhibit large seasonal as well as yearly variations in the same ecosystem (Salt and Mayes, 1993; Lembrechts et al., 1990; Burrough et al., 1999). Also, they vary largely between ecosystems, soils, plants, and microbial activity.

3.5 Chemistry of cesium

Cesium is a very reactive metal; it combines quickly with oxygen in the air and reacts violently with water. In the reaction with water, hydrogen gas is released; hydrogen gas ignites immediately as a result of the heat given off the reaction. Cesium must be stored in kerosene or mineral oil to protect it from reacting with oxygen and water vapor in the air. Cesium also reacts vigorously with acids, and with the halogens sulfur and phosphorus (Avery, 1995).

In comparison with many of the most commonly studied metals of environmental concern, e.g., Cd^{2+} and Cu^{+}, Cs^{+} is a very weak Lewis acid; i.e., the ion has a very low charge/radius ratio, and consequently has a low polarizing power and is seen by neighboring anions as a center of low charge density (Hughes and Poole, 1991). Thus, Cs^{+}, like other monovalent cations, interacts very weakly with ligands, and analytical problems relating to metal speciation can essentially be disregarded. Any Cs-ligand complexation will be primarily ionic/electrostatic in nature and therefore weaker than the more covalent interactions that occur between nitrogen- or sulfur-donor ligands and softer metals, such as Cu^{2+} and the monovalent ion Ti^{+} (Gadd, 1992; Hughes and Poole, 1991).

The toxicity of certain metals is related to a strong coordinating ability (Ochiai, 1987). Therefore, soft metals are generally toxic towards microorganisms, whereas hard metals are often nontoxic and in many cases are essential macronutrients for microbial growth (Avery and Tobin, 1993; Hughes and Poole, 1991). No essential biological role of stable Cs has been elucidated, although trace quantities have been detected in many microorganisms (Gadd, 1992). Moreover, the chemical similarity of Cs^{+} and other alkali monovalent cations, in particular K^{+}, is the key factor that governs the high mobility of Cs^{+} in biological systems (Komarov and Bennett, 1983). K^{+} is an essential cation for microbial growth and is accumulated intracellularly to cytosolic concentrations of several hundred millimolar, while Na^{+} is maintained at a low intracellular level (Padan and Vitterbo, 1988; Walderhaug et al., 1987). Cesium (Cs^{+}) may act as an analogue for K^{+} and enter cells via metabolism-dependent transport processes.

3.6 Microbial remediation of cesium

Microorganisms play a significant role in the transformation of radionuclides and affect their stability, mobility, and bioavailability in the environment. Microbial processes have been extensively investigated for the remediation of contaminated materials, soils, sediments, wastes, and wastewater effluents. Microbial activities affect the mobilization and immobilization of Cs by biosorption/bioaccumulation and dissolution/remobilization by metabolic and organic matter decomposition products (Strandberg et al., 1981; West et al., 1991; Avery and Tobin, 1993; Avery, 1995; Ohnuki et al., 2003; Lan et al., 2014; Mukhopadhyay et al. (2007); Vinichuk et al., 2013; Masao et al., 2019; Saskai et al., 2013; Koji and Ayumi, 2018; Wendling et al., 2004, 2005; Takuro et al., 2020). After the Chernobyl accident, the bioaccumulation of ^{137}Cs in various organisms, such as a cyanobacterium (Avery et al., 1991), fungi (Haselwandter et al., 1988), mushrooms (Korky and Kowalski, 1989), and mosses (Elstner et al., 1987), was studied as an indicator of radioactive contamination. Several types of microbes such as bacteria, fungi, mycorrhizae, algae, and lichens isolated from the natural and radionuclide-contaminated environments have been tested for their ability to concentrate and bioaccumulate Cs (Table 3.1). Likewise, the activities of naturally occurring microbial communities in the ecosystem to solubilize Cs during organic matter degradation and/or by their metabolic products were investigated (Wendling et al., 2004, 2005; Takuro et al., 2020). Studies of the bioaccumulation by the microbial cells or biomass could be useful for the removal and monitoring of cesium in the wastewater of nuclear facilities and liquid wastes (Plato and Denovan, 1974; Strandberg et al., 1981). An important factor in the design of a bioaccumulation process study is the isolation of the appropriate microorganisms.

3.6.1 Bacteria

Bacteria accumulate radionuclides and toxic metals extracellularly and/or intracellularly and concentrate them. Consequently, there has been great interest in using microbial biomass to remove radionuclides and toxic metals from waste streams. Strandberg et al. (1981) investigated the ability of *Pseudomonas aeruginosa*, *Saccharomyces cerevisiae*, and a mixed culture of denitrifying bacteria to accumulate cesium for the potential treatment of radioactively contaminated waste solutions. They observed very low levels of Cs bioaccumulation by these cultures. In contrast, Tomioka et al. (1992) isolated from soil Cs accumulating rod-coccus shaped bacteria *Rhodococcus erythropolis* CS98 and *Rhodococcus* sp. strain CS402. The cesium concentration in the culture medium decreased with cell growth. The strain CS98 showed the maximum removal efficiency of 90% at 24h of incubation, whereas strain CS402 exhibited the maximum removal efficiency of 47% at 48h of incubation. In the case of strain CS98, the cesium content in the

TABLE 3.1 Radiocesium accumulation by selected microorganisms.

Microorganism	Reference
Bacteria	
Pseudomonas aeruginosa *Rhodococcus erythropolis* strain CS98 *Rhodococcus sp.* strain CS402 *Halomonas israelensis* *Bacillus pumilus, Bacillus licheniformis* *Nocardiopsis sp.* 13H strain	Strandberg et al. (1981); Tomioka et al. (1992); Tomioka et al. (1994); Tomioka et al. (1998). Asad and Haggai (1998) Abdel-Razek et al. (2015) Sivaperumal et al. (2018)
Fungi	
Saccharomyces cerevisiae *Rhodosporidium fluviale* strain UA2 *Pleurotus citrinopileatus* *Pleurotus ostreatus*	Strandberg et al. (1981); Avery and Tobin (1993); Ohnuki et al. (2003) Lan et al. (2014) Mukhopadhyay et al. (2007) Bazala et al. (2008)
Mycorrhizal fungi:	
Lactarius rufus *Inocybelongcystis* *Arbuscular Mycorrhizal Fungi* (families including): *Glomeraceae, Gigasporaceae, Paraglomeraceae, Claroideoglomeraceae, Acaulosporaceae, Archeosporaceae, Ambisporaceae, Diversisporaceae, and uncultured Glomeromycotina*	Dighton and Horrill (1988) Vinichuk et al. (2013); Masao et al. (2019)
Algae	
Microcoleus vaginatus (Cyanobacteria) *Draparnaldia plumosa* (Green algae) *Navicula seminulum* *Synechocystis strain* PCC 6803 *Euglena intennedia,* *Chlorella pyrenoidosa* *Nostoc commune* *Vacuoliviride crystalliferum* *Galdieria sulphuraria* *Enteromorpha torta* (Marine green algae) *Padina australis* *Sargassum glaucescens* *Cystoseria indica* *Haematococcus pluvialis* and *Chlorella vulgaris* *Desmodesmus armatus* SCK *Synechococcus* PCC7002	Harvey and Patrick (1967) Avery et al. (1991) Williams (1960) Saskai et al. (2013) Koji and Ayumi (2018) Omar et al. (2010) Jalali-Rada et al. (2004) Lee et al. (2019) Kim et al. (2019) Yu et al. (2020)
Lichens	
Cladonia fimbriata, Cladonia squamosa, Pseudevernia furfuracea, Hypogymnia physodes *Cetraria islandica, Cladonia fimbriata, Usnea barbata*	Iurian et al. (2011) Nedić et al. (2006)
Azolla	
Azolla filiculoides	Mashkani and Ghazvini (2009)

cells increased rapidly, decreased slowly, and then became stable. The cesium content in CS402 increased over 48 h, decreased, and then became stable. The maximum values of cesium content of CS98 and CS402 were 52.0 μmol/g (dry weight) of cells at 17 h and 18.8 μmol/g (dry weight) of cells at 48 h, respectively (Tomioka et al., 1992). The pH increased from 7.0 to 9.1 with the progression of growth for all the bacteria tested. The total amount of cesium in the cells and in the medium ranged from 95% to 106% of the added cesium throughout the cultivation. Tomioka et al. (1992) concluded that the *Rhodococcus erythropolis* CS98 and *Rhodococcus* sp strain CS402 have a high-level ability to accumulate cesium. However, the cesium accumulated by strains CS98 and CS402 was released after 24 and 48 h of incubation, respectively. These bacteria are known to produce mesodiaminopimelic acid, mycolic acids, and tuberculostearic acids and what role if any, these acids play in Cs accumulation is not known. Strandberg et al. (1981)

estimated the values of ^{137}Cs concentration factors by *Pseudomonas aeruginosa* and *Saccharomyces cerevisiae* to be at 16 and 37, respectively, based on the wet weight. However, the values of the concentration factors of CS98 and CS402 were much higher than those of *P. aeruginosa* and *S. cerevisiae*.

Asad and Haggai (1998) reported that the cells of *Halomonas israelensis* grown in medium containing different salt concentrations (0.2–4 M) and incubated with 25 mM CsCl without potassium showed Cs uptake. NMR studies indicated that the intracellular cesium signal has shifted from the cesium signal in the medium and the shift depended on the salt concentration in growth medium. Intracellular cesium shift showed a lesser dependence on the concentration of salts in the medium, suggesting that the transport of Cs through the cell membrane was mostly by an active transport system. The cesium concentration in the cell was higher than that of the medium, ~ 100 mM intracellular concentration compared with 25 mM in the medium (Asad and Haggai, 1998).

Several bacterial species isolated from the hazardous liquid wastes at Hot Laboratories and waste Management Center, Cairo, Egypt, were investigated for their ability to remove ^{137}Cs from waste solutions. The biosorption capabilities of the free and immobilized biomass were studied by batch experiments. Different immobilized matrices such as calcium alginate (CA), chitosan (CTS), chitosan-alginate (CTS/CA), and polyvinyl alcohol-alginate (PVA/CA) were tested for the application in the biosorption system. The immobilized system showed the maximum biosorption capacities at ^{137}Cs solution activity of 15,000 Bq/mL with CA and CA-immobilized *Bacillus pumilus* and *Bacillus licheniformis* beads. The CA-immobilized system could be used for more than one cycle with a removal efficiency of about 50% at the first cycle (Abdel-Razek et al., 2015).

3.6.2 Fungi

Several species of fungi, including yeast cells (Strandberg et al., 1981; Avery and Tobin, 1993; Ohnuki et al., 2003), the mycelia of saprophytic fungi (Dighton et al., 1991; Lan et al., 2014; Seeprasert et al., 2016; El-Sayyad et al., 2018), the hyphae and fruit bodies of mushrooms, and mycorrhizae, accumulate radiocesium and play a major role in translocating Cs in forest ecosystem (Dighton and Horrill, 1988; Muhopadhyay et al., 2007; Vinichuk et al., 2013; Ohnuki et al., 2016; Masao et al., 2019). The mycelium of *Aspergillus niger* accumulated the Cs-137 from spiked simulated aqueous solutions (El-Sayyad et al., 2018) and *Fusarium* sp., *Trichoderma* sp., and *Aspergillus* sp. from decomposing plant residues (Seeprasert et al., 2016).

Significant accumulation of radiocesium by a wide variety of wild and edible mushrooms from contaminated wood, litter, and soils has been observed (Table 3.1). They have been extensively examined for radiocesium transfer in the food chain as well as their potential application in the bioremediation of the contaminated soils. These studies also showed that the accumulation of radioactive Cs varied considerably with the fungal species. However, the mechanisms of accumulation from contaminated materials to the fruit bodies is not fully understood.

3.6.2.1 *Mycorrhizae*

Mycorrhizae is the mutual symbiotic association of a fungus and plant particularly in the rhizosphere of the root zone. Arbuscular mycorrhiza (AM) refers to mycorrhizas whose hyphae have extensively penetrated into the plant cells. Mycorrhiza play an important role in plant nutrient availability, particularly phosphorous. Elevated levels of radiocesium have been observed in fruiting bodies of several mycorrhizal fungal species at Chernobyl (Dighton and Horrill, 1988). The arbuscular mycorrhizal fungi (AMF) are beneficial for the remediation of ^{137}Cs-contaminated soils (Masao et al., 2019). However, no information was available on species diversity and the composition of AMF communities in soil contaminated with ^{137}Cs after the Fukushima-Daiichi Nuclear Power Plant (NPP) disaster. The AMF families identified include *Glomeraceae, Gigasporaceae, Paraglomeraceae, Claroideoglomeraceae, Acaulosporaceae, Archeosporaceae, Ambisporaceae, Diversisporaceae,* and *uncultured Glomeromycotina*. Among them, *Glomeraceae* was the most abundant in both grassland and paddy field, followed by *Paraglomeraceae* (Masao et al., 2019).

3.6.3 Lichens and mosses

Lichens consist of a simple photosynthetic organism, a green alga typically a cyanobacterium, living among filaments of multiple fungi species in a mutualistic symbiotic relationship. The exudates from lichen have a chelating ability. Mosses are small, nonvascular flowerless plants that form dense green clumps or mats. The individual plants are usually composed of simple leaves. Cyanobacteria colonize moss and receive shelter in return for providing fixed nitrogen.

Lichens are well known to accumulate radiocesium and retain for several years. Numerous studies after the Chernobyl nuclear accident in 1986 showed that lichens and mosses are suitable as bioindicators of the nuclear fallout. Iurian et al. (2011) measured ^{137}Cs activities in different mosses, epigenic and epiphytic lichens, collected from three sampling sites from Koppl, in the Salzburg Province. The concentrations of Cs-137 ranged from 1765 ± 13

to $7345 \pm 33\,Bq\,kg^{-1}$ dry weight in selected lichens (*Cladonia fimbriata, Cladonia squamosa, Pseudevernia furfuracea, Hypogymnia physodes*) and from 1145 ± 16 to $14092 \pm 46\,Bq\,kg^{-1}$ dry weight in moss species (*Leucobryum glaucum, Sphagnum papillosum, Dicranodontium denudatum, Polytrichum strictum, Sphagnum fallax*) (Iurian et al., 2011). After 20 years since the Chernobyl nuclear accident, the cesium levels in lichens and mosses from the mountain region of province were still very high. These nonvascular plants are suitable and inexpensive bioindicators of the radioactive fallout. Differences in ^{137}Cs levels were observed between different species of lichens and mosses, mainly due to the morphological characteristics or the growth rate. The moss species *Leucobryum glaucum* showed the highest activity of Cs-137 (14,092 $\pm$ 46 Bq/kg dry weight), but it cannot be concluded that all mosses have higher capacity for the storage of radionuclide than the lichens. Among the species studied, the moss had the highest biological half-life for Cs-137 (22.5 $\pm$ 0.7 years) (Iurian et al., 2011).

The mechanism of uptake by lichens may involve surface complexation with the functional groups on the biomass surface, or by the precipitation on the cell wall surface. Intracellular uptake involving metabolic processes has been proposed, but has not been confirmed. The nature of Cs association in moss whether it is with the nonvascular plant, the cyanobacterium, or both, has not fully been elucidated.

A 5% solution of both ammonium oxalate and phosphoric acid solubilized 77.5% of ^{137}Cs from *Cetraria islandica*, 47.6% from *Cladonia fimbriata*, and 46.4% from *Usnea barbata*. The tested lichen species have similar specific radioactivities (i.e., amount of ^{137}Cs); the difference is attributed to the existence of different types of bonds between radiocesium and the binding sites. The crystals precipitated from the extracts contained most of the soluble ^{137}Cs, and the amount and the specific radioactivity of crystals varied between the lichen species (Nedić et al., 2006).

3.6.4 Algae

Studies of cesium accumulation by algae have been investigated due to their importance in aquatic ecosystems. Harvey and Patrick (1967) calculated the ^{137}Cs concentration factors in various algae on the basis of the dry weight of biomass. The highest values of the concentration factors in cyanobacteria, green algae, and diatoms were 9.5×10^2 (*Microcoleus vaginatus*), 1.5×10^3 (*Draparnaldia plumosa*), and 1.6×10^3 (*Navicula seminulum*), respectively. Williams (1960) reported that the values of the concentration factors of ^{137}Cs for *Euglena intennedia* and *Chlorella pyrenoidosa* were 7.1×10^2 and 1.1×10^2, respectively. The average value of the concentration factor of ^{137}Cs for the freshwater algae was 4.6×10^2 (Harvey and Patrick, 1967).

The heterocystous blue green alga *Nostoc commune* grows under high radiation exposure environment, and it forms jelly-like clumps of polysaccharides, which are known to accumulate the radionuclides deposited on them, thus simplifying their removal from the contaminated environment. Saskai et al. (2013) monitored ^{134}Cs and ^{137}Cs in Nihonmatsu City, Fukushima Prefecture, and reported that *N. commune* accumulated 415,000 Bq/kg dry weight ^{134}Cs and 607,000 Bq/kg dry weight ^{137}Cs. These results were confirmed by a controlled cultivation experiment with the blue green alga from radiocesium-contaminated soil (Saskai et al., 2013). The decontamination of polluted soil and water can be achieved by using *N. commune* because this species forms jelly-like clumps and can be easily collected. In addition, when the biomass dried, the weight loss was ~ 90%, resulting in less volume of radioactive waste generated for disposal.

The eustigmatophycean algae *Vacuoliviride crystalliferum*, the cyanophytes *Stigonema ocellatum*, and *Nostoc commune* showed the highest bioaccumulation activity for the removal of cesium, strontium, and iodine from the environment (Koji and Ayumi, 2018). Besides these strains, the extremophilic unicellular red algae *Galdieria sulphuraria* removed high levels of dissolved cesium from media in mixotrophic growth conditions (Koji and Ayumi, 2018).

The marine green algae *Enteromorpha torta* (Wulfen) J. Agardh collected from the Western Alexandria Coast removed ^{134}Cs from the aqueous solutions (Omar et al., 2010). Different dry weights of *E. torta* (0.2, 0.15, 0.1, 0.05, and 0.025 g) were immobilized with 4% calcium alginate. Immobilized algal biomass (IAB) enhanced the adsorption capability of ^{134}Cs from solution (Omar et al., 2010).

Microalgae *Haematococcus pluvialis* and *Chlorella vulgaris* removed 95% and 90% of soluble ^{137}Cs, respectively, from aqueous medium (Lee et al., 2019). Kim et al. (2019) evaluated the potential sequestration of cesium (Cs^+) by nine microalgae under heterotrophic growth conditions in order to develop a system for the treatment of radioactive wastewater. They examined the effects of initial Cs^+ concentration (100–500 μM), pH (5–9), K^+ and Na^+ concentrations (0–20 mg/L), and different organic carbon sources (acetate, glycerol, glucose) on Cs^+ removal. Of the nine microalgae tested, *Desmodesmus armatus* SCK removed the most Cs^+ under various environmental conditions. Addition of organic substrates significantly enhanced Cs^+ uptake by *D. armatus*, even in the presence of a competitive cation (K^+). Magnetic nanoparticles coated with a cationic polymer (polyethylenimine) were used to separate ^{137}Cs-containing microalgal biomass under a magnetic field. Combining both bioaccumulation and magnetic separation removed more than 90% of the radioactive ^{137}Cs from an aqueous medium. They suggested that the use of microalgae-magnetic particles with an inexpensive organic substrate appears to have great potential for bioremediation of ^{137}Cs-polluted wastewater (Kim et al., 2019).

3.6.5 Azolla

Azolla, an aquatic fern, is a very invasive plant. It lives in a symbiotic relationship with the cyanobacterium *Anabaena azollae*. The cyanobacterium fixes atmospheric nitrogen and provides the essential nutrient to the plant. Azolla has been extensively studied as a biofertilizer among others as a principle source of nitrogen to rice paddies. The mechanism of Cs bioaccumulation neither by the fern nor by the cyanobacterium has been determined.

Cesium bioaccumulation studies by *Azolla filiculoides* showed that it accumulated up to 195 mg/g Cs (Mashkani and Ghazvini, 2009). The best Cs removal results were observed when *A. filiculoides* was treated with 2M $MgCl_2$ and 30 mL H_2O_2 8 mM at pH 7 for 12 h, and it was then washed with NaOH solution at pH 10.5 for 6 h. Pretreatment of *Azolla* presumably has modified the surface characteristics, which improved Cs biosorption. The binding of Cs on the cell wall of *Azolla* was characterized by micro-PIXE and FT-IR (Mashkani and Ghazvini, 2009).

3.6.6 Mechanisms of cesium accumulation in microorganisms

Microorganisms living or dead possess an abundance of functional groups on the cell surface capable of binding metals. Biosorption of metals occurs by complexation of positively charged metal ions with negatively charged reactive sites such as amino, carboxyl, hydroxyl, and phosphate groups on the cell surface. Extracellular polymeric substances (EPS) secreted by microorganisms are also known to sequester metals. Intracellular accumulation of metals in all microorganisms occurs via energy-dependent transport systems. Several studies on the mechanism of bioaccumulation and release of cesium by bacteria and fungi have been reported (Avery et al., 1991; Bossemeyer et al., 1989; Johnson et al., 1991; Tomioka et al., 1994; Ohnuki et al., 2003; Lan et al., 2014). Avery et al. (1991) showed that it is primarily intracellular uptake and the cesium accumulation was directly proportional to the extracellular cesium concentration over the range of 0.2 to 2.0 mM cesium. The accumulated cesium was released after 24 h of incubation by *Synechocystis strain* PCC6803. They considered that this release might be a response to an increased internal osmotic pressure. The accumulation of cesium was also markedly influenced by external pH, with increasing cesium accumulation at greater (alkaline) pHs in *Synechocystis strain* PCC 6803 (Avery et al., 1991).

The mechanism of Cs accumulation by bacteria is energy dependent like monovalent cation K^+ transport system and not simple adsorption (Tomioka et al., 1994). Ohnuki et al. (2003) determined Cs accumulation and distribution in yeast cells (*S. cerevisiae*) by micro-PIXE (particle-induced X-ray emission) and EDS (energy-dispersive spectroscopy) coupled to a SEM (scanning electron microscope), and found that cells grown in the presence of Cs showed that Cs was uniformly distributed in the cells. Lan et al. (2014) used transmission electron microscopy (TEM), PIXE, and enhanced proton back scattering Spectroscopy (EPBS) to understand the structure surface morphology and the mechanisms of Cs biosorption by *Rhodosporidium fluviale* strain UA2 isolated from Cs solution. Cesium association with *R. fluviale* strain UA2 occurred via passive and active mechanisms; Cs ions were sorbed to the cell surface by electrostatic attraction and transported by an active energy-dependent transport system through the cell membrane and deposited in the cytoplasm.

Marine actinobacterium *Nocardiopsis sp.* 13H strain isolated from nuclear power plant sites in India removed 88.6% ± 0.72% of Cs^+ from the test solution containing 10 mM $CsCl_2$. Biosorption was confirmed by scanning electron microscopy (SEM) coupled with energy-dispersive spectroscopy (EDS). Most of the bound cesium was associated with extracellular polymeric substances (EPS), suggesting its interaction with the surface active groups. Fourier transform infrared (FTIR) spectroscopic analysis revealed biosorption of Cs with the carboxyl, hydroxyl, and amide groups on the cell surface (Sivaperumal et al., 2018).

Yu et al. (2020) investigated the biosorption mechanism of Cs by marine cyanobacteria *Synechococcus* PCC7002 by Fourier transform infrared spectroscopy, Raman spectroscopy, scanning electron microscopy, transmission electron microscopy, energy-dispersive X-ray spectrometry, and three-dimensional excitation-emission matrix fluorescence spectroscopy. They observed passive biosorption of low concentrations of Cs bound to the extracellular proteins in EPS and with the amino, hydroxyl, and phosphate functional groups on the cell walls, as well as an active intracellular accumulation of high concentration Cs presumably due to intracellular transport mechanisms.

The radioactive Cs is known to accumulate in fungal hyphae, and it is believed that radioactive Cs is transported to the fruiting body through hyphae (Vinichuk et al., 2005). The hyphae function in the uptake and transport of the radioactive Cs dissolved in interstitial water into fruit bodies (Bazala et al., 2008). Cesium accumulated in the hyphae of *Pleurotus ostreatus* is associated with intercellular materials of polyphosphate. Sugiyama et al. (2008) examined the characteristics of Cs accumulation and localization in the mycelia of edible mushroom *Pleurotus ostreatus*-Y1. Scanning electron microscopic images showed the presence of white spots, and energy-dispersive X-ray microanalyzer analysis indicated larger amounts of Cs and P in these spots in mycelia cultured on medium containing 25 mM CsCl. The ^{137}Cs activities in the mycelia were about 4 – 6 times higher than those in water used. Higher Cs concentrations were associated with the sediment fraction containing vacuolar pellets and that Cs in the mycelia was trapped by polyphosphate in vacuoles or other organelles.

Ohnuki et al. (2016) reported the presence of a direct pathway of radioactive Cs accumulation into the fruit body from the contaminated wood logs. Another study on the accumulation paths of radioactive Cs in hyphae using inhibitors of uptake suggested a possible indirect pathway of Cs accumulation in fruit bodies by extracellular transport via interhyphal cavities (Bystrzejewska-Piotrowska and Bazala, 2008). In the forest, the fruiting bodies of edible and inedible mushrooms tend to grow after rain events. The rainwater dissolves radioactive Cs in the litter zone (Sakai et al., 2015). A small portion of the dissolved radioactive Cs is transported directly to the fruiting body, which causes the excess accumulation of radioactive Cs in the fruit body rather than through hyphae. Thus, the direct accumulation pathway of radioactive Cs from the contaminated wood, litter, and soil facilitates the migration of radioactive Cs in forest. The fungal hyphae strongly retain radioactive Cs in organic layers in the forest ecosystem (Dighton et al., 1991; Seeprasert et al., 2016).

The presence of minerals in agar medium inhibits the accumulation of radioactive Cs by unicellular fungus (Ohnuki et al., 2015). Likewise, the presence of the minerals at the position near to the fruit body decreases the concentration of radioactive Cs in the fruit body. Addition of minerals such as zeolite and vermiculite onto the litter zone of the contaminated forest is expected to reduce the accumulation of radioactive Cs by edible as well as inedible mushrooms. It is likely that radioactive Cs transported through hyphae is released outside of the hyphae and is sorbed by the minerals (Ohnuki et al., 2016).

The mechanism of Cs uptake by lichens is generally considered as an abiotic process, i.e., surface complexation with functional groups on lichen biomass surface, or by the precipitation reactions with solid phases on the exteriors of cell walls. Intracellular exchange mechanisms involving metabolic processes facilitating Cs uptake have been proposed but not documented (Pipíška et al., 2005).

3.7 Remobilization of cesium due to microbial activity

Microbial activities under aerobic and anaerobic conditions can affect the mobilization and/or remobilization of Cs in soil and aquatic environments. The soil rhizosphere, and the soil zone influenced by plant roots, is chemically, biologically, and mineralogically distinct from the bulk soil (Kruyts and Delvaux, 2002). The soil solution composition and pH in the rhizosphere can be quite different from those of the surrounding bulk soil due to the presence of plant roots, organic exudates, and microbial communities. Many studies have documented the significance of plant residue decomposition products, exudates from plant roots, and rhizosphere microorganisms to phyllosilicate mineral weathering processes and remobilization of Cs in the soil environment and bioavailability (Berthelin and Leyval, 1982; Wendling et al., 2004, 2005; Kim et al., 1997; Robert and Berthelin, 1986; Berthelin, 1983; Boyle and Voigt, 1973; Johnson et al., 1991; Clint et al., 1992; Song et al., 2012). Wendling et al. (2005) studied Cs desorption from illite in the presence of exudates from the rhizosphere bacteria of crested wheatgrass (*Agropyron desertorum*). Bacterial exudates significantly increased Cs desorption from high-Cs illite (120 mmol kg^{-1}) after 16 days. Cesium desorption from illite in the presence of 0.2% *Bacillus*, 1% *Ralstonia*, and 2% *Enterobacter* exudate solutions was significantly enhanced compared with Cs desorption with the cell-free bacterial growth media. More Cs was desorbed from illite in the presence of 1% *Ralstonia* and 2% *Enterobacter* exudate solutions than in the presence of 0.2% *Bacillus* exudate solution, suggesting differences in their chemical composition and/or concentration (Wendling et al., 2005). Anaerobic degradation of radiocesium-contaminated crop residues released about 82% of plant bound radiocesium into the water phase. The radiocesium from the water phase was removed with 90% efficiency using zeolite (30 g/L) as an adsorbent; however, Prussian blue beads at 3 g/L removed 99% of radiocesium (Takuro et al., 2020). These studies suggest that by enhancing the microbial activity in soils and in plant residues, radiocesium can be extracted and recovered from the contaminated environment.

Although microbial processes in the contaminated environment regulate the mobility and bioavailability of cesium, we have very little understanding of the nature of the association of radiocesium and the chemical form of the association with wood and organic matter, and the identity of the microbial degradation products involved. Furthermore, the mechanisms of release Cs bound from freshly contaminated vs aged wood, leaf litter, and soil are not fully understood in order to develop an appropriate bioremediation strategy. One would presume that the freshly contaminated soils are much more amenable to bioremediation by increasing microbial activity than the aged soils wherein Cs is irreversibly bound with frayed edge sites.

It is evident that several of these studies demonstrate the potential use of microorganisms such as bacteria, fungi, and algae by biosorption and bioaccumulation processes and/or selective extraction by microbial metabolic products in the remediation of radiocesium-contaminated liquid effluent wastes and soils.

3.8 Conclusions

The fate of radiocesium in the environment is strongly dependent on the adsorption properties of the solid phase, including mineralogy, pH, ionic strength, competing cations, and organic matter. Cesium interactions with phyllosilicate mineral surfaces are relatively strong due to the formation of inner sphere complexes. Cesium bound to interlayer sites is not readily exchanged by other cations and is generally considered fixed and irreversible. Cesium has very small hydration energy; thus, the electrostatic attraction of Cs ions by clay particles is large and therefore is preferentially sorbed. Microorganisms affect radiocesium mobility and bioavailability. Further research is needed to understand the cesium accumulation mechanisms in various microorganisms and in particular the role of microbial processes in regulating radiocesium mobility from various contaminated soils, wood, and plant residues. Microbial exudates are naturally occurring, promising remedial agent for radiocesium decontamination that warrent further studies.

Acknowledgment

The work was supported in part by the National Research Foundation of Korea funded by the Ministry of Education, Science and Technology under the World Class University (WCU) and BK21+ program.

References

Abdel-Razek, A.S., Shaaban, M., Mahmoud, S., Kandeel, E.M., 2015. Bioaccumulation of ^{137}Cs by immobilized bacterial species isolated from radioactive wastewater. J. Appl. Environ. Microbiol. 3, 112–118.

Alexakhin, R., Firsakova, S., Rauret, I., Dalmau, G., Arkhipov, N., Vandecasteele, C., et al., 1996. Fluxes of radionuclides in agricultural environments: main results and still unsolved problems. In: Karaoglou, A., Desmet, G., Kelly, G.N., Menzel, H.G. (Eds.), The Radiological Consequences of the Chernobyl Accident. European Commission, Brussels, p. 39. EUR 16544 EN.

Asad, S., Haggai, G., 1998. Nuclear magnetic resonance studies of cesium-133 in the halophilic halotolerant bacterium Ba_1. Chemical shift and transport studies. NMR Biomed. 11, 80–86.

Auerbach, S.I., 1986. Comparative behaviour of three long-lived radionuclides in forest ecosystems. In: Annual Seminar on the Cycling of Long-lived Radionuclides in the Biosphere. CEC-CIEMAT meeting, Sept. 1-19, Madrid. vol. I.

Avery, S.V., 1995. Cesium accumulation by microorganisms: uptake mechanisms cation competition, compartmentalization and toxicity. J. Ind. Microbiol. 14, 76–84.

Avery, S.V., Tobin, J.M., 1993. Mechanism of adsorption of hard and soft metal ions to *Saccharomyces cerevisiae* and influence of hard and soft anions. Appl. Environ. Microbiol. 59, 2851–2856.

Avery, S.V., Codd, G.A., Gadd, G.M., 1991. Caesium accumulation and interactions with other monovalent cations in the cyanobacterium Synechocystis PCC 6803. J. Gen. Microbiol. 137, 405–413.

Bazala, M.A., Goda, K., Bystrzejewska-Piotrowska, G., 2008. Transport of radiocesium in mycelium and its translocation to fruitbodies of a saprophytic *macromycete*. J. Environ. Radioact. 99, 1200–1202.

Beckmann, C., Faas, C., 1992. Radioactive contamination of soils in lower Saxony, Germany, after the Chernobyl accident. Analyst 117, 525–527.

Bellenger, J.P., Staunton, S., 2008. Adsorption and desorption of ^{85}Sr and ^{137}Cs on reference minerals, with and without inorganic and organic surface coatings. J. Environ. Radioact. 99, 831–840.

Berthelin, J., 1983. Microbial weathering processes. In: Krumbein, W.E. (Ed.), Microbial Geochemistry. Blackwell Scientific Publications, Oxford, pp. 223–262.

Berthelin, J., Leyval, C., 1982. Ability of symbiotic and non-symbiotic rhizospheric microflora of maize (*Zea mays*) to weather micas and to promote plant growth and plant nutrition. Plant Soil 68, 369–377.

Bossemeyer, D., Schlosser, A., Bakker, E.P., 1989. Specific cesium transport via the *Escherichia coli* Kup (TrkD) K^+ uptake system. J. Bacteriol. 171, 2219–2221.

Bostick, B.C., Vairavamurthy, M.A., Karthikeyan, K.G., Chorover, J., 2002. Cesium adsorption on clay minerals: an EXAFS spectroscopic investigation. Environ. Sci. Technol. 36, 2670–2676.

Boyle, J.R., Voigt, G.K., 1973. Biological weathering of silicate minerals: implications for tree nutrition and soil genesis. Plant Soil 38, 191–201.

Bunzl, K., Foster, H., Kracke, W., Schimmack, W., 1994. Residence times of fallout $^{239+240}$Pu, ^{238}Pu, ^{241}Am and ^{137}Cs in the upper horizons of an undisturbed grassland soil. J. Environ. Radioact. 22, 11–27.

Burrough, P.A., Van der Perk, M., Howard, B.J., Prister, B.S., Sansone, U., Voitsekhovitch, O.V., 1999. Environmental mobility of radiocesium in the Pripyat catchment Ukraine/Belarus. Water Air Soil Poll. 110, 35–55.

Bystrzejewska-Piotrowska, G., Bazala, M.A.A., 2008. Study of mechanisms responsible for incorporation of cesium and radiocesium into fruitbodies of king oyster mushroom (*Pleurotus eryngii*). J. Environ. Radioact. 99, 1185–1191.

Cha, H.J., Kang, M.J., Chung, G.H., Choi, G.S., Lee, C.W., 2006. Accumulation of ^{137}Cs in soils on different bedrock geology and textures. J. Radioanal. Nucl. Chem. 267, 349–355.

Clint, G.M., Harrison, A.F., Howard, D.M., 1992. Rates of leaching of ^{137}Cs and potassium from different plant litters. J. Environ. Radioact. 16, 65–76.
Comans, R.N.J., Haller, M., De Preter, P., 1991. Sorption of cesium on illite: non-equilibrium behavior and reversibility. Geochim. Cosmochim. Acta 55, 433–440.
Dahu, D., Zhenya, Z., Zhongfang, L., Yingnan, Y., Tianming, C., 2016. Remediation of radiocesium-contaminated liquid waste, soil, and ash: a mini review since the Fukushima Daiichi nuclear power plant accident. Environ. Sci. Pollut. Res. 23, 2249–2263.
De Vries, W., Van der Kooy, A., 1986. Radioactivity Measurements Arising From Chernobyl, IRI Report, 190-86-01. Interfaculty Reactor Institute, Delft.
Dighton, J., Horrill, A.D., 1988. Radiocesium accumulation in the mycorrihal fungi *Lactorius rufus* and *Inocybelongcystis*. In Upland Britain. Trans. Mycol. Soc. 91, 335–337.
Dighton, J., Clint, G.M., Poskitt, J., 1991. Uptake and accumulation of Cs-137 by upland grassland soil fungi: a potential pool of Cs immobilization. Mycol. Res. 95, 1052–1056.
Dumat, C., Staunton, S., 1999. Reduced adsorption of cesium on clay minerals caused by various humic substances. J. Environ. Radioact. 46, 187–200.
El-Sayyad, H.E., Eskander, S.B., Bayuomi, T.A., 2018. Capability of *Aspergillus niger* to bioconcentrate cesium-137 and cobalt- 60 from medium and low-level radioactive waste solution simulates. Indian J. Microbiol. Res. 5, 318–325.
Elstner, E.F., Fink, R., Holl, W., Lengfelder, E., Ziegler, H., 1987. Natural and Chernobyl-caused radioactivity in mushrooms, mosses and soil-samples of defined biotops in SW Bavaria. Oecologia 73, 553–558.
Fan, Q.H., Xu, J.Z., Niu, Z.W., Li, P., Wu, W.S., 2012. Investigation of Cs(I) uptake on Beishan soil combined batch and EDS techniques. Appl. Radiat. Isot. 70, 13–19.
Fowler, S.W., Buat-Menard, P., Yokoyama, Y., Ballestra, S., Holm, E., Nguyen, H.V., 1987. Rapid removal of Chernobyl fallout from Mediterranean surface waters by biological activity. Nature 329, 56–58.
Francis, C.W., Brinkley, F.S., 1976. Preferential adsorption of ^{137}Cs to micaceous minerals in contaminated freshwater sediment. Nature (London) 260, 511–513.
Frissel, M.J., Noordijk, H., Van Bergeijk, K.E., 1990. The impact of extreme environmental conditions as occurring in natural ecosystems, on the soil-plant transfer of radionuclides. In: Desmet, G., Nassimbeni, P., Belli, M. (Eds.), Transfer of Radionuclides in Natural and Semi-Natural Environments. Elsevier Applied Science, London, p. 40.
Gadd, G.M., 1992. Metals and microorganisms: a problem of definition. FEMS Microb. Lett. 100, 197–204.
Gerzabek, M.H., Strebl, F., Temmel, B., 1998. Plant uptake of radionuclides in lysimeter experiments. Environ. Poll. 99, 93–103.
Giannakopoulou, F., Haidouti, C., Chronopoulou, A., Gasparatos, D., 2007. Sorption behavior of cesium on various soils under different pH levels. J. Hazard. Mater. 149, 553–556.
Gommers, A., Thirty, Y., Vandenhove, H., Vandecasteele, C.M., Smolders, E., Merckx, R., 2000. Radiocesium uptake by one-year-old willows planted as short rotation coppice. J. Environ. Qual. 29, 1384–1390.
Grutter, A., Von Gunten, R., Kohle, M., Rossler, E., 1990. Sorption, desorption and exchange of cesium on glaciofluvial deposits. Radiochem. Acta 50, 177–184.
Hakem, N., Mahamid, I., Apps, J., Moridis, G., 1999. Sorption of cesium and strontium on Hanford soil. J. Radio Nucl. Chem. 246, 275–278.
Harvey, R.S., Patrick, R., 1967. Concentration of ^{137}Cs, ^{65}Zn, and ^{85}Sr by fresh-water algae. Biotechnol. Bioeng. 9, 449–456.
Haselwandter, K., Berreck, M., Brunner, P., 1988. Fungi as bioindicators of radiocaesium contamination: pre- and post-Chernobyl activities. Trans. Br. Mycol. Soc. 90, 171–174.
Howard, B.J., Howard, D.C., 1997. The radioecological significance of semi-natural ecosystems. In: Health Impacts of Large Releases of Radionuclides. Ciba. Found. Symp, 203, pp. 21–37.
Hughes, M.N., Poole, R.K., 1991. Metal speciation and microbial growth-the hard (and soft) facts. J. Gen. Microbiol. 137, 725–734.
Ikemoto, T., Magara, Y., 2011. Measures against impacts of nuclear disaster on drinking water supply systems in Japan. Water Pract. Technol. 6. wpt20110078.
Iurian, A.R., Hofmann, W., Lettner, H., Türk, R., Cosma, C., 2011. Long term study of Cs-137 concentrations in lichens and mosses. Rom. J. Phys. 56, 983–992.
Izrael, Y.A., De Cort, M., Jones, A.R., Nazarov, I.M., Fridman, S.D., Kvasnikova, E.V., et al., 1996. The atlas of cesium-137 contamination of Europe after the Chernobyl accident. In: Karaoglou, A., Desmet, G., Kelly, G.N., Menzel, H.G. (Eds.), The Radiological Consequences of the Chernobyl Accident. European Commission, Brussels, p. 1. EUR 16544 EN.
Jalali-Rada, R., Ghafouriana, H., Asefa, Y., Dalira, S.T., Sahafipoura, M.H., Gharanjik, B.M., 2004. Biosorption of cesium by native and chemically modified biomass of marine algae: introduce the new biosorbents for biotechnology applications. J. Hazar. Mater. 116, 125–134.
Johnson, E., O'Donnel, A.G., Ineson, P., 1991. An autoradiographic technique for selecting Cs-137-sorbing microorganisms from soil. J. Microbiol. Methods 13, 293–298.
Kim, Y., Kirkpatrick, R.J., 1997. ^{23}Na and ^{133}Cs NMR study of cation adsorption on mineral surfaces: local environments, dynamics, and effects of mixed cations. Geochim. Cosmochim. Acta 61, 5199–5208.
Kim, K.Y., McDonald, G.A., Jordan, D., 1997. Solubilization of hydroxyapatite by *Enterobacter agglomerans* and cloned *Escherichia coli* in culture medium. Biol. Fert. Soils 24, 347–352.
Kim, H., Yang, H.M., Park, C.W., Yoon, I.-H., Seo, B.-K., Eun-Kyung Kim, E.-K., et al., 2019. Removal of radioactive cesium from an aqueous solution via bioaccumulation by microalgae and magnetic separation. Sci. Rep. 9, 10149. https://doi.org/10.1038/s41598-019-46586-x.
Koji, I., Ayumi, M., 2018. Bioremediation of biophilic radionuclides by algae. Intech Open, 1–17. https://doi.org/10.5772/intechopen.81492.
Komarov, E., Bennett, B.G., 1983. Selected Radionuclides. World Health Organization, Geneva.
Konoplev, A.V., Viktorova, N.V., Virchenko, E.P., Popov, V.E., Bulgakov, A.A., Desmet, G.M., 1993. Influence of agricultural countermeasures on the ratio of different chemical forms of radionuclides in soil and soil solution. Sci. Tot. Environ. 137, 147–162.
Korky, J.K., Kowalski, L., 1989. Radioactive cesium inedible mushrooms. J. Agric. Food. Chem. 37, 568–569.
Kruyts, N., Delvaux, B., 2002. Soil organic horizons as a major source for radiocesium biorecycling in forest ecosystems. J. Environ. Radioact. 58, 175–190.

Lan, T., Feng, Y., Liao, J., Li, X., Ding, C., Zhang, D., et al., 2014. Biosorption behavior and mechanism of cesium-137 on *Rhodosporidium fluviale* strain UA2 isolated from cesium solution. J. Environ. Radioact. 134, 6–13.

Lee, K.Y., Lee, S.H., Lee, J.E., Lee, S., 2019. Biosorption of radioactive cesium from contaminated water by microalgae *Haematococcus pluvialis* and *Chlorella vulgaris*. J. Environ. Manag. 233, 83–88.

Lembrechts, J.F., Stoutjesdijk, J.F., Van Ginkel, J.H., Noordijk, H., 1990. Soil-to-grass transfer of radionuclides: local variation and fluctuations as a function of time. In: Desmet, G., Nassimbeni, P., Belli, M. (Eds.), Transfer of Radionuclides in Natural and Semi-Natural Environments. Elsevier Applied Science, London, p. 524.

Masao, H., Dong, J.K., Katsunori, I., 2019. First report of community dynamics of arbuscular mycorrhizal fungi in radiocesium degradation lands after the Fukushima-Daiichi Nuclear disaster in Japan. Sci. Rep. 9 (8240), 1–10.

Mashkani, S., Ghazvini, P., 2009. Biotechnological potential of *Azolla filiculoides* for biosorption of Cs and Sr: application of micro-PIXE for measurement of biosorption. Biores. Technol. 100, 1915–1921.

Muck, K., Gerzabek, M.H., 1996. In: Long-term reduction of root uptake of Cs-isotopes after nuclear fallout, *Mitt.d. Osterr. Bodenkundl.Ges.*, H. 53, S. 199–206. International Symposium on Radioecology.

Mukhopadhyay, B., Nag, M., Laskar, S., Lahiri, S., 2007. Accumulation of radiocesium by *Pleurotus citrinopileatus* species of edible mushroom. J. Radioanal. Nuclear Chem. 273, 415–418.

Myttenaere, C., Schell, W.R., Thiry, Y., Sombre, L., Ronneau, C., 1992. Modelling of ^{137}Cs cycling in forest: recent developments and research needed. Sci. Tot. Environ. 136, 77–91.

Nakamaru, Y., Ishikawa, N., Tagami, K., Uchida, S., 2007. Role of soil organic matter in the mobility of radiocesium in agricultural soils common in Japan. Colloid Surf. A 306, 111–117.

Nedić, O., Stanković, A., Stanković, S., 2006. Specificity of Lichen species in respect to ^{137}Cs binding. Int. J. Environ. Anal. Chem. 76, 311–318.

Nimis, P.L., 1996. Radiocesium in plants of forest ecosystems. Stud. Geobot. 15, 3–49.

Ochiai, E.I., 1987. General Principles of Biochemistry of the Elements. Plenum Press, New York.

Ohnuki, T., Sakamoto, F., Kozai, N., Ozaki, T., Narumi, I., Francis, A.J., et al., 2003. Application of micro-PIXE technique to uptake study of cesium by *Saccharomyces cerevisiae*. Nucl. Instrum. Meth. Phys. Res. B 210, 378–382.

Ohnuki, T., Fuminori, S., Shinya, Y., Naofumi, K., Hiroyuki, S., Satoshi, U., et al., 2015. Effect of minerals on accumulation of Cs by fungus *Saccharomyces cerevisiae*. J. Environ. Radioact. 144, 127–133.

Ohnuki, T., Yukitoshi, A., Fuminori, S., Naofumi, K., Tadafumi, N., Yoshito, S., 2016. Direct accumulation pathway of radioactive cesium to fruit bodies of edible mushroom from contaminated wood logs. Sci. Rep. 6, 1–6. https://doi.org/10.1038/srep298666.

Omar, H.A., Abdel-Razek, A.S., Sayed, M.S., 2010. Biosorption of cesium-134 from aqueous solutions using immobilized marine algae: equilibrium and kinetics. Nat. Sci. 8, 140–147.

Padan, E., Vitterbo, A., 1988. Cation transport in Cyanobacteria. Meth. Enzymol. 167, 561–577.

Pipíška, M., Koèiová, M., Horník, M., Augustín, J., Lesný, J., 2005. Influence of time, temperature, pH and inhibitors on bioaccumulation of radiocaesium – 137Cs by lichen *Hypogymnia physodes*. Nukleonika 50 (Suppl. 1), S29–S37.

Plato, P., Denovan, J.T., 1974. The influence of potassium on the removal of ^{137}Cs by live Chlorella from low level radioactive wastes. Radiat. Bot. 14, 37–41.

Poldini, L., 1990. Naturalness and artificiality. In: Desmet, G., Nassimbeni, P., Belli, M. (Eds.), Transfer of Radionuclides in Natural and Semi-Natural Environments. Elsevier Appl Science, London, p. 17.

Prister, B.S., Belli, M., Sanzharova, N.I., Fesenko, S.V., Bunzl, K., Petriaev, E.P., et al., 1996. Behaviour of radionuclides in meadows including countermeasures application. In: Karaoglou, A., Desmet, G., Kelly, G.N., Menzel, H.G. (Eds.), The Radiological Consequences of the Chernobyl Accident. European Commission, Brussels, p. 261. EUR 16544 EN.

Rayyes, A.H., Ronneau, C., Stone, W.E.E., Genet, M.J., Ladriere, J., Cara, J., 1993. Radiocesium in hot particles: solubility vs. chemical speciation. J. Environ. Radioact. 21, 143–151.

Rigol, A., Vidal, M., Rauret, G., 2002. An overview of the effect of organic matter on soil-radiocesium interaction: implications in root uptake. J. Environ. Radioact. 58, 191–216.

Robert, M., Berthelin, J., 1986. Role of biological and biochemical factors in soil mineral weathering. In: Schnitzer, P.M.H. (Ed.), Interactions of Soil Minerals With Natural Organics and Microbes SSSA Special Publication Number 17. Soil Science Society of America, Inc, Madison, pp. 453–495.

Sakai, M., Takashi, G., Risa, S.N., Junjiro, N.N., Michiko, S., Hiroto, T., et al., 2015. Radiocesium leaching from contaminated litter in forest streams. J. Environ. Radioact. 144, 15–20.

Salt, C.A., Mayes, R.W., 1993. Plant uptake of radiocesium on heather moorland grazed by sheep. J. Appl. Ecol. 30, 235–246.

Sanzharova, N.I., Fesenko, S.V., Alexakhin, R.M., Anisimov, V.S., Kuznetsov, V.K., Chernyayeva, L.G., 1994. Changes in the forms of ^{137}Cs and its availability for plants as dependent on properties of fallout after the Chernobyl nuclear power plant accident. Sci. Tot. Environ. 154, 9–22.

Saskai, H., Shirato, S., Tahara, T., Sato, K., Takenaka, H., 2013. Accumulation of radioactive cesium released from Fukushima Daichi nuclear power plant in terrestrial Cyanobacteria *Nostoc commune*. Microb. Environ. 28, 466–469.

Seeprasert, P., Yoneda, M., Shimada, Y., 2016. The influence of soil fungi on the sorption of cesium and strontium in the soil organic layer. Int. J. Environ. Sci. Develop. 7, 415–419.

Shender, M., Eriksson, A., 1993. Sorption behaviour of caesium in various soils. J. Environ. Radioact. 19, 41–51.

Sivaperumal, P., Kamala, K., Rajaram, R., 2018. Adsorption of cesium ion by marine actinobacterium Nocardiopsis sp. 13H and their extracellular polymeric substances (EPS) role in bioremediation. Environ. Sci. Pollut. Res. 25, 4254–4267.

Skuterud, L., Travnikova, I.G., Balonov, M.I., Strand, P., Howard, B.J., 1997. Contribution of fungi to radiocesium intake by rural populations in Russia. Sci. Tot. Environ. 193, 237–242.

Song, N., Zhang, X., Wang, F., Zhang, C., Tang, S., 2012. Elevated CO_2 increases Cs uptake and alters microbial communities and biomass in the rhizosphere of *Phytolacca americana* Linn (pokeweed) and *Amaranthus cruentus* L. (purple amaranth) grown on soils spiked with various levels of Cs. J. Environ. Radioact. 112, 29–37.

Staunton, S., Roubaud, M., 1997. Adsorption of ^{137}Cs on montmorillonite and illite: effect of charge compensating cation, ionic strength, concentration of Cs, K, and fulvic acid. Clays Clay Miner. 45, 251–260.

Strand, P., Balanov, M., Skuterud, L., Hove, K., Howard, B.J., Prister, B.S., et al., 1996. Exposures from consumption of agricultural and seminatural products. In: Karaoglou, A., Desmet, G., Kelly, G.N., Menzel, H.G. (Eds.), The Radiological Consequences of the Chernobyl Accident. European Commission, Brussels, p. 261. EUR 16544 EN.

Strandberg, G.W., Shumate, S.E.I.I., Parrott, J.R., 1981. Microbial cells as biosorbents for heavy metals: accumulation of uranium by *Saccharomyces cerevisiae* and *Pseudomonas aeruginosa*. Appl. Environ. Microbiol. 41, 237–245.

Strebl, F., Gerzabek, M.H., Bossew, P., Kienzl, K., 1999. Distribution of radiocesium in an Austrian forest stand. Sci. Tot. Environ. 226, 75–83.

Sugiyama, H., Takahashi, M.N., Terada, H., Kuwahara, C., Maeda, C., Anzai, et al., 2008. Accumulation and localization of cesium in edible mushroom (*Pleurotus ostreatus*) mycelia. J. Agric. Food Chem. 56 (20), 9641–9646.

Takuro, K., Hidetoshi, K., Kai-Qin, X., Takao, A., 2020. Bioleaching and removal of radiocesium in anaerobic digestion of biomass crops: Effect of crop type on partitioning of cesium. Biotechnol. Rep. https://doi.org/10.1016/j.btre.2020.e00561.

Tomioka, N., Uchiyama, H., Yagi, O., 1992. Isolation and characterization of cesium-accumulating bacteria. Appl. Environ. Microbiol. 58, 1019–1023.

Tomioka, N., Uchiyama, H., Yagi, O., 1994. Cesium accumulation and growth characteristics of *Rhodococcus erythropolis* CS98 and *Rhodococcus* sp. strain CS402. Appl. Environ. Microbiol. 60, 2227–2231.

Tomioka, N., Uchiyama, H., Yagi, O., Kokufura, E., 1998. Revovery of ^{137}Cs by a bioaccumulation system using *Rhodococcus erythropolis* CS98. J. Ferment. Bioeng. 85, 604–608.

Tsukada, H., Shibata, H., Sugiyama, H., 1998. Transfer of radiocesium and stable cesium from substrata to mushrooms in a pine forest in Rokkasho-mura Aomori Japan. J. Environ. Radioact. 39, 149–160.

Tsukada, H., Hasegava, H., Hisamatsu, S., Yamasaki, S., 2002. Transfer of Cs and stable Cs from paddy soil to polished rice in Aomori. Jpn. J. Environ. Radioact. 59, 351–363.

Vinichuk, M.M., Johanson, K.J., Rosen, K., Nilsson, I., 2005. Role of the fungal mycelium in the retention of radiocaesium in forest soils. J. Environ. Radioact. 78, 77–92.

Vinichuk, M., Martensson, A., Ericsson, T., Rosen, K., 2013. Effect of arbuscular mycorrhizal (AM) fungi on 137Cs uptake by plants grown on different soils. Environ. Radioact. 115, 151–156.

Walderhaug, M.O., Dosch, D.C., Epstein, W., 1987. Potassium transport in bacteria. In: Rosen, B.P., Silver, S. (Eds.), Ion Transport in Prokaryotes. Academic Press, New York, pp. 85–130.

Wauters, J., Elsen, A., Cremers, A., Konoplev, A.V., Bulgakov, A.A., Comans, R.N.J., 1996. Prediction of solid/liquid distribution coefficients of radiocesium in soils and sediments. Part one: a simplified procedure for the solid-phase characterization. Appl. Geochem. 11, 589–594.

Wendling, L.A., Harsh, J.B., Palmer, C.D., Hamilton, M.A., Flury, M., 2004. Cesium sorption to illite as affected by oxalate. Clays Clay Miner. 52, 376–382.

Wendling, L.A., Harsh, J.B., Ward, T.E., Palmer, C.D., Hamilton, M.A., Boyle, J.S., Flury, M., 2005. Cesium desorption from illite as affected by exudates from rhizosphere bacteria. Environ. Sci. Technol. 39, 4505–4512.

West, J., Haigh, D., Hooker, P., Rowe, E., 1991. Microbial influence on the sorption of ^{137}Cs onto materials relevant to the geological disposal of radioactive waste. Experientia 47, 549–552.

White, P.J., Broadly, M.R., 2000. Tansley review no. 113: Mechanisms of cesium uptake by plants. New Phytol. 147, 241–256.

Williams, L.G., 1960. Uptake of cesium137 by cells and detritus of *Euglena and Chlorella*. Limnol. Oceanogr. 5, 301–311.

Wright, S.M., Strand, P., Sickel, M.A.K., Howard, B.J., Howard, D.C., Cooke, A.I., 1997. Spatial variation in the vulnerability of Norwegian Arctic counties to radiocesium deposition. Sci. Tot. Environ. 202, 173–184.

WSD, HSB, MHLW, 2011. Interim Report For Measures Against Radioactive Materials in Drinking Water (in Japanese). accessed on 28.04.12 http://www.mhlw.go.jp/stf/shingi/2r9852000001g9fq-att/2r9852000001g9jp.pdf.

Yoshida, S., Muramatsu, Y., Dvornik, M., Zhuchenko, A., Linkov, I., 2004. Equilibrium of radiocaesium with stable caesium within biological cycle of contaminated forest ecosystems. J. Environ. Radioact. 75, 301–313.

Yu, R., Chai, H., Yu, Z., Wu, X., Liu, Y., Shen, L., Li, J., Ye, J., Liu, D., Ma, T., Gao, F., Zeng, W., 2020. Behavior and mechanism of cesium biosorption from aqueous solution by living *Synechococcus* PCC7002. Microorganisms 2020 (8), 491.

CHAPTER

4

Management of biodegradable waste through the production of single-cell protein

Rajeswari Uppala and Azhaguchamy Muthukumaran

Department of Biotechnology, Kalasalingam Academy of Research and Education, Krishnankoil, Tamil Nadu, India

OUTLINE

4.1 Introduction

Earth is the reservoir of natural products that are beneficial to all animals, but human beings exploited it for increasing the industrial revolution. Solid waste management includes collecting the waste, treatment, and then disposal because it was no longer useful in further. There will be chances of unsanitary conditions due to improper disposal of municipal solid waste, which can lead to pollution of the environment. Solid waste treatment is always a mounting concern in both large and small cities. Valorization of food organic waste needs to be adapted for the proper disposal of urban waste. There are some conventional methods for waste disposal available, such as conventional landfill, incineration, and composting methods. Solid wastes produced by municipal origin can be managed by composting and anaerobic digestion (AD) (Abdel-Shafy and Mansour, 2018). The primary aim of waste management is to reduce or nullify the harmful impacts of disposed waste on human health and the environment.

Agricultural waste has been increasing rapidly throughout the world in recent years. Due to agricultural waste, there is a chance of creating environmental problems and other negative impacts on the environment. Thus, agricultural waste has drawn more important attention for the disposable methods. Therefore, it has developed to adopt proper approaches to reduce and reuse agricultural waste (Xue et al., 2016). The world population also increased the food wastage accumulation that developed a critical problem for food wastage management. The increase in the accumulation of food waste leads to various issues, such as pollution and health problems associated with it. It is necessary to adopt a standard management technique and then to reduce the burden of food waste accumulation

https://doi.org/10.1016/B978-0-12-822933-0.00016-4

(Paritosh et al., 2017). Food wastage contains various materials that can be decomposed and contaminated. The increase in the food waste problem is due to the involvement of many sectors such as the collection and disposal section, agriculture and industry areas, and finally, retail and consumer sectors. In the waste management hierarchy, a series of solutions have to be implemented. One of the best solutions is to donate edible portions to social service. Biofuels or biopolymers can be produced from food waste through industrial processes (Girotto et al., 2015).

Innovative recycling methods should be investigated to guard our community health. In this context, a more exciting and economically friendly practice is to convert food waste into value-added products. In a study, the authors reported the fermentation process using raw materials like citrus wastes (Gervasi et al., 2018).

Single-cell protein (SCP) can be adopted as a supplement to proteins in the diet. Protein scarcity in the diet can be eliminated by replacing the expensive traditional foods such as fishmeal and soya meal with SCP. In addition to this, wastes generated from industries and agriculture practices can be bio-converted to cheaper protein-rich foods and fodder stocks. Bioconversion of wastes can balance the negative cost value of wastes, which are being used as starting material in SCP production (Anupama and Ravindra, 2000).

Another potential source of SCP production is secondary clarifiers of pulp mills (Kellems et al., 1981). The protein scarcity problem can be solved partially by the application of resource-efficient and novel practices for SCP production. The SCP could be potentially produced by the conversion of various low-value waste streams as protein-rich food. Recent studies evaluated the coupling of anaerobic digestion and SCP production with a possibility of urban biowaste valorization (Khoshnevisan et al., 2019).

4.2 Biodegradable wastes and environmental hazards

Microorganisms like bacteria and fungi are useful to break down various biodegradable materials such as rice straw, disposable cups, and utensils, and finally produce methane. Methane is useful as a valuable energy source, but it causes the greenhouse effect when it is released into the atmosphere.

4.2.1 Types of biodegradable wastes

Polystyrene waste: Wide usage of polystyrene causes soil pollution due to its nonbiodegradable nature, and it also causes the accumulation of plastic-associated contaminants in various water bodies, including rivers, lakes, and oceans. Polystyrenes (PS) are generally resistant to the biodegradation process and are considered to be durable. In a study, PS product, Styrofoam, was feed to mealworms. These larvae (*Tenebrio molitor Linnaeus*) chew and eat Styrofoam and within 24 h, degraded the Styrofoam efficiently within the gut (Yang et al., 2015).

Municipal solid wastes (MSW): Improper urban management of municipal solid wastes insisted the world search for alternatives to treat these wastes efficiently. Rapidly increasing population and modern lifestyle are also one of the causes for the generation of large amounts of MSW. Vermicomposting is one of the methods to promote safe, sustainable management of biodegradable MSW to a safe and hygienic form. Certain MSWs collected from cities and houses, including kitchen, garden, paper, feces, and vegetable, can be transformed into organic fertilizers or vermicompost. This has a positive impact on agricultural practices and plants. Sustainability of earthworms requires good ventilation, proper temperature, moisture content, and pH, due to their sensitive nature to the environment for the optimal vermicomposting process. Along with the maintenance of these ideal conditions, the effective precomposting and proper microbial inoculum are required for the vermicomposting of MSW. Remodeling of traditional vermireactor can improve the quality of the municipal solid waste vermicomposting method (Sim and Wu, 2010).

Agricultural wastes: Pharmaceutical, food, and beverage industries use biovanillin as the prominent flavor. There is narrow accessibility of vanilla pods in the market due to the high cost, and it created a marginal approach towards biovanillin. Natural vanillin is the chief organic ingredient obtained from the vanilla beans. In addition, vanillin can also be made available through microbial-mediated synthesis using a natural precursor such as ferulic acid. In the future, it can be produced on a large scale by microbial fermentation in an economical manner. Many researchers explained about biotransformation of ferulic acid into biovanillin using agricultural waste resources (Zamzuri and Abd-Aziz, 2013).

Papermill wastes: Papermill industries release the sludge from the pulp, which causes the soil and water pollution. The resistant organic and inorganic parts of paper mill sludge are converted into high-value soil ameliorating material with less investment and duration (Hazarika and Khwairakpam, 2018). Plant and plant-based natural resources can be utilized as an alternative to remediate papermill waste. The naturally available source *Moringa oleifera* can be employed as an ecofriendly coagulant with nontoxic and biodegradable nature. It was not exhibited any effect on the pH of water. Compared with conventional coagulants like alum, it showed more advantages (Boulaadjoul et al., 2018).

4.2.2 Health hazards of inorganic waste

Industrialization all over the globe leads to the production of a large number of wastes like a greenhouse gas, solid waste, and releases wastes into water bodies. Large amounts and uncontrollable release of wastes into the environment cause health hazards and also trouble human life (Yadav et al., 2019). Municipal solid wastes were composed of various chemical and biological agents, and finally, form compost as a final product. The final product, compost, should be harmless to the environment. All these contaminants, when exposed to different populations, lead to health hazards, from plant workers to consumers. Compost produced from the municipal solid wastes contains a large number of metals, fungal toxins, and also some persistent organic pollutants. Compost that develops health risk for the population should not be commercially produced (Domingo and Nadal, 2009).

During the paper manufacturing, the pulping and bleaching processes release a variety of pollutants irrespective of inorganic and organic nature, into water bodies. Gaseous pollutants such as sulfur, chlorine dioxide, methyl mercaptan, sodium sulfide, and hydrogen sulfide, etc., cause skin irritation, eye irritation, heart problems, nausea, headache, and respiratory disorders. Several inorganic and organic pollutants such as hexadecanoic acids, octacosane, β-sitosterol, trimethylsilyl ether, chlorocatechols, terpenes, methanol, phenol, alkylated phenols, etc., were found to be involved in disrupting the endocrine system. Inorganic chemicals such as iron, copper, magnesium were reported for their neurotoxicity to juvenile channel catfish (*Ictalurus punctatus*) due to their accumulation in various body parts like gill, liver, ovary, and muscle (Singh and Chandra, 2019).

Paper and pulp industries widely use $NaClO_3$ as a nonselective herbicide. This sodium chlorate leads to the disintegration of genetic material and enhanced cross-linking between DNA and protein. Histological studies reported a significant damage to the kidney from $NaClO_3$ in tested animals. It also leads to $NaClO_3$-induced nephrotoxicity due to redox imbalance. It eventually causes damage to DNA and membrane, and alters metabolism and dysfunction of brush border membrane enzyme (Ali et al., 2018).

4.2.3 Reuse of degradable wastes

Water contaminated with the drainage produces some amount of sludge. Sludge obtained from drainage is accountable for the issues of stability, clearance, storage, social impacts, and environmental problems. Various applications of the sludge include its importance as a reusable stabilizer in polluted soil, as fertilizer in agro-ecosystem practices, as a substitute raw material in construction procedures, as a cover over tailings for acid mine drainage inhibition, for carbon sequestration, and also in cement and pigment industries (Rakotonimaro et al., 2017).

4.3 Recycling methods for biodegradable wastes

The paper and pulp industry (PPI) is considered as the most polluting industry due to the production of large amounts of solid and liquid discharge. It shows the adverse effects on the environment and human beings. Treatment methods of PPI sludge and waste materials should be modified on a priority basis. PPI sludge contains several organic chemicals such as lignin, cellulose, and hemicellulose. Substrates for bioenergy and value-added products can be obtained from every 10 million tons of waste generated each year (Chakraborty et al., 2019). There are various methods available for the recycling process such as shearing, high-pressure homogenization procedures such as the physical method, aerobic and anaerobic treatment involved in biological techniques, and whole cell-based and enzyme-based methods such as biochemical techniques.

From the past 30 years, water contaminated with chemical and biological contaminants has become a significant concern for the public authorities and industry. Water is contaminated with toxic pollutants in wastewaters due to the activities of domestic and industries. Recently, wastewater treatment involves the combination of physical, chemical, and biological methods. It helps to remove the insoluble particles and soluble contaminants from effluents (Crini and Lichtfouse, 2019).

4.3.1 Physical methods

Sand filtration, electrokinetics, adsorption, coagulation, electrosorption/CDI, reverse osmosis, electrodialysis, and membrane methods come under the physical control of hazardous wastes. Management of hazardous wastes using physical methods such as photoelectrochemical oxidation, photocatalysis, nonthermal plasma, supercritical fluid, electrochemical reduction processes, sonochemical, and electrochemical oxidation was well reported (Wang et al., 2019).

Antibiotic pollution is one of the major problems for the environment due to induced antibiotic resistance. It is responsible for human health hazards. Adsorption and biodegradation are the main ways of removing antibiotics from sewage. The analysis of biodegradation of antibiotics in a biological sewage treatment process involves biodegradability, degrading bacteria, and degradation products (Zhang et al., 2018). Aeration and sand filtration are commonly used for drinking water production. Pesticide residues could not be removed with simple physical methods. Other physical methods, such as bioaugmentation of sand filter units, have been suggested as another potential "green" technology for the removal of pesticides (Hylling et al., 2019).

4.3.2 Chemical methods

Ozonization is the commonly used chemical method to remove the microorganisms. Various other chemical approaches for the control of hazardous waste include hydrogen peroxide-based, persulfate-based, Fenton and Fenton-like, and potassium permanganate processes. Chemicals applied in the wastewater management help to inactivate or kill the microbes. The chemical methods that induce chemical reactions are called chemical unit processes.

Chlorination is the other chemical method in drinking water treatment plants, in which contaminated waters are treated with Cl_2, ClO_2, or NaOCl reagents. In this process, there is a chance of advancement in the development of organic and inorganic by-products. In the breakpoint chlorination method, approximately 10 times more chlorine has to be used than the disinfection (Stefán et al., 2019).

There was a study about the efficacy of several oxidants or redox buffers (ferric chloride, sodium hypochlorite, potassium permanganate, hydrogen peroxide, and potassium nitrate) for governing the emission of sulfur gases during the sewage treatment. When sewage sludge was treated with potassium permanganate and hydrogen peroxide, there was less emission of sulfur gases like hydrogen sulfide, carbonyl sulfide, carbon sulfide, and methanethiol, from sewage sludge collected from the wastewater treatment plant. Sulfur gas like hydrogen sulfide emission can be reduced by using sodium hypochlorite (Devai and Delaune, 2002).

4.4 Biological methods

4.4.1 Various biological methods involved in the management of biodegradable waste

There was a considerable production of MSW in developing countries (India), contributed by a rapid growth in the population and industries. Various methods of disposal and treatment of MSW are composting, landfills, incineration, bio-methanation, and pyrolysis. Among these, the most popular, economically feasible, and effective disposal method is aerobic/anaerobic bioreactor landfill (Fig. 4.1) (Patil et al., 2017). These aerobic or anaerobic bioreactor landfills assist in the enhancement of operating lifetime and reduce the treatment costs (Borglin et al., 2004).

4.4.1.1 Anaerobic digestion

This method involves the production of biogas from organic waste materials assisted by appropriate microorganisms. This treatment method is commonly practiced for the control of animal wastes produced from large animal

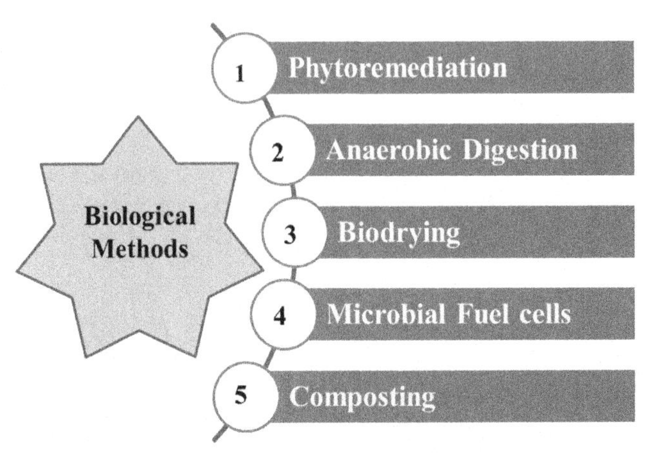

FIG. 4.1 Biological methods employed for the management of biodegradable waste.

farms. In this process, the high content of methane is produced and acts as a combustible energy source (Shaw et al., 2017). Due to current environmental issues like global warming, recently, anaerobic digestion of solid wastes plays the most crucial role in biological remediation. Biogas produced in this treatment process acts as a renewable energy source (Muhammad Nasir et al., 2012).

The municipal solid waste consists of a blend of primary and secondary municipal sludge. In the anaerobic digestion method, sludge is treated using the first-stage mesophilic anaerobic and a second-stage aerobic digester. This study helps to monitor the aerobic digester and identify the steps involved in the efficient removal of solids and nitrogen. In the first phase, continuous aeration was provided to aerobic digester with a batch feeding. When aerobic retention time increased, more solids and ammonia were eliminated. The retention time of 3 days removed about 90% or greater ammonia. Anaerobic digestion was followed for 3–5 days in the second phase. Anaerobic digested sludge when continuously fed with different intermittent aeration times helps in removing the solids (Kim and Novak, 2017).

4.4.1.2 *Phytoremediation*

Ecological restoration, as well as remediation, is possible only by the phytoremediation using native plants. It serves as a perfect residence for the accumulation of microorganisms in the rhizosphere region. It also holds natural enzymatic activities and accumulates and biodegrades various contaminants of organic and inorganic forms. Native species are the best and effective at the phytoremediation process than the non-native plants. A native plant is easier to cultivate and aesthetically pleasing, socially acceptable, and also ecologically safe. Researchers studied about phytoremediation of polycyclic aromatic hydrocarbons by *Chromolaena odorata*, obtained from crude oil-polluted land against *Medicago sativa*. *Chromolaena odorata* showed an improved degradation of hydrocarbon chains compared with the reference plant. Phytoremediation contributed by native plants is best associated with its natural contaminated area and is solar driven. It prevents erosion, and abolishes secondary airborne and waterborne waste (Futughe et al., 2020).

4.4.1.3 *Biodrying*

Biodrying is a convective evaporation process. Biodrying exploits the biological heat produced by the aerobic reactions of organic compounds. Biodrying is a versatile practice as it uses the heat to enable the moisture content and volume to decrease and increase the bulk density (Tom et al., 2016).

4.4.1.4 *Microbial fuel cell*

Microorganisms are useful to convert the chemical energy of organic compounds into electricity in the microbial fuel cell (MFC) technology. By employing a variety of microbes that utilize a wide array of carbon sources from wastes, electric energy can be produced. Therefore, the conversion of wastes mediated by microbes with new bioremediation approaches is recommended. MFC is considered as an effective and ecosystem-friendly tactic for the production of energy (Chaturvedi and Verma, 2016).

4.4.1.5 *Composting*

The composting process requires the composition of solid wastes and also microbial diversity at that particular landfill. Several genera such as *Stenotrophomonas*, *Pseudomonas*, *Xanthomonas*, *Klebsiella*, *Caulobacter*, *Alcaligenes*, and *Achromobacter* are the most prominent free-living and nitrogen-fixing microorganisms. Nitrogen fixation by microorganisms augments the quality of compost. Enzyme-producing microorganisms such as pectinolytic, amylolytic, and aerobic cellulolytic organisms were observed in the inner and exterior parts of the heaps in large quantities. The nitrogen-fixing ability of these organisms was further investigated (Pepe et al., 2013).

4.5 Biodegradable waste into value-added products

Several reusable substances can be produced from biological wastes that include soluble sugars and fiber. When solid wastes are disposed of directly into the soil, they will create environmental hazards. Therefore, the synthesis of potential value-added products from biodegradable wastes is a highly attractive process. Solid-state fermentation offers several advantages. This includes the production of commercially valuable metabolites using biodegradable wastes, which enables the control of waste accumulation in an ecofriendly manner (Rodríguez Couto, 2008).

Several kinds of organic wastes were used for the production of value-added compounds, which include enzymes, pharmaceuticals, organic acids, polysaccharides, flavors, mushrooms, biodegradable plastics, gibberellic acid, dietary fibers, animal feed, biomass (for bioremediation), acetone, glycerol butanol, and fuel like ethyl alcohol (Stabnikova et al., 2010).

In the solid waste treatment process, microbial electrolysis cells (MECs) are applied as an efficient method for the production of energy and resource recovery. MECs are able to convert any biodegradable waste into biofuels, H_2, and valuable products. When different substrates are used and also operated in various conditions, the system efficacy could be changed (Lu and Ren, 2016).

One of the challenging threats that are facing globally now is food wastage. Several types of food wastes are discharged from food processing companies, hospitals, and houses. These wastes consist of proteins, lipids, carbohydrates, and inorganic compounds. Conversion of food waste to value-added products is one of the efficient management tasks of food waste. This food waste can be transformed into several value-added products like biopigments, colorants, bioactive compounds, phytochemicals, livestock feed, food supplements, dietary fibers, emulsifiers, edible oils, essential oils, biofertilizers, biopreservatives, biofuels, and single-cell proteins through efficient methods. Converting waste into value-added products is the most useful and ecofriendly procedure (Gunjal et al., 2019).

Another useful strategy to treat organic waste is through vermicomposting. Vermicomposting using earthworms helps to recover the nutrients and enriches the soil, most useful for agricultural purposes. Researchers worked on the possibility of treatment of onion waste through vermicomposting for agricultural purposes. In this procedure, the cow dump was taken and mixed with the onion waste from the food industry. This mixture was kept for composting at the pilot scale for 154 days. Thus, one of the feasible methods to treat organic waste for agriculture purposes is the vermicomposting of onion waste along with cow dump. This can enhance soil fertility through vermicomposting (Pellejero et al., 2020).

Dried biomass of several useful microorganisms such as yeast, bacteria, algae, and fungi can be used as single-cell protein. This dried biomass is enriched with proteins, amino acids, vitamins, and lipid content. The microorganisms utilize any waste materials, agricultural and industrial waste, and feedstock, and increase their quantity and quality of natural protein concentrates. Single-cell protein can be produced by using a variety of substrates as raw material. For example, among the various microorganisms, commonly yeast can produce 250 tons of proteins within a span of 24h. Single-cell proteins are useful to reduce the problem of malnutrition and can be utilized as a vital supplement of proteins.

4.5.1 Single-cell protein (SCP)

Dried biomass of useful microorganisms, mostly yeast, algae like spirulina are rich in nutrients that can be used as single-cell proteins. They are rich in protein and can be used as a protein supplement in human foods or animal feeds. Algae, fungi, yeast, and bacteria are able to utilize inexpensive feedstock and agricultural wastes rich in carbon source and act as substrates for their growth to produce the biomass, protein concentrate, or amino acids. The single-cell protein produced from microbial biomass is rich in protein and can act as a natural protein concentrate (Nasseri et al., 2011).

4.5.2 Production of SCP from biodegradable waste

Production of SCP from the animal or agricultural wastes is an interesting choice, which can decrease the cost of protein feedstock for animals. For example, chicken manure can be used for SCP production by exploiting the photosynthetic bacteria for submerged fermentation (Saejung and Salasook, 2020).

Industrial processing units generate several food products as by-products, and they can be reprocessed as elements for the human diet (Fig. 4.2). All food products are often reused and studied. But few reports suggest the insertion of such by-products into the new food products and their acceptability. Solid-state fermentation of guava peels and cashew bagasse for protein improvement was studied and prepared as comprised cereal bars for the human diet. Solid-state fermentation of fruit by-products were carried out at 30°C (temperature), with an equilibrium humidity of 70% and aw of 0.9 (water activity). Researchers maintained the inoculum size of *Saccharomyces cerevisiae* (3% concentration) for the cashew bagasse and the guava peels (5% concentration). A different formulation of three cereal bars was used based on protein-rich by-products. Also, the analysis of physicochemical, sensorial, and instrumental parameters was accomplished. In order to analyze the results, a multivariate analysis was employed. The result of the instrumental analysis showed improved hardness and cohesiveness with high dietary fiber. When compared

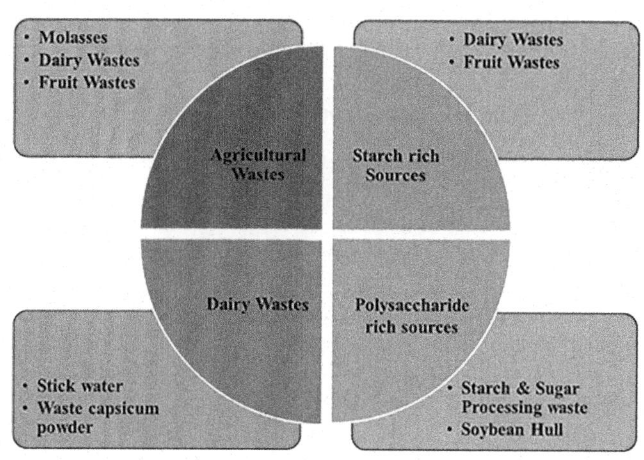

FIG. 4.2 Various substrates used in the production of single-cell protein.

with control bars and cereal bars, increased textural parameters in cereal bars were noticed. Textural parameters of cereal bars increased throughout the storage period compared with the controls. Altogether, the compositional stability of all the samples was discovered during the test period of 28 days. Protein-enriched by-products were substituted to enhance the nutritional value as well as economic value to cereal bars (Muniz et al., 2020).

To minimize the problem of protein scarcity, resources and novel practices should be improved to manufacture proteinaceous food and feed sources. The most important solution for this problem is the transformation of low-value waste streams into SCP. It also assessed the chance to valorizate urban biowaste by combining anaerobic digestion and synthesis of SCP. For this, methanotroph mixed culture was fed with raw and upgraded biogas. Then, mixed culture allowed to grow well by adding nutrients through the direct addition of pasteurized, centrifuged, and filtered digest (Khoshnevisan et al., 2019).

When starch is used as a substrate, first, it was enzymatically removed and transformed into SCP. Pruksari and colleagues worked with the extraction and partial precipitation of rice bran proteins. Protein and mineral fractions rich in soluble dietary fibers were obtained by ultrafiltration (3 kDa). Around 69% of rice bran has been converted to valuable products such as protein fractions and minerals. The same procedure continued with 12% of the rice bran to obtain soluble dietary fiber. However, soluble dietary fiber accounts for 2% of total rice bran, and the remaining section accounts for the mixture of minerals and monomeric sugars (Pruksasri et al., 2019).

Capsicum powder mixture (CPM) is a rich source of protein and other nutrients. Thus, the inclusion of capsicum powder to animal feed can enhance the protein need and also reduce ecosystem pollution. SCP can also be produced using capsicum powder. The highest SCP production was observed with a yeast strain. The ability of *Candida utilis* to produce SCP from CPM was studied where the microorganism employed was served as the producer of biomass. Capsicum powder is employed as the culture medium to grow four different yeast strains. The sugar, nitrogen, and phosphorous contents in CPM were 16.3, 3.7, and 785.4 mg/L, respectively, which are useful for the growth of the yeast, and the obtained biomass can be utilized as protein feed to animals (Zhao et al., 2010).

Brewery process water can be used as a source to produce SCP. Microbial biomass serves as an alternative to being the protein base for animal feed when global fisheries decline. Bacteria such as rhizosphere diazotrophs of alpha and beta proteobacteria in the bioreactor can efficiently produce the SCP. Commercially produced feed diazotrophs can be used as SCP (Lee et al., 2015).

Fungi are able to grow on the rubber waste as the substrate can be employed to produce the proteinaceous food. Natural rubber waste serum acts as a potential substrate for SCP production. According to Mahat and colleagues, *Rhizopus oligosporus* grown on natural rubber waste serum showed the maximum production of mycelium at 28 °C and a pH of 4.0 (Mahat and MacRae, 1992).

R. oligosporus is an edible filamentous fungus that can be grown on organic wastes and helpful to achieve the growing demand for single-cell protein. Anaerobic digestion of a large quantity of organic solid waste can be converted to value-added products (SCP) with the help of *R. oligosporus*. It can be grown on volatile fatty acids (VFAs) as favorable potential substrates to produce a large quantity of biomass and can be used as animal feed (Wainaina et al., 2020). Citrus pulp could be used to produce the single-cell protein (SCP) and crude pectinolytic enzymes. Single-cell protein was produced from lemon juice clarification pulp using *Aspergillus niger* and *Trichoderma viride*. *T. viride* showed higher amounts of SCP and improved enzymatic property in their biomass and supernatants (De Gregorio, 2002).

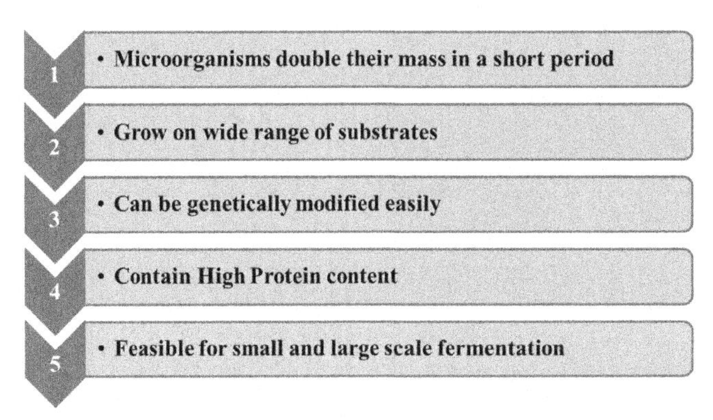

FIG. 4.3 Importance of microorganisms in single-cell protein production.

4.5.3 Advantages of single-cell protein

Microorganisms are able to produce large amounts of biomass due to their fast multiplication in less time. They can produce single-cell protein in a large scale and have multiple advantages over conventional food production practices (Fig. 4.3). Microbial genetic material modifications are easy in the laboratory with advances in recombinant DNA technology to change the amino acid composition and to achieve high protein content. Another advantage of single-cell production is that a wide range of raw materials such as organic solid wastes can be used as a substrate. SCP production helps in decreasing the number of pollutants, is cost-effective and ecofriendly, and can be operated at various climatic conditions. SCP is having various applications apart from food and as a feed supplement. SCP can be used in cosmetics and helps to maintain healthy hair. SCP is blended in the cosmetics to produce various herbal face creams and other herbal beauty products such as bio-lipsticks and hair dyes.

4.6 Applications of single-cell protein

Single-cell protein (SCP) can be applied in different industries such as food, fodder, and also in cosmetic industries. Different groups of microorganisms such as bacteria, marine microalgae, yeasts, and molds can be used as SCP. Production of raw materials such as various kitchen wastes, and industrial wastes like rubber wastes, are the cheap resources and can be used as a substrate to produce SCP. Thus, it is an economically feasible source of protein for use in animal feed or processed food for human consumption. It often offers the dietary requirements of protein. Global protein shortage problem could be solved with SCP production by applying many microorganisms that have been used to convert various substrates into biomass. Biotransformation of low-value products to value-added products with rich nutritive value involves various strategies and is regarded as an excellent strategy. SCP production is more profitable because it provides the cheapest protein products in the market (Srividya et al., 2013).

Untreated industrial waste mostly causes environmental hazards and is a cause for the substantial quantity of human-made waste streams. The harmful substance can be neutralized by microbiological treatment to reduce environmental pollution. The transformation of solid wastes into value-added products such as SCP is an ecofriendly and economically feasible method (Spalvins et al., 2018).

Single-cell protein is the best proteinaceous food supplement for malnutrition people and also extremely good for healthy eyes and skin. Microalgae species are rich in protein and are used in the production of SCP. It acts as a good source of minerals, amino acids, vitamins, crude fibers, etc. *Spirulina platensis* contains the richest protein to supplement the food for human beings. *S. platensis* contains high levels of minerals, vitamins, phenolics, amino acids, essential fatty acids, and pigments, which benefits human health (Lupatini et al., 2017).

Single-cell protein could be used in therapeutic and natural medicine to control obesity. It also lowers the blood sugar level in diabetic patients. It enhances the stress relief in people. It inhibits the accumulation of cholesterol in the body. Isogenic cells show difference in their abundance of cellular proteins due to the variability in SCP levels and these proteins have regulatory roles such as governing the cell fate during apoptosis (Levy and Slavov, 2018). SCP comprises trehalose, which can be hydrolyzed by trehalase and is useful for the absorption by the human intestine. Bergoz and colleagues studied the enzymatic properties of hundred intestinal biopsies. The study revealed two cases with a very low concentration of trehalase level and anticipated the clinical intolerance to even with less concentration of trehalose. It was concluded that 98% of healthy people were tolerant of high amounts of trehalose (Bergoz et al., 1982).

Protein is the main component of the food pyramid, and due to which they need for protein supply boosted in recent years. One of the very important issues that need to be addressed during food processing is to enhance the supply of protein. That is, during food processing, the by-products have to be reduced, and also waste-derived proteins have to be transformed into the "gutter to gold." Various industries are using waste protein, which is the best way to explore novel applications in the pharmaceutical and cosmetic industries. Waste proteins are synthesized from various sources such as visceral proteins from dead animals, plant by-products, dairy waste, and can be transformed to produce animal feed (Luo and Wang, 2016).

Researchers explained the solid-state fermentation of waste mixtures obtained from various food industries with the help of particular industrially important microorganisms. The industrially important microorganisms are *Kluyveromyces marxianus*, and *S. cerevisiae* used in the kefir preparation was studied for SCP production. The raw material fermented by *K. marxianus* showed the maximum amount of fat and protein. To analyze the protein content, nutrients, minerals content, volatile compounds, and aroma, GC/MS was used. Due to the presence of high fat and protein contents in the fermented product, it could be used as livestock feed (Aggelopoulos et al., 2014).

In the field of biotechnology, a prominent topic is the bioconversion of sugar industry wastes into value-added products. *Rhodopseudomonas faecalis* was used to produce single-cell protein rich in carotenoids. When the bacteria cultivated on wastewater sources, it boosted the growth and dehydrogenase activity of *R. faecalis* and significantly reduced the total sugar concentration (Saejung and Salasook, 2020).

Increasing production demand and limited production of animal feed are the major problems facing the livestock and fish farming industries where the rapid production rate of proteins by the microbial conversion of cheap-value nonfood wastes into single-cell protein is the best method to be followed (Mahan et al., 2018). Good quality and nutrient-rich food is always the need of the hour, especially when there is a rapid growth in the population. Aquaculture has developed to accomplish some of these needs and is thus one of the rapidly growing businesses dealing with animal protein. It acts as a source for the sustainable, renewable protein ingredient. To fulfill the protein demand, microbial-based single-cell protein (SCP) and its derivatives are more important (Jones et al., 2020).

4.7 Future perspectives

For many decades, humans and animals are consuming microbial-based food products, which is not a revolutionary new idea. Human beings have started consuming alcoholic beverages, cheese, yogurt, and soya sauce for thousands of years. Microbial biomass production is one of the best methods to produce rich proteinaceous products with cheap resources. Microorganisms are used for SCP production owing to their quick growth rate and high amount of protein. In addition, they can exploit cheap feedstock as carbon and energy sources for their growth. SCP produced with the help of microorganisms is applied in both the food and protein supplements of humans and other animals. Previously according to the protein demand, only a limited number of commercially feasible SCP were observed. Recently, special interest in SCP has resulted in the introduction of recombinant DNA technology (Tusé and Miller, 1984).

Perspectives in nutrition analysis and quality would lead to the maximum usage of SCP in the coming years. The latest technological advances in biotechnology should address complications associated with SCP, such as high purine content and the possible presence of allergens.

Finally, it can be concluded that the production of SCP must reach the challenge of increased protein needs required to prevent malnutrition in humans with an increased global population rate. SCP, when coupled with aquaculture production, is supposed to ensure future protein needs and sufficient without any direct competition with food for people (Tlusty et al., 2017). Protein in the diet is highly important as it constitutes the building blocks of living organisms. SCP production contributes substantially to the world food and protein needs and reduces the problem of malnutrition. SCP also supports the needs of animal husbandry as fodder or feed to livestock. In the future, it acts as a source of protein or protein supplement and a very promising source with the potential to satisfy the world shortage of food.

References

Abdel-Shafy, H.I., Mansour, M.S.M., 2018. Solid waste issue: Sources, composition, disposal, recycling, and valorization. Egypt. J. Pet. 27, 1275–1290. https://doi.org/10.1016/j.ejpe.2018.07.003.

Aggelopoulos, T., Katsieris, K., Bekatorou, A., Pandey, A., Banat, I.M., Koutinas, A.A., 2014. Solid state fermentation of food waste mixtures for single cell protein, aroma volatiles and fat production. Food Chem. 145, 710–716. https://doi.org/10.1016/j.foodchem.2013.07.105.

Ali, S.N., Arif, H., Khan, A.A., Mahmood, R., 2018. Acute renal toxicity of sodium chlorate: redox imbalance, enhanced DNA damage, metabolic alterations and inhibition of brush border membrane enzymes in rats. Environ. Toxicol. 33, 1182–1194. https://doi.org/10.1002/tox.22624.

Anupama, Ravindra, P., 2000. Value-added food: single cell protein. Biotechnol. Adv. https://doi.org/10.1016/S0734-9750(00)00045-8.

Bergoz, R., Vallotton, M.-C., Loizeau, E., 1982. Trehalase deficiency. Prevalence and relation to single-cell protein food. Ann. Nutr. Metab. 26, 291–295. https://doi.org/10.1159/000176576.

Borglin, S.E., Hazen, T.C., Oldenburg, C.M., Zawislanski, P.T., 2004. Comparison of aerobic and anaerobic biotreatment of municipal solid waste. J. Air Waste Manage. Assoc. 54, 815–822. https://doi.org/10.1080/10473289.2004.10470951.

Boulaadjoul, S., Zemmouri, H., Bendjama, Z., Drouiche, N., 2018. A novel use of Moringa oleifera seed powder in enhancing the primary treatment of paper mill effluent. Chemosphere 206, 142–149. https://doi.org/10.1016/j.chemosphere.2018.04.123.

Chakraborty, D., Shelvapulle, S., Reddy, K.R., Kulkarni, R.V., Puttaiahgowda, Y.M., Naveen, S., Raghu, A.V., 2019. Integration of biological pre-treatment methods for increased energy recovery from paper and pulp biosludge. J. Microbiol. Methods 160, 93–100. https://doi.org/10.1016/j.mimet.2019.03.015.

Chaturvedi, V., Verma, P., 2016. Microbial fuel cell: a green approach for the utilization of waste for the generation of bioelectricity. Bioresour. Bioprocess. 3, 38. https://doi.org/10.1186/s40643-016-0116-6.

Crini, G., Lichtfouse, E., 2019. Advantages and disadvantages of techniques used for wastewater treatment. Environ. Chem. Lett. https://doi.org/10.1007/s10311-018-0785-9.

De Gregorio, A., 2002. SCP and crude pectinase production by slurry-state fermentation of lemon pulps. Bioresour. Technol. 83, 89–94. https://doi.org/10.1016/S0960-8524(01)00209-7.

Devai, I., Delaune, R.D., 2002. Effectiveness of selected chemicals for controlling emission of malodorous sulfur gases in sewage sludge. Environ. Technol. 23, 319–329. https://doi.org/10.1080/09593332508618412.

Domingo, J.L., Nadal, M., 2009. Domestic waste composting facilities: a review of human health risks. Environ. Int. https://doi.org/10.1016/j.envint.2008.07.004.

Futughe, A.E., Purchase, D., Jones, H., 2020. Phytoremediation using native plants. In: Shmaefsky, B.R. (Ed.), Phytoremediation. Concepts and Strategies in Plant Sciences, Springer International Publishing, Cham, pp. 285–327.

Gervasi, T., Pellizzeri, V., Calabrese, G., Di Bella, G., Cicero, N., Dugo, G., 2018. Production of single cell protein (SCP) from food and agricultural waste by using Saccharomyces cerevisiae. Nat. Prod. Res. 32, 648–653. https://doi.org/10.1080/14786419.2017.1332617.

Girotto, F., Alibardi, L., Cossu, R., 2015. Food waste generation and industrial uses: a review. Waste Manag. https://doi.org/10.1016/j.wasman.2015.06.008.

Gunjal, A.B., Waghmode, M.S., Patil, N.N., Bhatt, P. (Eds.), 2019. Global Initiatives for Waste Reduction and Cutting Food Loss. IGI Global.

Hazarika, J., Khwairakpam, M., 2018. Evaluation of biodegradation feasibility through rotary drum composting recalcitrant primary paper mill sludge. Waste Manag. 76, 275–283. https://doi.org/10.1016/j.wasman.2018.03.044.

Hylling, O., Nikbakht Fini, M., Ellegaard-Jensen, L., Muff, J., Madsen, H.T., Aamand, J., Hansen, L.H., 2019. A novel hybrid concept for implementation in drinking water treatment targets micropollutant removal by combining membrane filtration with biodegradation. Sci. Total Environ. 694, 133710. https://doi.org/10.1016/j.scitotenv.2019.133710.

Jones, S.W., Karpol, A., Friedman, S., Maru, B.T., Tracy, B.P., 2020. Recent advances in single cell protein use as a feed ingredient in aquaculture. Curr. Opin. Biotechnol. 61, 189–197. https://doi.org/10.1016/j.copbio.2019.12.026.

Kellems, R.O., Aseltine, M.S., Church, D.C., 1981. Evaluation of single cell protein from pulp mills: laboratory analyses and in vivo digestibility. J. Anim. Sci. 53, 1601–1608. https://doi.org/10.2527/jas1982.5361601x.

Khoshnevisan, B., Tsapekos, P., Zhang, Y., Valverde-Pérez, B., Angelidaki, I., 2019. Urban biowaste valorization by coupling anaerobic digestion and single cell protein production. Bioresour. Technol. https://doi.org/10.1016/j.biortech.2019.121743.

Kim, J., Novak, J.T., 2017. Nitrogen removal in a staged anaerobic-aerobic sludge digestion system. Water Environ. Res. 89, 17–23. https://doi.org/10.2175/106143016X147336816962.

Lee, J.Z., Logan, A., Terry, S., Spear, J.R., 2015. Microbial response to single-cell protein production and brewery wastewater treatment. Microb. Biotechnol. 8, 65–76. https://doi.org/10.1111/1751-7915.12128.

Levy, E., Slavov, N., 2018. Single cell protein analysis for systems biology. Essays Biochem. 62, 595–605. https://doi.org/10.1042/EBC20180014.

Lu, L., Ren, Z.J., 2016. Microbial electrolysis cells for waste biorefinery: a state of the art review. Bioresour. Technol. 215, 254–264. https://doi.org/10.1016/j.biortech.2016.03.034.

Luo, Y., Wang, T., 2016. Pharmaceutical and cosmetic applications of protein by-products. In: Protein Byproducts. Elsevier, pp. 147–160.

Lupatini, A.L., Colla, L.M., Canan, C., Colla, E., 2017. Potential application of microalga Spirulina platensis as a protein source. J. Sci. Food Agric. 97, 724–732. https://doi.org/10.1002/jsfa.7987.

Mahan, K.M., Le, R.K., Wells, T., Anderson, S., Yuan, J.S., Stoklosa, R.J., Bhalla, A., Hodge, D.B., Ragauskas, A.J., 2018. Production of single cell protein from agro-waste using Rhodococcus opacus. J. Ind. Microbiol. Biotechnol. 45, 795–801. https://doi.org/10.1007/s10295-018-2043-3.

Mahat, M.S., MacRae, I.C., 1992. Rhizopus oligosporus grown on natural rubber waste serum for production of single cell protein: a preliminary study. World J. Microbiol. Biotechnol. 8, 63–64. https://doi.org/10.1007/BF01200687.

Muhammad Nasir, I., Mohd Ghazi, T.I., Omar, R., 2012. Production of biogas from solid organic wastes through anaerobic digestion: a review. Appl. Microbiol. Biotechnol. 95, 321–329. https://doi.org/10.1007/s00253-012-4152-7.

Muniz, C.E.S., Santiago, Â.M., Gusmão, T.A.S., Oliveira, H.M.L., Conrado, L.d.S., de Gusmão, R.P., 2020. Solid-state fermentation for single-cell protein enrichment of guava and cashew by-products and inclusion on cereal bars. Biocatal. Agric. Biotechnol. 25, 101576. https://doi.org/10.1016/j.bcab.2020.101576.

Nasseri, A.T., Rasoul-Ami, S., Morowvat, M.H., Ghasemi, Y., 2011. Single cell protein: production and process. Am. J. Food Technol. 6, 103–116. https://doi.org/10.3923/ajft.2011.103.116.

Paritosh, K., Kushwaha, S.K., Yadav, M., Pareek, N., Chawade, A., Vivekanand, V., 2017. Food waste to energy: an overview of sustainable approaches for food waste management and nutrient recycling. Biomed. Res. Int. 2017, 1–19. https://doi.org/10.1155/2017/2370927.

Patil, B.S., Anto, C.A., Singh, D.N., 2017. Simulation of municipal solid waste degradation in aerobic and anaerobic bioreactor landfills. Waste Manag. Res. 35, 301–312. https://doi.org/10.1177/0734242X16679258.

Pellejero, G., Rodriguez, K., Ashchkar, G., Vela, E., García-Delgado, C., Jiménez-Ballesta, R., 2020. Onion waste recycling by vermicomposting: nutrients recovery and agronomical assessment. Int. J. Environ. Sci. Technol. 17, 3289–3296. https://doi.org/10.1007/s13762-020-02685-1.
Pepe, O., Ventorino, V., Blaiotta, G., 2013. Dynamic of functional microbial groups during mesophilic composting of agro-industrial wastes and free-living (N2)-fixing bacteria application. Waste Manag. 33, 1616–1625. https://doi.org/10.1016/j.wasman.2013.03.025.
Pruksasri, S., Wollinger, K.K., Novalin, S., 2019. Transformation of rice bran into single-cell protein, extracted protein, soluble and insoluble dietary fiber, and minerals. J. Sci. Food Agric. 99, 5044–5049. https://doi.org/10.1002/jsfa.9747.
Rakotonimaro, T.V., Neculita, C.M., Bussière, B., Benzaazoua, M., Zagury, G.J., 2017. Recovery and reuse of sludge from active and passive treatment of mine drainage-impacted waters: a review. Environ. Sci. Pollut. Res. 24, 73–91. https://doi.org/10.1007/s11356-016-7733-7.
Rodríguez Couto, S., 2008. Exploitation of biological wastes for the production of value-added products under solid-state fermentation conditions. Biotechnol. J. 3, 859–870. https://doi.org/10.1002/biot.200800031.
Saejung, C., Salasook, P., 2020. Recycling of sugar industry wastewater for single-cell protein production with supplemental carotenoids. Environ. Technol. 41, 59–70. https://doi.org/10.1080/09593330.2018.1491633.
Shaw, G.T.-W., Liu, A.-C., Weng, C.-Y., Chou, C.-Y., Wang, D., 2017. Inferring microbial interactions in thermophilic and mesophilic anaerobic digestion of hog waste (J.-H. Shin, Ed.). PLoS One 12. https://doi.org/10.1371/journal.pone.0181395, e0181395.
Sim, E.Y.S., Wu, T.Y., 2010. The potential reuse of biodegradable municipal solid wastes (MSW) as feedstocks in vermicomposting. J. Sci. Food Agric. 90, 2153–2162. https://doi.org/10.1002/jsfa.4127.
Singh, A.K., Chandra, R., 2019. Pollutants released from the pulp paper industry: aquatic toxicity and their health hazards. Aquat. Toxicol. 211, 202–216. https://doi.org/10.1016/j.aquatox.2019.04.007.
Spalvins, K., Zihare, L., Blumberga, D., 2018. Single cell protein production from waste biomass: comparison of various industrial by-products. Energy Procedia 147, 409–418. https://doi.org/10.1016/j.egypro.2018.07.111.
Srividya, A., Vishnuvarthan, V., Murugappan, M., Dahake, P.G., 2013. Single cell protein—a review. Int. J. Pharm. Res. Sch. 2, 472–485.
Stabnikova, O., Wang, J.-Y., Ivanov, V., 2010. Value-added biotechnological products from organic wastes. In: Environmental Biotechnology. Humana Press, Totowa, NJ, pp. 343–394.
Stefán, D., Erdélyi, N., Izsák, B., Záray, G., Vargha, M., 2019. Formation of chlorination by-products in drinking water treatment plants using breakpoint chlorination. Microchem. J. https://doi.org/10.1016/j.microc.2019.104008.
Tlusty, M., Rhyne, A., Szczebak, J.T., Bourque, B., Bowen, J.L., Burr, G., Marx, C.J., Feinberg, L., 2017. A transdisciplinary approach to the initial validation of a single cell protein as an alternative protein source for use in aquafeeds. PeerJ 5. https://doi.org/10.7717/peerj.3170, e3170.
Tom, A.P., Pawels, R., Haridas, A., 2016. Biodrying process: a sustainable technology for treatment of municipal solid waste with high moisture content. Waste Manag. 49, 64–72. https://doi.org/10.1016/j.wasman.2016.01.004.
Tusé, D., Miller, M.W., 1984. Single-cell protein: current status and future prospects. CRC Crit. Rev. Food Sci. Nutr. 19, 273–325. https://doi.org/10.1080/10408398409527379.
Wainaina, S., Kisworini, A.D., Fanani, M., Wikandari, R., Millati, R., Niklasson, C., Taherzadeh, M.J., 2020. Utilization of food waste-derived volatile fatty acids for production of edible Rhizopus oligosporus fungal biomass. Bioresour. Technol. 310, 123444. https://doi.org/10.1016/j.biortech.2020.123444.
Wang, J., Shih, Y., Wang, P.Y., Yu, Y.H., Su, J.F., Huang, C.P., 2019. Hazardous waste treatment technologies. Water Environ. Res. https://doi.org/10.1002/wer.1213.
Xue, L., Zhang, P., Shu, H., Wang, R., Zhang, S., 2016. Restoration and improvement of rural environment of China. Agricult. Waste. Water Environ. Res. 88, 1334–1369. https://doi.org/10.2175/106143016X14696400495019.
Yadav, G., Dash, S.K., Sen, R., 2019. A biorefinery for valorization of industrial waste-water and flue gas by microalgae for waste mitigation, carbon-dioxide sequestration and algal biomass production. Sci. Total Environ. 688, 129–135. https://doi.org/10.1016/j.scitotenv.2019.06.024.
Yang, Y., Yang, J., Wu, W.-M., Zhao, J., Song, Y., Gao, L., Yang, R., Jiang, L., 2015. Biodegradation and mineralization of polystyrene by plastic-eating mealworms: part 2. Role of gut microorganisms. Environ. Sci. Technol. 49, 12087–12093. https://doi.org/10.1021/acs.est.5b02663.
Zamzuri, N.A., Abd-Aziz, S., 2013. Biovanillin from agro wastes as an alternative food flavour. J. Sci. Food Agric. 93, 429–438. https://doi.org/10.1002/jsfa.5962.
Zhang, X.-Y., Li, R.-Y., Ji, M., 2018. Mechanisms and influencing factors of antibiotic removal in sewage biological treatment. Huan jing ke xue= Huanjing kexue 39, 5276–5288. https://doi.org/10.13227/j.hjkx.201803190.
Zhao, G., Zhang, W., Zhang, G., 2010. Production of single cell protein using waste capsicum powder produced during capsanthin extraction. Lett. Appl. Microbiol. 50, 187–191. https://doi.org/10.1111/j.1472-765X.2009.02773.x.

CHAPTER

5

Application of plant-based natural coagulants in water treatment

Arunkumar Patchaiyappan[a,b] *and Suja Purushothaman Devipriya*[c]

[a]Department of Social Sciences, French Institute of Pondicherry, Pondicherry, India [b]Department of Ecology and Environmental Sciences, School of Life Sciences, Pondicherry University, Pondicherry, India [c]School of Environmental Studies, Cochin University of Science and Technology, Cochin, Kerala, India

OUTLINE

5.1 Introduction

Many communities worldwide use plant-based materials to treat turbid water sources for centuries (Fuglie, 1999). In the past, plant products are used as a desirable alternative through the point-of-use techniques that are cheaper in developing economies because of the lack of technical advancement (Miller et al., 2008). More than one billion poor populations in Asian and African countries rely heavily on these plant coagulants to clean up turbid water for drinking purpose. Several historical anecdotes explain the usage of plant products in treating turbid water for drinking water purposes. Sushruta Samhita, an ancient Indian text, quotes that Indian rural women used numerous plant sources in purifying water during early times. Women in rural India have utilized various plant seeds such as nirmali, moringa, and pongamia to clarify turbid water. In tropical African countries like Sudan, *M.oleifera* indigenous to India was utilized for water clarification purposes. In rain-fed ponds and wells, the burnt coconut shells and the wood of *Phyllanthus emblica* (amla) were used to treat turbid water. In addition to antibacterial and insecticidal effects, tulsi (*Ocimum sanctum*) is an ideal water purifier. In India, the rural communities used copper and bronze pots to store water, while plastic containers used now are known to breed bacteria (Shiva, 1988). The plants used as natural coagulants in other tropical regions were *Moringa oleifera, Strychnos potatorum*, mesquite bean, and *cactus latifaria*. Šćiban et al. (2009) also tested coagulation efficiency with horse chestnut, common oak, turkey oak, and northern red oak utilized by the traditional European communities. When used, these seed materials serve as a principal coagulant. Research performed on natural coagulants demonstrated that coagulant action is incorporated within seeds' cotyledons and is defined by the seed's virtue and its geographical region. Different authors have explored the potential of the active fraction present in the seeds. The seed coagulant ability depends on various parameters used for seed extraction and seed purification (Ghebremichael et al., 2005). The researchers proved that regarding *Moringa oleifera*, the cationic protein of molecular weight ranging from 6 to 66 KDa was responsible for initiating excellent

Cost Effective Technologies for Solid Waste and Wastewater Treatment
https://doi.org/10.1016/B978-0-12-822933-0.00012-7

coagulation with promising disinfectant properties (Gassenschmidt et al., 1995; Okuda et al., 2001; Narasiah et al., 2002). These cationic proteins embedded within the seeds will readily react with negative particles in the suspension, greatly enhancing the formation of flocs to settle. Yet overdose coagulant does not induce coagulation; rather, it increases the solutiońs turbidity by reversing the suspensiońs zeta potential.

The harvest of seeds is the crucial element in deciding the use of these coagulants. The indigenous communities use these methods as a sociocultural response and treat turbid water with these indigenous seeds. Sudanese women in their household swirl the cloth bag in a container filled with turbid water containing the crushed seeds of *Moringa oleifera* and encourage it for clarification (Madsen et al., 1987). In arid regions of Southern Tamil Nadu, the seeds of *Strychnos potatorum* are scraped mostly along the walls of the earthenware comprising turbid water to accomplish a clarification effect (Patchaiyappan et al., 2016). This indigenous water treatment method includes a course of action from grinding the seeds, gathering fully grown quality seeds, separating husk, and finally point of use to treat turbid water. The introduction of technological interventions has revolutionized coagulant preparation in water treatment.

The crude extract with primary processing contains more than the active component necessary for coagulation; other macromolecules such as lipids have an intimidating effect on the coagulation potential by intensifying the organic load and depreciating the quality of treated water (Arnoldsson et al., 2008). The active component extraction is normally carried out with organic solvents after having removed the lipid content of the seeds. The defatted seeds were then treated with water or salt solution to isolate the coagulant protein. The most effective and straightforward approach is believed to be the extraction of coagulant protein with water, as they are mainly water-soluble proteins. The extract produced in the saltwater exhibited improved coagulant action than that of aqueous extract (Ghebremichael, 2007; Sánchez-Martín et al., 2012). Due to salting in process, the salt extraction process removes significant amounts of soluble protein. Besides, tertiary manufacturing methods such as gel permeation chromatography, dialysis, and lyophilization have been tested to resolve the crude extract's drawbacks. The coagulant protein obtained with tertiary-level methods was found to have an elevated coagulant activity than the crude extract (Okuda et al., 2001). While the coagulant potential is enhanced by the tertiary-level processing method, it is not possible in the water treatment technology due to the additional processing expense and professional operating procedure. In field-level programs, this stage of coagulant processing is scarcely applied and it is still functional only in research laboratories and academic disciplines. Consequently, the scientists concentrate on cost-effective, reasonable treatment and extraction process for commodification at the industrial scale of coagulant-mediated water treatment ventures.

Water pollution is indeed a significant public health problem in underdeveloped and developing nations. Fecal contamination found in the polluted surface water and sediments contributes to enteric diseases and related morbidity and mortality. The natural coagulants utilized for clarifying turbidity also impart disinfection properties that make these organic alternatives of paramount importance. These natural coagulants have also been environmentally safe alternatives for chemical equivalents such as chlorine and chloramine. The experiments were constructed at ground level as well as laboratory scale to evaluate the microbial growth along with sediments. Other records of plant-based natural coagulants in reducing bacterial load and a linear association with turbidity reduction have also been found. The researchers also had antibacterial polypeptides isolated from *Moringa oleifera* that allowed waterborne pathogens to be sedimented and disinfected (Eilert et al., 1981). Moreover, the flocculating protein was used to achieve improved organic load reduction and disinfection besides metal ion reduction (Poumaye et al., 2012). To assess the efficacy of *Moringa oleifera* in the elimination of helminth larvae, field-based laboratory experiments have been taken up. The natural coagulant was observed to decrease helminth eggs by 94%–95.5% with a simultaneous reduction in turbidity of around 80%–96%. Other than disinfection, removing sediment and bacteria from the water at the end of the coagulation is a method for lowering the microbial population count in turbid water.

Coagulation is the process in which the electrically double-layered colloidal particles reduce their repulsive ability. The optimal physical characteristics such as pH, alkalinity, dosage, and ionic strength need to be tailored to specific conditions for realizing the coagulation potential. Natural polymers made up of polyelectrolytes comprise ionic active sites such as carboxyl, hydroxyl, or amino groups which involve in coagulation process. These polyelectrolytes either possess anionic, cationic, or amphoteric properties. It is believed that the coagulation of suspended particles in the suspension is carried out by four types of coagulation systems, viz. charge neutralization, interparticle bridging, double-layer compression, and sweep flocculation (Crittenden et al., 2012). Usually, the coagulation process operates according to various rules and one or more strategies for successful destabilization. The double-layer compression process is mainly dependent on the inclusion of an insensitive polyelectrolyte in more significant amounts. If a current system is introduced with a robust ion solution, it alters the total ion concentration. A subsequent rise in ion concentration will compress the double layer covering the colloidal particle and result in the formation of flocs. This coagulation mechanism's efficiency is disputed and is not generally favored, whereas the presence of bivalent cations such as calcium and magnesium causes this type of coagulation mechanism (Duan and Gregory, 2003). The polymers such as proteins and polysaccharides improve this form of the coagulation process. When a long chain of polyelectrolytes is inserted into the suspension, it bonds and holds many colloids nearby. The bridging process will be further enhanced when higher molecular polymer chains are being used for coagulation. This form of coagulation process has been reported with anionic polysaccharides extracted from *Opuntia ficus indica* mucilage (Miller et al., 2008).

The coagulant dose heavily influences the charge neutralization type of mechanism and, if the optimum dosage is surpassed, it is easy to restabilize the particles. The negatively charged colloidal particle will be attracted to the positively charged coagulant. Zeta potential measurements are essential for the effectiveness of coagulation. The zeta potential indicates the optimum coagulant dosage required for net charge neutralization. Cationic protein isolated from the seeds of *Moringa oleifera* follows a charge neutralization coagulation pathway (Ndabigengesere et al., 1995). Sweep flocculation mechanism is typically found in plant-based coagulants; alum usually employs this coagulation process method. The coagulation is caused by a large number of coagulant particles being introduced into the coagulation system. The sludge produced in this method makes it unfavorably difficult.

Adsorption is a natural method used to eliminate contaminants and nutrients from wastewater. Natural biomass like agrowaste, yeast, and bacteria has been extensively studied for adsorption efficiency. The polymer material obtained from the biomass is often utilized as adsorbents owing to their porosity, reactivity, and higher surface area (Srinivasan, 2013; Kozlowski and Walkowiak, 2002; Rio and Delebarre, 2003). Current findings have shown that natural coagulants are a safer natural adsorbent in the disposal of toxic metals from metal smelters and unsafe disposal of municipal solid waste (Van der Oost et al., 2003). These toxic heavy metals have a prolonged biological life span and persist in the environment. The presence of toxic heavy metals often generates complications in humans, such as cancer, dermatitis, and respiratory syndromes (Bagheri and Saraji, 2001; Lee et al., 2003; Van der Bruggen and Vandecasteele, 2003). The natural coagulant undergoes trace metal sorption process through amino acid interacting with metal ions or metal oxides at a particular pH. The biomass's sorption capacity is known to be enhanced when they are pretreated with alkalis, surfactants, and acids. The sorption phenomenon in biomass is governed by the active binding sites on the natural coagulant surface and varies with the metal ions properties to be adsorbed. Henceforth, the plant coagulant is now being popularly regarded as a potential material for eliminating dyes, effluents, and surfactants by the academic community (Hwang et al., 2000; Pearce et al., 2003; McMullan et al., 2001; Fu and Viraraghavan, 2001). Many researchers are looking at natural polysaccharides like cellulose, sugar, gums, and mucus. Chitosan, a remnant of crustaceans' exoskeleton, was commonly used as a flocculant in extracting organic matter from wastewater. Even xanthan gum and guar gum have been widely used in the wastewater treatment industries. Real polymers contain flocculating agents, which are reinforced by grafting polymers with synthetic monomers, including acrylamide.

The production of alternative products like coagulants is always related to a few limits and restrictions. There are several explanations why there has not been much research into this issue. Among these are lack of knowledge and regulatory approval. Despite all these limitations, it is desirable to discuss the bright potential of these alternative and safe technologies for water treatment. The science community should investigate the natural chemical processes to find a suitable commercial method to decentralize its usage. The current use within academic and R&D laboratories should be encouraged to a wide-scale application by the continuous discovery of the chemistry and these natural coagulants' mechanism.

5.2 Moringa oleifera

Moringa oleifera is a fast-growing drought-resistant deciduous tree, which is most commonly used by the indigenous communities in Asia and Africa and has been rightly regarded as the most promising natural coagulant by pioneering studies (Ghebremichael, 2007). Ndabigengesere et al. (1995) isolated dimeric cationic protein with a 13 KDa, which effectively produces lesser sludge volume than alum. The proposed coagulation of the aforementioned study was presumed to be adsorption and charge neutralization. Okuda et al. (1999) contrasted with ordinary distilled water extraction methods by trying the novel approach of coagulant extraction. The technique involved salt-induced extraction method utilizing sodium chloride and resulted in effective protein yielding 7.4 times more effective than protein extracted from pure water. The enhanced coagulant activity was apparently due to the enhanced ionic strength resulting from the salting mechanism. Okuda et al. (2001) presumed that the coagulation mechanism is carried out by the enmeshment of kaolin particles with insoluble matters of the netlike structure produced by the salt-extracted protein. For the coagulation process that binds the active coagulant molecule to form a net structure, the bivalent ions are necessary. The research outlined that the coagulation process like double-layer trapping, interparticle bridging, or charge neutralization are not responsible for clarification process.

Ndabigengesere and Narasiah (1998) experimented to determine the capacity of *Moringa oleifera* and alum to treat turbid water. The experiments involved assaying varied dosages of the crude extract, in both powder and liquid forms of *Moringa oleifera* seeds. After treatment, several physical parameters were measured including pH, conductivity, alkalinity, and ion concentration. In the treated water, the organic compounds' concentration gradually improved after the treatment process, but the other physical parameters did not improve significantly. From the experiment results, it was deduced that only when the active protein fraction was purified, the seeds might be feasible for use in the wastewater and the water.

Pritchard et al. (2010a) investigated the performance of *Moringa oleifera* in studies with chemical coagulants. The experiments proved that the seed extract of moringa seeds is known to reduce more than 80% of turbidity and coliforms in simulated synthetic water. Despite alum's better performance, the outcomes were adequate to support its more widespread use. An experiment was also conducted by Pritchard et al. (2010b) to define the suitable dosing method, the optimum dosages that can be used, the impact of pH, temperature, and shelf life. Cold temperatures hindered the coagulation mechanisms, and the shellfish with a life span of 18 months was found to have the best reduction in turbidity.

Nurfahasdi (2012) investigated *Moringa oleifera*'s coagulant fraction, which was the active component in sediment removal from turbid water. The active anticoagulant fraction was isolated using NaCl and KCl. The morphological characterization of the extractant solution indicated the presence of amine and ketone moieties and the order of clarification was KCl > NaCl > distilled water. The higher the KCl concentration, the higher the level of peptide content present in the extract. Kawo and Daneji (2011) used *Moringa oleifera* to treat the water samples obtained from Challawa River. The coagulant obtained from moringa seeds lowered the turbidity to 14 NTU, and it also exhibited a significant decline in other dissolved salts. The statistical analysis revealed that the probability of bacterial recovery was dependent on the coagulant dosage amount administered. Masamba et al. (2010) studied removing heavy metals by moringa seeds as a coagulant in jar test apparatus. The varying parameters including initial concentration, ionic strength, and agitation time were tried to assess the amount of cadmium sorbed. The cadmium sorption was in terms of the Freundlich isotherm and the adsorption energy at 30°C was consistent with physisorption.

Kwaambwa et al. (2010) carried out an experimental investigation of moringa protein adsorption onto the silica surface to evaluate its flocculation mechanism. The analysis indicated that high molecular weight protein combines firmly with silica at even lower concentrations. Though the amino acid present in the protein possesses a net positive charge at neutral pH, the amino acid played a major role in either hydrophobic or electrostatic association. The adsorbed protein with the ability to form destabilized particulates can provide clean water.

5.3 Strychnos potatorum

Strychnos potatorum L.f is a deciduous tree with a wide distribution in India and Burma. The hard, ripe seeds of this tree are called "nirmali" and also known as "thethankottai" in local parlance. Sushruta Samhita, an old Sanskrit text, describes that these seeds profoundly clarify turbid water. The seeds are pummeled and scraped against the earthenware container walls to produce a water-purifying effect. NATFLOC, polyelectrolyte derived from *Strychnos potatorum* seeds, has been developed by the indigenous communities of Andhra Pradesh for water purification (Saif et al., 2012). Tripathi et al. (1976) found that the anionic polyelectrolytes embedded in the seed extract cause particles in water to aggregate through bridging mechanism. Adinolfi et al. (1994) claimed that the galactomannans and galactan present in the nirmali seed extract reduce more than 80% of water turbidity in kaolin-simulated synthetic turbid water. The polysaccharide possesses many hydroxyl functional groups that provide a weak adsorption site and easy interparticle bonding.

Senthamarai et al. (2013) explored how methylene blue adsorbs onto Strychnos potatorum seeds' surfaces in an aqueous solution. The experiment model claimed that the seed surface adsorption behavior follows pseudo-second-order kinetics. It was also determined that adsorption was carried out by inter- and intracellular protein transporters. The calculated isothermal model best suits the Freundlich equation, which demonstrated the adsorption of methylene blue onto the seed powder. The experiments also proved that the seed powder had an adsorption potential of 78.84 mg of a dye molecule per gram of powder. Thermodynamic studies have indicated that the adsorption mechanism produces heat.

Jayaram et al. (2009) examined *Strychnos potatorum* seeds' metal adsorption of lead from an aqueous suspension. The study involved a variety of experimental parameters in adsorption efficiency. Lead is most absorbed when in a 5.0 pH, and the equilibrium reaches at 360 min. The adsorption process was best described by pseudo-second-order kinetic model and also, the Freundlich and Langmuir isotherms described the adsorption process best. The monolayer adsorption capacity was determined to be 16.420 mg/g. Lakshmipathiraj et al. (2013) found that *Strychnos potatorum* seeds effectively adsorbed chromium (Cr VI). The adsorption equilibrium was attained within 4 min, and the amount of material absorbed was determined to be 59 mg/g. The anions, just like the cations, also affected the adsorption efficiency to a small degree. The removal rate was governed by second-order kinetics and adsorption was explained by electrostatic and chemical interactions. Saif et al. (2012) determined how bound proteins from *Strychnos potatorum* retain Cd in aqueous solutions. The protein from the seeds was analyzed for its cadmium binding ability by fractional precipitation of calcium sulfate. The cadmium removal rate was pH-dependent, with maximum removal at pH around 5 with a period of around 6 h.

Patchaiyappan et al. (2020) explored the efficiency of the nirmali seeds in clarifying the synthetic turbid water. In addition to the clarifying effect, the seed extracts were tested for their disinfection efficacy by spiking *E. coli* and MS-2 bacteriophage in synthetic turbid water. The seed extract of *S. potatorum* resulted in a 3–4 log reduction of *E. coli* and 1.3 log reduction for MS-2 phages. The optimization of operating parameters like dosage, alkalinity, pH, and ionic concentration on turbidity reduction were also presented. With aqueous extract showing disinfection efficacy, a bioactive compound methyl palmitoleate isolated from the seed extracts also displayed the ability to kill bacteria. Also, Arunkumar et al. (2019) examined the turbidity reduction potential and isolated the respective active constituents like proteins and polysaccharides from *Strychnos potatorum* seeds. The isolated 12-kDa protein exhibited better coagulation efficiency than the isolated galactomannan residue in a small-scale coagulation assay.

Similarly, Patchaiyappan et al. (2016) separated titanium dioxide particles from aqueous suspension, which is a significant limitation in the photocatalytic water treatment process. The studies showed that the plant-based coagulant offers a solution to less efficient immobilization strategies that exist. The aqueous extract extracted from *Strychnos potatorum* seeds showed that it was capable of effectively sedimenting TiO_2. The potential recovery of sedimented titanium dioxide photocatalysts has been researched thoroughly by carrying out various experiments. The recovered catalyst after repeated photocatalytic experiments pursued pseudo-first-order kinetics and displayed greater photocatalytic activity. Compared to pure TiO_2, none of the recovered catalysts was found to have a substantial loss in surface and morphological properties; also, the recovered catalysts were highly suitable for recyclability and reusability.

5.4 Leguminous species

Šćiban et al. (2005) studied the coagulation efficiency of some leguminous species, such as *Phaseolus vulgaris, Robinia pseudoacacia, Ceratonia siliqua,* and *Amorpha fruticosa.* With *Phaseolus vulgaris* exhibiting higher coagulation ability, the purified protein expressed the best coagulation efficiency at 35 NTU, 22 times better than crude extract, and reducing the organic load up to 16 times. Partial purification of these protein coagulants has also been observed and compared to the coagulation activity of crude proteins (Antov et al., 2010). This additional pretreatment phase is crucial to minimize the accumulation of organic loadings (Ghebremichael, 2007), which is a frequent issue in the usage of natural coagulants to reduce the dosage of coagulants while ensuring optimum coagulation processes. Their seeds generated an even lower proportion of lipids. It removes the need for delipidation and is hugely advantageous in sustainability issues (Antov et al., 2010). Centrifugation and coagulation using plant coagulants were compared for different wastewater processes. The coagulation process in treating wastewater sample resulted in COD removal of up to 69% under optimal condition compared to centrifugation (Prodanović et al., 2011).

5.5 Other plant coagulants

Jatropha curcas is a genus of flowering plants in the spurge family, Euphorbiaceae, widespread in tropical and subtropical parts of the world. *Jatropha curcas* seed and its press cake have an inherent coagulant property which is used to treat wastewater (Makkar et al., 1997). The disinfectant property is as effective as alum, in addition to its coagulant activity, projecting it as a possible coagulant. It removes 99% turbidity at pH3 and 120 ml/L dosage (Abidin et al., 2011). The seed's active coagulant is expected to be a soluble cationic protein when extracted with 0.5 M NaCl. Many cactus species in Central American nations are reported to contain anionic polyelectrolyte, which is used to clarify turbid water and can be used as a coagulant aid. Bratby (2006) has shown the efficacy of cactus as a flocculant with alum as a main coagulant. Turbidity removal efficiency could achieve 94% at a pH of around 10, and the temperature is found to have a minor influence on its coagulation efficiency (Zhao et al., 2007). The possible coagulation mechanism in Opuntia species is adsorption and bridging, which is often caused by hydrogen bonding or dipole interactions

(Crittenden et al., 2012). The divalent cations embedded in them react with anionic polymers to form insoluble precipitates (Miller et al., 2008).

The extraction of coagulant protein from the coconut endosperm was demonstrated by Fatombi et al. (2013). The mechanism of coagulation is assumed to be heteroaggregation between oppositely charged colloids; the coagulant protein is likely to develop higher order complexes of greater complexity as the coagulant concentration is increased. The coagulant isolated from the coconut is effective over a broad pH range. The coagulation efficiency of the mustard seed extracts was investigated by Bodlund et al. (2014). The work yielded coagulant protein with thermal resistance properties even after heat treatment at 95°C for 5h. They described coagulant proteins with a molecular mass between 6.5 and 9 KDa. In the peptide sequence analysis of the isolated mustard protein, Moringa MO2.1 protein and napin-3 protein isolated from those seeds were structurally similar.

Furthermore, it has been shown that these isolated coagulant proteins are also effective in the removal of microorganisms. In Tanzania, indigenous community used the seeds of *Vigna unguiculata* and *Parkinsonia aculeata* for clarification of turbid water. Marobhe et al. (2007) analyzed the lectins isolated from the conventional coagulants and found that cationic proteins of molecular weight 6 KDa have more significant coagulant activity than their crude extract with a broad pH ranging from 5.5 to 8. Also, the coagulant proteins were found to be heat resistant. In addition to turbidity removal, it is also found to reduce hardness and heavy metals to an extent. It also contains antiviral and antifungal compounds imparting disinfectant properties to them (Lim, 2012).

The slimy mucilage obtained from the seedpods of *Hibiscus esculentus* is utilized as an alternative coagulant for treating synthetic turbid water and effluent. The performance of the mucilage is verified with synthetic turbid water up to 3000 NTU. When used as a coagulant aid, it increased the efficiency of alum to 97.1%. The coagulation efficiency of alum in turbid water was improved with this seed extract's addition, resulting in higher turbidity removal up to 97.1%. The alum consumption was reduced by 50% and it is efficient for many applications (Al-Samawi and Shokralla, 1996). Further studies on effluent treatment have revealed that the mucilage obtained from the seedpods has removed more than 80% of suspended solids from tannery and sewage effluents (Agarwal et al., 2003). The coagulation effect was primarily due to the direct bonding between the waste and the chelators and concurrent agglomeration into colloidal particles (Agarwal et al., 2001, 2003). The polysaccharides and proteins had a strong interaction with one another during the floc formation (El-Mahdy and El-Sebaiy, 1984). Like other plant coagulants, higher efficiency was achieved with lower dosages, and the increase in dosage contributed to the organic load (Anastasakis et al., 2009). The mucilage exhibited promising disinfection properties against common surface water pathogens like *Escherichia coli* and *Staphylococcus aureus* (de Carvalho et al., 2011). Hence, this plant-based coagulant may fit into the ambit of treating raw surface water.

The mucilage of *Coccinia indica* fruit is used to clarify turbid water and found that 94% turbidity was eliminated in 100 NTU turbidity (Asrafuzzaman et al., 2011). The principal component involved in the turbidity reduction was found to be a polysaccharide. The turbidity clarification of surface water with this plant coagulant is more desirable as it kills harmful bacteria like Salmonella, causing waterborne diseases. The extract of *Cicer arietinum* was also tested for turbidity reduction, and it was found that 95% reduction was obtained with turbid water of 95 NTU. The turbidity reduction efficiency was comparable to chemical coagulant like alum and the inherent coagulant fraction involved in the turbidity reduction was presumed to be carbohydrate. The legume was also known to reduce total coliforms indicating their disinfection efficacy. The legume also contains bioactive compounds like cicerin and arietin, antifungal peptides with N-terminal sequences (Lim, 2012). The other grain, horse gram, was also examined for its coagulation efficiency. The natural coagulant removed nearly 95% of turbidity with initial turbidity ranging from 25 to 1600mg/L and the optimum pH was found to be 7.5 (Bhole, 1995).

5.6 Conclusion

A number of natural compounds had been found to clarify turbidity water, but only a handful of compounds have the ability to clean the waterborne pathogens and reduce the turbidity in the water. The development of alternative products like plant-based coagulants is often linked to a few limitations and constraints. The limitations can be linked to the lack of scientific validation and development of technical feasibility. Since plant-based coagulants address the environmental concerns, further research and commercialization of these alternative sustainable solutions should be promoted. Large-scale application and popularization of such natural coagulants are feasible only with continuous exploration of these indigenous coagulants' mechanism and chemistry. India has a vast natural plant diversity that harbors many potential solutions for many water treatment problems. These materials' usage is documented, but formal elucidation of their properties and mechanism needs to be elaborated in scientific and robust manner. The growing interest in green chemistry and sustainable management endeavors and such works are being initiated in a steady phase. While the plant-based coagulants' properties have yet to be fully understood, the chemical coagulants are largely used. Gradual adoption may be made until commodifications shortfalls, and the drawbacks of natural coagulants have been resolved. It is vital that life-cycle assessments should be accomplished for analyzing the various socioeconomic implications of plant-based coagulants.

References

Abidin, Z.Z., Ismail, N., Yunus, R., Ahamad, I.S., Idris, A., 2011. A preliminary study on Jatropha curcas as coagulant in wastewater treatment. Environ. Technol. 32 (9), 971–977.

Adinolfi, M., Corsaro, M.M., Lanzetta, R., Parrilli, M., Folkard, G., Grant, W., Sutherland, J., 1994. Composition of the coagulant polysaccharide fraction from Strychnos potatorum seeds. Carbohydr. Res. 263 (1), 103–110.

Agarwal, M., Srinivasan, R., Mishra, A., 2001. Study on flocculation efficiency of okra gum in sewage wastewater. Macromol. Mater. Eng. 286 (9), 560–563.

Agarwal, M., Rajani, S., Mishra, A., Rai, J.S.P., 2003. Utilization of okra gum for treatment of tannery effluent. Int. J. Polym. Mater. 52 (11–12), 1049–1057.

Al-Samawi, A.A., Shokralla, E.M., 1996. An investigation into an indigenous natural coagulant. J. Environ. Sci. Health A 31 (8), 1881–1897.

Anastasakis, K., Kalderis, D., Diamadopoulos, E., 2009. Flocculation behavior of mallow and okra mucilage in treating wastewater. Desalination 249 (2), 786–791.

Antov, M.G., Šćiban, M.B., Petrović, N.J., 2010. Proteins from common bean (Phaseolus vulgaris) seed as a natural coagulant for potential application in water turbidity removal. Bioresour. Technol. 101 (7), 2167–2172.

Arnoldsson, E., Bergman, M., Matsinhe, N., Persson, K.M., 2008. Assessment of drinking water treatment using Moringa oleifera natural coagulant. Vatten 64 (2), 137.

Arunkumar, P., Sadish Kumar, V., Saran, S., Bindun, H., Devipriya, S.P., 2019. Isolation of active coagulant protein from the seeds of Strychnos potatorum–a potential water treatment agent. Environ. Technol. 40 (12), 1624–1632.

Asrafuzzaman, M., Fakhruddin, A.N.M., Hossain, M.A., 2011. Reduction of turbidity of water using locally available natural coagulants. ISRN Microbiol. 2011.

Bagheri, H., Saraji, M., 2001. New polymeric sorbent for the solid-phase extraction of chlorophenols from water samples followed by gas chromatography–electron-capture detection. J. Chromatogr. A 910 (1), 87–93.

Bhole, A.G., 1995. Relative evaluation of a few natural coagulants. Aqua-J. Water Supply: Res. Technol. 44 (6), 284–290.

Bodlund, I., Pavankumar, A.R., Chelliah, R., Kasi, S., Sankaran, K., Rajarao, G.K., 2014. Coagulant proteins identified in mustard: a potential water treatment agent. Int. J. Environ. Sci. Technol. 11 (4), 873–880.

Bratby, J., 2006. Coagulation and Flocculation in Water and Wastewater Treatment. IWA Publishing.

Crittenden, J.C., Trussell, R.R., Hand, D.W., Howe, K.J., Tchobanoglous, G., 2012. MWH's Water Treatment: Principles and Design. John Wiley & Sons.

de Carvalho, C.C., Cruz, P.A., da Fonseca, M.M.R., Xavier-Filho, L., 2011. Antibacterial properties of the extract of Abelmoschus esculentus. Biotechnol. Bioprocess Eng. 16 (5), 971.

Duan, J., Gregory, J., 2003. Coagulation by hydrolyzing metal salts. Adv. Colloid Interf. Sci. 100, 475–502.

Eilert, U., Wolters, B., Nahrstedt, A., 1981. The antibiotic principle of seeds of Moringa oleifera and Moringa stenopetala. Planta Med. 42 (05), 55–61.

El-Mahdy, A.R., El-Sebaiy, L.A., 1984. Preliminary studies on the mucilages extracted from okra fruits, Taro tubers, Jew's mellow leaves and fenugreek seeds. Food Chem. 14 (4), 237–249.

Fatombi, J.K., Lartiges, B., Aminou, T., Barres, O., Caillet, C., 2013. A natural coagulant protein from copra (Cocos nucifera): isolation, characterization, and potential for water purification. Sep. Purif. Technol. 116, 35–40.

Fu, Y., Viraraghavan, T., 2001. Fungal decolorization of dye wastewaters: a review. Bioresour. Technol. 79 (3), 251–262.

Fuglie, L.J., 1999. The Miracle Tree: *Moringa oleifera*, Natural Nutrition for the Tropics. Church World Service.

Gassenschmidt, U., Jany, K.D., Bernhard, T., Niebergall, H., 1995. Isolation and characterization of a flocculating protein from Moringa oleifera lam. Biochim. Biophys. Acta (BBA)-Gen. Subj. 1243 (3), 477–481.

Ghebremichael, K., 2007. Overcoming the drawbacks of natural coagulants for drinking water treatment. Water Sci. Technol. Water Supply 7 (4), 87–93.

Ghebremichael, K.A., Gunaratna, K.R., Henriksson, H., Brumer, H., Dalhammar, G., 2005. A simple purification and activity assay of the coagulant protein from Moringa oleifera seed. Water Res. 39 (11), 2338–2344.

Hwang, H.M., Slaughter, L.F., Cook, S.M., Cui, H., 2000. Photochemical and microbial degradation of 2, 4, 6-trinitrotoluene (TNT) in a freshwater environment. Bull. Environ. Contam. Toxicol. 65 (2), 228–235.

Jayaram, K., Murthy, I.Y.L.N., Lalhruaitluanga, H., Prasad, M.N.V., 2009. Biosorption of lead from aqueous solution by seed powder of Strychnos potatorum L. Colloids Surf. B: Biointerfaces 71 (2), 248–254.

Kawo, A.H., Daneji, I.A., 2011. Bacteriological and physcio–chemical evaluation of water treated with seed powder of Moringa oleifera. Bayero J. Pure Appl. Sci. 4 (2), 208–212.

Kozlowski, C.A., Walkowiak, W., 2002. Removal of chromium (VI) from aqueous solutions by polymer inclusion membranes. Water Res. 36 (19), 4870–4876.

Kwaambwa, H.M., Hellsing, M., Rennie, A.R., 2010. Adsorption of a water treatment protein from Moringa oleifera seeds to a silicon oxide surface studied by neutron reflection. Langmuir 26 (6), 3902–3910.

Lakshmipathiraj, P., Umamaheswari, S., Raju, G.B., Prabhakar, S., Caroling, G., Kato, S., Kojima, T., 2013. Studies on adsorption of Cr (vi) onto Strychnos potatorum seed from aqueous solution. Environ. Prog. Sustain. Energy 32 (1), 35–41.

Lee, K.B., Gu, M.B., Moon, S.H., 2003. Degradation of 2, 4, 6-trinitrotoluene by immobilized horseradish peroxidase and electrogenerated peroxide. Water Res. 37 (5), 983–992.
Lim, T.K., 2012. Edible Medicinal and Non-Medicinal Plants. Vol. 1 Springer, Dordrecht, The Netherlands, pp. 285–292.
Madsen, M., Schlundt, J., El Fadil, E.O., 1987. Effect of water coagulation by seeds of *Moringa oleifera* on bacterial concentrations. J. Trop. Med. Hyg. 90, 101–109.
Makkar, H.P.S., Becker, K., Sporer, F., Wink, M., 1997. Provenances of *Jatropha curcas*. J. Agric. Food Chem. 45 (8), 3152–3157.
Marobhe, N.J., Dalhammar, G., Gunaratna, K.R., 2007. Simple and rapid methods for purification and characterization of active coagulants from the seeds of *Vigna unguiculata* and *Parkinsonia aculeata*. Environ. Technol. 28 (6), 671–681.
Masamba, W.R.L., Mataka, L.M., Mwatseteza, J.F., Sajidu, S.M.I., 2010. Cadmium sorption by *Moringa stenopetala* and *Moringa oleifera* seed powders: batch, time, temperature, pH and adsorption isotherm studies. Acad. J. 2 (3), 50–59.
McMullan, G., Meehan, C., Conneely, A., Kirby, N., Robinson, T., Nigam, P., Banat, I.M., Marchant, R., Smyth, W.F., 2001. Microbial decolourization and degradation of textile dyes. Appl. Microbiol. Biotechnol. 56 (1), 81–87.
Miller, S.M., Fugate, E.J., Craver, V.O., Smith, J.A., Zimmerman, J.B., 2008. Toward understanding the efficacy and mechanism of Opuntia spp. as a natural coagulant for potential application in water treatment. Environ. Sci. Technol. 42 (12), 4274–4279.
Narasiah, K.S., Vogel, A., Kramadhati, N.N., 2002. Coagulation of turbid waters using *Moringa oleifera* seeds from two distinct sources. Water Sci. Technol. Water Supply 2 (5–6), 83–88.
Ndabigengesere, A., Narasiah, K.S., 1998. Quality of water treated by coagulation using *Moringa oleifera* seeds. Water Res. 32 (3), 781–791.
Ndabigengesere, A., Narasiah, K.S., Talbot, B.G., 1995. Active agents and mechanism of coagulation of turbid waters using *Moringa oleifera*. Water Res. 29 (2), 703–710.
Nurfahasdi, M., 2012. Extraction and Characterization of Coagulation Active Agent from *Moringa oleifera* for Turbidity and Hardness Removal in Raw Water Sources (Doctoral dissertation). Universiti Malaysia Perlis (UniMAP).
Okuda, T., Baes, A.U., Nishijima, W., Okada, M., 1999. Improvement of extraction method of coagulation active components from Moringa oleifera seed. Water Res. 33 (15), 3373–3378.
Okuda, T., Baes, A.U., Nishijima, W., Okada, M., 2001. Coagulation mechanism of salt solution-extracted active component in *Moringa oleifera* seeds. Water Res. 35 (3), 830–834.
Patchaiyappan, A., Saran, S., Devipriya, S.P., 2016. Recovery and reuse of TiO2. Korean J. Chem. Eng. 33 (7), 2107–2113.
Patchaiyappan, A., Sarangapany, S., Saksakom, Y.A., Devipriya, S.P., 2020. Feasibility study of a point of use technique for water treatment using plant-based coagulant and isolation of a bioactive compound with bactericidal properties. Sep. Sci. Technol. 55 (1), 112–122.
Pearce, C.I., Lloyd, J.R., Guthrie, J.T., 2003. The removal of colour from textile wastewater using whole bacterial cells: a review. Dyes Pigments 58 (3), 179–196.
Poumaye, N., Mabingui, J., Lutgen, P., Bigan, M., 2012. Contribution to the clarification of surface water from the *Moringa oleifera*: case M'Poko River to Bangui, Central African Republic. Chem. Eng. Res. Des. 90 (12), 2346–2352.
Pritchard, M., Craven, T., Mkandawire, T., Edmondson, A.S., O'neill, J.G., 2010a. A comparison between Moringa oleifera and chemical coagulants in the purification of drinking water–an alternative sustainable solution for developing countries. Phys. Chem. Earth, A/B/C 35 (13), 798–805.
Pritchard, M., Craven, T., Mkandawire, T., Edmondson, A.S., O'neill, J.G., 2010b. A study of the parameters affecting the effectiveness of Moringa oleifera in drinking water purification. Phys. Chem. Earth, A/B/C 35 (13), 791–797.
Prodanović, J.M., Šćiban, M.B., Antov, M.G., Dodić, J.M., 2011. Comparing the use of common bean extracted natural coagulants with centrifugation in the treatment of distillery wastewaters. Rom. Biotechnol. Lett. 16 (5), 6638–6647.
Rio, S., Delebarre, A., 2003. Removal of mercury in aqueous solution by fluidized bed plant fly ash. Fuel 82 (2), 153–159.
Saif, M.M.S., Kumar, N.S., Prasad, M.N.V., 2012. Binding of cadmium to *Strychnos potatorum* seed proteins in aqueous solution: adsorption kinetics and relevance to water purification. Colloids Surf. B: Biointerfaces 94, 73–79.
Sánchez-Martín, J., Beltrán-Heredia, J., Peres, J.A., 2012. Improvement of the flocculation process in water treatment by using moringa oleifera seeds extract. Braz. J. Chem. Eng. 29 (3), 495–502.
Šćiban, M.B., Klašnja, M.T., Stojimirović, J.L., 2005. Investigation of coagulation activity of natural coagulants from seeds of different leguminose species. Acta Period. Technol. 36, 81–90.
Šćiban, M., Klašnja, M., Antov, M., Škrbić, B., 2009. Removal of water turbidity by natural coagulants obtained from chestnut and acorn. Bioresour. Technol. 100 (24), 6639–6643.
Senthamarai, C., Kumar, P.S., Priyadharshini, M., Vijayalakshmi, P., Kumar, V.V., Baskaralingam, P., Sivanesan, S., 2013. Adsorption behavior of methylene blue dye onto surface modified *Strychnos potatorum* seeds. Environ. Prog. Sustain. Energy 32 (3), 624–632.
Shiva, V., 1988. Staying Alive: Women, Ecology and Development. Zed Books.
Srinivasan, R., 2013. Natural polysaccharides as treatment agents for wastewater. RSC Green Chem. Ser., 51–81.
Tripathi, P.N., Chaudhuri, N., Bokil, S.D., 1976. Nirmali seed a naturally occurring coagulant. Indian J. Environ. Health 18 (4).
Van der Bruggen, B., Vandecasteele, C., 2003. Removal of pollutants from surface water and groundwater by nanofiltration: overview of possible applications in the drinking water industry. Environ. Pollut. 122 (3), 435–445.
Van der Oost, R., Beyer, J., Vermeulen, N.P., 2003. Fish bioaccumulation and biomarkers in environmental risk assessment: a review. Environ. Toxicol. Pharmacol. 13 (2), 57–149.
Zhao, M., Yang, N., Yang, B., Jiang, Y., Zhang, G., 2007. Structural characterization of water-soluble polysaccharides from Opuntia monacantha cladodes in relation to their anti-glycated activities. Food Chem. 105 (4), 1480–1486.

CHAPTER

6

Recent applications of downflow hanging sponge technology for decentralized wastewater treatment

Mahmoud Nasr[a,b], *Michael Attia*[c], *Hani Ezz*[b], *and Mona G. Ibrahim*[b,d]

[a]Sanitary Engineering Department, Faculty of Engineering, Alexandria University, Alexandria, Egypt [b]Environmental Engineering Department, Egypt-Japan University of Science and Technology (E-JUST), Alexandria, Egypt [c]Irrigation Engineering and Hydraulics Department, Faculty of Engineering, Alexandria University, Alexandria, Egypt [d]Environmental Health Department, High Institute of Public Health, Alexandria University, Alexandria, Egypt

OUTLINE

6.1 Introduction

Recently, "Clean Water and Sanitation" has been considered "Goal 6" of the sustainable development goals (SDGs), affecting the economic, social, political, and environmental aspects in most countries. Unfortunately, billions of people worldwide have lack access to adequate domestic wastewater treatment due to the limited financial and administrative supports (Nasr et al., 2014a, b). Moreover, some countries have an inadequate connection to sanitation, which could be associated with spreading infectious diseases and the occurrence of various health risk issues (Hamdy et al., 2018). As a result, several efforts and actions have been performed to provide proper sanitation systems, including the application of decentralized wastewater treatment works (Chernicharo and Almeida, 2011). The decentralized systems represent the technologies of wastewater collection, treatment, and dispersal/reuse for individual units (industry, cluster of homes, rural and remote communities, or institutional facility) (El-Tabl et al., 2013). The decentralized wastewater treatment systems should be appropriately designed and implemented to obey the stringent laws for effluent discharges into the environment (Awolusi et al., 2016). Accordingly, comprehensive researches and studies on the decentralized wastewater treatment approach should be conducted, providing a suitable option for protecting the environment in low-income countries (Mahmoud et al., 2019).

https://doi.org/10.1016/B978-0-12-822933-0.00014-0

Rural domestic wastewater is usually difficult to collect due to the lack of sewerage systems (infrastructure) that transfer wastewater into centralized wastewater treatment facilities. Hence, several technologies such as constructed wetlands, septic tanks, soil trench systems, sand filters, and lagoons have been considered for rural sewage treatment (Nasr et al., 2014a, b). However, these technologies may suffer from the large land requirements and low removal efficiency of pollutants, limiting their broad applications (Jiang et al., 2005). Recently, downflow hanging sponge (DHS) has emerged as a promising and cost-effective technology for decentralized wastewater treatment, especially in countries with a shortage of energy resources (Fleifle et al., 2013). The DHS unit can be used as either a stand-alone treatment process or a posttreatment stage (i.e., especially for the case of highly polluted sources) (Racho et al., 2011). The DHS technology depends on the application of sponge material to provide a large surface area for bacterial biomass growth and accumulation (Jiang et al., 2005). In turn, the bacterial community attached to the sponge media tends to utilize the organic substrate in the feed, reducing the COD, nitrogen, and phosphorus pollutants (Matsuura et al., 2010). The sponge carrier is characterized by a high void ratio, which improves the ability to capture more biomass both inside and on the surface of the sponge piece (Matsushita et al., 2018). The dissolved oxygen (DO) profile of the sponge element (i.e., high DO on the surface and low DO in the deep zone) allows for enriching aerobic, anoxic, and anaerobic microorganisms (Araki et al., 1999). Moreover, several openings (windows) are distributed along the reactor height to permit the air diffusion from the surrounding environment into the reactor (Ismail et al., 2020). The DHS reactor is constructed in segments connected in series, and each of the two segments is separated by a space to facilitate the air diffusion mechanism (Hatamoto et al., 2018). This pattern is known as a self- or natural ventilation system, reducing the costs required for installing aeration equipment (Ismail et al., 2020). Because of its wide applications to treat various wastewater sources, the DHS technology and its modifications should be comprehensively summarized, giving beneficial knowledge to researchers, decision makers, and public and private stakeholders.

Hence, this chapter represents DHS technology as a suitable decentralized wastewater treatment option in developing and low-income countries. This objective is comprehensively reported regarding (i) a literature survey in the SCOPUS database within 1997–2020, (ii) concept and idea of DHS, (iii) DHS configuration and modifications, (iv) advantages and limitations, and (v) recommended future studies.

6.2 Sanitation and hygiene in decentralized communities

The idea of decentralized wastewater treatment deals with relatively small communities (i.e., single households or groups of dwellings) that are not served by central sewer systems (Chernicharo and Almeida, 2011). The treatment facility should be located *on-site* or near (<3 km) the pollution source to avoid huge pump stations and long-distance wastewater conveyance. The capacity of the decentralized systems can be flexibly expanded to track the population growth patterns (Machdar et al., 1997). Moreover, decentralized systems can be installed to receive gray water with low levels of pathogens, which can be simply treated by separating it from blackwater (toilets). Treated wastewater from such systems can be employed for local water reuse schemes, mainly for nondrinking purposes (Tawfik et al., 2015). Local collection systems are installed, requiring smaller area and lower costs compared with those used for regional wastewater treatment plants (WWTPs). The decentralized facilities should be properly designed, installed, and operated to achieve the best performance and benefit.

6.3 Downflow hanging sponge (DHS) in SCOPUS database (1997–2020)

The literature survey of this chapter was performed using the keywords "Downflow," "Hanging," and "Sponge" in the SCOPUS library (https://www.scopus.com) for the 1997–2020 period. The obtained information was evaluated regarding the number and type of publications, subject area, and most contributing countries. This systematic and comprehensive survey depicts that the DHS system was originated by a research group in Japan in the late 1990s for treating sewage via an upflow anaerobic sludge blanket (UASB) + DHS combined system (Agrawal et al., 1997; Machdar et al., 1997). Further, the number of publications increased and reached up to 41 articles by 2020. However, the actual number of publications is expected to be more than 41, because other authors use alternative expressions such as "Down-flow hanging sponge" (Chuang et al., 2014), "DHS" (Furukawa et al., 2016), "Continuous-flow sponge bioreactor" (Hatamoto et al., 2017), and "Self-aerated sponge tower" (Ismail et al., 2020). This assessment showed that the DHS technology could be appropriately employed for managing the wastewater treatment systems, regarding pollutant removal, economic/cost attributes, energy utilization, and wastewater reuse. The studies were conducted based on laboratory-scale, pilot-plant (Tandukar et al., 2005; Tanikawa et al.,

2016; Watari et al., 2017a, b), and full-scale (Chernicharo and Almeida, 2011; Pattananuwat et al., 2013) systems. Most of the publications related to the DHS technology were spread over the following journals: *Water Science and Technology* (36.6%), *Applied Microbiology and Biotechnology* (7.3%), *Bioresource Technology* (7.3%), *Environmental Science and Pollution Research* (4.9%), and *Water Research* (4.9%). The scopes of these journals include *Water and Wastewater Treatment Technologies; Waste Management; Water Chemistry and Microbiology;* and *Cleaner Production*. Moreover, Japan was the top country contributing to the DHS-related publication (65.9%) and followed by Egypt (12.2%), Viet Nam (9.8%), China (7.3%), Thailand (7.3%), Brazil (4.9%), India (4.9%), and Indonesia (4.9%). This observation assigns that DHS is an applicable technology for both developed and developing countries. The types of these publications were articles (78.0%), conference papers (14.6%), reviews (4.9%), and others (2.4%). The bibliometric analysis also demonstrated that the main funding sponsors were Japan Society for the Promotion of Science (26.8%) followed by Engineering and Physical Sciences Research Council (4.9%) and Ministry of the Environment, Government of Japan (4.9%). The subject fields covering these documents included environmental science (85.4%), biochemistry, genetics and molecular biology (14.6%), chemical engineering (14.6%), energy (9.8%), engineering (9.8%), immunology and microbiology (7.3%), and agricultural and biological sciences (4.9%). There is also cooperation among authors with different affiliations (i.e., partnerships), implying the interaction between environmental, civil, mechanical, and electrical engineering fields.

A summary of this literature survey from 1997 to 2020 can be illustrated as follows: In 1997, an integrated UASB + DHS system was operated with a total HRT of 8.3h, achieving removal efficiencies of ≈100%, 94%, 81%, and ≈ 100% for BOD_{total}, COD_{total}, $COD_{soluble}$, and suspended solids (SS), respectively, from sewage (Machdar et al., 1997). Moreover, the system was able to attain an adequate nitrification efficiency of 73%–78%. In the same year, Agrawal et al. (1997) used the DHS unit for the postdenitrification of UASB effluent at 13–30°C and HRT < 1 h, achieving removal efficiency of 84% for N (NO_3 + NO_2). After 2 years, Araki et al. (1999) investigated the simultaneous carbon removal and nitrification process by DHS operated with UASB effluents. Their study revealed the presence of nitrifying bacteria (*Nitrosomonas* and *Nitrobacter*) due to maintaining the aerobic condition even within the interior space of sponge cubes. Further, Machdar et al. (2000) used a curtain-type DHS unit to treat the UASB effluent with an entire HRT of 8h. The combined system achieved removal efficiencies of 94%–97%, 81%–84%, 52%–61%, and 63%–79% for $BOD_{unfiltered}$, $COD_{unfiltered}$, ammonia-nitrogen, and SS, respectively. During the subsequent 2 years, Uemura et al. (2002) used the DHS technique to remove pathogenic microorganisms, including total coliphages and fecal coliforms, from the UASB effluent. In 2005, Tandukar et al. (2005) represented the fourth generation of DHS combined with a UASB reactor to treat municipal wastewater as a pilot-scale system. This generation of DHS was developed to reduce the clogging issues within the sponge pores, enhancing the air dissolution in the wastewater. The UASB + DHS system reported satisfactory removals of $BOD_{unfiltered}$, nutrient, and *E. coli*. Further, Yamaguchi et al. (2006) used the UASB + DHS system to treat low-strength wastewater under low-temperature conditions based on the sulfur-redox action process. The proposed system achieved an overall CODcr removal efficiency greater than 90% at HRT = 12 h and 8 °C. In 2008, Tawfik et al. (2008) investigated the effect of HRT (16, 11, and 8h) on the UASB + DHS performance to treat domestic wastewater. Their study demonstrated that prolonging the HRT to 16h significantly ($P < 0.05$) improved the removal efficiencies of BOD_5, COD, ammonia-nitrogen, total suspended solids (TSS), and coliform. Sludge samples collected from the UASB + DHS system were partially stabilized, and their settling properties were acceptable. Mahmoud et al. (2009) found that the upper segments were responsible for the organic matter removal, whereas the nitrification process was achieved within the bottom segments. Their study also depicted that the DHS unit had an average sludge yield of 0.08 g TSS/g $COD_{removed}$, equivalent to a solid retention time (SRT) of 121 d. Watanabe et al. (2009) reported that the integrated UASB + DHS system could be used after a flocculent yeast, i.e., Hansenula anomala J224 PAWA, application to eliminate phosphorus from sweet potato and barley shochu wastewater types. In 2010, Matsuura et al. (2010) found that a two-stage closed DHS system was able to eliminate more than 99% of dissolved methane from the UASB effluent within HRT = 2 h. Wichitsathian and Racho (2010) depicted that the two modules, viz. fungal-DHS and bacterial-DHS, were capable of treating UASB effluent of tapioca starch wastewater. Racho et al. (2011) also used the two DHS-based fungal (FDHS) and bacterial (BDHS) cultures as a posttreatment step for the UASB effluent of tapioca starch wastewater. Their study demonstrated that the BDHS showed higher nitrogen removal compared with the FDHS unit, which could be assigned to the ability of bacteria to maintain the simultaneous nitrification and denitrification processes. Fleifle et al. (2013) investigated the effect of reducing the HRT from 5.26 to 1.50 h on the DHS performance treating agricultural drainage water (ADW). Their study revealed that the operational HRT should not exceed 2.63 h to have a final effluent complying with the regulation standards for reuse in irrigation. Additional research was conducted to treat molasses-based bioethanol wastewater via a full-scale combined UASB + DHS system (Pattananuwat et al., 2013). In another study (Cao et al., 2015), the DHS unit was used

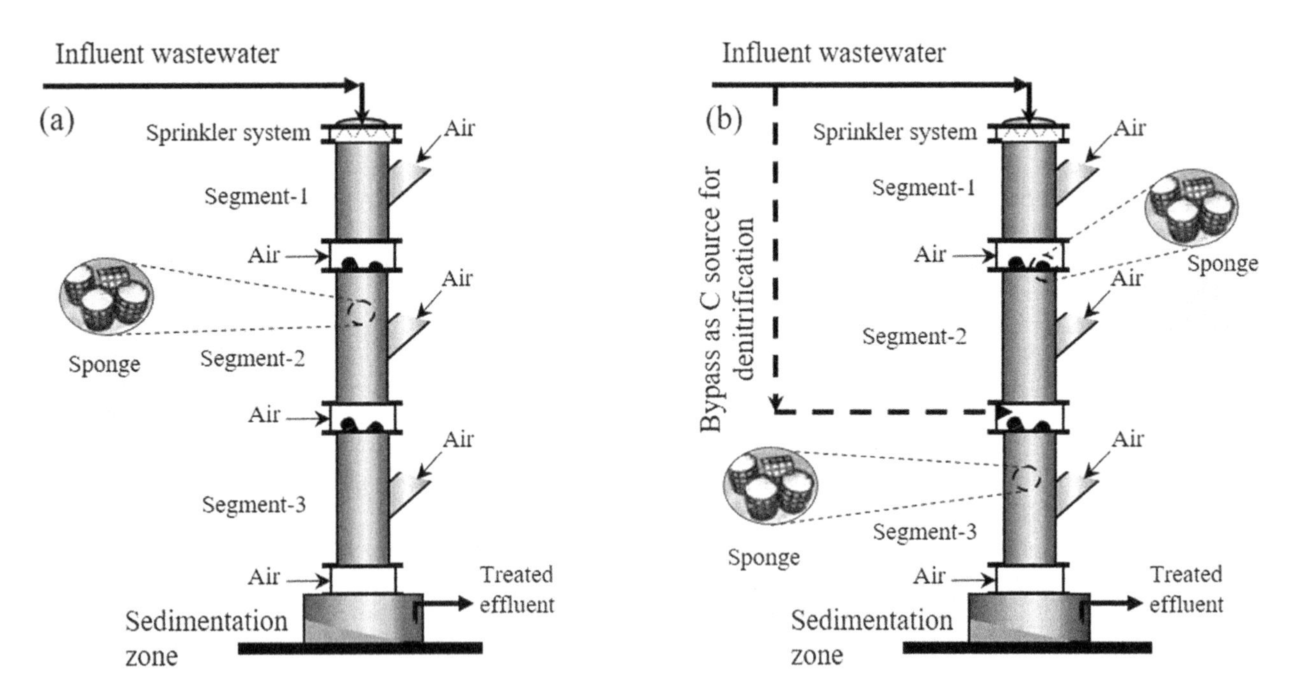

FIG. 6.1 Downflow hanging sponge (DHS) reactor for (A) C removal and (B) C and N removal.

to cultivate manganese-oxidizing bacteria (MnOB) that were used to recover minor metals (e.g., Ni and Co) from wastewater. Matsuura et al. (2015) used two-stage closed DHS units to eliminate dissolved methane from the UASB effluent, achieving CH_4 removal efficiency greater than 99%. Tawfik et al. (2015) employed a pilot-scale up-flow anaerobic hybrid (AH) reactor + DHS to treat domestic wastewater within 9.2 h. Their results showed that the main portions of total coliform (TC), fecal coliform (FC), and fecal streptococci (FS) were eliminated by the DHS stage. Further, Beas et al. (2015) revealed that the DHS effluent, particularly for organic compounds, complied with the United States Environmental Protection Agency standards (USEPA, 2004) for cultivating nonfood crops. In 2016, the DHS unit was used after a baffled reactor (BR) and UASB to treat natural rubber processing wastewater (Watari et al., 2016). The combined system achieved removal efficiencies of 98.6% ± 1.2% and 98.0% ± 1.4% for COD_{total} and TSS, respectively. Tanikawa et al. (2016) employed a pilot-scale of a two-stage UASB and DHS system to treat natural rubber processing effluents. Their study demonstrated that the COD removal efficiency reached up to 95.7% ± 1.3%, and the recovered methane was equivalent to the electricity needed to operate the entire treatment system. Watari et al. (2017a, b) used a DHS unit to treat the anaerobic effluent of natural rubber processing wastewater, recording removal efficiencies of 64.2 ± 7.5% and 55.3 ± 19.2% for COD_{total} and total nitrogen (TN), respectively. The DHS unit operated at HRT = 4.8 h revealed adequate treatment performances compared with the 30-day HRT algal system. Further, more researches were conducted during 2017. For instance, DHS was combined with anaerobic and settling tanks to treat effluents of a natural rubber processing factory. It was able to reduce 92% of greenhouse gas emissions compared with the existing treatment systems (Watari et al., 2017a, b). Matsuura et al. (2017) reported the presence of methanotrophic communities within the DHS column, owing to the recovery of dissolved CH_4 from the UASB effluent. Moreover, modified aerobic/anoxic DHS bioreactors showed adequate C and N removal from wastewater, suggesting its applicability for decentralized systems (Bundy et al., 2017). A closed-type DHS could also be employed after anaerobic treatment for nitrogen and dissolved methane removal (Hatamoto et al., 2017). In addition, DHS could be used for toluene gas treatment due to the enrichment of *Pseudoxanthomonas spadix* and *Pseudomonas* sp. within the sponge media (Yamaguchi et al., 2018). Jong et al. (2018) found that the denitrification process could be achieved via bypassing a portion of the influent discharge to the lower submerged layers for supplying the carbon source (Fig. 6.1), achieving 71% removal of TN from domestic wastes. In 2018, Hatamoto et al. (2018) reviewed the development of the six DHS configurations, including the full-scale applications in India, Egypt, and Japan. After 2 years, the DHS reactor was used for treating heavy metal-containing wastewater via enriching manganese-oxidizing bacteria (MnOB) (Matsushita et al., 2020). Further, the anaerobic baffled reactor (ABR) + DHS combined system was employed to treat natural rubber (NR) processing wastewater, in which the genera *Gordonia* (i.e., rubber-degrading bacteria) were detected in the sludge retained within the sponge (Tanikawa et al., 2020). Furthermore, heavy metal (Zn, Cu, Cr, Pb, and Cd) removal has been observed by the UASB and DHS combined system via binding the metal ions with some anions such as carbonates, hydroxyls, and bicarbonates (Kumar et al., 2020). Iwano et al. (2020) demonstrated that the DHS effluent was appropriate for producing duckweed biomass due to nutrients availability, especially phosphorus. The combined ABR and DHS system was also employed to eliminate azo dye from synthetic dyeing wastewater at a total HRT = 23.2 h (Nguyen et al., 2020). The integrated system achieved COD and color removal efficiencies of 90% and 58%, respectively, in which *Clostridium* sp. was responsible for the azo dye degradation. In 2020, a review article was published by Centeno-Mora et al. (2020), representing details of the closed-DHS reactor to eliminate the CH_4 and H_2S gasses from the UASB effluent. The DHS reactor could also be operated under an anaerobic condition for enhancing N removal and eliminating N_2O gas (Tran et al., 2020). Further, still more searches regarding the DHS applications in wastewater treatment are ongoing.

6.4 Downflow hanging sponge (DHS) concept and configuration

The DHS concept was proposed similar to the criteria of the trickling filter technology, but with using sponge pieces as the support media instead of rocks (Watari et al., 2017a, b). Polyurethane sponge has a void ratio greater than 95%, making it an excellent site for the growth and attachment of active biomass. In these systems, wastewater is supplied from the top of the packing media (Fig. 6.1A) and trickles down toward the reactor's bottom (Racho et al., 2011). A fraction of the influent wastewater can be introduced to the lower segments to supply the C source required for the denitrification process (Fig. 6.1B). Air transfers from the outside environment and diffuses into wastewater during the wastewater distribution onto the supporting media surface. Although this design does not require external aeration equipment, additional aerators may be added in some cases for highly polluted waste or if the reactor's height exceeds 3.0 m (Watari et al., 2017a, b). The upper segments of the reactor with high volumetric organic load are responsible for removing the major portion of the biodegradable organic matter (e.g., BOD). The lower portion of the reactor represents a suitable environment for the dominance of nitrifying bacteria for nitrogen removal. As wastewater trickles downward through the media, the microbial biomass feeds on the available substrate and grows to form biofilms attached to the sponge surface (Nguyen et al., 2020). Biomass can retain both inside and outside the sponge matrix. By increasing the biofilm thickness, the attached biomass sloughs off, falls downward, and precipitates in a clarifier positioned at the reactor's bottom. The carbonaceous matter and nutrients are transformed into inorganic products, surplus biomass, and evolved gas (i.e., due to anaerobic activities within the interior sponge pores). The variation of oxygen levels within the sponge profile implies the occurrence of aerobic, anoxic, and anaerobic pathways, providing suitable conditions for removing multiple pollutant elements (Yamaguchi et al., 2018). For instance, the outer portion of the sponge contains aerobic microbes because of the presence of a sufficient amount of oxygen (DO 4–6 mg/L). The unique properties of the sponge media also ensure the accomplishment of the nitrification/denitrification reactions.

6.5 Downflow hanging sponge (DHS) generations

The sponge media design has been upgraded throughout the 1997–2020 years, named as the first to the sixth generation of DHS reactor (G1–G6). These generations can be illustrated as follows:

The first generation of DHS (DHS-G1) consisted of sponge material (cube shape 1.5 cm × 1.5 cm × 1.5 cm) connected using nylon string (2 m height). This system was designed, implemented, and operated as a posttreatment stage for the UASB anaerobic unit at HRT = 2.4 h and ambient temperature (Agrawal et al., 1997; Machdar et al., 1997).

The second generation of DHS (DHS-G2) contained sponge pieces having the shape of a triangular prism (length = 75 cm and triangular sides = 3 cm, i.e., 3 cm × 3 cm × 75 cm) arranged to form a curtain structure. The DHS-G2 unit was also used for the posttreatment of UASB fed with real sewage at HRT = 2.0 h and 25 °C (Araki et al., 1999; Machdar et al., 2000).

The third generation of DHS (DHS-G3) was designed and implemented similar to the trickling filter system. Small sponge pieces (Ø2.7 cm × 2.7 cm) were packed in polypropylene plastic net and distributed randomly in the reactor. The system was able to illuminate both organic matter and fecal coliform from wastewater at HRT = 2 h and 25 °C (Uemura et al., 2002; Tawfik et al., 2010).

Similar to DHS-G3, the fourth generation of the DHS reactor (DHS-G4) was filled with sponge pieces covered with rigid plastic, but the sponge length increased to 50 cm. Hence, the sponge size was 2.5 cm × 2.5 cm × 50 cm. Multiple sponge elements were organized in a row architecture and stacked one above to form an arrayed sponge shape. This structure formed more gaps that would facilitate the transfer of air from outside into wastewater (Tandukar et al., 2005; Okubo et al., 2016).

The concept of the fifth generation of DHS (DHS-G5) was established based on improving the DHS-G2 configuration by aligning several DHS curtains. Multiple curtains were assembled to construct a pilot-scale DHS reactor (85 cm × 49.5 cm × 49.5 cm) (Tandukar et al., 2007).

The sixth DHS generation (DHS-G6) was designed similar to the trickling filter type of DHS-G3. However, the DHS-G6 unit was occupied with rigid sponge material manufactured by introducing epoxy resin into the polyurethane network structure. The hard sponge pieces with no support materials were able to enhance the interactions between wastewater, air, and biomass without using the plastic cover (Onodera et al., 2014).

6.6 Downflow hanging sponge (DHS) advantages

The DHS system has several advantages over other decentralized wastewater treatment facilities, including:

- **i)** The DHS module can be easily upgraded by modifying the carriers' shape, size, and material, the reactor's ventilation openings and segment number, and operational conditions.
- **ii)** The flexibility in the DHS design allows for its application as a single biological treatment step, especially when the wastewater source has a low-to-medium pollution load strength. Moreover, the DHS unit can be emerged as a posttreatment step for the cases of high pollution load with increased amounts of N and P components (e.g., industrial wastewater source).
- **iii)** The sponge media can capture large amounts of bacterial species and retain them within the pores for extended durations that can reach up to several months. This increase in the solid retention time (SRT) reduces the generation of sludge biomass and the amount of biowaste for disposal.
- **iv)** The DHS reactor is a simple technology that does not contain complicated electromechanical parts (e.g., neither air diffusers nor mixing tools), making the system has low operational and maintenance costs.
- **v)** The DHS reactor can perform reasonably well for the long term under a continuous mode of operation. In addition, it can offer stable and almost constant effluent characteristics (e.g., pH), even when the composition of the influent is highly fluctuating. Hence, it has been reported that the DHS effluent would meet the desired standards and regulations (Tawfik et al., 2008).
- **vi)** The DHS reactor has advantages over conventional treatment systems, regarding simple configuration, easy maintenance, appropriate operational and maintenance costs, low excess sludge production, and small land requirement.
- **vii)** The energy requirement is assigned mainly for the pumping process, with no need for force aeration.

Based on the aforementioned advantages, the DHS system proposes a cost-effective sewage treatment approach for decentralized applications, which is highly recommended in developing countries.

6.7 Downflow hanging sponge (DHS) challenges for future researches

Although the DHS system has been broadly employed for treating different wastewater sources, some points should be considered during design, implementation, and operation:

- **i)** The DHS operation has observed reduced clogging issues compared with conventional attached growth wastewater treatment systems. However, the accumulation of biosolids (concentrated biomass) within the sponge voids, increasing media weight, should be regularly monitored and recorded.
- **ii)** Still, the DHS technology requires further studies for improving the operational strategies, regarding artificial ventilation devices, influent wastewater bypass, recirculation scheme, and sponge densities (fine or coarse), especially for highly varying hydraulic loads.
- **iii)** Like most biological-based systems, the DHS technology should be adapted to face challenges from operating under cold climate conditions or treating unbalanced C/N ratio wastewater.
- **iv)** The DHS performance to treat wastewater is improved under operation with a uniform flow distribution onto the sponge surface. For instance, the DHS unit can work more efficiently as UASB's post-treatment to handle shock hydraulic and organic loads.

6.8 Conclusions and recommendations

This chapter represents an appropriate wastewater treatment technique, namely DHS, which has been employed as a decentralized system in different countries, mainly Japan, Egypt, and India. Based on the chapter's objectives, it was concluded that:

- The DHS concept is similar to the trickling filter, in which stones and blocks are replaced with sponge (polyurethane) material. Hence, the sponge pores, with a voids' percentage >90%, act as homes for biomass to settle, grow, and accumulate.
- The DHS unit has a long SRT (several months), providing enough time for sludge to reach the endogenous respiration phase (i.e., self-degradation of any attached biomass). Hence, the sludge disposal issue can be avoided.
- The aerobic condition in the reactor is maintained via the air diffusion from the surrounding environment, and hence, the cost of adding an external ventilation source is excluded.
- Efficient removal efficiencies of organic matter (BOD and COD), nutrients (N and P), and pathogenic microorganisms by DHS were demonstrated in several studies.

Although the DHS system shows adequate carbonaceous matter removal, nitrification/denitrification process, and pathogenic microorganisms' reduction, some recommendations should be considered in future studies:

- The DHS system should be modified by providing a carbon source for the bottom layers to achieve the denitrification process.
- More research on the modeling and simulation of DHS should be conducted to understand the C and N conversions along with the reactor height and determine the optimum operating conditions relevant to the organic and nutrient loads.
- Further researches should proceed in the direction of DHS applications for decentralized communities in emerging economies.

Acknowledgments

The first author would link to acknowledge Nasr Academy for Sustainable Environment (NASE). The third author thanks Egyptian Ministry of Higher Education (MoHE) and Japan International Cooperation Agency (JICA).

References

Agrawal, L.K., Ohashi, Y., Mochida, E., Okui, H., Ueki, Y., Harada, H., Ohashi, A., 1997. Treatment of raw sewage in a temperate climate using a UASB reactor and the hanging sponge cubes process. Water Sci. Technol. 36 (6-7), 433–440.

Araki, N., Ohashi, A., Machdar, I., Harada, H., 1999. Behaviors of nitrifiers in a novel biofilm reactor employing hanging sponge-cubes as attachment site. Water Sci. Technol. 39 (7), 23–31.

Awolusi, O., Nasr, M., Kumari, S., Bux, F., 2016. Artificial intelligence for the evaluation of operational parameters influencing nitrification and nitrifiers in an activated sludge process. Microb. Ecol. 72 (1), 49–63.

Beas, R., Kujawa-Roeleveld, K., Zeeman, G., Van Lier, J., 2015. A downflow hanging sponge (DHS) reactor for faecal coliform removal from an upflow anaerobic sludge blanket (UASB) effluent. Water Sci. Technol. 72 (11), 2034–2044.

Bundy, C.A., Wu, D., Jong, M.-C., Edwards, S.R., Ahammad, Z.S., Graham, D.W., 2017. Enhanced denitrification in downflow hanging sponge reactors for decentralised domestic wastewater treatment. Bioresour. Technol. 226, 1–8.

Cao, L.T.T., Kodera, H., Abe, K., Imachi, H., Aoi, Y., Kindaichi, T., Ozaki, N., Ohashi, A., 2015. Biological oxidation of Mn(II) coupled with nitrification for removal and recovery of minor metals by downflow hanging sponge reactor. Water Res. 68, 545–553.

Centeno-Mora, E., Fonseca, P.R., Andreão, W.L., Brandt, E.M.F., de Souza, C.L., de Lemos Chernicharo, C., 2020. Mitigation of diffuse CH_4 and H_2S emissions from the liquid phase of UASB-based sewage treatment plants: challenges, techniques, and perspectives. Environ. Sci. Pollut. Res. 27 (29), 35979–35992.

Chernicharo, C., Almeida, P., 2011. Feasibility of UASB/trickling filter systems without final clarifiers for the treatment of domestic wastewater in small communities in Brazil. Water Sci. Technol. 64 (6), 1347–1354.

Chuang, H.-P., Wu, J.-H., Ohashi, A., Abe, K., Hatamoto, M., 2014. Potential of nitrous oxide conversion in batch and down-flow hanging sponge bioreactor systems. Sustain. Environ. Res. 24 (2), 117–128.

El-Tabl, A., Wahaab, R., Younes, S., 2013. Downflow hanging sponge (DHS) reactor as a novel post treatment system for municipal wastewater. Life Sci. J. 10 (3), 409–414.

Fleifle, A., Tawfik, A., Saavedra, O., Elzeir, M., 2013. Treatment of agricultural drainage water via downflow hanging sponge system for reuse in agriculture. Water Sci. Technol. Water Supply 13 (2), 403–412.

Furukawa, A., Matsuura, N., Mori, M., Kawamata, M., Kusaka, J., Hatamoto, M., Yamaguchi, T., 2016. Development of a DHS-USB recirculating system to remove nitrogen from a marine fish aquarium. Aquacult. Eng. 74, 174–179.

Hamdy, A., Mostafa, M., Nasr, M., 2018. Zero-valent iron nanoparticles for methylene blue removal from aqueous solutions and textile wastewater treatment, with cost estimation. Water Sci. Technol. 78 (2), 367–378.

Hatamoto, M., Sato, T., Nemoto, S., Yamaguchi, T., 2017. Cultivation of denitrifying anaerobic methane-oxidizing microorganisms in a continuous-flow sponge bioreactor. Appl. Microbiol. Biotechnol. 101 (14), 5881–5888.

Hatamoto, M., Okubo, T., Kubota, K., Yamaguchi, T., 2018. Characterization of downflow hanging sponge reactors with regard to structure, process function, and microbial community compositions. Appl. Microbiol. Biotechnol. 102 (24), 10345–10352.

Ismail, S., Nasr, M., Abdelrazek, E., Awad, H.M., Zhaof, S., Meng, F., Tawfik, A., 2020. Techno-economic feasibility of energy-saving self-aerated sponge tower combined with up-flow anaerobic sludge blanket reactor for treatment of hazardous landfill leachate. J. Water Process Eng. 37, 101415.

Iwano, H., Hatohara, S., Tagawa, T., Tamaki, H., Li, Y.-Y., Kubota, K., 2020. Effect of treated sewage characteristics on duckweed biomass production and microbial communities. Water Sci. Technol. 82 (2), 292–302.

Jiang, Z.-W., Fang, X., Zhang, Y.-L., Zhao, J.-F., 2005. Recent achievements of anaerobic treatment of sewage. Tongji Daxue Xuebao/J. Tongji Univ. 33 (4), 489–493.

Jong, M.-C., Su, J.-Q., Bunce, J.T., Harwood, C.R., Snape, J.R., Zhu, Y.-G., Graham, D.W., 2018. Co-optimization of sponge-core bioreactors for removing total nitrogen and antibiotic resistance genes from domestic wastewater. Sci. Total Environ. 634, 1417–1423.

Kumar, M., Gogoi, A., Mukherjee, S., 2020. Metal removal, partitioning and phase distributions in the wastewater and sludge: performance evaluation of conventional, upflow anaerobic sludge blanket and downflow hanging sponge treatment systems. J. Clean. Prod. 249, 119426.

Machdar, I., Harada, H., Ohashi, A., Sekiguchi, Y., Okui, H., Ueki, K., 1997. A novel and cost-effective sewage treatment system consisting of UASB pre-treatment and aerobic post-treatment units for developing countries. Water Sci. Technol. 36 (12), 189–197.

Machdar, I., Sekiguchi, Y., Sumino, H., Ohashi, A., Harada, H., 2000. Combination of a UASB reactor and a curtain type DHS (downflow hanging sponge) reactor as a cost-effective sewage treatment system for developing countries. Water Sci. Technol. 42 (3–4), 83–88.

Mahmoud, M., Tawfik, A., Samhan, F., El-Gohary, F., 2009. Sewage treatment using an integrated system consisting of anaerobic hybrid reactor (AHR) and downflow hanging sponge (DHS). Desalin. Water Treat. 4 (1–3), 168–176.

Mahmoud, A., Mostafa, M., Nasr, M., 2019. Regression model, artificial intelligence, and cost estimation for phosphate adsorption using encapsulated nanoscale zero-valent iron. Sep. Sci. Technol. 54 (1), 13–26.

Matsushita, S., Komizo, D., Cao, L.T.T., Aoi, Y., Kindaichi, T., Ozaki, N., Imachi, H.D., Ohashi, A., 2018. Production of biogenic manganese oxides coupled with methane oxidation in a bioreactor for removing metals from wastewater. Water Res. 130, 224–233.

Matsushita, S., Hiroe, T., Kambara, H., Shoiful, A., Aoi, Y., Kindaichi, T., Ozaki, N., Imachi, H., Ohashi, A., 2020. Anti-bacterial effects of MnO_2 on the enrichment of manganese-oxidizing bacteria in downflow hanging sponge reactors. Microb. Environ. 35 (4, ME20052), 1–8.

Matsuura, N., Hatamoto, M., Sumino, H., Syutsubo, K., Yamaguchi, T., Ohashi, A., 2010. Closed DHS system to prevent dissolved methane emissions as greenhouse gas in anaerobic wastewater treatment by its recovery and biological oxidation. Water Sci. Technol. 61 (9), 2407–2415.

Matsuura, N., Hatamoto, M., Sumino, H., Syutsubo, K., Yamaguchi, T., Ohashi, A., 2015. Recovery and biological oxidation of dissolved methane in effluent from UASB treatment of municipal sewage using a two-stage closed downflow hanging sponge system. J. Environ. Manage. 151, 200–209.

Matsuura, N., Hatamoto, M., Yamaguchi, T., Ohashi, A., 2017. Methanotrophic community composition based on pmoA genes in dissolved methane recovery and biological oxidation closed downflow hanging sponge reactors. Biochem. Eng. J. 124, 138–144.

Nasr, M., Elreedy, A., Abdel-Kader, A., Elbarki, W., Moustafa, M., 2014a. Environmental consideration of dairy wastewater treatment using hybrid sequencing batch reactor. Sustain. Environ. Res. 24 (6), 449–456.

Nasr, M., Moustafa, M., Seif, H., El-Kobrosy, G., 2014b. Application of fuzzy logic control for Benchmark simulation model. Sustain. Environ. Res. 24 (4), 235–243.

Nguyen, T.H., Watari, T., Hatamoto, M., Sutani, D., Setiadi, T., Yamaguchi, T., 2020. Evaluation of a combined anaerobic baffled reactor–downflow hanging sponge biosystem for treatment of synthetic dyeing wastewater. Environ. Technol. Innov. 19, 100913.

Okubo, T., Kubota, K., Yamaguchi, T., Uemura, S., Harada, H., 2016. Development of a new non-aeration-based sewage treatment technology: performance evaluation of a full-scale down-flow hanging sponge reactor employing third-generation sponge carriers. Water Res. 102, 138–146.

Onodera, T., Tandukar, M., Sugiyana, D., Uemura, S., Ohashi, A., Harada, H., 2014. Development of a sixth-generation down-flow hanging sponge (DHS) reactor using rigid sponge media for post-treatment of UASB treating municipal sewage. Bioresour. Technol. 152, 93–100.

Pattananuwat, N., Aoki, M., Hatamoto, M., Nakamura, A., Yamazaki, S., Syutsubo, K., Araki, N., Takahashi, M., Harada, H., Yamaguchi, T., 2013. Performance and microbial community analysis of a full-scale hybrid anaerobic-aerobic membrane system for treating molasses-based bioethanol wastewater. Int. J. Environ. Res. 7 (4), 979–988.

Racho, P., Jindal, R., Wichitsathian, B., 2011. Posttreatment of UASB effluents of tapioca starch wastewater using downflow hanging sponge system. J. Hazard. Toxic Radioact. Waste 16 (1), 9–17.

Tandukar, M., Uemura, S., Machdar, I., Ohashi, A., Harada, H., 2005. A low-cost municipal sewage treatment system with a combination of UASB and the "fourth-generation" downflow hanging sponge reactors. Water Sci. Technol. 52 (1–2), 323–329.

Tandukar, M., Ohashi, A., Harada, H., 2007. Performance comparison of a pilot-scale UASB and DHS system and activated sludge process for the treatment of municipal wastewater. Water Res. 41 (12), 2697–2705.

Tanikawa, D., Syutsubo, K., Hatamoto, M., Fukuda, M., Takahashi, M., Choeisai, P.K., Yamaguchi, T., 2016. Treatment of natural rubber processing wastewater using a combination system of a two-stage up-flow anaerobic sludge blanket and down-flow hanging sponge system. Water Sci. Technol. 73 (8), 1777–1784.

Tanikawa, D., Kataoka, T., Sonaka, H., Hirakata, Y., Hatamoto, M., Yamaguchi, T., 2020. Evaluation of key factors for residual rubber coagulation in natural rubber processing wastewater. J. Water Process Eng. 33, 101041.

Tawfik, A., El-Gohary, F., Ohashi, A., Harada, H., 2008. Optimization of the performance of an integrated anaerobic-aerobic system for domestic wastewater treatment. Water Sci. Technol. 58 (1), 185–194.

Tawfik, A., Ohashi, A., Harada, H., 2010. Effect of sponge volume on the performance of down-flow hanging sponge system treating UASB reactor effluent. Bioprocess Biosyst. Eng. 33 (7), 779–785.

Tawfik, A., El-Zamel, T., Herrawy, A., El-Taweel, G., 2015. Fate of parasites and pathogenic bacteria in an anaerobic hybrid reactor followed by downflow hanging sponge system treating domestic wastewater. Environ. Sci. Pollut. Res. 22 (16), 12235–12245.

Tran, P.T., Hatamoto, M., Tsuba, D., Watari, T., Yamaguchi, T., 2020. Positive impact of a reducing agent on autotrophic nitrogen removal process and nexus of nitrous oxide emission in an anaerobic downflow hanging sponge reactor. Chemosphere 256, 126952.

Uemura, S., Takahashi, K., Takaishi, A., Machdar, A., Ohashi, A., Harada, H., 2002. Removal of indigenous coliphages and fecal coliforms by a novel sewage treatment system consisting of UASB and DHS units. Water Sci. Technol. 46 (11–12), 303–309.

USEPA, 2004. Guidelines for Water Reuse. Environmental Protection Agency, Camp Dresser & McKee, Inc., Washington, DC, p. 480.

Watanabe, T., Masaki, K., Iwashita, K., Fujii, T., Iefuji, H., 2009. Treatment and phosphorus removal from high-concentration organic wastewater by the yeast Hansenula anomala J224 PAWA. Bioresour. Technol. 100, 1781–1785.

Watari, T., Thanh, N.T., Tsuruoka, N., Tanikawa, D., Kuroda, K., Huong, N.L., Tan, N.M., Hai, H.T., Hatamoto, M., Syutsubo, K., Fukuda, M., Yamaguchi, T., 2016. Development of a BR-UASB-DHS system for natural rubber processing wastewater treatment. Environ. Technol. (UK) 37 (4), 459–465.

Watari, T., Mai, T.C., Tanikawa, D., Hirakata, Y., Hatamoto, M., Syutsubo, K., Fukuda, M., Nguyen, N.B., Yamaguchi, T., 2017a. Development of downflow hanging sponge (DHS) reactor as post treatment of existing combined anaerobic tank treating natural rubber processing wastewater. Water Sci. Technol. 75 (1), 57–68.

Watari, T., Mai, T.C., Tanikawa, D., Hirakata, Y., Hatamoto, M., Syutsubo, K., Fukuda, M., Nguyen, N.B., Yamaguchi, T., 2017b. Performance evaluation of the pilot scale upflow anaerobic sludge blanket—downflow hanging sponge system for natural rubber processing wastewater treatment in South Vietnam. Bioresour. Technol. 237, 204–212.

Wichitsathian, B., Racho, P., 2010. Quantification of organic and nitrogen removal in downflow hanging sponge (DHS) systems as a post-treatment of UASB effluent. Water Sci. Technol. 62 (9), 2121–2127.

Yamaguchi, T., Bungo, Y., Takahashi, M., Sumino, H., Nagano, A., Araki, N., Imai, T., Yamazaki, S., Harada, H., 2006. Low strength wastewater treatment under low temperature conditions by a novel sulfur redox action process. Water Sci. Technol. 53 (6), 99–105.

Yamaguchi, T., Nakamura, S., Hatamoto, M., Tamura, E., Tanikawa, D., Kawakami, S., Nakamura, A., Kato, K., Nagano, A., Yamaguchi, T., 2018. A novel approach for toluene gas treatment using a downflow hanging sponge reactor. Appl. Microbiol. Biotechnol. 102 (13), 5625–5634.

CHAPTER

7

Assessment of biochar application in decontamination of water and wastewater

Alaa El Din Mahmoud[a,b] *and Srujana Kathi*[c]

[a]Environmental Sciences Department, Faculty of Science, Alexandria University, Alexandria, Egypt [b]Green Technology Group, Faculty of Science, Alexandria University, Alexandria, Egypt [c]ICSSR Postdoctoral Research Fellow, Department of Applied Psychology, Pondicherry University, Puducherry, India

OUTLINE

7.1 Introduction

Biochar (BC) has received significant attention in recent years especially for the removal of toxic elements or compounds present in water and wastewater. BC-based adsorbents exhibit vital potential for waste and wastewater remediation (Chaukura et al., 2020). The preferable precursors of producing BC are agricultural wastes. Therefore, BC is rich in carbon and contains a certain amount of minerals. The presence of silica (Si) and phosphorous (P) minerals can enhance the stability of biochar as reported in Xu et al. (2017).

The application of biochar-based materials such as catalysts, catalyst supports, and adsorbents for environmental remediation has increased considerably because biochars can be modified for a specific purpose depending on the preparation methods and feedstock (Lee and Park, 2020). The production of biochar is a sustainable option for waste management as well as remediation of pollutants from polluted water (Gope and Saha, 2020). Wastes to biochar via pyrolysis can be an effective option for biosolids management. It helps reducing biosolid volume as well as adding value to them by converting them into high-quality biochar (Patel et al., 2020).

Contaminant removal is based on the functional groups available on the surface. Surface modification of biochars is also recommended to enhance removal efficiency. Effective surface modification of BC ensures the removal of heavy metals from wastewater as well as the recyclability of HM-loaded BC (Wang et al., 2019). Moderate capacity wood biochar is an environmentally superior alternative to coal-based PAC for micropollutant removal from wastewater, and its use can offset a wastewater facility's carbon footprint (Thompson et al., 2016). It can be widely used to enhance soil fertility in a financially viable manner and also maintain ecosystem balance (Yaashikaa et al., 2019).

Cost Effective Technologies for Solid Waste and Wastewater Treatment
https://doi.org/10.1016/B978-0-12-822933-0.00009-7

Continuing concerns regarding possible population-level impacts of pharmaceuticals from wastewater effluents has contributed to a search for sustainable and cheap technologies that will result in the effective removal of pharmaceuticals from reclaimed water. Recently, biochar has been explored as a potential material for the adsorption of pharmaceuticals from aqueous solutions (Ndoun et al., 2020). Biochar provides a promising alternative to activated carbon for the removal of organic contaminants of emerging concerns in various wastewater and concentrate streams (Lin et al., 2017).

This chapter presents the latest developments on the use of BC in a wastewater treatment process. The chapter intends to discuss BC production and properties, factors influencing the quality of BC, environmental applications of BC, and removal of organic pollutants and heavy metals. Also, the possible challenges of BC application in water and wastewater treatment are emphasized in the hope of helping the researchers to develop empathy toward the interactions between BC-based materials and contaminants and to explore the application of BC-based materials in environmental pollution management.

7.2 Biochar production and properties

The BCs are usually produced from biomass feedstock by pyrolysis of biowaste under conditions of limited oxygen (Xiao et al., 2020). Feedstock materials vary from corn stover, peanut hulls, macadamia nut shells, wood chips, and turkey manure plus wood chips (Spokas and Reicosky, 2009). The characteristics of the feedstock impact the conversion process and product properties (Weber and Quicker, 2018). The performance of BC is determined by preparation factors and loading processes (Pan et al., 2020).

The BCs can be prepared by carbonization technologies such as pyrolysis, gasification, flash carbonization, and hydrothermal carbonization. BC produced at high pyrolysis temperatures (600–700 °C) contains fewer oxygen- and hydrogen-containing functional groups, on its surface possessing a low ion-exchange capacity presents an excellent platform for further activation and it has been applied to contaminant removal, through sorption processes (Panwar and Pawar, 2020). The degree and magnitude of decomposition of each of the biomass components are dependent on process variables such as feedstock type, reaction temperature, and heating rate. H/C and O/C atomic ratios are important indicators for biochar quality (Jafri et al., 2018).

Generally, nanomaterials have a high surface that is reflected in the number of surface-active sites to improve the physicochemical properties of the material (Mahmoud, 2020a, b). Biochar-based nanocomposites combine the advantages of biochar with nanomaterials. Nanoparticles improve biochar's physicochemical properties and serve as active sites (Tan et al., 2016). In the process of the development of more cost-effective materials, several types of biochar-based hybrid materials, such as magnetic BC composites (MBCs), nanometal/nanometallic oxides/hydroxide BC composites, and layered nanomaterial-coated BCs, as well as physically/chemically activated BCs, have also been developed (Huang et al., 2019).

In high-temperature pyrolysis, pyrogenic nanopores contribute minimally to total porosity. However, these pores provide the majority of biochar surface area that is critical for contaminant sorption (Gray et al., 2014). Depending on the different types of biochar-based materials, organic contaminants can be removed by different mechanisms, such as physical adsorption, electrostatic interaction, π–π interaction, and Fenton process, as well as photocatalytic degradation (Huang et al., 2019).

7.3 Factors influencing quality of biochar

The physicochemical properties of biochar vary according to the modification synthesis methodology, such as surface area, pH, molar ratios (H/C, O/C, N/C), surface charge, mineral content, elementary compositions, and binding sites (Panwar and Pawar, 2020). The selection of the activation method mainly depends on its environmental application. Therefore, there are physical and chemical activation techniques.

Physical/steam activation technique generates pores in the produced biochar at high temperatures. Besides, gas purging increases its surface area and pore volume. However, this technique is not applicable to enhance the surface functional groups of BCs. On the contrary, chemical/impregnation activation technique is more favorable for synthesizing the required BCs. Metal oxide-modified biochar had better surface functionalities than did physical- and chemical-activated biochar and better sorption of organic and inorganic contaminants from potable water and wastewater (Panwar and Pawar, 2020). Contaminant removal is mainly based on the presence of functional groups and charges on the surface of the BCs. Alkaline activation improves the surface area and oxygen-containing functional group in the activated BCs.

Consequently, BCs can remove various contaminants that are mainly based on the presence of functional groups and charges on their surface. Acidic treatment provides more oxygenated functional groups on BC surfaces (Godwin et al., 2019). Tang et al. (2018) reported that the sorption performance of alkali-acid-modified magnetic biochar, derived from sewage sludge, was better than either alkali- or acid-modified magnetic biochar for tetracycline removal. Furthermore, biochar derived from rice husk can be used as a potential activator of peroxymonosulfate (Huong et al., 2020). Zhang et al. (2019a) applied biochar-TiO_2 to remove sulfamethoxazole and compared the catalytic efficiency with bare TiO_2. Compared to bare TiO_2, the biochar-supported TiO_2 showed improved activity for the degradation of sulfamethoxazole, 91.27% from 58.47% in the system using TiO_2.

7.4 Environmental application of biochar

Adsorption has proven to be an effective and economically viable method of removing various contaminants compared to other treatment techniques (Mahmoud et al., 2016; Badr et al., 2020).

Since biochar has many advantages such as various types of raw materials, low cost, and recyclability, it can achieve the effect of turning waste into treasure when used for environmental treatment (Dai et al., 2019). BC has many environmental applications such as contaminant removal where we focus here on it as well as carbon sequestration, and soil amelioration (Oliveira et al., 2017). BC composites with metal oxides and carbonaceous materials alter the physicochemical properties of bare BCs and make them more applicable in a wide range of contaminant removal (Premarathna et al., 2019). Table 7.1 summarizes the potential of different biochars in treating contaminants from wastewater.

Biochar or modified biochar materials were utilized for the removal of phosphate (PO_4^{3-}) and nitrate (NO_3^-) that are responsible for the eutrophication in the wake of phytoplankton blooming and rapid decline of dissolved oxygen in water (Lee and Park, 2020). The coating of nanoparticles on carbon structure could also increase the adsorption sites to attract the organic contaminants from water (Tan et al., 2016). Among the various metal oxides, TiO_2 is widely used as a photocatalyst for environmental remediation because of its nontoxicity, low cost, and high stability. The activity of TiO_2 as a photocatalyst can be improved by various surface modifications, such as doping and the addition of effective components to TiO_2 as biochar (Lee and Park, 2020).

It is also applicable to use locally available wastes such as sewage sludge because sludge pyrolysis could be one of the efficient methods for the valorization of these waste materials as an adsorbent (Regkouzas and Diamadopoulos, 2019). In soil remediation, sewage sludge is combined with wood biochar as an alternative for reusing and aggregating value to these wastes.

7.5 Removal of organic pollutants

Application of biochar for the removal of organic pollutants from soils and aquatic systems have been widely under research, including: hydrophobic organics such as polycyclic aromatic hydrocarbons (PAHs), phthalic acid esters (PAEs), and estrogen contaminants, and ionizable organics such as sulfonamides, pesticides, and phenolic

TABLE 7.1 Potential of different biochars in treating wastewater.

Precursor of prepared biochar	Contaminant	Removal percentage or update capacity	References
Date-palm	Phosphate and nitrate	The maximum adsorption capacities of phosphate and nitrate were 177.97 and 28.06 $mg\,g^{-1}$, respectively	Alagha et al. (2020)
Peanut shell	Cadmium:Cd(II) and Copper:Cu(II)	The maximum adsorption capacities of Cd(II) and Cu(II) were 29.9 and 34.1 $mg\,g^{-1}$, respectively	Li et al. (2020)
Napa cabbage and rape	Phosphate	> 92% at a pH range of 2–10	Zhang et al. (2019b)
Loofah	Chromium:Cr(VI) and Copper:Cu(II)	After three reuse cycles, Cr(VI) and Cu(II) adsorptions by magnetic loofah biochar were 23.34 and 42.6 $mg\,g^{-1}$	Xiao et al. (2019)
Corncob	Mercury:Hg(II)	The maximum Hg(II) adsorption capacity for the optimized aminobiochar was 14.1 $mg\,g^{-1}$, which was almost twice as that for pristine biochar	Bao et al. (2018)
Yak manure	Lead: Pb(II)	The maximum adsorption capacity of Pb(II) increased from 76.41 to 169.57 $mg\,g^{-1}$ when amended with H_2O_2	Wang and Liu (2018)

compounds (Han et al., 2016; Hu et al., 2020). The sorption of organic contaminants is depending on the adsorbent type and mostly governed by π–π electron donor-acceptor interactions, hydrophobic interactions, H-bonding, and electrostatic attraction (Mahmoud et al., 2018, 2020a; Hu et al., 2020).

Highly polar biochar, pyrolyzed at temperature ≤ 400 °C, includes aromatic π-configuration, and electron donor, and acceptor functional groups. These π electron-rich biochar functional groups (− ve) have electron donor-acceptor interactions with π electron-deficiency (+ ve) of organic dyes as an example. Hence, these interactions have resulted in an electrostatic repulsion and promoted pollutant adsorption via H-bonding between biochar and organic dyes (Teixidó et al., 2011). Rice husk biochar activation peroxy monosulfate showed excellent activity toward organic pollutants that retained high stability even after five cycles (Huong et al., 2020).

Qin et al. (2017) conducted degradation experiments of 1,3-dichloropropene using rice-husk-derived biochar prepared at temperatures in the range of 300–700 °C. The best catalytic activity of the prepared BCs was for BC prepared at 500 °C. Another study proved the successful application of BC in the removal of phenanthrene (100% removal efficiency) and achieved 76% removal efficiency for naphthalene. On the contrary, BC was not efficient in the removal of *E. coli* (Reddy et al., 2014).

Although different nonconventional low-cost materials have been investigated and adsorbents based on clays and metal-organic frameworks have been studied (de Andrade et al., 2018), BCs have more potential in the removal of pharmaceuticals and minimize their exposure to the environment and human. Biochar is a promising medium for pharmaceutical removal in on-site sewage systems. Carbamazepine removal was high using BC with no biofilm due to the adsorption process (Dalahmeh et al., 2018). Adsorption of pharmaceutical compounds was dominated by a complex interplay of three mechanisms: hydrophobic partitioning, hydrogen bonding, and π–π electron donor-acceptor (EDA) interactions.

The adsorption capacity of pharmaceuticals by biochar was affected by their physicochemical properties including ash content, specific surface area, charge, pore volume, as well as hydrophobicity, π-energy, and speciation of pharmaceuticals. The adsorption of pharmaceuticals in concentrate was pH-dependent, the kinetic rate constant increased with decreasing pH due to the electrical interactions between pharmaceutical molecules and adsorbents (Lin et al., 2017).

7.6 Removal of heavy metals

Heavy metals are nonbiodegradable and may be passed along the food chain through bioaccumulation (Mahmoud et al., 2016, 2020b). Therefore, BC is increasingly being considered as an alternative agent in water treatment technologies for the removal of metal ions (Inyang et al., 2016). BC is used as a low-cost adsorbent to remove heavy metals from the aqueous solution, which is a promising and an emerging wastewater treatment technology. About 99% of Cd(II), Pb(II), and Zn(II) were removed from aqueous solution using BCs (Patra et al., 2017). It was noticed that the bioavailable heavy metal fractions were transformed into the relatively stable fractions so the bioavailability and ecotoxicity of the heavy metals in biochar were significantly reduced (Devi and Saroha, 2014). BC-chitosan after cross-linking can be cast into membranes that can be utilized as an adsorbent for metal ion uptake (Hussain et al., 2017). The concentration of Cd(II), Cr(VI), Cu(II), Pb(II), Ni(II), and Zn(II) is decreased by 18%, 19%, 65%, 75%, 17%, and 24%, respectively. The variation can be explained in terms of the chemical properties of the metal ions (Mahmoud, 2020a) as well as the properties of the biochar (Reddy et al., 2014). When BC was impregnated by hydrous ferric oxide (HFO), HFO-BC exhibited excellent performance to two representative heavy metals, i.e., Cd(II) and Cu(II), with maximal experimental sorption capacities of $29.9\,mg\,g^{-1}$ for Cd(II) and $34.1\,mg\,g^{-1}$ for Cu(II) (Li et al., 2020). Understanding how to synthesize BC nanocomposites and how they remove HMs from the solution is crucial for their applications as reported in Ho et al. (2017).

7.7 Challenges of biochar application in water and wastewater treatment

Despite the excellent sorption ability of BC for heavy metals, it is difficult to separate and reuse after adsorption when applied to wastewater treatment process (Son et al., 2018). Separation and recycle usage are also challenging when using powder-type BC during the wastewater treatment. The nanoscale metals have been considered as a high-quality adsorbent toward HM ions due to their high specific surface area, fast reactive rate, high reactivity, and magnetism (Ho et al., 2017). The binding of metals by sewage sludge is unstable, which may in the future result in desorption of metals into the environment (Bogusz et al., 2019).

7.8 Conclusions and recommendations

The applicability of low-cost raw material as a precursor for producing BC appears to be promising. An ecofriendly prepared biochar is applicable in water and wastewater treatment process. It depends on the basic properties of the used biochar and the type of the pollutants either inorganic or organic. In addition, the tenability of biochar's functional groups can enhance the uptake capacity of various heavy metals from aqueous solutions. Using acids or alkali in BC's production can make it more applicable to specific pollutants if required. Recently, biochar amendments with nanomaterials are on focus to enhance the uptake capacity of various pollutants. The economic cost of using BC in water and wastewater should be further analyzed.

Acknowledgments

Srujana Kathi would like to thank the ICSSR for providing postdoctoral research fellowship, as well as Pondicherry University, India, for the outreach and support. Alaa El Din Mahmoud would like to thank the support of Science, Technology, and Innovation Funding Authority (STDF-STIFA), Egypt, for the Project ID: 42961.

References

Alagha, O., Manzar, M.S., Zubair, M., Anil, I., Mu'azu, N.D., Qureshi, A., 2020. Comparative adsorptive removal of phosphate and nitrate from wastewater using biochar-MgAl LDH nanocomposites: coexisting anions effect and mechanistic studies. Nanomaterials 10 (2), 336.

Badr, N.B., Al-Qahtani, K.M., Mahmoud, A.E.D., 2020. Factorial experimental design for optimizing selenium sorption on *Cyperus laevigatus* biomass and green-synthesized nano-silver. Alex. Eng. J.

Bao, J., Zheng, H., Tufail, H., Irshad, S., Du, J., 2018. Adsorption-assisted decontamination of Hg (ii) from aqueous solution by multi-functionalized corncob-derived biochar. RSC Adv. 8 (67), 38425–38435.

Bogusz, A., Oleszczuk, P., Dobrowolski, R., 2019. Adsorption and desorption of heavy metals by the sewage sludge and biochar-amended soil. Environ. Geochem. Health 41 (4), 1663–1674.

Chaukura, N., Masilompane, T.M., Gwenzi, W., Mishra, A.K., 2020. Biochar-based adsorbents for the removal of organic pollutants from aqueous systems. In: Emerging Carbon-Based Nanocomposites for Environmental Applications, pp. 147–174.

Dai, Y., Zhang, N., Xing, C., Cui, Q., Sun, Q., 2019. The adsorption, regeneration and engineering applications of biochar for removal organic pollutants: a review. Chemosphere 223, 12–27.

Dalahmeh, S., Ahrens, L., Gros, M., Wiberg, K., Pell, M., 2018. Potential of biochar filters for onsite sewage treatment: adsorption and biological degradation of pharmaceuticals in laboratory filters with active, inactive and no biofilm. Sci. Total Environ. 612, 192–201.

de Andrade, J.R., Oliveira, M.F., da Silva, M.G., Vieira, M.G., 2018. Adsorption of pharmaceuticals from water and wastewater using nonconventional low-cost materials: a review. Indus. Eng. Chem. Res. 57 (9), 3103–3127.

Devi, P., Saroha, A.K., 2014. Risk analysis of pyrolyzed biochar made from paper mill effluent treatment plant sludge for bioavailability and eco-toxicity of heavy metals. Bioresour. Technol. 162, 308–315.

Godwin, P.M., Pan, Y., Xiao, H., Afzal, M.T., 2019. Progress in preparation and application of modified biochar for improving heavy metal ion removal from wastewater. J. Bioresour. Bioprod. 4 (1), 31–42.

Gope, M., Saha, R., 2020. Removal of heavy metals from industrial effluents by using biochar. In: Intelligent Environmental Data Monitoring for Pollution Management. Academic Press, pp. 25–48.

Gray, M., Johnson, M.G., Dragila, M.I., Kleber, M., 2014. Water uptake in biochars: the roles of porosity and hydrophobicity. Biomass Bioenergy 61, 196–205.

Han, L., Ro, K.S., Sun, K., Sun, H., Wang, Z., Libra, J.A., Xing, B., 2016. New evidence for high sorption capacity of hydro char for hydrophobic organic pollutants. Environ. Sci. Technol. 50, 13274–13282.

Ho, S.H., Zhu, S., Chang, J.S., 2017. Recent advances in nanoscale-metal assisted biochar derived from waste biomass used for heavy metals removal. Bioresour. Technol. 246, 123–134.

Hu, B., Ai, Y., Jin, J., Hayat, T., Alsaedi, A., Zhuang, L., Wang, X., 2020. Efficient elimination of organic and inorganic pollutants by biochar and biochar-based materials. Biochar, 1–18.

Huang, Q., Song, S., Chen, Z., Hu, B., Chen, J., Wang, X., 2019. Biochar-based materials and their applications in removal of organic contaminants from wastewater: state-of-the-art review. Biochar 1 (1), 45–73.

Huong, P.T., Jitae, K., Al Tahtamouni, T.M., Tri, N.L.M., Kim, H.H., Cho, K.H., Lee, C., 2020. Novel activation of peroxy monosulfate by biochar derived from rice husk toward oxidation of organic contaminants in wastewater. J. Water Process Eng. 33, 101037.

Hussain, A., Maitra, J., Khan, K.A., 2017. Development of biochar and chitosan blend for heavy metals uptake from synthetic and industrial wastewater. Appl. Water Sci. 7 (8), 4525–4537.

Inyang, M.I., Gao, B., Yao, Y., Xue, Y., Zimmerman, A., Mosa, A., Cao, X., 2016. A review of biochar as a low-cost adsorbent for aqueous heavy metal removal. Crit. Rev. Environ. Sci. Technol. 46 (4), 406–433.

Jafri, N., Wong, W.Y., Doshi, V., Yoon, L.W., Cheah, K.H., 2018. A review on production and characterization of biochars for application in direct carbon fuel cells. Process Saf. Environ. Protect. 118, 152–166.

Lee, J.E., Park, Y.K., 2020. Applications of modified biochar-based materials for the removal of environment pollutants: a mini review. Sustainability 12 (15), 6112.

Li, Y., Gao, L., Lu, Z., Wang, Y., Wang, Y., Wan, S., 2020. Enhanced Removal of Heavy Metals From Water by Hydrous Ferric Oxide-Modified Biochar. ACS Omega.

Lin, L., Jiang, W., Xu, P., 2017. Comparative study on pharmaceuticals adsorption in reclaimed water desalination concentrate using biochar: Impact of salts and organic matter. Sci. Total Environ. 601, 857–864.

Mahmoud, A.E.D., 2020a. Nanomaterials: green synthesis for water applications. In: Handbook of Nanomaterials and Nanocomposites for Energy and Environmental Applications, pp. 1–21.

Mahmoud, A.E.D., 2020b. Graphene-based nanomaterials for the removal of organic pollutants: insights into linear versus nonlinear mathematical models. J. Environ. Manage. 270, 110911.

Mahmoud, A.E.D., Fawzy, M., Radwan, A., 2016. Optimization of Cadmium (CD^{2+}) removal from aqueous solutions by novel biosorbent. Int. J. Phytoremediation 18 (6), 619–625.

Mahmoud, A.E.D., Stolle, A., Stelter, M., Braeutigam, P., 2018. Adsorption Technique for Organic Pollutants Using Different Carbon Materials, Abstracts of Papers of the American Chemical Society. Amer Chemical Soc, Washington, DC.

Mahmoud, A.E.D., Franke, M., Stelter, M., Braeutigam, P., 2020a. Mechanochemical versus chemical routes for graphitic precursors and their performance in micropollutants removal in water. Powder Technol., 366.

Mahmoud, A.E.D., Fawzy, M., Hosny, G., Obaid, A., 2020b. Equilibrium, kinetic, and diffusion models of chromium (VI) removal using *Phragmites australis* and *Ziziphus spina-christi* biomass. Int. J. Environ. Sci. Technol., 1–12.

Ndoun, M.C., Elliott, H.A., Preisendanz, H.E., Williams, C.F., Knopf, A., Watson, J.E., 2020. Adsorption of pharmaceuticals from aqueous solutions using biochar derived from cotton gin waste and guayule bagasse. Biochar, 1–16.

Oliveira, F.R., Patel, A.K., Jaisi, D.P., Adhikari, S., Lu, H., Khanal, S.K., 2017. Environmental application of biochar: current status and perspectives. Bioresour. Technol. 246, 110–122.

Pan, X., Gu, Z., Chen, W., Li, Q., 2020. Preparation of biochar and biochar composites and the application in a Fenton-like process for wastewater decontamination: a review. Sci. Total Environ., 142104.

Panwar, N.L., Pawar, A., 2020. Influence of activation conditions on the physicochemical properties of activated biochar: a review. Biomass Convers. Bior., 1–23.

Patel, S., Kundu, S., Halder, P., Ratnnayake, N., Marzbali, M.H., Aktar, S., Sharma, A., 2020. A critical literature review on biosolids to biochar: an alternative biosolids management option. Rev. Environ. Sci. Biotechnol., 1–35.

Patra, J.M., Panda, S.S., Dhal, N.K., 2017. Biochar as a low-cost adsorbent for heavy metal removal: a review. Int. J. Res. Biosci. 6 (1), 1–7.

Premarathna, K.S.D., Rajapaksha, A.U., Sarkar, B., Kwon, E.E., Bhatnagar, A., Ok, Y.S., Vithanage, M., 2019. Biochar-based engineered composites for sorptive decontamination of water: a review. Chem. Eng. J. 372, 536–550.

Qin, J., Chen, Q., Sun, M., Sun, P., Shen, G., 2017. Pyrolysis temperature-induced changes in the catalytic characteristics of rice husk-derived biochar during 1,3-dichloropropene degradation. Chem. Eng. J. 330, 804–812.

Reddy, K.R., Xie, T., Dastgheibi, S., 2014. Evaluation of biochar as a potential filter media for the removal of mixed contaminants from urban storm water runoff. J. Environ. Eng. 140 (12), 04014043.

Regkouzas, P., Diamadopoulos, E., 2019. Adsorption of selected organic micro-pollutants on sewage sludge biochar. Chemosphere 224, 840–851.

Son, E.B., Poo, K.M., Chang, J.S., Chae, K.J., 2018. Heavy metal removal from aqueous solutions using engineered magnetic biochars derived from waste marine macro-algal biomass. Sci. Total Environ. 615, 161–168.

Spokas, K.A., Reicosky, D.C., 2009. Impacts of sixteen different biochars on soil greenhouse gas production. Annals of Environmental Science. 3 U.S. Department of Agriculture.

Tan, X.F., Liu, Y.G., Gu, Y.L., Xu, Y., Zeng, G.M., Hu, X.J., Li, J., 2016. Biochar-based nano-composites for the decontamination of wastewater: a review. Bioresour. Technol. 212, 318–333.

Tang, L., Yu, J., Pang, Y., Zeng, G., Deng, Y., Wang, J., Feng, H., 2018. Sustainable efficient adsorbent: alkali-acid modified magnetic biochar derived from sewage sludge for aqueous organic contaminant removal. Chem. Eng. J. 336, 160–169.

Teixidó, M., Pignatello, J.J., Beltrán, J.L., Granados, M., Peccia, J., 2011. Speciation of the ionizable antibiotic sulfamethazine on black carbon (biochar). Environ. Sci. Technol. 45 (23), 10020–10027.

Thompson, K.A., Shimabuku, K.K., Kearns, J.P., Knappe, D.R., Summers, R.S., Cook, S.M., 2016. Environmental comparison of biochar and activated carbon for tertiary wastewater treatment. Environ. Sci. Technol. 50 (20), 11253–11262.

Wang, Y., Liu, R., 2018. H_2O_2 treatment enhanced the heavy metals removal by manure biochar in aqueous solutions. Sci. Total Environ. 628, 1139–1148.

Wang, L., Wang, Y., Ma, F., Tankpa, V., Bai, S., Guo, X., Wang, X., 2019. Mechanisms and reutilization of modified biochar used for removal of heavy metals from wastewater: a review. Sci. Total Environ. 668, 1298–1309.

Weber, K., Quicker, P., 2018. Properties of biochar. Fuel 217, 240–261.

Xiao, F., Cheng, J., Cao, W., Yang, C., Chen, J., Luo, Z., 2019. Removal of heavy metals from aqueous solution using chitosan-combined magnetic biochars. J. Colloid Interface Sci. 540, 579–584.

Xiao, L., Feng, L., Yuan, G., Wei, J., 2020. Low-cost field production of biochars and their properties. Environ. Geochem. Health 42 (6), 1569–1578.

Xu, X., Zhao, Y., Sima, J., Zhao, L., Mašek, O., Cao, X., 2017. Indispensable role of biochar-inherent mineral constituents in its environmental applications: a review. Bioresour. Technol. 241, 887–899.

Yaashikaa, P.R., Kumar, P.S., Varjani, S.J., Saravanan, A., 2019. Advances in production and application of biochar from lignocellulosic feedstocks for remediation of environmental pollutants. Bioresour. Technol. 292, 122030.

Zhang, S., Lyu, H., Tang, J., Song, B., Zhen, M., Liu, X., 2019a. A novel biochar supported CMC stabilized nano zero-valent iron composite for hexavalent chromium removal from water. Chemosphere 217, 686–694.

Zhang, Z., Yan, L., Yu, H., Yan, T., Li, X., 2019b. Adsorption of phosphate from aqueous solution by vegetable biochar/layered double oxides: fast removal and mechanistic studies. Bioresour. Technol. 284, 65–71.

Further reading

Cha, J.S., Park, S.H., Jung, S.C., Ryu, C., Jeon, J.K., Shin, M.C., Park, Y.K., 2016. Production and utilization of biochar: a review. J. Indus. Eng. Chem. 40, 1–15.

CHAPTER

8

In situ chemical oxidation (ISCO) remediation: A focus on activated persulfate oxidation of pesticide-contaminated soil and groundwater

Rama Mohan Kurakalva[a,b]

[a]Hydrogeochemistry Group, CSIR-National Geophysical Research Institute (CSIR-NGRI), Hyderabad, Telangana, India
[b]Faculty of Physical Sciences, Academy of Scientific & Innovative Research (AcSIR), Ghaziabad, India

OUTLINE

8.1 Introduction

Remediation of the contaminated soil and groundwater is indispensable as it poses a significant risk to public health and the environment. Many different remediation technologies by means of either in situ or ex situ mode have developed for the last few decades to mitigate the risk imposed due to soil and groundwater contamination. The selection of remediation technologies (in situ/ex situ) or in combination of technologies is dependent on the contaminant type. Further, remediation technologies would be useful and helpful in preventing further spreading of contamination. Remediation methods involve three major types of mechanisms in their processes that include (i) containment, (ii) removal, and (iii) treatment, thereby achieving remediation (Fig. 8.1). *Containment* is the process where the contaminant is restricted to a specified domain to prevent further spreading. *Removal* is the process where the contaminant is transferred from an open to a controlled environment. *Treatment* is the process where the contaminant is transferred into a nonhazardous substance (Brusseau, 2019). Of the three methods, *treatment* is a preferred approach for remediation of contaminated soil and groundwater as contaminants completely eliminate and convert to nontoxic substances. However, containment and removal techniques are also essential when it is not feasible with the only treatment process of the contaminants. Hence, all the three types of remedial actions mentioned already, in turn, would be useful.

https://doi.org/10.1016/B978-0-12-822933-0.00011-5

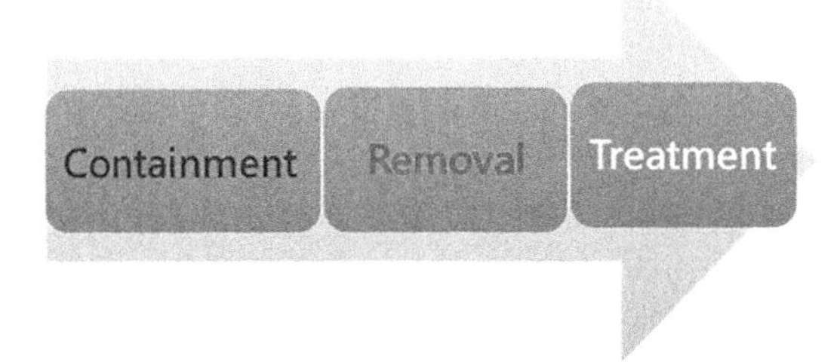

FIG. 8.1 Different actions occur in the processes of remediation.

In situ treatment technology allows "in place clean up" of contaminated field sites, as they are cheap in some cases than the other remedial methods. Further, the risk associated with the contaminants by promoting the destruction of the contaminants via transforming reactions is eliminated. Two major types of in situ treatment methods frequently used are based on the biological process called in situ bioremediation, and the chemical process involved is termed in situ chemical oxidation remediation. The chemical oxidation methods are based on injecting a reagent into the subsurface to promote transformation reactions. Advanced oxidation processes (AOPs) are becoming increasingly popular as an alternative for treating organic contaminants in soil and groundwater system.

8.2 Chemical oxidation

The chemical oxidation process involves the injection or emplaces a highly reactive substance to break apart the bonds in a contaminant compound, and that destroys the molecule. It involves one-half of a redox reaction (Fig. 8.2); it means one of the reactants in a reaction process becomes oxidized or loses electrons, while another reactant becomes reduced or gains electrons.

If contaminant remediation accomplished in place using a "chemical oxidation process" called "in situ chemical oxidation" (ISCO). It is also one of the AOPs involving the use of strong chemical oxidants for destructing or degrading a variety of organic contaminants (Tsai et al., 2008; Tsitonaki et al., 2010; Watts et al., 1990, 1991). Chemical oxidation technologies are preferable as this process is typically carried out at mild temperatures and room pressures, making it more appropriate for soil and groundwater remediation. The chemical oxidants are usually introducing either by direct injection or by other means into the contaminated medium such as soil, groundwater, solid waste, and wastewater, as shown in Fig. 8.3.

The AOPs have been successfully used the processes like ozonation (Andreozzi et al., 2003; Ikehata et al., 2006, 2008; Nakada et al., 2007; Rosal et al., 2010), photocatalysis (Miranda-Garcia et al., 2010; Nienow et al., 2009; Pereira et al., 2007; Shemer et al., 2006; Sires et al., 2007; Yang et al., 2008; Antoniou et al., 2010), electrochemical (Murugananthan et al., 2007; Torriero et al., 2006), and Fenton's reagent (Tekin et al., 2006; Li et al., 2012) for abatement of priority pollutants. In recent years, a relatively new oxidant "persulfate (PS)" has considerable attention as an alternative to oxidants mentioned earlier due to its capability of destructing many contaminants at mild conditions.

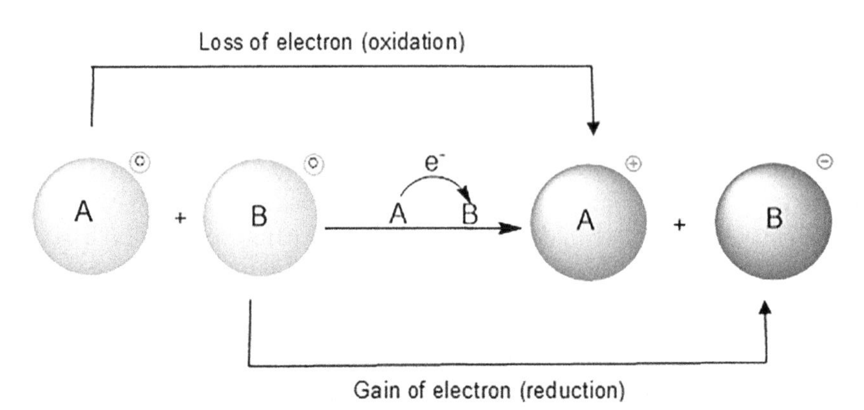

FIG. 8.2 Schematic of chemical oxidation reaction.

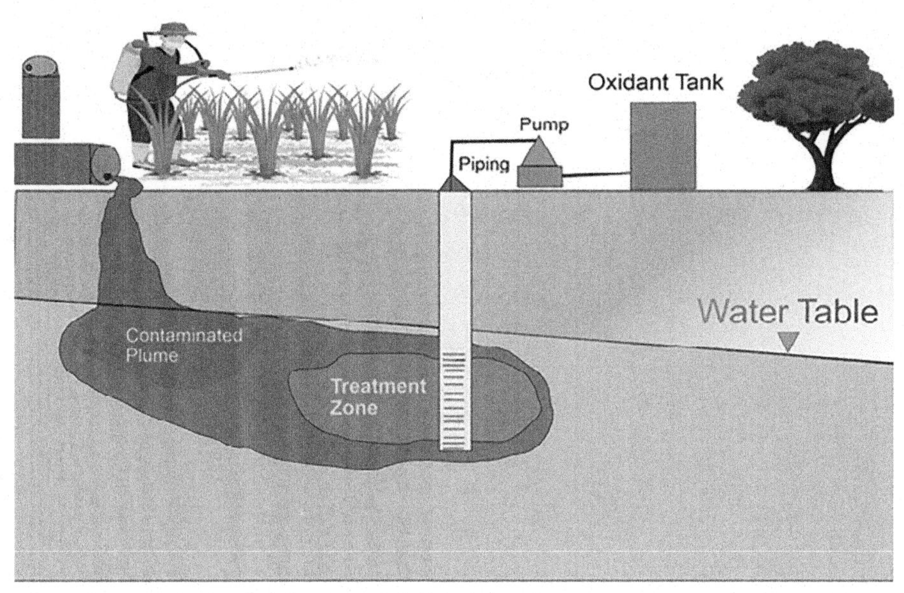

FIG. 8.3 Schematic diagram of in situ chemical oxidation (ISCO) remediation.

8.3 ISCO remediation

In situ chemical oxidation remediation reasonably established treatment technology for contaminated soil and groundwater, including source zones and plumes. The basic concept involved in remediation is oxidizing compounds that give electrons away to other compounds in the reaction used to reduce or degrade the contaminants into harmless compounds (In situ Chemical Reduction, 2020). ISCO has been used mainly for the treatment of chlorinated organic solvents and petroleum hydrocarbons with a target of reducing contaminant mass in a source zone so as to attain maximum level of contaminants in a plume (Siegrist et al., 2011). USEPA designated ISCO as the most frequently used in situ method for source remediation (U.S. Environmental Protection Agency (US EPA), 2017).

Currently, the ISCO process is utilizing several common oxidants, and their oxidation potential is shown in Table 8.1. Persulfate as an oxidant have depicted significant awareness as sustainable technology and also it became an alternative to oxidants such as hydrogen peroxide/Fentońs reagent in water and wastewater treatment (Pac et al., 2019; Ikea et al., 2018). Initially, the persulfate-based AOPs were introduced in late 1990 for soil and groundwater remediation. Chemical oxidation was readily adapted in remediation methods through tank-based reactors for ex situ treatment of organic contaminants in groundwater (Lee et al., 2020; Barbeni et al., 1987; Bellamy et al., 1991; Bowers et al., 1989; Venkatadri and Peters, 1993; Watts et al., 1993).

8.4 Mechanism of generation of sulfate radicals

Persulfate (peroxydisulfate, $S_2O_8{}^{2-}$) ion (Fig. 8.4) is one of the most potent oxidants discovered by M. Berthelot in 1878 (Kolthoff and Miller, 1951).

Persulfate (PS) ion has standard redox potential ($E°$) of 2.01 V (Eq. 8.1). It has the property of nonselective reactive, relatively stable storage and handling at room temperature (Latimer, 1952; Liu et al., 2012). In an aqueous solution, persulfate anion is kinetically slow for the oxidation reactions.

$$\underset{\text{Persulfate anion}}{S_2O_8{}^{2-}(aq)} + 2e^- \rightarrow \underset{\text{Sulfate radical}}{2SO_4^{2-}} \quad E° = 2.01\text{eV} \tag{8.1}$$

Activation of persulfate by various agents can achieve a higher oxidative potential through the formation of sulfate radicals (Berlin, 1986; Sergio et al., 2012), as shown in Fig. 8.5. A comprehensive equation for the activation of persulfate is shown in Eq. (8.2).

$$\underset{\text{Persulfate anion}}{S_2O_8{}^{2-}} + \text{activator} \rightarrow \underset{\text{Sulfate radical}}{SO_4^{\cdot-} + (SO_4^{\cdot-}\,\text{or}\,SO_4^{2-})} \quad E° = 2.6\text{eV} \tag{8.2}$$

TABLE 8.1 Oxidation potential of some most popular oxidants.

Oxidant	Formula	Structure	Standard reduction potential (*E*) in V
Fluorine	F_2	F—F	3.03
Hydroxyl radical	OH	$^{-}$O—H	2.80
Sulfate radical	SO_4^-	$^{-}$O—S(=O)(=O)—O^{-}	2.60
Ozone	O_3	$^{-}$O—O^{+}=O	2.07
Persulfate anion	$S_2O_8^{2-}$	$^{-}$O—S(=O)(=O)—O—O—S(=O)(=O)—O^{-}	2.01
Hydrogen peroxide	H_2O_2	HO—OH	1.78
Per hydroxyl radical	HO_2	$^{-}$O—OH	1.70
Permanganate anion	MnO_4^-	O=Mn(=O)(=O)—O^{-}	1.68
Oxygen	O_2	O=O	1.23

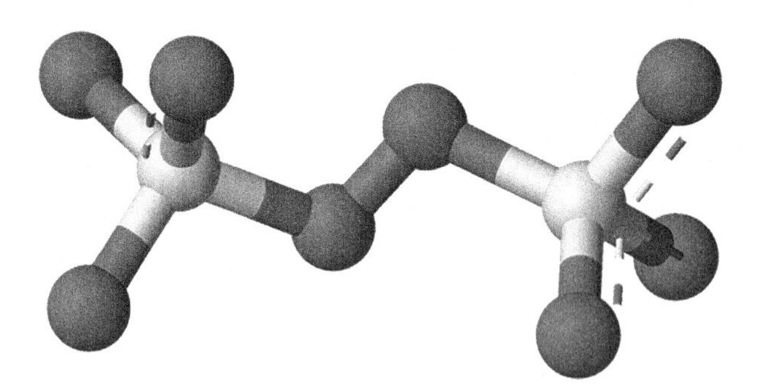

FIG. 8.4 Persulfate ion.

where the activator (Eq. 8.2) could be heat or transition metal ions or UV/photon or ultrasound. These activator(s) imparts energy to persulfate anions causing direct cleavage of the peroxide bond (O-O) and produces two sulfate radicals in accordance with Eqs. (8.3), (8.4).

$$S_2O_8^{2-} + \text{energy input} \rightarrow 2SO_4^{\bullet-} \tag{8.3}$$

$$S_2O_8^{2-} + e^- \rightarrow SO_4^{2-} + 2SO_4^{\bullet-} \tag{8.4}$$

8.5 Activation of persulfate

Activation of persulfate enhances the degree of removal efficiency as it attains higher oxidation potential, which is responsible for increasing contaminant degradation in situ in the soil and groundwater. The following discussions are the most popular activation methods.

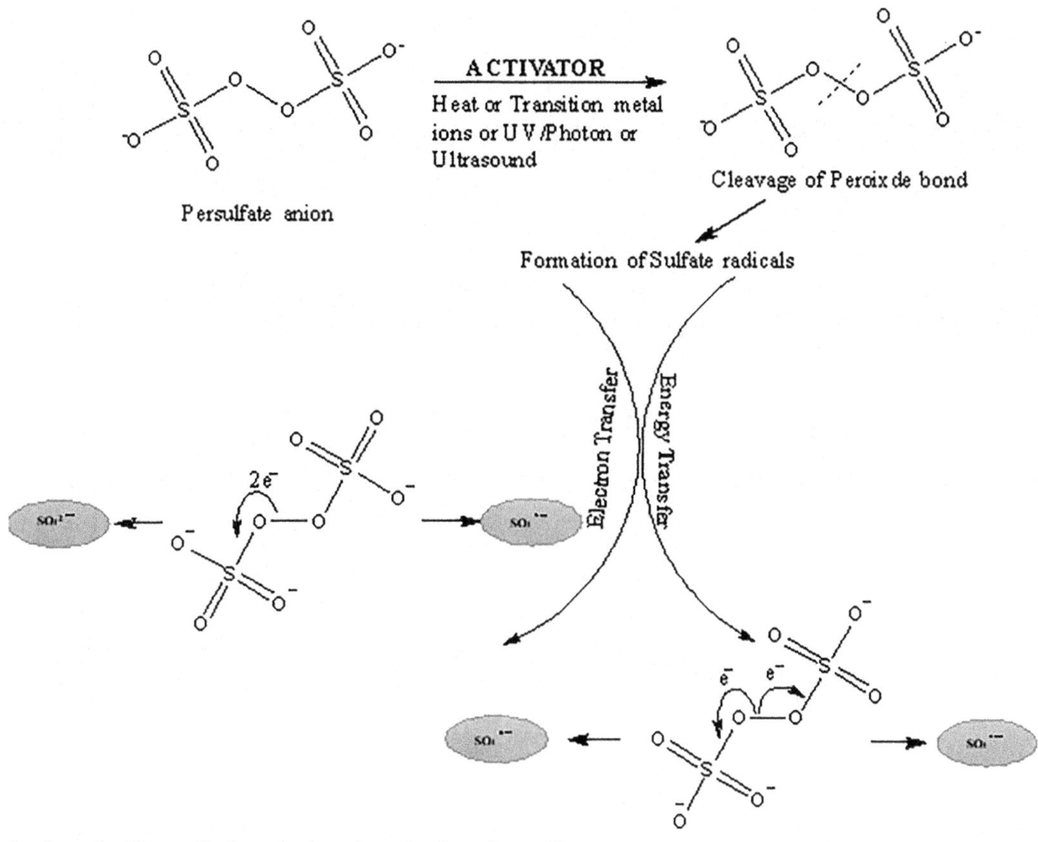

FIG. 8.5 Mechanism of sulfate radical generation via activation of persulfate.

8.5.1 Heat

Persulfate can be activated through the supply of heat and achieved sulfate radicals as per Eq. (8.5)

$$S_2O_8^{2-} + \text{Heat} \rightarrow 2SO_4^{\cdot-} \tag{8.5}$$

8.5.2 Acidic catalysis

From the literature (Sergio et al., 2012; Eberson, 1987), it shows that persulfate can produce sulfate radicals under acidic conditions as shown in Eqs. (8.6), (8.7):

$$S_2O_8^{2-} + H^+ \rightarrow HSO_8^- \tag{8.6}$$

$$HSO_8^- \rightarrow H^+ + SO_4^{\cdot-} + SO_4^{2-} \tag{8.7}$$

8.5.3 Transition metals

Activation of persulfate with transition metal ions, particularly using divalent metal ions, serves as an electron donor (i.e., electron transfer from transition metal ions) and forms sulfate-free radicals deduced as per Eq. (8.8).

$$S_2O_8^{2-} + Me^{n+} \rightarrow SO_4^{\cdot-} + Me^{(n+1)+} + SO_4^{2-} \tag{8.8}$$

where Me is the metal catalyst.

Iron is the most widely investigated activator among the other transition metal ions for contaminant destruction (Anipsitakis and Dionysiou, 2004; Liang et al., 2004a, 2004b, 2007; Jiang et al., 2013; Wu et al., 2015; Rastogi et al., 2009a; Romero et al., 2010) as shown in Eqs. (8.9)–(8.11).

$$S_2O_8^{2-} + Fe^{2+} \rightarrow Fe^{3+} + SO_4^{2-} + SO_4^{\cdot-} \tag{8.9}$$

$$SO_4^{\cdot -} + Fe^{2+} \rightarrow Fe^{3+} + SO_4^{2-} \quad (8.10)$$

$$SO_4^{\cdot -} + O_p \rightarrow O_p^{ox} \quad (8.11)$$

where O_p is the organic pollutant p, O_p^{ox} is the organic pollutant oxidized.

In fact, all the activation processes discussed earlier generate sulfate radicals, $SO_4^{\cdot -}$, which are responsible for the destruction of organic contaminants, as shown in Eq. (8.11), that are typically considered. Nevertheless, an undesirable reaction has occurred between sulfate radicals (Eq. 8.9) and Fe^{2+} through Eq. (8.10), thereby yielding both radicals (sulfate and sulfate-free radical) through Fe^{2+} consumption (Liang et al., 2004, 2007; Jiang et al., 2013; Wu et al., 2015; Rastogi et al., 2009a; Romero et al., 2010). It was deduced that the oxidation rate of Fe^{2+} by $SO_4^{\cdot -}$ is much higher than the oxidation of the organic compounds by the persulfate radicals (Lee et al., 2009; Vicente et al., 2011).

8.6 Pesticides in soil and groundwater

The pesticide is defined as any substance or mixture of substances intended for preventing, destroying, repelling, or mitigating any pest (EPA, 2020). Pesticides have advantages such as increased food production and control of vector-borne diseases. At the same time, they have resulted in serious implications on human health and the environment (Lari et al., 2014). Among the South Asian and African countries, India is one of the largest manufacturers of pesticides (Lari et al., 2014). Drinking water samples from hand pumps and wells in India have been reported that the organochlorine pesticides (OCPs) were found beyond the permissible limits of EPA Standards (Hundal et al., 2006). Among the various pesticides, some of these chemicals, like OCPs, do pose a potential risk to humans and other life forms and undesirable side effects to the environment (Kole and Bagchi, 1995; Forget, 1993). OCPs are a vital group of environmental contaminants that have been of great concern due to their persistent nature and chronic adverse effects on wildlife and humans (Igbedioh, 1991; Andersson et al., 2001). Due to widespread occurrence of OCPs contamination in soil and groundwater, studies on eco-friendly remediation such as iron activated persulfate oxidation (APO) has drawn great environmental concern. In this connection, chlorinated pesticide (e.g., Aldrin chosen as a model compound) remediation contaminated aquifers through iron-APO was investigated.

8.7 Investigation on iron-activated persulfate oxidation of Aldrin

Aldrin (CAS No. 309-00-2) is a priority toxic pollutant (Fig. 8.6) and has found in at least 207 of 1613 of the most serious hazardous waste sites recognized by USEPA on the National Priorities List (Nair et al., 1991). United Nations Environment Programme (UNEP) adopted Stockholm Convention on POPs, a global agreement in the year 2004. The permissible limit of Aldrin concentration in drinking water is 0.03 μg/L as per the WHO guideline values (ASTDR, 2002). Hence, there is a need to remediate contaminated soil and groundwater.

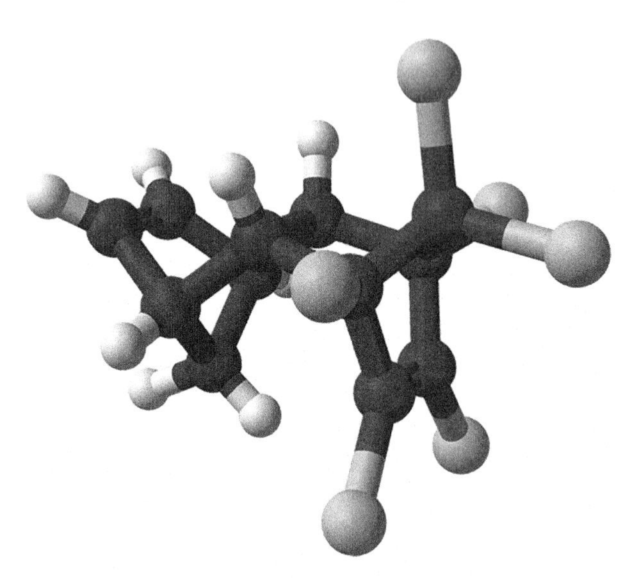

FIG. 8.6 Structure of aldrin.

On the contrary, among the various ISCO remediation methods, persulfate activated with iron is a widely studied metal among the other transition metals, as it is effective in the degradation of organic compounds. It is relatively nontoxic, environmentally friendly, and cost-effective (WHO, 2003; Kurakalva, 2016a). Therefore, the present investigations focused on the iron-activated persulfate remediation of pesticide that contaminates soil and groundwater are proposed. The analytical methodologies and of the batch experimental details were used for the ferrous ion-activated Aldrin degradation as described elsewhere (Kurakalva, 2016b).

Effect of pH on Aldrin degradation. Effect of pH on the formation of active radical species such as SO_4^{-} and $OH^{\cdot}$ and the reaction between the active radical species were performed at a variable pH range of 3.0–11.0. The data showed that at pH 3.0, nearly 99.9% of Aldrin was degraded at $S_2O_8^{2-}/Fe^{2+}$/Aldrin molar ratio of 50/50/1 (180 min).

However, for the study persulfate degradation of Aldrin, the experiments were carried out without adjusting the initial pH of the solution. In the aqueous solution of the contaminant (Aldrin), the initial pH of the solution is 6.5, which is observed before the addition of Fe^{2+}. In the first addition of Fe^{2+}, pH is decreased to 3.0; and in further sequential additions, it reached 2.6 with an optimized oxidant/activator/contaminant ratio as mentioned previously. This has been observed that persulfate anion can produce sulfate radicals as per Eqs. (8.6), (8.7) and rapidly attacks the organic contaminants degraded completely. This further verified that persulfate anion could also hydrolyze to form hydrogen peroxide (Rastogi et al., 2009), under acidic conditions as Eq. (8.12) possibly acts as a dual oxidant (i.e., persulfate and hydrogen peroxide) in the system.

$$S_2O_8^{2-} + 2H_2O \rightarrow H_2O_2 + 2HSO_4^{-} \quad (8.12)$$

Also, in the presence of aqueous solutions, sulfate radicals ($SO_4^{\cdot-}$) proceed to radical interconversion reactions to produce the hydroxyl radical (Eq. 8.13).

$$SO_4^{\cdot-} + H_2O \rightarrow SO_4^{2-} + {}^{\cdot}OH + H^{+} \quad (8.13)$$

A decrease in the degradation of Aldrin observed at > pH 4.0 due to Fe^{2+} oxidation of persulfate to form precipitation of Fe^{3+}, which is insoluble as there is no sulfate radical generation:

$$2Fe^{2+} + S_2O_8^{2-} \rightarrow 2Fe^{3+} + 2SO_4^{2-} \quad (8.14)$$

$$Fe^{3+} + 3H_2O \rightarrow Fe(OH)_3 + 3H^{+} \downarrow \quad (8.15)$$

8.8 Influence of Fe^{2+} on Aldrin degradation

Experiments were carried out with different ratios of Fe^{2+}/PS and obtained an optimized ratio as 50/1 for the maximum degradation of Aldrin, and it followed immediately and then increased very slowly (Fig. 8.7).

This might be due to its pH that is decreased from 6.5 to 3.0 upon adding Fe^{2+} to the contaminant aqueous solution and rapidly degraded 70.3% of the Aldrin as the reaction is progressed according to Eqs. (8.6), (8.7).

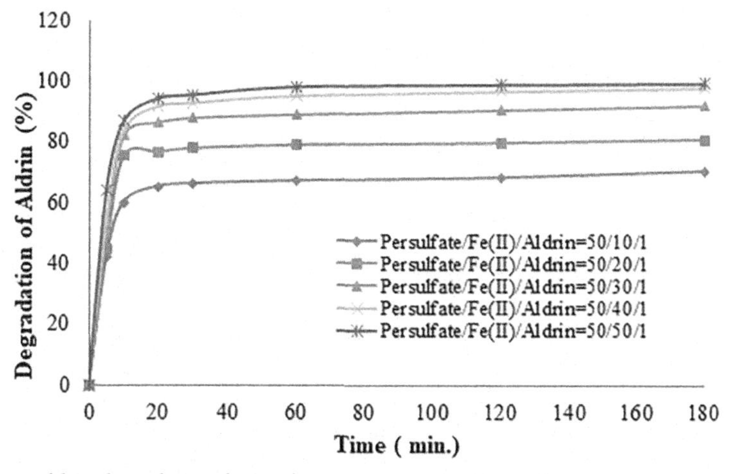

FIG. 8.7 Effect of Fe^{2+} dosage on aldrin degradation during ferrous ion activated persulfate oxidation. $[Aldrin]_o$ = 0.001 μM; $[PS]_o$ = 0.05 μM; $[Fe^{2+}]_o$ = 0.01–0.05 μM; initial pH 6.5; reaction time = 180 min.

TABLE 8.2 Summary of pseudo-first-order rate constant (K_{obs}) between different molar ratios in persulfate-Fe^{2+} oxidation system.

$S_2O_8^{2-}/Fe^{2+}$/Aldrin molar ratios	Initial $S_2O_8^{2-}$ concentration (μM)	Initial Fe^{2+} concentration (μM)	C_t/C_0	Pseudo-first-order rate constant (K_{obs})
50/10/1	0.05	0.01	0.40	0.0 9
50/20/1	0.05	0.02	0.25	0.14
50/30/1	0.05	0.03	0.18	0.17
50/40/1	0.05	0.04	0.16	0.18
50/50/1	0.05	0.05	0.13	0.20
1/10/1	0.001	0.01	0.73	0.03
5/10/1	0.005	0.01	0.62	0.05
10/10/1	0.010	0.01	0.48	0.08
15/10/1	0.015	0.01	0.45	0.08
20/10/1	0.020	0.01	0.41	0.08

Also, sequential mode of addition of Fe^{2+}/Aldrin (20/1 to 50/1) has resulted in a further 29.6% of Aldrin degradation, as the availability of Fe^{2+} ions increasing sulfate radicals that attack Aldrin. As it can be proved through the reported investigation of Fentońs reaction (PeroxygenTalk, 2010), gradual addition of Fe^{2+} could minimize the oxidation of Fe^{2+} by OH. In conclusion, $S_2O_8^{2-}/Fe^{2+}$/Aldrin molar ratio of 50/50/1 degraded 99.9% of Aldrin.

Kinetics of oxidation. Aldrin oxidation in the PS-Fe^{2+} system proceeds initially much slower as detailed earlier. The degradation of Aldrin in the PS-Fe^{2+} system can be approached as the pseudo-first order less than 10 min. The data on rate constants (K_{obs}) for different molar ratios are presented in Table 8.2. It postulated persulfate that followed pseudo-first-order reaction kinetics and determined by

$$-d[\text{Aldrin}]/dt = k[\text{Aldrin}] \tag{8.16}$$

It can be written as follows:

$$\text{Ln}(C_t/C_o) = -kt \tag{8.17}$$

where C_t and C_o are the concentrations of Aldrin (μg/L) at t and t_0, respectively. k is the pseudo-first-order rate constant (min^{-1}).

Degradation of Aldrin in aqueous solutions is well fitted to the pseudo-first-order kinetic reaction.

Effect of oxidant and activator concentration. Fig. 8.8 demonstrates that variable oxidant concentration affects the degradation of Aldrin accordingly. It was noticed that ~ 99% of oxidants are decomposed with all the molar ratios investigated.

With regard to activator (Fe^{2+}), concentration with various molar ratios instigated an increase in Aldrin degradation about 29%, relatively after initially added Fe^{2+} to the system. This further verified the degradation of Aldrin occurring rapidly with the initial dose of activator. On the contrary, it is found that oxidant decomposes gradually while increasing the activator content. Thus, the oxidant and activator properties are critical in the determination of the rate of reaction and the degree of contaminant mineralization.

8.9 Challenges and opportunities in field applications

Persulfate-based ISCO remediation method is much faster than most other remediation methods; it has been taken care of in field applications. As the most popular and successful investigations, it is one of the best methods for in situ remediation of contaminated groundwater (Pignatello and Baher, 1994). The subsurface properties affect reactivity, and maintaining persulfate/activator levels in the subsurface is critical for successful implementation (Matzek and Carter, 2016; Boni and Sbaffoni, 2012; Ahmad et al., 2013). For instance, soil organic matter under basic conditions enhanced activated persulfate degradation of nitrobenzene and hexachloroethane (Sra et al., 2014). Besides, the properties of contaminants like hydrophobic compounds sorbed to soil and subsurface solids decrease the efficiency of destruction while using activated persulfate (Teel et al., 2009). ISCO generally improves the biodegradability of the

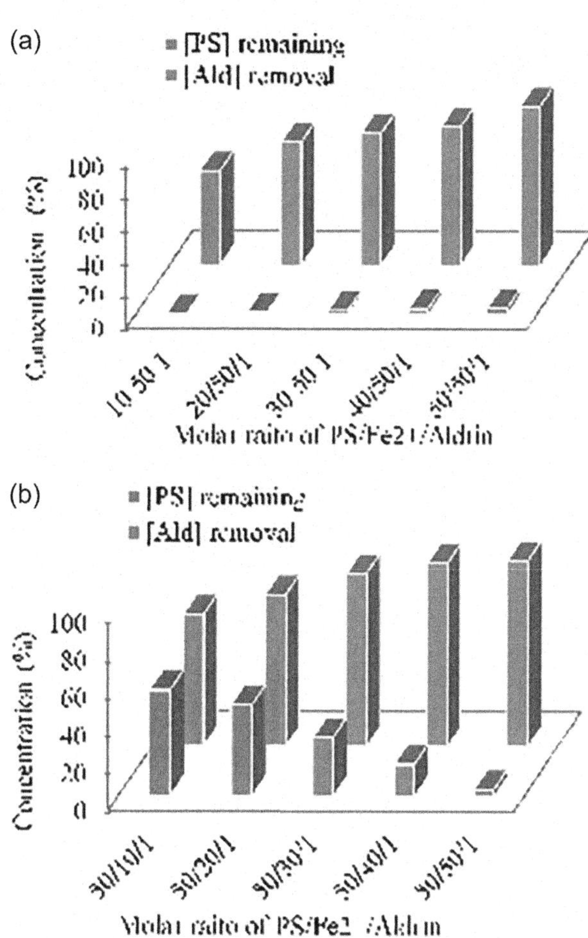

FIG. 8.8 Aldrin removal and persulfate remaining in the ferrous ion activated oxidation system. Increase of persulfate concentration $[PS]_o$ = 0.01–0.05 μM; $[Fe^{2+}]_o$ = 0.05 μM; initial pH 6.5; reaction time = 180 min. Increase of ferrous ion concentration $[PS]_o$ = 0.05 μM; $[Fe^{2+}]_o$ = 0.01–0.05 μM; initial pH 6.5; reaction time = 180 min.

contaminant pool; however, its impacts on bioavailability should be studied and optimized before its implementation on a field scale. Besides, chemical oxidant quantity and concentrations should be optimized as these would influence subsequent biological degradation of contaminants.

Insufficient Fe^{2+} concentrations generally lead to inefficient persulfate usage, whereas too much Fe^{2+} scavenges the sulfate radicals, as shown in Eq. (8.18) that, in turn, decreases degradation efficiency in the system (Romero et al., 2010; Lee et al., 2009).

$$SO_4^{\cdot -} + 2Fe^{2+} \rightarrow Fe^{3+} + SO_4^{2-} \quad (8.18)$$

However, the use of hydroxylamine is useful to maintain iron levels in situ via a quick cycle of Fe^{3+} back to Fe^{2+} that enhances the degradation of trichloroethylene as reported (Vicente et al., 2011). Besides, iron chelator functionality will help in improving the longevity of iron, which enhances the remediation efficiency via radical sustainability. Further, subsurface parameters like mineralogy and natural organic matter (NOM) would consequently enhance APO (Boni and Sbaffoni, 2012). Therefore, geochemical characterization of subsurface of the contaminated site provides vital information in filed applications and its safety concern for sustainable management.

8.10 Conclusions

ISCO provides highly effective and relatively inexpensive strategies for the remediation of recalcitrant organic contaminants. The iron-APO improves with increased persulfate addition; however, this approach is found to be not sustainable due to the depletion of reactants, limiting degradation of targeted contaminants. Among the different methods of activation of PS, transition metal activation has economic advantages compared to thermal activation, radiation in terms of energy consumption. If remediation site contain natural minerals like goethite, ferrihydrite, and limonite enhances the effectiveness for activation of persulfate oxidation. The development of methods/techniques to maintain ratios of oxidants and activator concentrations in situ will improve utilization at the field scale.

Acknowledgments

Rama Mohan Kurakalva (RMK) would like to thank the Director, CSIR-NGRI, for permitting to publish this research work. The author acknowledges the Science Engineering Research Board (SERB), Department of Science and Technology (DST), Government of India, for providing financial assistance through the Fast Track Scheme for Young Scientists (Grant No. SR/FTP/ES-29/2008). He also acknowledges the support rendered from Environmental Geochemistry Laboratory, CSIR-NGRI, for providing GC-MS analysis, and finally, he thanked Dr. Prabhakar Sripadi, CSIR-IICT, Hyderabad, India, for his support in drawing the chemical structures in this manuscript.

References

Ahmad, M., Teel, A.L., Watts, R.J., 2013. Mechanism of persulfate activation by phenols. Environ. Sci. Technol. 47 (11), 5864–5871.

Andersson, P., Berg, A.H., Bjerselius, R., Norrgren, L., Olsén, H., Olsson, P.E., Orn, S., Tysklind, M., 2001. Bioaccumulation of selected PCBs in Zebrafish, three-spined stickleback and arctic char after three different routes of exposure. Arch. Environ. Contam. Toxicol. 40 (4), 519–530.

Andreozzi, R., Caprio, V., Marotta, R., Vogna, D., 2003. Paracetamol oxidation from aqueous solutions by means of ozonation and H_2O_2/UV system. Water Res. 37, 993–1004.

Anipsitakis, G.P., Dionysiou, D.D., 2004. Radical generation by the interaction of transition metals with common oxidants. Environ. Sci. Technol. 38 (13), 3705–3712.

Antoniou, M.G., de la Cruz, A.A., Dionysiou, D.D., 2010. Degradation of microcystin-LR using sulfate radicals generated through photolysis, thermolysis and e-transfer mechanisms. Appl. Catal. B: Environ. 96 (3–4), 290–298.

ASTDR, 2002. Toxicological profile for aldrin/dieldrin. In: Services, U. S. D. o. H. a. H (Ed.), Agency for Toxic Substances and Disease Registry.

Barbeni, M., Nfinero, C., Pelizzetti, E., Borgarello, E., Serpon, N., 1987. Chemical degradation of chlorophenols with Fenton's reagent. Chemosphere 16, 2225–2237.

Bellamy, W.D., Hickman, P.A., Ziemba, N., 1991. Treatment of VOC-contaminated ground water by hydrogen peroxide and ozone oxidation. Res. J. Water Pollut. Control Federat. 63, 120–128.

Berlin, A.A., 1986. Kinetics of radical-chain decomposition of persulfate in aqueous solutions of organic compounds. Kinet. Catal. 27, 34–39.

Boni, M.R., Sbaffoni, S., 2012. Chemical oxidation by sodium persulphate for the treatment of contaminated groundwater. Laboratory tests. Chemical Engineering Transactions 28, 157–162.

Bowers, A.R., Gaddipati, P., Eckenfelder, W.W., Monsen, R.M., 1989. Treatment of toxic or refractory wastewaters with hydrogen peroxide. Water Sci. Technol. 21, 477–486.

Brusseau, M.L., 2019. Soil and groundwater remediation. Environ. Pollut. Sci., 329–354. https://doi.org/10.1016/b978-0-12-814719-1.00019-7.

Eberson, L., 1987. Electro Transfer Reactions in Organic Chemistry. Springer-Verlag, Berlin.

EPA, 2020. Basic Information About Pesticides. https://www.epa.gov/ingredients-used-pesticide-products/basic-information-about-pesticide-ingredients. (Accessed 28 November 2020).

Forget, G., 1993. Balancing the need for pesticides with the risk to human health: impact of pesticide use on health in developing countries. In: Proceedings of a Symposium Held in Ottawa, 17–20 September 1990, pp. 2–16.

Hundal, B.S., Anand, Singh, R., 2006. Pesticide marketing: the Indian Scenario. IUP J. Manage. Econ. 4 (2), 32–37.

Igbedioh, S.O., 1991. Effects of agricultural pesticides on humans, animals, and higher plants in developing countries. Arch. Environ. Health 46 (4), 218–224.

Ikea, I.A., Lindenb, K.G., Orbella, J.D., Duke, M., 2018. Critical review of the science and sustainability of persulfate advanced oxidation processes. Chem. Eng. J. 338, 651–669.

Ikehata, K., Naghashkar, N.J., Gamal Ei-Din, M., 2006. Degradation of aqueous pharmaceuticals by ozonation and advanced oxidation processes: a review. Ozone Sci. Eng. 28 (6), 353–414.

Ikehata, K., Gamal El-Din, M., Snyder, S.A., 2008. Ozonation and advanced oxidation treatment of emerging organic pollutants in water and wastewater. Ozone Sci. Eng. 30 (1), 21–26.

Anon., 2020. In situ Chemical Reduction. https://clu-in.org/techfocus/default.focus/sec/In_Situ_Chemical_Reduction/cat/Overview/. (Accessed 28 November 2020).

Jiang, X.X., Wu, Y.L., Wang, P., Li, H.J., Dong, W.B., 2013. Degradation of bisphenol A in aqueous solution by persulfate activated with ferrous ion. Environ. Sci. Pollut. Res. 20, 4947–4953.

Kole, R.K., Bagchi, M.M., 1995. Pesticide residues in the aquatic environment and their possible ecological hazards. J. Inland Fish. Soc. India 27, 79–89.

Kolthoff, I.M., Miller, I.K., 1951. The chemistry of persulfate. I. The kinetics and mechanism of decomposition of persulfate in aqueous medium. J. Am. Chem. Soc. 73, 3055.

Kurakalva, R.M., 2016a. In situ chemical oxidation (ISCO) technology for remediation of pesticide contaminated soil and groundwater. In: First Networking Conference.

Kurakalva, R.M., 2016b. In situ remediation of aldrin via activated persulfate oxidation. Geochicago 302-318. https://doi.org/10.1061/9780784480168.031.

Lari, S.Z., Khan, N.A., Gandhi, K.N., Meshram, T.S., Thacker, N.P., 2014. Comparison of pesticide residues in surface water and ground water of agriculture intensive areas. J. Environ. Health Sci. Eng. 12 (11). https://doi.org/10.1186/2052-336X-12-11.

Latimer, W.M., 1952. The Oxidation States of the Elements and Their Potentials in Aqueous Solutions. Prentice-Hall, Inc., Englewood Cliffs, NJ.

Lee, J., Gunten, U.V., Kim, J.H., 2020. Persulfate-based advanced oxidation: critical assessment of opportunities and roadblocks. Environ. Sci. Technol. 54 (6), 3064–3081.

Lee, Y.C., Lo, S.L., Chiueh, P.T., Chang, D.G., 2009. Efficient decomposition of perfluorocarboxylic acids in aqueous solution using microwave-induced persulfate. Water Res. 43 (11), 2811–2816.

Li, W., Venkateswarlu, N., Zhou, Q., Korshin, G.V., 2012. Effects of Fenton treatment on the properties of effluent organic matter and their relationships with the degradation of pharmaceuticals and personal care products. Water Res. 46 (2), 403–412.

Liang, C., Bruell, C.J., Marley, M.C., Sperry, K.L., 2004a. Persulfate oxidation for in situ remediation of TCE. I. Activated by ferrous ion with and without a persulfate-thiosulfate redox couple. Chemosphere 55, 1213–1223.

Liang, C., Bruell, C.J., Marley, M.C., Sperry, K.L., 2004b. Persulfate oxidation for in situ remediation of TCE. II. Activated by chelated ferrous ion. Chemosphere 55 (9), 1225–1233.

Liang, C., Huang, C.F., Mohanty, N., Lu, C.J., Kurakalva, R.M., 2007. Hydroxypropyl beta cyclodextrin-mediated iron-activated persulphate oxidation of trichloroethylene and tetrachloroethylene. Ind. Eng. Chem. Res. 46 (20), 6466–6479.

Liu, C.S., Higgins, C.P., Wang, F., Shih, K., 2012. Effect of temperature on oxidative transformation of perfluorooctanoic aid (PFOA) by persulfate activation in water. Sep. Purif. Technol. 91, 46–51.

Matzek, L.W., Carter, K.E., 2016. Activated persulfate for organic chemical degradation: a review. Chemosphere 151, 178–188.

Miranda-Garcia, N., Maldonado, M.I., Coronado, J.M., Malato, S., 2010. Degradation study of 15 emerging contaminants at low concentration by immobilized TiO_2 in a pilot plant. Catal. Today 151 (1–2), 107–113.

Murugananthan, M., Yoshihara, S., Rakuma, T., Uehara, N., Shirakashi, T., 2007. Electrochemical degradation of 17b-estradiol (E2) at boron-doped diamond (Si/BDD) thin film electrode. Electrochim. Acta 52 (9), 3242–3249.

Nair, A., Dureja, P., Pillai, M.K.K., 1991. Levels of aldrin and dieldrin in environmental samples from Delhi, India. Sci. Total Environ. 108 (3), 255–259.

Nakada, N., Shinohara, H., Murata, A., Kiri, K., Managaki, S., Sato, N., Takada, H., 2007. Removal of selected pharmaceuticals and personal care products (PPCPs) and endocrine disrupting chemicals (EDCs) during sand filtration and ozonation at a municipal sewage treatment plant. Water Res. 41, 4373–4382.

Nienow, A.M., Hua, I., Poyer, I.C., Bezares-Cruz, J.C., Jafvert, C.T., 2009. Multifactor statistical analysis of H_2O_2-enhanced photodegradation of nicotine and phosphamidon. Ind. Eng. Chem. Res. 48, 3955–3963.

Pac, T.J., Baldock, J., Brodie, B., et al., 2019. In situ chemical oxidation: lessons learned at multiple sites. Remediat. J. 29 (2), 75–91.

Pereira, V.J., Linden, K.G., Weinberg, H.S., 2007. Evaluation of UV irradiation for photolytic and oxidative degradation of pharmaceutical compounds in water. Water Res. 41, 4413–4442.

PeroxygenTalk, 2010. Activated Persulfate Chemistry: Combined Oxidation and Reduction Mechanisms. pp. 1–4.

Pignatello, J., Baher, K., 1994. Ferric complexes as catalysts for "Fenton" degradation of 2,4-D and metolachlor in soil. J. Environ. Qual. 23, 365–370.

Rastogi, A., Al-Abed, S.R., Dionysiou, D.D., 2009. Effect of inorganic, synthetic and naturally occurring chelating agents on Fe(II) mediated advanced oxidation of chlorophenols. Water Res. 43 (3), 684–694.

Rastogi, A., Al-Abed, S.R., Dionysiou, D.D., 2009a. Sulphate radical-based ferrous peroxymonosulphate oxidative system for PCBs degradation in aqueous and sediment systems. Appl. Catal. B Environ. 85 (3-4), 171–179.

Romero, A., Santos, A., Vicente, F., Gonzalez, C., 2010. Diuron abatement using activated persulphate: effect of pH, Fe(II) and oxidant dosage. Chem. Eng. J. 162 (1), 257–265.

Rosal, R., Rodriguez, A., Perdigon-Melon, J.A., Petre, A., Garcia-Calvo, E., Gomez, M.J., Aguera, A., Fernandez-Alba, A.R., 2010. Occurrence of emerging pollutants in urban wastewater and their removal through biological treatment followed by ozonation. Water Res. 44, 578–588.

Sergio, R., Aurora, S., Arturo, R., Fernando, V., 2012. Kinetic of oxidation and mineralization of priority and emerging pollutants by activated persulfate. Chem. Eng. J. 213, 225–234.

Shemer, H., Kunukcu, Y.K., Linden, K.G., 2006. Degradation of the pharmaceutical metronidazole via UV, Fenton and photo-Fenton processes. Chemosphere 63, 269–276.

Siegrist, R.L., Crimi, M., Simpkin, T.J., 2011. In Situ Chemical Oxidation for Groundwater Remediation. New York, Springer.

Sires, I., Arias, C., Cabot, P.L., Centellas, F., Garrido, J.A., Rodriguez, R.M., Brillas, E., 2007. Degradation of clofibric acid in acidic aqueous medium by electro-Fenton and photoelectro-Fenton. Chemosphere 66, 1660–1669.

Sra, K.S., Thomson, N.R., Barker, J.F., 2014. Stability of activated persulfate in the presence of aquifer solids. Soil Sediment Contam. 23 (8), 820–837.

Teel, A.L., Cutler, L.M., Watts, R.J., 2009. Effect of sorption on contaminant oxidation in activated persulfate systems. J. Environ. Sci. Health A 44 (11), 1098–1103.

Tekin, H., Bilkay, O., Ataberk, S.S., Balta, T.H., Ceribasi, I.H., Sanin, F.D., Dilek, F.B., Yetis, U., 2006. Use of Fenton oxidation to improve the biodegradability of a pharmaceutical wastewater. J. Hazard. Mater. 136 (2), 258–265.

Torriero, A.A.J., Tonn, C.E., Sereno, L., Raba, J., 2006. Electrooxidation mechanism of non-steroidal anti-inflammatory drug piroxicam at glassy carbon electrode. J. Electroanal. Chem. 588 (2), 218–225.

Tsai, T., Kao, C.M., Yeh, T.Y., Lee, M.S., 2008. Chemical oxidation of chlorinated solvents in contaminated groundwater: a review. Pract. Period. Hazard. Toxic Radioactive Waste Manage. 12 (2), 116–126.

Tsitonaki, A., Petri, B., Crimi, M., Mosbek, H., Siegrist, R.L., Bjerg, P.P., 2010. In situ oxidation of contaminated soil and groundwater using persulfate: a review. Crit. Rev. Environ. Sci. Technol. 40, 55–91.

U.S. Environmental Protection Agency (US EPA), 2017. Superfund Remedy Report, fifteenth ed. US EPA, Washington, DC.

Venkatadri, R., Peters, R.W., 1993. Chemical oxidation technologies: ultraviolet light/hydrogen peroxide, Fenton's reagent, and titanium dioxide assisted photocatalysis. J. Hazard. Waste Hazard. Mater. 10 (2), 107–149.

Vicente, F., Santos, A., Romero, A., Rodriguez, S., 2011. Kinetic study of diuron oxidation and mineralization by persulphate: effects of temperature, oxidant concentration and iron dosage method. Chem. Eng. J. 170 (1), 127–135.

Watts, R.J., Udell, M.D., Rauch, P.A., Leung, S.W., 1990. Treatment of pentachlorophenol-contaminated soils using Fenton's reagent. Hazard. Waste Hazard. Mater. 7 (4), 335–345.

Watts, R.J., Smith, B.R., Miller, G.C., 1991. Catalyzed hydrogen peroxide treatment of octachlorodibenzo-p-oxin (OCCD) in surface soils. Chemosphere 23 (7), 949–955.

Watts, R.J., Udell, M.D., Monsen, R.M., 1993. Use of iron minerals in optimizing the peroxide treatment of contaminated soils. Water Environ. Res. 65, 839–844.
WHO, 2003. Aldrin and Dieldrin in Drinking-Water, Background Document for Preparation of WHO Guidelines for Drinking-Water Quality. World Health Organization Geneva.
Wu, X.L., Gu, X.G., Lu, S.G., Qiu, Z.F., Sui, Q., Zang, X.K., Miao, Z.W., Xu, M.H., 2015. Strong enhancement of trichloroethylene degradation in ferrous ion activated persulfate system by promoting ferric and ferrous ion cycles with hydroxylamine. Sep. Purif. Technol. 147, 186–193.
Yang, L., Yu, L.E., Ray, M.B., 2008. Degradation of paracetamol in aqueous solutions by TiO_2 photocatalysis. Water Res. 42 (13), 3480–3488.

CHAPTER

9

Composting of food waste: A novel approach

M. Anand, T.B. Anjali, K.B. Akhilesh, and John K. Satheesh

School of Environmental Studies, Cochin University of Science and Technology, Cochin, Kerala, India

OUTLINE

9.1 Introduction

India has drawn the world's attention due to the high-paced growth of industrialization, urbanization, and population. However, another aspect of higher economic development has resulted in increased waste generation and consumption of natural resources, hence ecological degradation and pollution. People are becoming more aware about the detrimental effects of currently used waste disposal methods on the environment, therefore it is high time to bring out an effective waste management system in the country. According to the Central Pollution Control Board (CPCB, 2016), urban India generated 62 Mt of municipal solid waste (MSW) in 2015, or 169,864 t/day or 450 g/capita/day. Approximately 82% (50 Mt) of MSW was collected, and the remaining 18% (12 Mt) consisted of litter. Waste treated was only 28% (14 Mt) of the collected waste, and the remaining 72% (36 Mt) was openly dumped (CPCB, 2016; MNRE, 2016). In 1947, 2001, and 2011, the total urban solid waste was 6, 31, and 48 Mt, respectively (Rawat et al., 2013; Singh et al., 2011). At this rate, the total urban MSW will be 165 Mt by 2030, 230 Mt by 2041, and 436 Mt by 2050 (Annepu, 2012; WtR (Waste to Resources), 2014).

MSW composition in India is approximately 40%–60% compostable, 30%–50% inert, and 10%–30% recyclable. According to the National Environmental Engineering Research Institute (NEERI), Indian waste consists of 0.64% ± 0.8% nitrogen, 0.67% ± 0.15% phosphorus, and 0.68% ± 0.15% potassium and has a 26 ± 5 C/N ratio (Gupta et al., 2015; Joshi and Ahmed, 2016).

Approximately 1.3 billion tonnes of food waste (FW) is produced worldwide (Pham et al., 2015). In India, FW constitutes ~ 50% of the total MSW (FW generation = ~ 30 million tonnes) (Sharma and Jain, 2019a, b). Out of a total of 62 million tonnes of MSW, 82% of waste is collected while three-fourth of the collected waste is landfilled. Organic waste which comprises of nearly 52% of the nation's municipal solid waste poses around 15065 Gg CO^2 -eq of greenhouse gas emissions to our climate through disposal sites (MoEFCC, 2018). However, as per the Planning

Cost Effective Technologies for Solid Waste and Wastewater Treatment
https://doi.org/10.1016/B978-0-12-822933-0.00007-3

Commission's report of 2014, the potential of our country to convert this waste to wealth is about 5.4 million tonnes of compost and about 72 MW of electricity from biogas (Planning Commission, 2014). The two promising technologies identified under Nationally Appropriate Mitigation Action phase 1 project for India targeting organic waste are also biomethanation and composting (GIZ-India, 2015).Conventionally, biological processes such as composting and anaerobic digestion (AD) are suggested for the treatment of wet biodegradable solid waste fraction (Sharma and Jain, 2019a, b; Manu et al., 2017; Manu et al., 2019). Composting is one of the most preferred options for the organic fraction of MSW due to its cost-effectiveness, production of nutrient-rich material, and mitigation of greenhouse gas emission (Onwosi et al., 2017). But there is a lack of environmentally sound facilities for the treatment and disposal of MSW. However, large-scale centralized composting suffers from some drawbacks such as long processing time, uneven passive air distribution, and high odor potential, whereas AD has issues pertaining to slow hydrolysis, foaming, and safe disposal of slurry. According to reports when composting and anaerobic digestion methods are compared, composting is more environment-friendly and recirculates the organic matter back into the nature. Therefore, there is an immediate need to study and develop more advanced composting technologies like in-vessel composting systems. In urban areas, large quantities of compostable wastes are produced within a small geographical area, which creates opportunity for collection and disposal. However, the density of urban areas also presents certain challenges. For one, large-scale composting requires space, which is scarce and expensive. In addition, given the likely proximity of neighbors to any site selected, the composting process must be strictly controlled in order to avoid odors and pests. For composting to be accepted as a viable alternative to landfilling and to other man-made methods like incineration, pyrolysis, thermal gasification, plasma arc, sanitary landfill, palletization, microwave, and laser waste destruction, effective separation of the organic fraction needs to be achieved. If effective waste management practices are adopted, it is possible to practice the separation of the degradable materials at the source of generation, i.e., the individual houses. Zurbrügg et al. (2004) said that "early separation of the waste flows and their decentralized recycling could contribute to reducing the waste quantities to be transported by the city and to free capacities which can be used elsewhere."

The traditional and current methods of waste disposal like large-scale centralized composting units are becoming more and more inadequate and unhealthy in pier-urban villages and rural towns in India, as we know it from our past and present experiences. Therefore, to reduce the environmental impacts of composting methods, highly sophisticated enclosed composting systems have been developed worldwide. The advantages of in-vessel systems are the following: it requires less space, provides better control of the process, and is highly efficient. In-vessel systems share the common feature that the material being composted is contained and usually enclosed. In most cases, enclosure means that the composting materials are not affected by the external environment like temperature, rainfall, and vectors. In addition, odors and leachate can be monitored, collected, and treated. Moreover, presently available composting technologies in the western countries are not suitable to handle a wide variety of source material especially for the type of food waste found in the regional settings in India, which is high in organic content and moisture with less physical structure. Secondly, these highly automated systems are not economically viable for a country like India and the maintenance costs for these units are very high due to import of spare parts. Therefore, there is a need in the art with the composting apparatus and system that is optimized and customized for the source material in the regional settings, economically viable, simple to operate, and highly efficient.

In the present study, an in-vessel composting machine was designed, developed, and optimized to be used at household and community levels. The system was made using locally available materials and indigenized technology.

9.1.1 Potential factors for in-vessel composting of food waste in India

Developing countries like India generate more food waste compared to developed countries. The characteristics of the municipal waste in India indicate a high biodegradable fraction of more than 50%. Moreover, Visvanathan et al. (2004) described that the MSW stream in most Asian countries is dominated by organic portion composed of food wastes, yard wastes, and mixed papers. The biodegradable portion of the waste mainly remained in the waste stream. The putrefying nature of food waste makes it less viable for storage and transportation. It also hinders the recovery of recyclable materials. The average moisture content is relatively high and the best disposal solution for this type of waste is an aerobic composting method. This situation has urged the need to develop and study various alternative composting technologies like in-vessel composting with locally available materials, best suited for Indian conditions. Composting has emerged as an attractive option for treating food wastes due to less environmental pollution and beneficial use of the final product (Filippi et al., 2002; Das et al., 2003; Benito et al., 2006; Zenjari et al., 2006). Food waste composting in a variety of composting systems has gained considerable attention in the last decade (Richard, 1998; Donahue et al., 1998; Nakasaki et al., 1998; Nakasaki and Ohtaki, 2002; Lemus and Lau, 2002; Das et al., 2003; Kwon and Lee, 2004; Nakasaki et al., 2004; Seo et al., 2004; Cekmecelioglu et al., 2005; Komilis and

Ham, 2006; Chang et al., 2006). Food waste has unique properties as a raw compost agent, because of high moisture content and low physical structure. However, it is important to mix food waste with a bulking agent like sawdust, yard waste, etc., which will absorb some of the excess moisture as well as add structure to the mix. Bulking agents with a high C/N ratio such as sawdust and yard waste are good choices (Chynoweth et al., 2003). In recent years, several small-to-medium-sized in-vessel composting systems for on-site application have been developed and used by schools, restaurants, and hotels. An indigenously developed in-vessel composting system has been successfully demonstrated and proved to be an efficient, eco-friendly, cost-effective, and nuisance-free solution for the management of household wastes according to (Iyengar and Bhave, 2006). Recently, Kim et al. (2008) evaluated a pilot-scale in-vessel composting for food waste treatment and found out that the final compost produced along with the bulking agent was satisfactory for agricultural application in terms of electrical conductivity as a salt content index and heavy metal contents.

9.2 Methods

9.2.1 Feedstock

The food waste was obtained from cafeteria inside the university campus. The collection timing was fixed at 11.00 a.m. in order to get a uniform composition of food waste that mainly consists of breakfast items like rice cake, dosa, sambar, chutney, poori, and also vegetable peelings. The cooked food wastes and vegetable peelings were collected separately and later blended into 50:50 percentage proportions before subjecting it to composting process, which is essential for comparison and process evaluation of different trial runs. Sawdust was used as a bulking agent to supplement carbon, absorb moisture, and maintain air space. The physical and chemical properties of kitchen waste, sawdust, and blended food waste are presented in Table 9.1.

9.2.2 Experimental design

One of the first tasks in developing a successful composting activity is getting the right combination of ingredients. Two parameters are particularly important in this regard: moisture content and the carbon-to-nitrogen (C/N) ratio. Moisture is essential to all living organisms. For most compost mixtures, 55%–60% is the recommended upper limit for moisture content; because composting is usually a drying process (through evaporation due to microbial generated heat), starting moisture contents are usually in this upper range.

The usual recommended range for C/N ratios at the start of the composting process is about 30/1, but this ideal may vary depending on the bioavailability of the carbon and nitrogen. As carbon gets converted to CO_2 (and assuming minimal nitrogen losses), the C/N ratio decreases during the composting process, with the ratio of finished compost typically close to 20/1. The desired values of moisture percentage and C/N ratio for starting the composting operation were obtained by substituting the values of mass, moisture, carbon, and nitrogen percentage of food waste and sawdust into the formula formulated by Richard and Trautmann (1996). Here the moisture percentage goal was set at 60% and the C/N ratio obtained after mixing 50 kg of food waste with 4 kg of sawdust was in the range of 30:1 to 40:1.

Initially, several bench-scale trial runs were carried out for a period of 2 years, to evaluate and optimize the composting reactor and aeration duty cycles. The final design was subjected to three consecutive experimental runs with actual kitchen waste, loaded on daily basis, for which the results are interpreted and discussed in the following sections.

TABLE 9.1 Physical and chemical properties of feedstock.

S no.	Composition	Kitchen waste	Sawdust	Blended food waste
1	Moisture %	75–85	36–40	70–78
2	pH	6.5–8.1	5.2–6.0	6.8–8.4
3	Conductivity (mV)	57–72	–	62–68
4	Carbon %	35–48	47–48	41–44
5	Nitrogen %	1.50–3.26	0.42–0.48	1.43–2.60
6	Organic matter %	31–45	50–70	30–42
7	C/N ratio	12.8–28.1	100–140	15.7–20.20

9.2.3 Aeration strategy

Intermittent aeration with positive and negative pressure aeration systems controlled by a programmable logic control unit with temperature feedback mechanism was the strategy adopted in this study. The upper temperature set point is 70°C, above which the blower and suction device will work continuously to bring down the temperature below the set point, see Table 9.2. The temperature feedback logarithm is written in LADDER LOGIC language that analyzes the chamber and ambient temperature continuously and activates the blower and suction devices according to the preprogrammed duty cycle. The schematic diagram of aeration control system is represented in Figs. 9.1 and 9.2.

9.2.4 Reactor design

The principle behind an in-vessel composting system is to provide air and moisture at a level that optimizes microbial activity as rapidly as possible and then maintains it for the desired period. In the present study, the reactor was designed and constructed based on the aforementioned principle, in order to provide optimum levels of oxygen,

TABLE 9.2 Temperature range and duty cycle.

	Duty cycle				
Temperature range (°C)	**Blower on (s)**	**Blower off (s)**	**Suction on (s)**	**Suction off (s)**	**Process control and set point**
Room temp–45	4	600	3	600	
45–70	3	900	2	900	Temperature feedback 70°C
70–room temp	2	1800	1	1800	

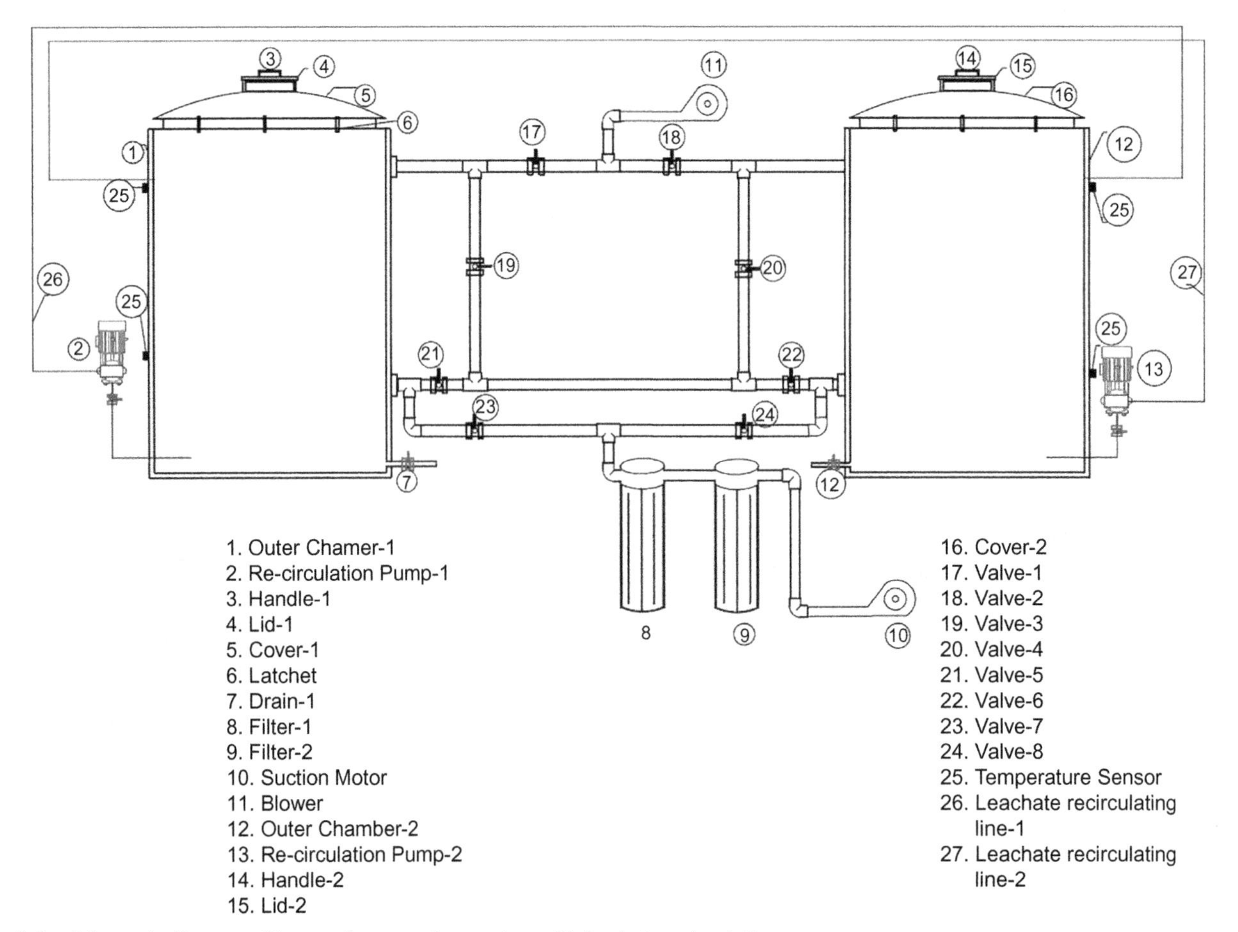

FIG. 9.1 Schematic diagram of in-vessel composting system with leachate recirculation.

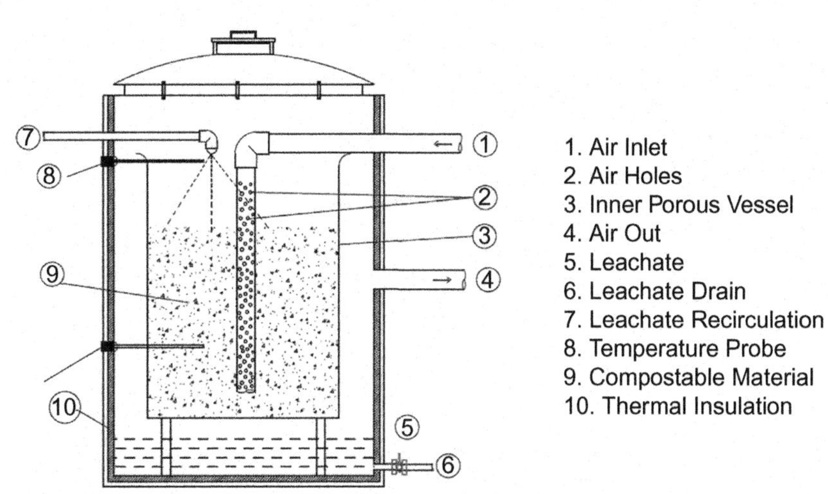

FIG. 9.2 Sectional drawing of the in-vessel composting chambers.

moisture, and temperature inside the vessel so as to encourage the growth of aerobic microorganisms for the bioconversion of organic matter in an effective way.

The innovative in-vessel composter system underwent five developmental stages, starting from a simple model made up of a conventional plastic bucket and a computer exhaust fan to the latest twin-chambered in-vessel composter with an integrated aeration system. Design I was a double-walled single-chamber system with simple aeration. This system did not meet the required operational parameters of an in-vessel composting method and it became anaerobic. Design II was a twin-chambered apparatus coupled with the negative pressure system. There were problems in maintaining a uniform aeration inside the compost matrix, and the required temperature was not attained. In Design III, the aeration method was enhanced by coupling an integrated positive and negative aeration system. This design showed good operational indices, but the temperature rise was not satisfactory. This was followed by Design IV that was provided with thermal insulation on the inner side of the reactor vessel to rectify the problem mentioned earlier. Finally, Design V, a programmable logic control unit, was added to the aeration system to synchronize the aeration duty cycle to that of microbial degradation process. The sequence of developments of the novel in-vessel composter is shown in Fig. 9.3.

9.2.5 Reactor configuration

Composting is done in perforated vessels encased by insulated and airtight vertical container made up of FRP. The unique double-walled arrangement of the unit with positive and negative aeration systems enhances the composting process. Air inlet is provided at the center of each of the perforated vessels, and the chambers are interconnected in such a way that the exhaust air from the first chamber is diverted through the second chamber that contains mature

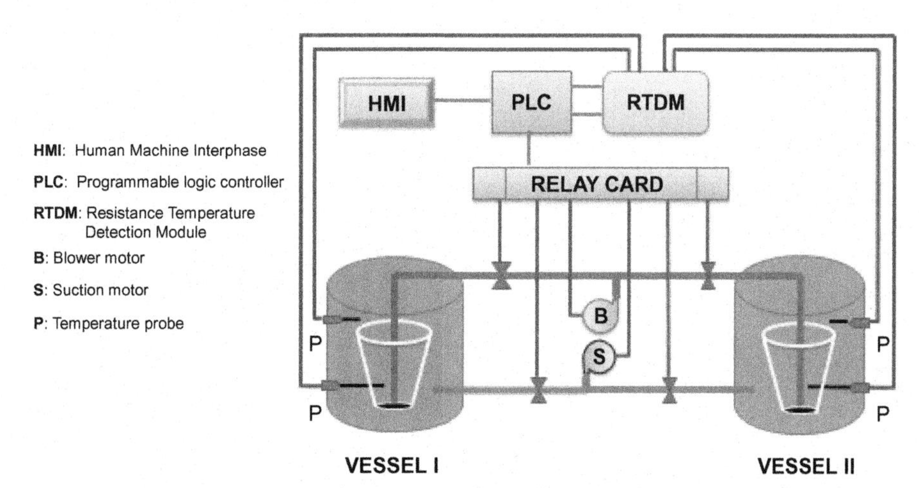

FIG. 9.3 Schematic diagram of aeration control system.

compost so as to capture noxious gases. Additionally, two external filters, one filled with activated carbon and the other with 5-micron polypropylene fibers, are provided at the suction end of the aeration system for added safety. With the twin-chamber system, the waste material can be loaded intermittently in the first chamber and vice versa for continuous stabilization and curing activity, after which the matured compost can be harvested. The schematic diagram of newly developed in-vessel composter is shown in Fig. 9.4.

The aeration system consists of interconnected pipings with manual and solenoid valves and controlled by a programmable logic control system, working on the temperature feedback method. The aeration system is programmed and operated based on the initial characteristics of the feedstock and bulking agent, in three-stage aeration duty cycles, to enhance and speed up the pre-/postmesophyllic and thermophilic stages of composting. Moisture is maintained mainly by aeration and recirculation of leachate and the excess leachate is removed through the outlet valve provided at the bottom of each chamber.

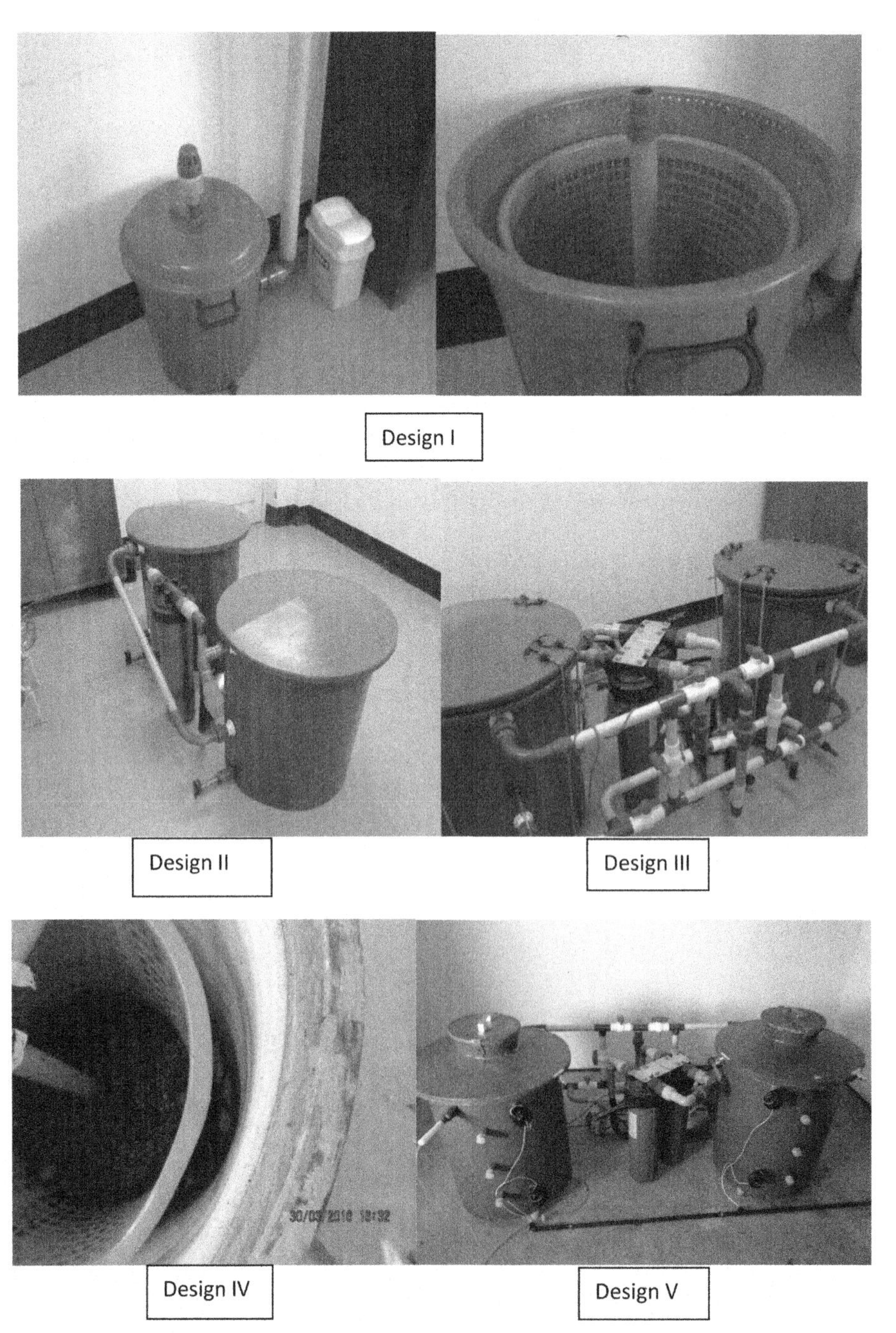

FIG. 9.4 Sequence of developments of novel in-vessel composter.

9.2.6 Operational procedure

The apparatus is self-contained to provide for a continuous input of raw waste, generating a bulk compost material of significantly less total volume and weight than the input material. About 2–3 kg of raw compost mixer was transferred to the first chamber on daily basis until it gets filled, after which it is transferred to the second chamber and vice versa. During the degradation process, the temperature build-up inside the chambers was monitored individually by the control unit and the blower and suction devices are switched on and off, as per the programmed duty cycles for respective temperature ranges. This method helps to synchronize the stabilization and curing phases of composting in the system and the finished compost can be withdrawn from the respective chambers on monthly basis.

9.2.7 Process monitoring and physicochemical analysis

The process to convert organic waste into compost is affected by many parameters to influence the composting rate, compost quality, and the generation of odorous gasses. Parameters used to monitor the progress of a composting process and the product quality include temperature, moisture content, pH, electrical conductivity, C/N ratio, organic matter content, methane, ammonia, pathogens, germination tests, and nutrient analysis.

Samples were collected from three vertical sampling ports provided on the reactor chambers, one on the upper, middle, and lower portions of the compost matrix. Samples were collected on every alternate day from the start of the experiment until 30–35 days of composting period. Samples were collected, stored, and analyzed according to the Handbook of Test Methods for the Examination of Composting and Compost by US Composting Council, 2002. Statistical analysis consists of the calculation of arithmetic means as well as standard deviations as a measure of variability of results. All the analyses were done in triplicate and the mean values were entered into spreadsheets using Microsoft Excel.

About 50 g of the wet sample is weighed and subjected for physicochemical and microbiological analyses. 25 g of the sample just mentioned is dried in a hot air oven at 105°C for 24 h for the determination of moisture, C/N ratio, and organic matter. The remaining 25 g of the sample is diluted in the ratio of 1:10 using distilled water and stirred for 2 h, filtered, and subjected for the pH, conductivity, and microbiological analyses.

Changes in temperature were measured at the intervals of 30 min through thermocouples made by RADIX, one fixed above the compost matrix inside the chamber and another one fixed at the lower middle portion that passes through the compost matrix, on both the chambers. Gas sampling was performed with a portable gas analyzer (BIOGAS5000, Geotechnical Instruments Ltd., UK) equipped with infrared absorption sensors to measure methane and carbon dioxide and with a galvanic cell-type sensor to measure oxygen. The measurements of the in situ oxygen and carbon dioxide contents at the headspace of the compost matrix inside the chambers were performed using a sampler. Acquisition time per measurement was approximately 30–60 s so as to achieve a constant reading by the gas analyzer. The oxygen and carbon dioxide concentration in the ambient air within the incubation room is measured prior to the initiation of a series of measurements. The ammonia gas was determined using UNIPHOS-282 (PM) ammonia monitor that has a 1 ppm low detection level.

9.3 Results and discussion

9.3.1 Operational indices

The operational indices like moisture content, pH, temperature, oxygen consumption, carbon dioxide evolution, and other off-gases were evaluated during a composting period of 35 days. The results indicate that moisture content remained almost constant throughout the composting period because of intermittent recirculation of leachate from the initial moisture content. Generally, the main mechanism of moisture loss is a result of produced metabolic heat and relates to aeration rate and bulking agent use. The initial moisture content at the start was 71%, which got reduced to 63 % and further to 53 % on the 10th and 21st days of composting, respectively. This reduction in moisture content is due to the evaporation loss associated with an increase in temperature as a result of metabolic activity. In the later stage, there was no much fluctuation and the composting process ended with 58 % moisture content. The results are represented in Fig. 9.5.

Fig. 9.6 indicates that the starting pH is 6.04 and dropped further during the initial weeks of composting to 5.01 due to the formation of organic acids. After this, the pH tends to move toward neutral again when these acids have been converted to carbon dioxide by the microbial action. This drop in pH value was observed for the days of loading due to the daily addition of biodegradable organic waste, and during the maturation phase, the pH subsequently rose and ended up with 6.8.

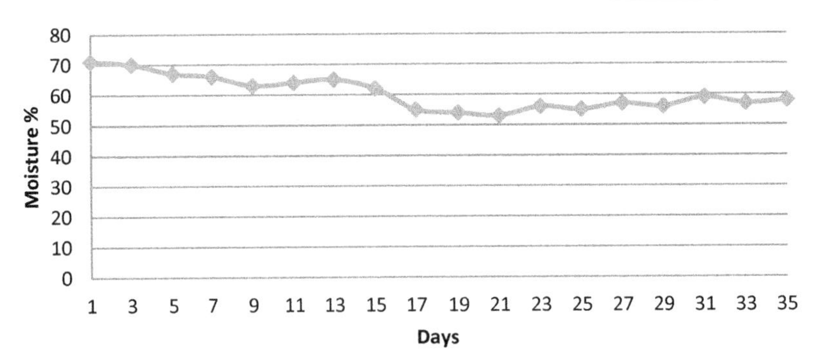

FIG. 9.5 Moisture content.

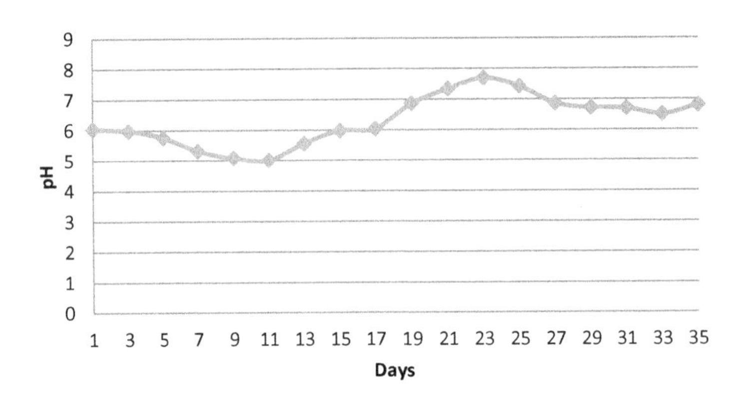

FIG. 9.6 pH.

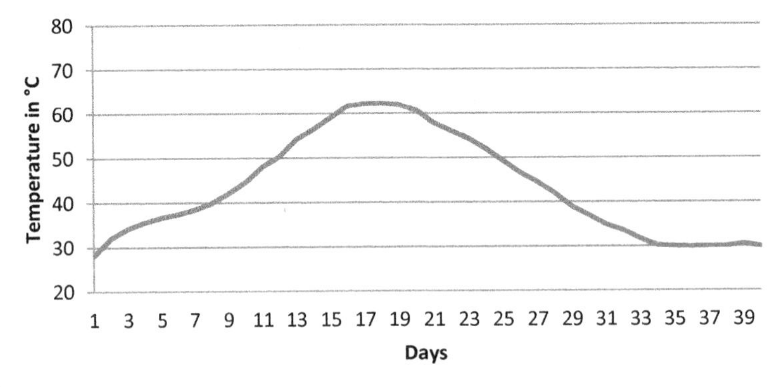

FIG. 9.7 Temperature profile.

The compost temperature (Fig. 9.7) increased from 28°C to 35°C in 2 days, and with a small lag, it increased consistently to above 60°C on 16th day, as the food waste was loaded on daily basis. The highest temperature remained for more than five days, which is sufficient to sterilize all the pathogenic microorganisms as stated by the international standards. Additionally, high temperature will facilitate the metabolism of organic matter to shorten the composting time with enhanced maturity of the final compost. The temperature started to decrease from the 21st day onward and reached 30°C on the 34th day of composting, indicating the complete stabilization of the compost material.

Fig. 9.8 shows that the oxygen consumption and carbon dioxide evolution inside the composting chambers remained almost in the ratio of 1:1, which indicates that complete aerobic reaction is taking place inside the chambers, and there are no anaerobic pockets in the compost matrix. The oxygen concentration was maintained above 10 % throughout the composting period without affecting the temperature rise and indicates the efficiency of the aeration system. The off-gases like methane and ammonia were observed in Designs I, II, and III, due to the inefficiency of the aeration system. Designs IV and V were coupled with integrated positive and negative aeration systems with fuzzy logic control system, in which these odorous gases were not detected. The details are provided in Table 9.3.

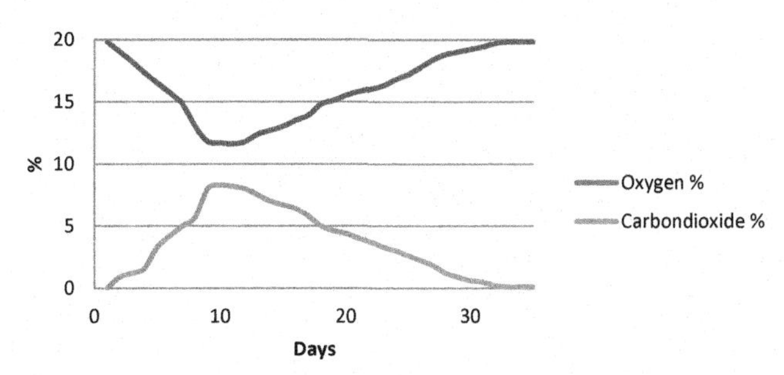

FIG. 9.8 O_2 consumption/CO_2 evolution.

TABLE 9.3 Comparison of analytical data.

Trial condition	Design I	Design II	Design III	Design IV	Design V
Feedstock in kg	20	50	50	50	50
Bulking agent in kg	2	4	4	4	4
Room temp in °C	29	28	27	27	28
pH	7.2	6.8	7.8	7.5	7.3
Initial C/N ratio	35:1	36:1	32:1	34:1	35.2
Initial moisture %	70	65	64	66	68
Max temp reached in °C	34	48	50	63	67
Max temp retained in days	–	1	1	2	4
Time for stabilization in days	–	35	33	29	30
Final C/N ratio	–	29:1	24:1	21:8	21:1
Final moisture in %	–	62	58	54	58
Methane %	3.6	3.2	ND	ND	ND
Ammonia in ppm	10	1.8	ND	ND	ND
Aeration	Fan	Suction motor	Blower and suction	Blower and suction	Blower and suction
Control	–	Analog	Analog	Analog	Logic

ND: Not detectable.

9.3.2 Compost maturity index

Compost maturity index is a very important parameter for compost production and application. Numerous maturity indices have been proposed, but no single method can be universally applied to all composts due to the variation in feedstock and composting technology (He et al., 1995). Organic matter and carbon-to-nitrogen ratios were investigated as compost maturity indices in this study.

Composting in Design V showed a higher reduction in organic matter content when compared to the previous designs. The starting organic matter content was around 60. 16% in the final run, and within 35 days, it got reduced to 19.82%. It appears that an improvement in the aeration strategy involving positive and negative aeration systems controlled by PLC favors a reduction in OM content because of higher microbial activity. The results are represented in Fig. 9.9.

In relation to the C/N ratios presented in Fig. 9.10, initial values obtained after mixing food waste with sawdust were in the range of 33.5–37.2:1, even though the ideal C/N ratio for composting is 15–30:1 as suggested by Haug (1993). In this study, a higher C/N value was preferred because of the presence of lignin content in sawdust that will reduce the available carbon for microbial degradation. After 35 days of composting, the final C/N ratio attained was 21.1, which indicates that the compost has matured.

9.3.3 Compost quality

To evaluate the quality of final compost, electrical conductivity (Fig. 9.11) as a salt content index and heavy metal content of the final compost produce were analyzed. The electrical conductivity indicates the total salt content in compost that indicates whether the salt content may affect the quality of compost to be used as a fertilizer.

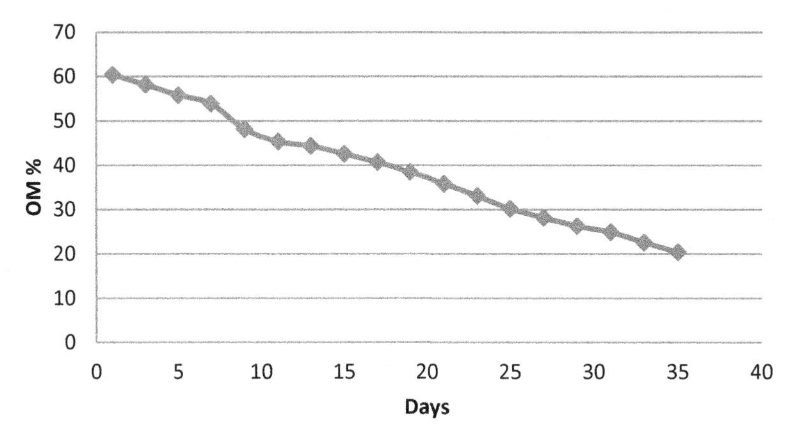

FIG. 9.9 Organic matter content (% dry matter).

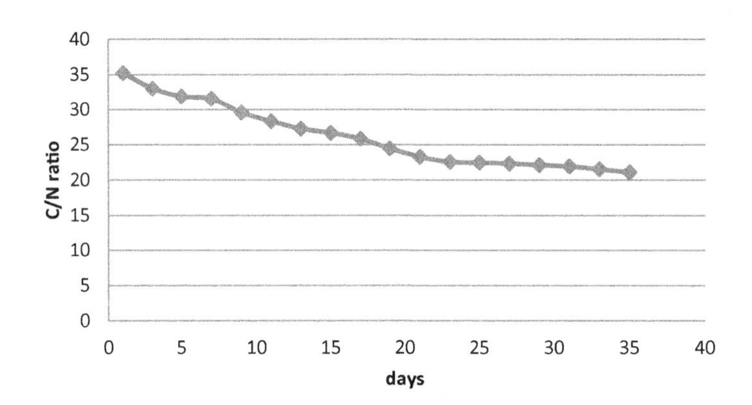

FIG. 9.10 C/N ratio.

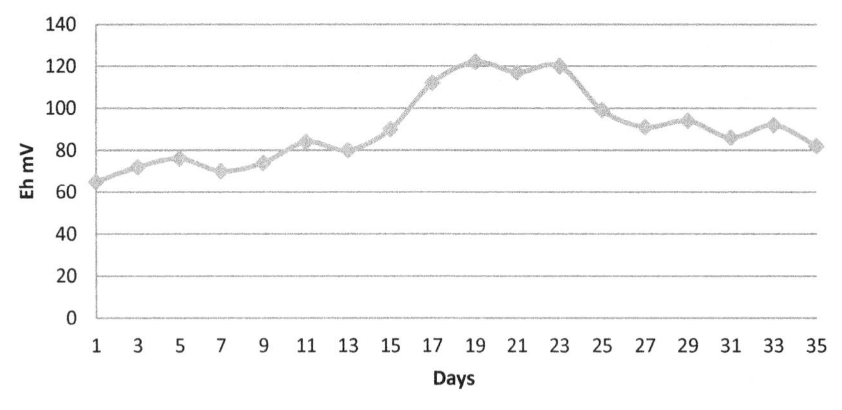

FIG. 9.11 Conductivity.

The electrical conductivity slightly increased in the initial stages of composting may be due to the release of decomposable compounds. It reached 122 mV on the 21st day from an initial value of 61 mV conductivity. The hike was there for 4 days and then started to decrease due to the decrease in microbial activity and ended up at 82 mV. Overall electrical conductivity was in the normal range.

After 35 days of composting process, the stabilized organic manure was spread over a plastic sheet for curing. Later on, the manure was subjected to nutrient analysis. The nutritive analyses are compared with the quality standards prescribed by the Dept. of Agriculture, Government of India, and all the parameters were within the limits. The comparisons are represented in Table 9.4.

The parameter to indicate the microbiological quality of water and food is "total coliforms." In this study, variations of total coliforms in the compost subject to fermentation and the final compost are monitored using the multiple-tube fermentation method. Since food waste comes from a variety of sources with no sanitary facilities to

TABLE 9.4 Compost quality evaluation.

Parameters	City/urban compost[a]	Experimental compost
Moisture, percent by weight, maximum	15.0–25.0	30
Color	Dark brown to black	Brown to black
Odor	Absence of foul odor	No foul smell
Particle minimum 90% material should pass through 4.0 mm	Minimum 90% material should pass through 4.0 mm, IS sieve	–
Bulk density (g/cc)	< 1.0	0.7
Total organic carbon percent by weight, minimum	12.0	35
Total nitrogen (as N) percent by weight, minimum	0.8	1.6
Total phosphate (as P_2O_5) percent by weight, minimum	0.4	0.1
Total potash (as K_2O) percent by weight, minimum	0.4	0.5
C/N ratio	< 20	21.1
pH (compost:water::1:2)	6.5–7.5	6.8
Conductivity (as dS m^{-1}) not more than	4.0	3.2
Pathogens	Nil	Nil
Heavy metal content (as mg/kg), maximum	Arsenic (as As_2O_3) 10.0, cadmium (as Cd) 5.0, chromium (as Cr) 50.0, copper (as Cu) 300.0, mercury (as Hg) 0.15, nickel (as Ni) 50.0, lead (as Pb) 100.0, zinc (as Zn) 1000.0	Nil

[a] *Quality Standards for City/Urban Compost as per FCO (2013) Dept. of Agriculture Cooperation, Govt. of India.*

store the material, the initial total coliforms were high. With a decrease in pH and an increasing temperature during the initial stage of composting process, the number of total coliforms is decreased. Stentiford (1996) pointed that the compost temperature maintained between 55 and 65°C would be sufficient for their total inactivation of pathogens. Lopez-Real and Baptista (1996) also reported that only a period of 3–4 days at 55°C was enough for their total elimination. The compost temperature in this study reaches as high as 60–65°C and is maintained for 5 days, which is sufficient to disinfect the compost. Phytotoxicity tests were conducted on ragi (*Eleusine coracana*) by growing the seeds on control soil substituted with 10%, 5%, and 2.5% of finished compost, and for leachate bioassay test 1:10 (w/v), compost extract was prepared and filtered, and a concentration series of 6.25%, 12.5%, 25%, 50%, and 100% was prepared using distilled water. The experiments are conducted and analyzed as per (208, Seedling Emergence and Seedling Growth Test as per **OECD** guidelines for the testing of chemicals) standard methods. Enhanced growth was observed in the shoot and root lengths at 5% compost substitution and 25% of leachate dilutions, respectively. The results are represented graphically in Fig. 9.12.

9.4 Conclusion

The management of MSW especially food waste is becoming a serious issue in the urban areas as the half-life of food waste is less than 4 h. There is an urgent need to treat this type of highly putrifiable waste on-site using an efficient technology. Here in the present study, the development of a suitable in-vessel composting system was aimed to solve the present-day solid waste management problem faced in Kerala. The preliminary results indicate that the system was found to be highly efficient, and with further research and development, the unit can be converted into a commercial model with reduced cost. Here 50 kg of food waste along with 4 kg sawdust was stabilized within 35 days on daily loading basis and after drying for 1–2 days, the final compost weighed approximately 11 kg which indicates a mass reduction of more than 80%. More studies have to be done in the area of odor and leachate management. The preference of various other bulking agents can be studied in detail based

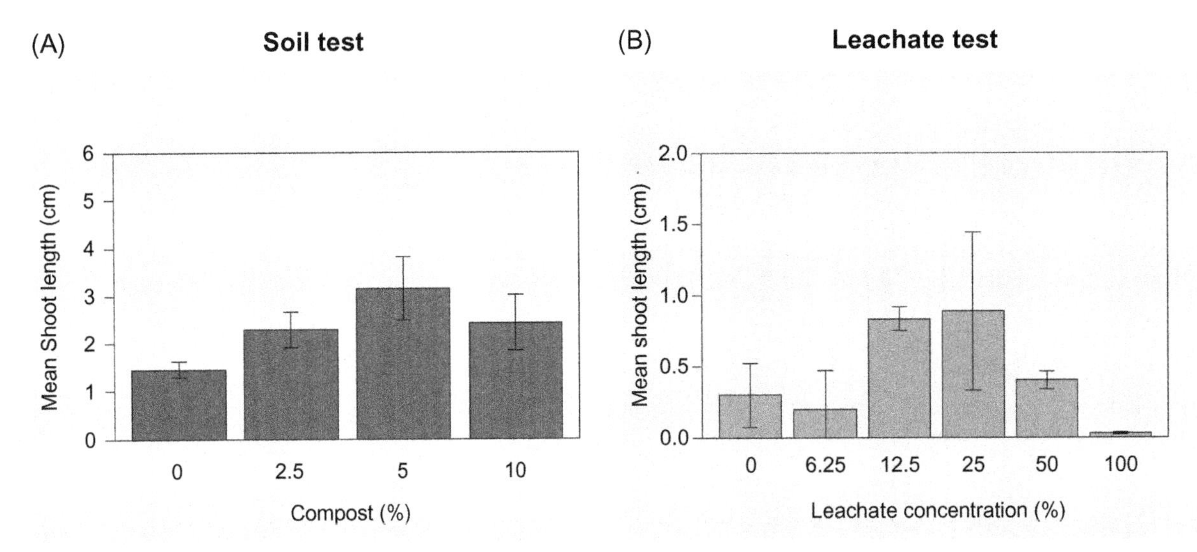

FIG. 9.12 Bioassay tests. (A) Soil test and (B) leachate test.

on their availability and available carbon content. In Kerala, composting and compost can enrich the soil, which is very important in today's scenario where all the top nutrient-rich soil is washed off during excessive rain and flooding. It is envisaged that a highly efficient in-vessel composting system can solve the present-day solid waste management problem in India.

References

Annepu, R.K., 2012. Sustainable solid waste management in India. Accessed December 29, 2016. http://www.seas.columbia.edu/earth /wtert/sofos/SustainableSolidWasteManagementinIndia_Final.pdf.

Benito, M., Masaguer, A., Moliner, A., De Antonio, R., 2006. Chemical and physical properties of pruning waste compost and their seasonal variability. Bioresour. Technol. 97 (16), 2071–2076.

Cekmecelioglu, D., Demirci, A., Graves, R.E., Davitt, N.H., 2005. Applicability of optimised in-vessel food waste composting for windrow systems. Biosyst. Eng. 91 (4), 479–486.

CPCB (Central Pollution Control Board), 2016. Central pollution control board (CPCB) bulletin, Government of India. Accessed March 14, 2017. http://cpcb.nic.in/openpdffile.php?id=TGF0ZXN0RmlsZS9MYXRlc3RfMTIzX1NVTU1BUllfQk9PS19GUy5wZGY.

Chang, J.I., Tsai, J.J., Wu, K.H., 2006. Thermophilic composting of food waste. Bioresour. Technol. 97 (1), 116–122.

Chynoweth, D.P., Sifontes, J.R., Teixeira, A.A., 2003. Sequential batch anaerobic composting of municipal and space mission waste and bioenergy crops. In: The Proceedings of ORBIT 2003, pp. 317–326.

Das, K.C., Tollner, E.W., Eiteman, M.A., 2003. Comparison of synthetic and natural bulking agents in food waste composting. Compost Sci. Util. 11 (1), 27–35.

Donahue, D.W., Chalmers, J.A., Storey, J.A., 1998. Evaluation of in-vessel composting of university postconsumer food wastes. Compost Sci. Util. 6 (2), 75–81.

Filippi, C., Bedini, S., Levi-Minzi, R., Cardelli, R., Saviozzi, A., 2002. Cocomposting of olive oil mill by-products: chemical and microbiological evaluations. Compost Sci. Util. 10 (1), 63–71.

GIZ-India, June 2015. Feasibility Study for a Waste NAMA in India.

Gupta, N., Yadav, K.K., Kumar, V., 2015. A review on current status of municipal solid waste management in India. J. Environ. Sci. 37, 206–217.

Haug, R., 1993. The Practical Handbook of Compost Engineering. Routledge.

He, X.T., Logan, T.J., Traina, S.J., 1995. Physical and chemical characteristics of selected US municipal solid waste composts. J. Environ. Qual. 24 (3), 543–552.

Iyengar, S.R., Bhave, P.P., 2006. In-vessel composting of household wastes. Waste Manag. 26 (10), 1070–1080. https://doi.org/10.1016/j.wasman.2005.06.011.

Joshi, R., Ahmed, S., 2016. Status and challenges of municipal solid waste management in India: a review. Cogent Environ. Sci. 2 (1), 1139434.

Kim, J.D., Park, J.S., In, B.H., Kim, D., Namkoong, W., 2008. Evaluation of pilot-scale in-vessel composting for food waste treatment. J. Hazard. Mater. 154 (1-3), 272–277.

Komilis, D.P., Ham, R.K., 2006. Carbon dioxide and ammonia emissions during composting of mixed paper, yard waste and food waste. Waste Manag. 26 (1), 62–70.

Kwon, S.H., Lee, D.H., 2004. Evaluation of Korean food waste composting with fed-batch operations. I. Using water extractable total organic carbon contents (TOCw). Process Biochem. 39 (10), 1183–1194.

Lemus, G.R., Lau, A.K., 2002. Biodegradation of lipidic compounds in synthetic food wastes during composting. Can. Biosyst. Eng. 44 (6), 6–33.

Lopez-Real, J., Baptista, M., 1996. A preliminary comparative study of three manure composting systems and their influence on process parameters and methane emissions. Compost Sci. Util. 4 (3), 71–82.

Manu, M.K., Kumar, R., Garg, A., 2017. Performance assessment of improved composting system for food waste with varying aeration and use of microbial inoculum. Bioresour. Technol. 234, 167–177.

Manu, M.K., Kumar, R., Garg, A., 2019. Decentralized composting of household wet biodegradable waste in plastic drums: effect of waste turning, microbial inoculum and bulking agent on product quality. J. Clean. Prod. 226, 233–241.

MNRE (Ministry of New and Renewable Energy), 2016. Power generation from municipal solid waste. Accessed December 15, 2017. http://164.100.47.193/lsscommittee/Energy/16_Energy_20.pdf.

MoEFCC, 2018. India: Second Biennial Update Report to the United Nations Framework Convention on Climate Change. Ministry of Environment, Forest and Climate Change, Government of India.

Nakasaki, K., Akakura, N., Atsumi, K., Takemoto, M., 1998. Degradation patterns of organic material in batch and fed-batch composting operations. Waste Manag. Res. 16 (5), 484–489.

Nakasaki, K., Nagasaki, K., Ariga, O., 2004. Degradation of fats during thermophilic composting of organic waste. Waste Manag. Res. 22 (4), 276–282.

Onwosi, C.O., Igbokwe, V.C., Odimba, J.N., Eke, I.E., Nwankwoala, M.O., Iroh, I.N., Ezeogu, L.I., 2017. Composting technology in waste stabilization: on the methods, challenges and future prospects. J. Environ. Manag. 190, 140–157.

Pham, T.P.T., Kaushik, R., Parshetti, G.K., Mahmood, R., Balasubramanian, R., 2015. Food waste-to-energy conversion technologies: current status and future directions. Waste Manag. 38, 399–408.

Planning Commission, May 2014. Report of the Task Force on Waste to Energy, Volume I. 2014. Report of the Task Force on Waste to Energy, Planning Commission.

Rawat, M., Ramanathan, A.L., Kuriakose, T., 2013. Characterisation of Municipal Solid Waste Compost (MSWC) From Selected Indian Cities—A Case Study for its Sustainable Utilisation.

Richard, T.L., 1998. The Kinetics of Solid-State Aerobic Biodegradation.

Tom L. Richard and Nancy M. Trautmann, 1996, Getting the Right Mix: Calculations for Thermophilic Composting. Cornell Waste Management Institute, Cornell University, Ithaca, NY, cwmi@cornell.edu. (Accessed 24 March 2012).

Seo, J.Y., Heo, J.S., Kim, T.H., Joo, W.H., Crohn, D.M., 2004. Effect of vermiculite addition on compost produced from Korean food wastes. Waste Manag. 24 (10), 981–987.

Sharma, K.D., Jain, S., 2019a. Overview of municipal solid waste generation, composition, and management in India. J. Environ. Eng. 145 (3), 04018143.

Sharma, K.D., Jain, S., 2019b. Overview of municipal solid waste generation, composition, and management in India. J. Environ. Eng. 145 (3), 04018143.

Singh, R.P., Tyagi, V.V., Allen, T., Ibrahim, M.H., Kothari, R., 2011. An overview for exploring the possibilities of energy generation from municipal solid waste (MSW) in Indian scenario. Renew. Sust. Energ. Rev. 15 (9), 4797–4808.

Stentiford, E.I., 1996. Composting control: principles and practice. In: Sequi, P., Lemmes, B., Papi, T., de Bertodli, M. (Eds.), The Science of Composting. Blackie Academic & Professional, London, pp. 49–59.

Visvanathan, C., Trankler, J., Joseph, K., Chiemchaisri, C., Basnayake, B.F.A., Gongming, Z., 2004. Municipal Solid Waste Management in Asia. Asian Regional Research Program on Environmental Technology (ARRPET). Asian Institute of Technology Publications.

WtR (Waste to Resources), 2014. Waste to Resources: A Waste Management Handbook. Energy and Resource Institute, New Delhi, India.

Zenjari, B., El Hajjouji, H., Baddi, G.A., Bailly, J.R., Revel, J.C., Nejmeddine, A., Hafidi, M., 2006. Eliminating toxic compounds by composting olive mill wastewater-straw mixtures. J. Hazard. Mater. 138 (3), 433–437.

Zurbrügg, C., Drescher, S., Patel, A., Sharatchandra, H.C., 2004. Decentralised composting of urban waste—an overview of community and private initiatives in Indian cities. Waste Manag. 24 (7), 655–662.

CHAPTER

10

Biological pretreatment for enhancement of biogas production

Aishiki Banerjee, Binoy Kumar Show*, Shibani Chaudhury, and S. Balachandran*

Department of Environmental Studies, Siksha-Bhavana, Visva-Bharati, Santiniketan, India

OUTLINE

10.1 Introduction

With the alarming rise in pollution due to the continuous use of fossil fuels and also escalating prices of petroleum and other non-renewable forms of energy required in our daily life, the quest for a promising, renewable, and recyclable source of energy has paced up in recent times (Mukhopadhyay and Chatterjee, 2010). Excessive use of fossil fuels has culminated in the accumulation of the greenhouse gas, adversely increasing the surface temperature of the earth by 5.2°C in the 21st century as reported by Sokolov et al. (2009). An immediate solution to the multifaceted problems caused by the non-renewable fuel energies requires a replacement to the less carbon-intensive fuels and substituting with renewable energy resources like solar, wind, and biomass (Chaemchuen et al., 2013).

Studies on waste management revealed that the increased generation of various solid wastes has directly affected the environment and economy in many emerging countries like India (Dhar et al., 2017). Handling of solid wastes

*These authors contributed equally to the work.

https://doi.org/10.1016/B978-0-12-822933-0.00020-6

is facing rising challenges with its improper management, making it hazardous to the society (Dhar et al., 2017). However, the organic matter in the wastes is the underlying code to a renewable energy currency (Singh et al., 2014). Hence, the surge to treat various forms of biowastes by aerobic or anaerobic degradation process for bioenergy has amplified with the demand to meet today's energy crisis in these developing nations (Evans and Wilkie, 2010). Solid organic waste management has entered a rapid developmental phase depending on factors like economic conditions, available resources, facilities, and manpower (Li et al., 2018a). The technology applied in treating these solid wastes mostly involves recycling, physical or mechanical, and biological treatments for waste-to-energy conversion (Psomopoulos et al., 2009). Biogas is considered as one of the most promising renewable sources of energy produced from the technology of converting organic wastes to methane, a universal energy carrier (Wirth et al., 2012), and has the potential to substitute the conventional source of energy. This renewable energy currency produced through anaerobic digestion recycles agricultural wastes, feedstocks, food wastes, and aquatic weeds resulting in by-products like manures and composts (Mathew et al., 2015).

Waste biomass has been established as a substitute for renewable energy, encircling several biochemical trails brought about by various biological and thermochemical conversions leading to the breakdown of complex chain carbohydrates (Iakovou et al., 2010). Plant biomass from infesting weeds can act as a promising starting material for obtaining energy. Noxious invasive weeds like *Eichhornia crassipes* (common name—water hyacinth) (Bhui et al., 2018; Priya et al., 2018; Mathew et al., 2015), *Salvinia* sp. (Syaichurrozi, 2018), *Pistia* sp. (Mthethwa et al., 2019), *Ipomoea* sp. (Stanislaus et al., 2017), *Lantana* sp. (Havilah et al., 2019), and *Ludwigia* sp. (Bhatia et al., 2020) with their fast replicating ability eventually grow into thick mats on land and waterbodies which get manifested and add to the menace causing severe ecological damage. Water hyacinth (*Eichhornia crassipes*) has been the most common aquatic macrophyte belonging to the Pontederiaceae family, the prime infesting weed in the waterbodies of many tropical and sub-tropical countries (Malik, 2007). Plant biomass is a vast repertoire for combating the recent energy crisis for fossil fuels (Chundawat et al., 2011). These plant biomasses composed of lignocellulosic material are a vast storehouse of bioenergy which can be a starting substrate for the major pathways of energy generation and power-rich chemicals.

Lignocellulosic feedstock has been the rewarding raw substrate for the production of biofuels at low costs via microbial fermentation (Herrera, 2006; Farrell et al., 2006; Malik, 2007). The lignocellulosic biofuel is cost-efficient, renewable, and eco-friendly, thus gaining more global popularity. These feedstocks can be sustainably transformed to fuels, energy-rich chemicals, and other materials by the biorefineries through a sequence of multistep processes (FitzPatrick et al., 2010). Significant amounts of polysaccharides like cellulose and hemicellulose and the complex biomolecule lignin can be converted to monomeric sugar compounds which are further valorized to produce the power chemicals (e.g., phenylpropanoids and levulinic acid) from the inexpensive waste biomass (Kawaguchi et al., 2016; Dutta et al., 2020; Abdolali et al., 2014). The main hurdles in breaking down the lignocellulose are due to the crystalline structure of cellulose surrounded by a sheath of hemicellulose having a protective outer covering composed of lignin (Horn et al., 2012; Sharma et al., 2019). Plant biomass from water weeds has attracted researchers for its high hemicellulose content which accounts for 35–55% of its dry weight which in turn can be exploited for various purposes like biogas, bioethanol, biohydrogen, biopolymer, high calorific fuel, biofertilizers, fish and animal feed, substrate for mushroom cultivation, and effluent treatment (Sharma et al., 1999; Nigam, 2002; Kumar et al., 2009; Sindhu et al., 2017; Bhui et al., 2018). The cellulose in plant biomass is hydrolyzed to reducing sugars which are then subjected to fermentation by yeast to yield alcohol. Other studies involve conversion of the hemicellulose to biogas or biomethane by the process of biomethanation (McInerney et al., 1980). In this intricate biochemical process, a great variety of diverse group of microbes participate in the gradual conversion of complex molecules to methane and carbon dioxide (Darke et al., 2002; Cirne et al., 2007; Shin and Youn, 2005). Although lignocellulosic biomass harnessing and conversion to energy faces many logistic and scientific challenges as reported by Chundawat et al. (2011), harvesting and transporting the collected biomass form a significant challenge due to the distance between refineries and pace of collection. Conversion to energy forms a significant technical challenge as carbohydrate chains of cellulose and hemicellulose contain a higher number of oxygen molecules than the crude oil; thus, reduction to form energy-rich biofuels at a low cost poses a vital test in current times (Kawaguchi et al., 2016). The biochemical processes associated with the various types of pretreatments of the lignocellulose matter have made the conversion more convenient facilitating the breakdown of lignin and the complex glycosidic bonds of cellulose and hemicellulose, making the process more economically viable (Kainthola et al., 2019). The various biochemical processes involved in generating an effective amount of biogas are discussed in the later part of this chapter. Thus, with the aim of using lignocellulosic material as a starting material for promoting long-term environmental sustainability, contemporary research is shifting toward lignocellulose depolymerization and its efficient breakdown.

10.2 Biochemical process

Methane fermentation is a complex process, which can be divided up into four phases: hydrolysis, acidogenesis, acetogenesis/dehydrogenation, and methanation (Fig. 10.1). The individual degradation steps are carried out by different consortia of microorganisms, which partly stand in syntrophic interrelation and place different requirements on the environment (Neshat et al., 2017).

10.2.1 Hydrolysis

Hydrolyzing and fermenting microorganisms are responsible for the initial attack on polymers and monomers and produce mainly acetate and hydrogen and varying amounts of volatile fatty acids such as propionate and butyrate (Fig. 10.1). Bacteria decompose long chains of complex carbohydrates, proteins, and lipids into small chains. Proteins are split into peptides and amino acids. A complex consortium of microorganisms participates in the hydrolysis and fermentation of organic material. Most of the bacteria are strict anaerobes such as *Bacteroides, Clostridia,* and *Bifidobacteria*. Furthermore, some facultative anaerobes such as *Streptococci* and *Enterobacteriaceae* also take part.

At the end of the degradation chain, two groups of methanogenic bacteria produce methane from acetate and hydrogen and CO_2. These bacteria are strict anaerobes. Only a few species are able to degrade acetate into CH_4 and CO_2, e.g., *Methanosarcina barkeri, Methanococcus mazei,* and *Methanothrix soehngenii*, whereas all the methanogenic bacteria are able to use hydrogen to form methane. The first and second groups of microbes, as well as the third and fourth

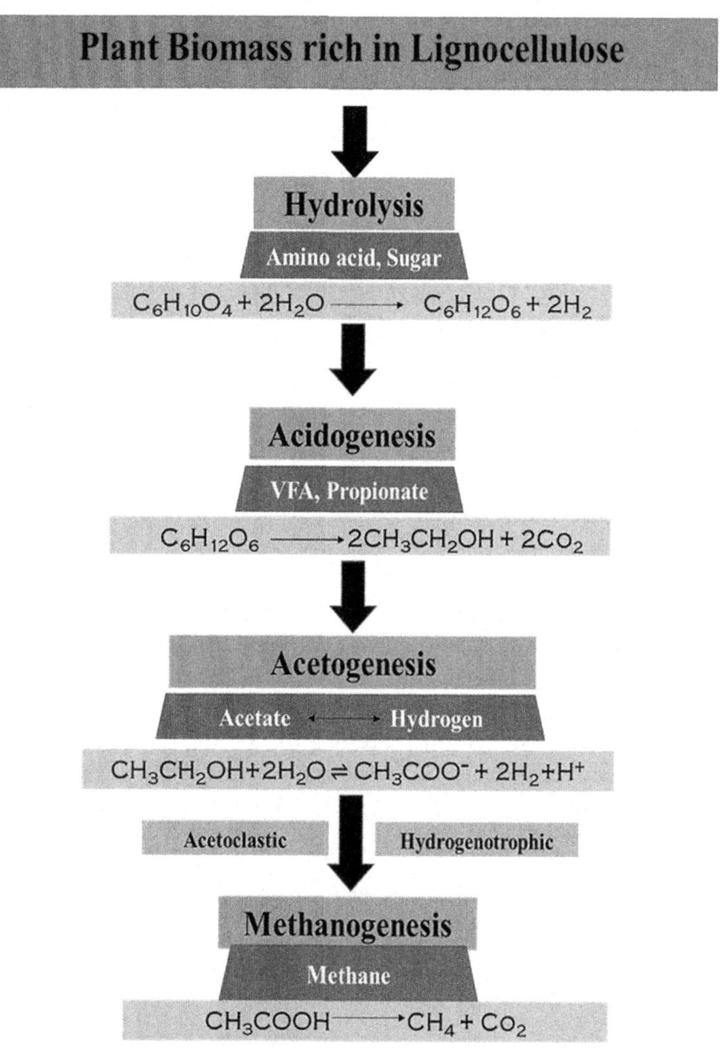

FIG. 10.1 Flowchart of anaerobic digestion.

groups, are linked closely with others (Schink, 1997). Therefore, the process can be accomplished in two stages. The effect of microbial enzymes in enhancement of anaerobic digestion (60%) by enzyme addition in batch digestion tests of sewage sludge was demonstrated by De Jonge et al. (2020). Eastman and Ferguson (1981) reported that the rate-limiting step during anaerobic digestion was the hydrolysis of waste from activated sludge.

$$C_6H_{10}O_4 + 2H_2O \Rightarrow C_6H_{12}O_6 + 2H_2$$

10.2.2 Acidogenesis

During acidogenesis, acid-producing bacteria convert the intermediates of fermenting bacteria into acetic acid, hydrogen, and carbon dioxide (Fig. 10.1). The fermentative bacteria which are involved in these steps are mainly *Hydrogenoanaerobacterium, Clostridium, Paraclostridium, Anaerosalibacter, Tissierella, Tepidanaerobacter, Ruminococcus, Selenomonas, Desulfomonas, Syntrophomonas, Syntrophobacter,* etc. (Chang et al., 2018; Uhlenhut et al., 2018). These bacteria are anaerobic or facultative anaerobic and can grow under acidic conditions. To produce acetic acid, they need oxygen and carbon. For this, they use dissolved O_2 or bound-oxygen. In a stable anaerobic digester, the primary degradation pathway leads to the formation of volatile fatty acids, acetate, alcohol, carbon dioxide, and hydrogen. The products from acidogenesis step can directly be utilized by methanogenic microorganisms (Schink et al., 2017). This step is usually considered as the fastest step in the anaerobic digestion process (Ma et al., 2013).

In the degradation pathway (Deublein and Steinhauser, 2011), the sugars are degraded to pyruvate leading to the formation of propionic acid by pyruvate degradation and the acetic acid is degraded to the formation of butyric acid.

$$\text{Acetic acid} > 2-\text{hydroxy butyrate} > \text{trans}-2-\text{butenic acid} > \text{butyric acid} > \text{butanol}.$$

The fatty acids are degraded by β-oxidation as they oxidize from coenzyme A, and two cations are separated in each step, which is set free as acetate. The amino acids are degraded by Stickland reaction by *Clostridium botulinum.* The microorganism takes two amino acids at the same time where one acts as a hydrogen donor and another as an acceptor and coupled it to acetate, ammonia, and CO_2. During the splitting of the amino acid cystine, hydrogen sulfide is formed.

$$C_6H_{12}O_6 \Leftrightarrow 2CH_3CH_2OH + 2CO_2$$

$$C_6\,H_{12}O_6 + 2H_2 \Leftrightarrow 2CH_3CH_2COOH + 2H_2O$$

$$C_6H_{12}O_6 \Rightarrow 3CH_3COOH$$

10.2.3 Acetogenesis

The process by which anaerobic bacteria like *Eubacterium, Acetogenium,* and *Clostridium* produce acetate by using a variety of energy and carbon as a substrate is known as acetogenesis. In this step, the biological oxygen demand (BOD) and the chemical oxygen demand (COD) of the biomass are reduced. By carbohydrate fermentation in acetogenesis, acetate is produced along with carbon dioxide and hydrogen. In this step, the long-chain fatty acids, which is the product of lipid hydrolysis, are oxidized into butyrate or propionate. Low partial pressure of hydrogen is essential for acetogenic reactions to be thermodynamically favorable (Schink, 1997). The reactions in acetogenesis step are endergonic. The products of acetogenesis step (acetate, CO2, H_2) are the substrates for the last step of the anaerobic digestion, methanogenesis.

$$H_3CH_2COO^- + 3H_2O \Leftrightarrow CH_3COO^- + H^+ + HCO_3^- + 3H_2$$

$$C_6H_{12}O_6 + 2H_2O \Leftrightarrow 2CH_3COOH + 2CO_2 + 4H_2$$

$$CH_3CH_2OH + 2H_2O \Leftrightarrow CH_3COO^- + 2H_2 + H^+$$

10.2.4 Methanogenesis

(Methane formation) Methane-producing bacteria, which were involved in the third step (Fig. 10.1), decompose compounds having low molecular weight. They utilize hydrogen, carbon dioxide, and acetic acid to form methane and carbon dioxide. Under natural conditions, CH_4-producing microorganisms occur to the extent that anaerobic conditions are provided, e.g., under water (e.g., in marine sediments) and in marshes. They are basically anaerobic and very sensitive to environmental changes if any occurs. The methanogenic bacteria like *Methanobacterium, Methanobrevibacter, Methanococcus, Methanomicrobium, Methanogenium,* and *Methanospirillum hungatei* belong to the archaebacter genus, i.e., to a group of bacteria with heterogeneous morphology and lot of common biochemical and molecular-biological properties that distinguish them from other bacteria. The main difference lies in the makeup of the bacteria's cell walls. During the methane production in methanogenesis, the acetate and H_2/CO_2 are converted to CH_4 and CO_2 by methanogenic archaea. The methanogenic archaea can utilize directly H_2/CO_2, acetate, and other one-carbon compounds, such as formate and methanol (Schink, 1997) for their growth and development.

Acetate-using methanogenic bacteria like *Methanosarcina barkeri, Methanobacterium söhngenii,* and *Methanobacterium thermoautotrophicum* grow in acetate very slowly, with a generation time of at least 100 h, where CO_2 has turned out to be essential for the growth. When a substrate which is rich in energy can be used, for example, methanol or methylamine, then the generation time is lower (40 h with *Methanosarcina* on methanol). However, the theoretically given generation times can be substantially longer under real conditions.

$$CO_2 + 4H_2 \Rightarrow CH_4 + 2H_2O$$

$$CH_3COOH \Rightarrow CH_4 + CO_2$$

10.3 Pretreatment of lignocellulosic biomass

Due to recalcitrance of lignocellulose to the enzymatic degradation, pretreatment is the necessary step in most biorefineries (Mosier et al., 2005; Principi et al., 2019). The nature of lignocellulose and its composition varies from one plant species to another and also with age and growing stages of the plant (Chen, 2014). The main focus of pretreating lignocellulosic biomass is to disintegrate the bonds among the cellulose hemicellulose and lignin and create smaller fragments increasing the surface area for further enzymatic digestion processes. The main barrier in lignocellulosic biomass degradation is its complex lignin structure. This complex structure can be altered by using various pretreatment techniques (Kainthola et al., 2019). According to previous studies, pretreatment can enhance the methane production greater than the untreated (da Silva Machado and Ferraz, 2017; Zhao et al., 2018). Pretreatment can be done in three ways—physical, chemical, and biological. Pretreatment helps in odor elimination, size reduction, inactivation of pathogen, increased energy recovery from the substrate, etc. (Pilli et al., 2011; Atelge et al., 2020). In this context, we mainly review the biological pretreatment part due to its cheap and environment-friendly nature.

10.3.1 Physical pretreatment

Physical pretreatment can be done in various ways like mechanical pretreatment (size reduction and separation by density) (Menardo et al., 2012), thermal pretreatment (liquid hot water pretreatment) (Atelge et al., 2020), and ultrasonic pretreatment (sonication) (Li et al., 2018b). Through physical pretreatment, the surface area of the substrate can be increased, which helps the microbes for degradation (Singh et al., 2016). The main drawback of physical pretreatment is its high energy consumption and maintenance expenses.

10.3.2 Chemical pretreatment

Chemical pretreatment is done by using various methods like (a) alkaline hydrolysis and alkaline peroxide (Bochmann and Montgomery, 2013; Kim et al., 2003); (b) organosolv process using wet oxidation (treatment with water and air or oxygen); (c) ozonolysis pretreatment method (treatment with ozone) (Salihu and Alam, 2016); and (d) acid hydrolysis pretreatment using sulfuric acid (Carlsson et al., 2012). Though the chemical pretreatment can enhance the methane production, it is not satisfactory in bioenergy industries due to its negative impact to the environment as well as an expensive procedure.

10.3.3 Biological pretreatment

Microbes which depend on lignocellulosic biomass for their nutrition help in biological pretreatment. Microbes secrete different types of enzymes which help in lignocellulose breakdown (Pérez et al., 2002). For cellulose degradation, endoglucanases, cellobiohydrolases, and β-glucosidases (Keller et al., 2020) play the critical role. Lignin can be degraded by peroxidase enzymes like lignin peroxidase (Lip) and manganese peroxidase (Mnp) (Sindhu et al., 2016), and for degradation of hemicellulose, xylanases play the critical role (Javier et al., 2007). Though the biological pretreatment method has several advantages like low energy and chemical requirement, it also has several disadvantages such as some extra storage to ensure the microbial attack after (Mansfield et al., 1999) enzyme loading is needed (Eriksson et al., 2002).

10.4 Microbial degradation of lignocellulosic biomass

Microbes play a crucial role in lignocellulose degradation in an intricate chain of biochemical pathways, reducing large complex molecules to its mineral form (Harmon et al., 1986; Strickland and Rousk, 2010; Sharma et al., 2019). All the physicochemical transitions including respiration, leaching, and fragmentation occur synergistically and are controlled by external factors like temperature, humidity, O_2/CO_2 concentration, and substrate quality including species, size, component, and position (Christy et al., 2014).

Lignocellulose is a complex biopolymer comprising 40%–60% of cellulose, 20%–40% of hemicellulose, and 10%–24% of lignin, respectively, which are strongly intermeshed by covalent and non-covalent linkages (Leonowicz et al., 1999; Pérez et al., 2002; Daorattanachai et al., 2012). The structure and function of the evident microbial communities greatly influence lignocellulose decomposition (Strickland and Rousk, 2010; Cleveland et al., 2014) and terrestrial carbon cycling greatly parameterize these processes (Allison, 2012). It is a renewable biomass that can be converted to different types of biofuels and other intermediate chemical compounds that are used as fuels or polymers (Fig. 10.2). Lignin plays a limiting factor in the process of decomposition. Due to the complexity of its structure, many challenges arise to efficiently yield biofuel with high energy density and physical and chemical characteristics similar to that of fossil fuels (Melero et al., 2012). The recalcitrant property of lignocellulose pretreatment is required before microbial degradation in order to improve its material accessibility and be used as the initial substrate for biofuel (Dai et al., 2019).

Microbial studies for the pretreatment of lignocellulosic material have been conducted by many research groups, but the use of microbes for lignocellulosic matter degradation is still far from industrial application due to the underlying problem of its complex structure (Putro et al., 2016). Most studies have approved the initial degradation of lignin to ease the process of enzymatic digestion of cellulose and must be carried out at low temperature in order to avoid carbohydrate disruption (Alvira et al., 2010; Zeng et al., 2014). The microbial degradation of lignin is known as biodelignification (Putro et al., 2016), and certain groups of bacteria and fungi that are good producers of lignocellulolytic enzymes aid in the bioconversion (Elisashvili et al., 2008; Bugg et al., 2011). In microbes, lignin is metabolized during the secondary stage, usually in the stationary phase, after the hemicellulose and water-soluble compounds being completely digested in the active growth phase (Moyson and Verachtert, 1991; Datta et al., 2017). Three main groups of fungus: white rot, brown rot, and soft rot fungi, and four classes of bacteria: actinomycetes, α-proteobacteria, β-proteobacteria, and ɣ-proteobacteria actively participate in this process (Alvira et al., 2010; Bugg et al., 2011).

10.4.1 Fungal decomposition of lignocellulose biomass

A collective group of fungi work simultaneously in wood and litter degradation, degrading celluloses and lignin present in lignocellulose (Table 10.1). While a large group of fungi are able to degrade cellulose and hemicellulose, only a smaller group of filamentous fungi have evolved to depolymerize lignin present in the complex lignocellulosic mass of plant cells (Sánchez, 2009; Datta et al., 2017). The fungal degradation of lignocellulose is directly proportional to its mycelial growth habit and takes place exocellularly either in association with the external cell envelope or completely outside the cell (Hammel, 1997; Sánchez, 2009).

Two-component enzyme systems of fungi help in the lignin degradation: (1) the hydrolases that degrade the polysaccharide chains and (2) a unique oxidative and extracellular ligninolytic system, which degrades lignin and opens phenyl rings. The most rapid degraders are the basidiomycetes (Ten Have and Teunissen, 2001; Bennet et al., 2002; Rabinovich et al., 2004), and *Phanerochaete chrysosporium* is the most studied white rot fungi with a lignocellulolytic system (Rodríguez et al., 1997; Kersten and Cullen, 2007; Quintero et al., 2006).

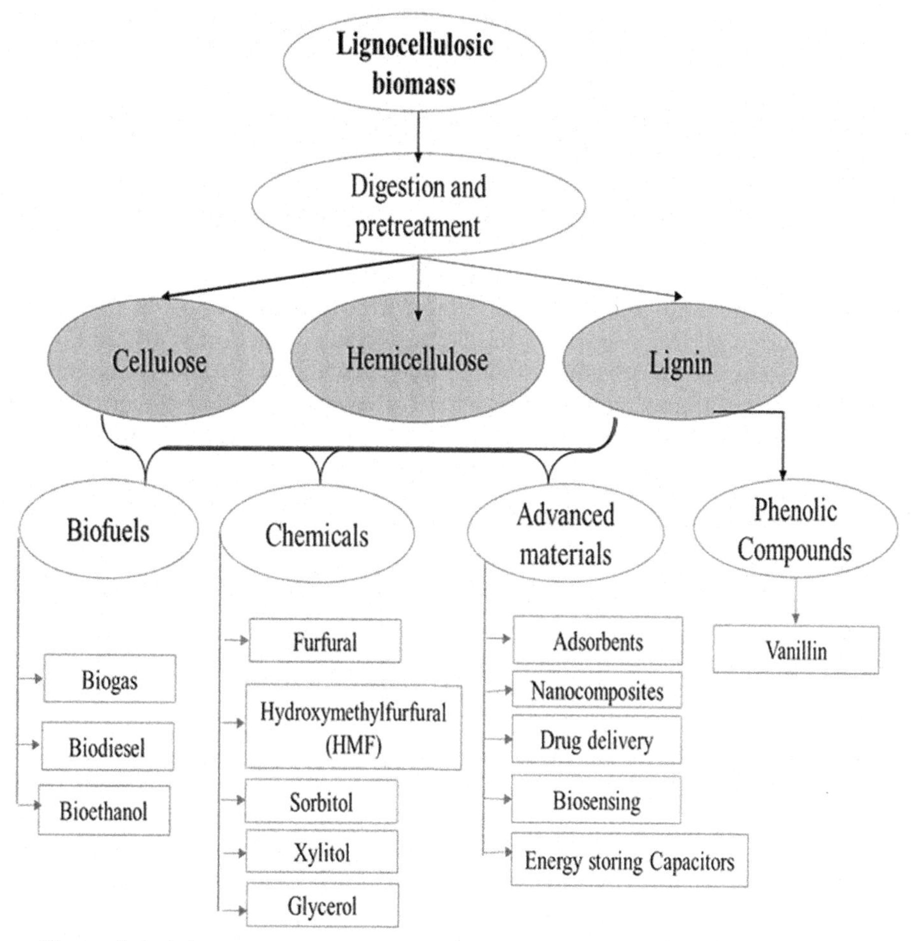

FIG. 10.2 Conversion of lignocellulosic biomass to various useful compounds.

TABLE 10.1 Examples of some lignocellulose-degrading microbes in recent works.

Lignocellulose-degrading microbes				
		Microbial Isolates	**Characteristic features**	**References**
Microbes	Bacteria	*Pseudomonas putida*, *Klebsiella pneumoniae*, and *Ochrobactrum tritici*	Kraft lignin degraders	Xu et al. (2018)
		Bathyarchaeota	Uses lignin as energy source	Yu et al. (2018)
		Brevibacillus parabrevis	Paper pulp degrader	Hooda et al. (2018)
		Planococcus sp.	Carotenoid producer and paper pulp effluent degrader	Majumdar et al. (2019)
	Fungus	*Strophariaceae*, *Tricholomataceae*, and *Bolbitiaceae*	Lignin degraders in soil	Datta et al. (2017)
		Marasmius cryptomeriae, *Phlebia tremellosa*, *Hymenochaete yasudai*, and *Stropharia rugosoannulata*	Lignin degraders of Japanese Cedar	Saito et al. (2018)
		Pleurotus ostreatus	Lignin degrader used in biological pretreatment	Mustafa et al. (2017)
		Ceriporiopsis subvermispora	Lignocellulose degrader	Liu et al. (2017)

Ligninolytic enzymes produced by the white rot fungi are (a) lignin peroxidase (LiP), (b) manganese peroxidase (MnP), (c) versatile peroxidases (VP), and (d) laccases (Howard et al., 2003; Martinez et al., 2004). All these enzymes target and actively degrade the phenolic and non-phenolic rings of lignin (Wong, 2009; Moreno et al., 2015). The effective breakdown of lignocellulose is carried out by many fungi like *Pleurotus ostreatus* and *Ceriporiopsis subvermispora* (Mustafa et al., 2017; Liu et al., 2017), but active lignin degradation is brought about only after complete hemicellulose and cellulose degradation (Kirk et al., 1976; Datta et al., 2017).

A study investigating the decomposition of water hyacinth (*Eichhornia crassipes*) biomass colonized by three species of white rot fungi (*Pleurotus ostreatus, Calocybe indica,* and *Pleurotus sajor-caju*) during solid-state fermentation proves all three organisms to be efficient degraders of the lignocellulose biomass in water hyacinth (Mukherjee et al., 2004). The weed biomass substrate was completely used for mycelial growth of the white oyster mushroom (*Pleurotus* florida) spawn prepared adding wheat grains. Commercial production of ethanol by *Saccharomyces cerevisiae* from the sugar obtained by the saccharification of water hyacinth biomass substrate was carried out in the same study (Raji et al., 2008). Significant enhancement in the yield of ethanol was obtained using pre-fermentation hydrolysis process using yeasts *S. cerevisiae* and *Pachysolen tannophilus* on water hyacinth substrate.

10.4.2 Bacterial decomposition of lignocellulose biomass

Bacterial ability of *co*-digestion of lignin along with the cellulosic polymers present in lignocellulose has been explored with certain limitations. The actinobacteria *Streptomyces* spp. have been demonstrated to *co*-metabolize lignocellulosic polymers (Vicuña, 1988) by various enzymatic pathways (Větrovský et al., 2014) while cellulose degraders have been shown to possess higher number of cellulases through a comparative genomic study (Medie et al., 2012). Lignin degradation by the strain of *Streptomyces* vir-*idosporusT7A* produces several extracellular peroxidases, which catalyze oxidative cleavage of β-aryl ether, biphenyl, diaryl propane, phenylcoumaran, and pinoresinol lignin model compounds (Ramachandra et al., 1988; Bugg et al., 2011; Putro et al., 2016). Species of *Pseudomonas, Alcaligenes, Nocardia,* and *Streptomyces* were studied to degrade single-ring aromatic compounds by Rüttimann et al. (1991). ^{14}C-labeled lignin molecule was used to study the targeted breakdown in strains of *Nocardia* and *Rhodococcus* (Zimmermann, 1990) while in the strain of *Sphingomonas paucimobilis* SYK-6 lignin breakdown catabolic pathways were extensively studied (Masai et al., 2007). Culture-dependent studies showed several pure cultured soil bacteria like *Enterobacter lignolyticus, Streptomyces viridosporus, Pseudomonas putida, Klebsiella pneumoniae,* and *Ochrobactrum tritici* to degrade lignin to low molecular weight compounds from different biomass suggesting their role in the catabolism and carbon cycling (Masai et al., 2007; DeAngelis et al., 2011a; Brown et al., 2012; Davis et al., 2013; Xu et al., 2018). The limitations in culture-dependent studies could identify mostly bacteria from three classes of Alpha- and Beta-Proteobacteria and Actinomycetes (DeAngelis et al., 2011b; Pold et al., 2015) which were much less efficient than the white rot fungal counterpart but were comparable with other lignin-degrading fungi (Ahmad et al., 2010).

However, our knowledge of the ecological decomposer's population is constrained in environmental contexts stemming from culture-dependent methods and complexity in bacterial populations. Only recently, many culture-independent studies target the lignocellulose conversion and the functional attributes of lignin-, cellulose-, and hemicellulose-degrading fungi and bacteria in coniferous forest soils with a cumulative bacterial degradation of the ^{13}C-labeled lignin model, particularly by the Gram-negative bacteria belonging to the family of Comamonadaceae and Caulobacteraceae (Wilhelm et al., 2019).

Bacterial decomposition studies of weed biomass substrate of water hyacinth and other potent aquatic weeds began in the late 80s and targeted mostly the culture-dependent approaches in identifying the isolates. In one such study, 13 bacteria were isolated from decaying litter of water hyacinth, 12 of which were reported as facultative anaerobes (Gaur et al., 1992). The bacterial decomposition initiated only after four days and then amplified exponentially thereafter. Bacterial dominance in the initial period of the macrophyte decay was well stated in the study (Singhal et al., 1991; Gamage and Asaeda, 2005). Recent studies to assess the possibilities of using mixed culture bacteria incorporated with various concentrations of sodium bicarbonate for hydrogen production from water hyacinth showed abundance of the class Firmicutes and order Clostridiales with biomass digestion speeding up with elevated levels of cellulose and xylanase (Wazeri et al., 2018). A generic survey of bacterial decomposition of water hyacinth plant parts showed leaves with a high rate of leaching at the beginning followed by petioles and roots in two sets of experiments with and without the addition of antibiotics under aerobic conditions (Gamage and Asaeda, 2005). In the current study of bacterial community from the rotary drum compost of green waste of water hyacinth, 12 bacterial isolates were obtained and identified using 16S rDNA phylogenetic study all belonging to *Bacillus* and *Enterobacter* family (Vishan et al., 2017).

10.5 Microbial enhancement of biogas

Many strains of fungi and bacteria have been found to augment the biogas production by utilizing specific enzyme activity for the target substrates (Attar et al., 1998; Christy et al., 2013). Biogas production from weed biomass mediated by the process of microbial decomposition is inexpensive when compared to the direct utilization of enzymes existing in the market. Bioenergy, a clean energy produced from the waste substances via aerobic or anaerobic decomposition, must be prioritized to usage of fossil fuels.

Aerobic decomposition is the process where oxygen acts as the terminal electron acceptor in microbial degradation of organic substrates to CO_2, nitrates, and sulfates. Aerobic decomposition occurs openly in nature and requires continuous addition of oxygen. Thus, composting in pits or stacks is possible with adequate inputs of oxygen (Christy et al., 2013).

Anaerobic decomposition is a universal technology applied for waste treatments, where the organic substrates are degraded in the absence of oxygen into biogas, and nutrients. Each step is a culmination of effective reaction and influenced by various factors. Two divisions of the kingdom Monera, the Eubacteria and the Archaebacteria, actively participate in the anaerobic digestion process (Veeken and Hamelers, 2000). Anaerobic digestion is by far the most suitable technology for waste treatment with advantages of sustainable energy production and improved environmental quality.

10.5.1 Microbes involved in the enhancement process

Microbes that are obligate or facultative anaerobes initiate the hydrolysis process belonging to the genera of *Clostridia, Micrococci, Bacteroides, Butyrivibrio, Fusobacterium, Selenomonas,* and *Streptococcus* that synthesize various hydrolyzing enzymes such as cellulase, cellobiase, and xylanase (Cirne et al., 2007). The lignocellulosic matter in the wastes is a rate-limiting step; hence, pretreatment of the lignocellulosic wastes using the suitable decomposing microorganism can be advantageous in this case (Mata-Alvarez et al., 2000). The addition of hemicellulolytic bacteria immobilized in activated zeolite showed enhanced biogas yield when compared to that of control (Weib et al., 2010). Fermentative microbes like *Streptococcus, Lactobacillus, Bacillus, Escherichia coli,* and *Salmonella* use the hydrolyzed substrates to produce organic acids such as acetic acid, propionic acid, butyric acid, and other short-chain fatty acids (Gujer and Zehnder, 1983). Synthesis of acetate, carbon dioxide, and hydrogen by the obligate hydrogen-producing acetogenic bacteria *Syntrophomonas wolfei* and *Syntrophobacter wolinii* has been identified as the key to biogas generation (Mah, 1982). Acetogen addition to the process enhances the successful biodegradation by coupling the oxidation of gaseous hydrogen to the reduction of carbon dioxide to acetate (Pester and Brune, 2007). Methanogens (*Methanospirillum hungatei, Methanoculles receptaculi, Methanosarcina thermophila*) are sensitive archaebacteria that catalyze the reaction to produce the end product methane by the process of methanogenesis. A consortium of cellulolytic and methanogenic bacteria enhanced the methanogenesis (Kalle et al., 1985) while immobilized methanogens isolated from the biodigester on a polyacrylonitrile-acrylamide support enhanced the process (Lalov et al., 2001). For improving biogas, eleven different microbial consortia with associated enzymatic activity were prepared by Sarkar et al. (2011) with significant results.

Cutting edge technology, the next-gen sequencing using SOLiD™ short-read DNA sequencing platform, was applied to study and decode sufficient useful information with functional contexts within a biogas-producing microbial community. The composition of the microbiological population and the controlling role of hydrogen metabolism appear to be the driving forces optimizing biogas-producing microbial communities (Wirth et al., 2012).

10.5.2 Microbial technologies to enhance biogas productivity

In recent studies, the enhancement of biogas production utilizing different microbial consortia has been aggravated as the need for a clean energy source has been upheaved. Extensive research is required to bring the biogas productivity to a commercial level. Scientific outlook toward the usage of microorganisms as efficient microbes is being searched to improvise and upgrade the process (Divya et al., 2015). Genetically engineered microbes were also designed in order to scale up the anaerobic digestion process (Loder et al., 2017). Several strains of yeasts were genetically modified to produce enhanced cellulolytic enzymes in order to speed up while eliminating the steps of cellulolysis. Recombinant strains of *Saccharomyces cerevisiae* were synthesized in the laboratory to simultaneously carry out saccharification and fermentation of cellulose and hemicellulose to 6 and 5 carbon sugars and finally to ethanol by producing extracellular enzymes like exoglucanase and glucosidase (Sedlak et al., 2003; Van Maris et al.,

2006). A modified *E. coli* mutant strain was engineered to consume two sugar substrates xylose and glucose that hold potential for the decomposition of variable sugar feed streams, such as lignocellulosic hydrolysates (Eiteman et al., 2008). Several approaches in cloning methanogenic bacterial genes in *E. coli* have long been targeted (Bertani and Baresi, 1987). Mutants of *Methanococcus voltae* PS, *Methanococcus maripaludis*, and *Methanosarcina* and transformed strains of *Clostridium* sp., *Klebsiella*, *E. coli*, *Lactobacillus* etc. have revealed maximum degradation activity than the wild counterparts (Divya et al., 2015). The field of recombinant DNA technology requires much expansion in terms of developing potent anaerobic microbial strains to allow the process to be expanded at a commercial level.

10.6 Future scope of research

Various areas in the biogas enhancement by microbes and the knowledge of microbial conversion of the nutrient biomass are still lacking. The effective utilization of the microbial digestion and the energy trapped in the process is still in its rudimentary stages. To make the bioenergy available to mankind, a broad area to explore is exposed for the researchers. Many challenges have been identified regarding the increase of methane content and purity of biogas that hinder the outcome and success of many projects. Pretreatment of biomass from aquatic macrophytes like water hyacinth is still in its budding stage. Although many studies claim to successfully generate and enhance biogas from various sources of lignocellulosic plant biomass, unified approaches should be targeted in a global context to bring about a revolution in this field.

10.7 Conclusion

The requirement of cleaner and sustainable source of energy has been the main target of many researchers due to the exhaustion of the fossil fuels, and also, biofuel would be an eco-friendly alternative to the otherwise harmful fossil fuels. The successful biogas production depends on its availability at low costs in various forms as electricity, gas, steam, and biodiesel. However, to bring the renewable biomass energy to the industrial level requires ample development, designing and stabilizing the process of anaerobic decomposition. The hidden knowledge in the microbial community requires decoding of the dynamic paradigms of biodegradability and various rate-limiting steps. In view of the current unprecedented pandemic situation, the global energy requirement scenario has seen a new direction. The annual IEA Global Energy Review meeting reported that global energy demand declined by 3.8% in the first quarter of 2020 with only a rise in needs for renewables due to larger installed capacity and priority dispatch. The demand for renewable energy sources is expected to increase in the days to come because of low operating costs and preferential access to many power systems.

Acknowledgment

The authors would like to acknowledge the grant received from BBSRC (BB/S011439/1) for financial support and research.

References

Abdolali, A., Guo, W.S., Ngo, H.H., Chen, S.S., Nguyen, N.C., Tung, K.L., 2014. Typical lignocellulosic wastes and by-products for biosorption process in water and wastewater treatment: a critical review. Bioresour. Technol. 160, 57–66.
Ahmad, A., Taylor, C.R., Pink, D., Burton, K., Eastwood, D., Bending, G.R., Bugg, T.D.H., 2010. Development of novel assays for lignin degradation: comparative analysis of bacterial and fungal lignin degraders. Molecular BioSystems *6*, 815–821.
Allison, S.D., 2012. A trait-based approach for modelling microbial litter decomposition. Ecol. Lett. 15, 1058–1070.
Alvira, P., Tomás-Pejó, E., Ballesteros, M., Negro, M.J., 2010. Pretreatment technologies for an efficient bioethanol production process based on enzymatic hydrolysis: a review. Bioresour. Technol. 101 (13), 4851–4861.
Atelge, M.R., Atabani, A.E., Banu, J.R., Krisa, D., Kaya, M., Eskicioglu, C., Kumar, G., Lee, C., Yildiz, Y.Ş., Unalan, S.E., Mohanasundaram, R., 2020. A critical review of pretreatment technologies to enhance anaerobic digestion and energy recovery. Fuel 270, 117494.
Attar, Y., Mhetre, S.T., Shawale, M.D., 1998. Biogas production enhancement by cellulytic strains of Actinomycetes. Biogas Forum I 72, 11–15.
Bennet, J.W., Wunch, K.G., Faison, B.D., 2002. Use of fungi biodegradation. In: Manual of Environmental Microbiology, pp. 960–971.
Bertani, G., Baresi, L., 1987. Genetic transformation in the methanogen *Methanococcusvoltae* PS. J. Bacteriol. 169 (6), 2730–2738.
Bhatia, P., Fujiwara, M., Ban, S., Toda, T., 2020. Effect of steam explosion pretreatment on methane generation from Ludwigia grandiflora. Biomass Bioenergy 142, 105771.
Bhui, I., Mathew, A.K., Chaudhury, S., Balachandran, S., 2018. Influence of volatile fatty acids in different inoculum to substrate ratio and enhancement of biogas production using water hyacinth and salvinia. Bioresour. Technol. 270, 409–415.

Bochmann, G., Montgomery, L.F., 2013. Storage and pretreatment of substrates for biogas production. In: The Biogas Handbook. Woodhead Publishing, pp. 85–103.
Brown, M.E., Barros, T., Chang, M.C., 2012. Identification and characterisation of a multifunctional dye peroxidase from a lignin-reactive bacterium. ACS Chem. Biol. 7 (12), 2074–2081.
Bugg, T.D., Ahmad, M., Hardiman, E.M., Rahmanpour, R., 2011. Pathways for degradation of lignin in bacteria and fungi. Nat. Prod. Rep. 28 (12), 1883–1896.
Carlsson, M., Lagerkvist, A., Morgan-Sagastume, F., 2012. The effects of substrate pretreatment on anaerobic digestion systems: a review. Waste Manag. 32 (9), 1634–1650.
Chaemchuen, S., Kabir, N.A., Zhou, K., Verpoort, F., 2013. Metal–organic frameworks for upgrading biogas via CO 2 adsorption to biogas green energy. Chem. Soc. Rev. 42 (24), 9304–9332.
Chang, S.E., Saha, S., Kurade, M.B., Salama, E.S., Chang, S.W., Jang, M., Jeon, B.H., 2018. Improvement of acidogenic fermentation using an acclimatized microbiome. Int. J. Hydrog. Energy 43 (49), 22126–22134.
Chen, H., 2014. Chemical composition and structure of natural lignocellulose. In: Biotechnology of Lignocellulose. Springer, Dordrecht, pp. 25–71.
Christy, P.M., Gopinath, L.R., Divya, D., 2013. A review on decomposition as a technology for sustainable energy management. Int. J. Plant Animal Environ. Sci. 3 (4), 44–50.
Christy, P.M., Gopinath, L.R., Divya, D., 2014. A review on anaerobic decomposition and enhancement of biogas production through enzymes and microorganisms. Renew. Sust. Energ. Rev. 34, 167–173.
Chundawat, S.P., Beckham, G.T., Himmel, M.E., Dale, B.E., 2011. Deconstruction of lignocellulosic biomass to fuels and chemicals. Annu. Rev. Chem. Biomol. Eng. 2, 121–145.
Cirne, D.G., Lehtomäki, A., Björnsson, L., Blackall, L.L., 2007. Hydrolysis and microbial community analyses in two-stage anaerobic digestion of energy crops. J. Appl. Microbiol. 103 (3), 516–527.
Cleveland, C.C., Reed, S.C., Keller, A.B., Nemergut, D.R., O'Neill, S.P., Ostertag, R., Vitousek, P.M., 2014. Litter quality versus soil microbial community controls over decomposition: a quantitative analysis. Oecologia 174 (1), 283–294.
da Silva Machado, A., Ferraz, A., 2017. Biological pretreatment of sugarcane bagasse with basidiomycetes producing varied patterns of biodegradation. Bioresour. Technol. 225, 17–22.
Dai, L., Wang, Y., Liu, Y., Ruan, R., Yu, Z., Jiang, L., 2019. Comparative study on characteristics of the bio-oil from microwave-assisted pyrolysis of lignocellulose and triacylglycerol. Sci. Total Environ. 659, 95–100.
Daorattanachai, P., Namuangruk, S., Viriya-empikul, N., Laosiripojana, N., Faungnawakij, K., 2012. 5-Hydroxymethylfurfural production from sugars and cellulose in acid-and base-catalysed conditions under hot compressed water. J. Ind. Eng. Chem. 18 (6), 1893–1901.
Darke, H.L., Kusel, K., Matthies, C., 2002. Ecological consequences of the phylogenetic and physiological diversities of acetogens. Antonie Van Leeuwenhook 81, 203–213.
Datta, R., Kelkar, A., Baraniya, D., Molaei, A., Moulick, A., Meena, R.S., Formanek, P., 2017. Enzymatic degradation of lignin in soil: a review. Sustainability 9 (7), 1163.
Davis, J.R., Goodwin, L., Teshima, H., Detter, C., Tapia, R., Han, C., Huntemann, M., Wei, C.L., Han, J., Chen, A., Kyrpides, N., 2013. Genome sequence of *Streptomyces viridosporus* strain T7A ATCC 39115, a lignin-degrading actinomycete. Genome Announc. 1 (4). e00416–13.
De Jonge, N., Davidsson, Å., la Cour Jansen, J., Nielsen, J.L., 2020. Characterisation of microbial communities for improved management of anaerobic digestion of food waste. Waste Manag. 117, 124–135.
DeAngelis, K.M., Allgaier, M., Chavarria, Y., Fortney, J.L., Hugenholtz, P., Simmons, B., Sublette, K., Silver, W.L., Hazen, T.C., 2011a. Characterisation of trapped lignin-degrading microbes in tropical forest soil. PLoS One 6 (4), e19306.
DeAngelis, K.M., D'Haeseleer, P., Chivian, D., Fortney, J.L., Khudyakov, J., Simmons, B., et al., 2011b. Complete genome sequence of *Enterobacter lignolyticus* SCF1. Stand. Genomic Sci. 5, 69–85.
Deublein, D., Steinhauser, A., 2011. Biogas from Waste and Renewable Resources: An Introduction. John Wiley & Sons, Second, Revised and expanded edition, Wiley-VCH Verlag GmbH & Co. KGaA, Weinheim, Germany.
Dhar, H., Kumar, S., Kumar, S., 2017. A review on organic waste to energy systems in India. Bioresour. Technol. 245, 1229–1237.
Divya, D., Gopinath, L.R., Christy, P.M., 2015. A review on current aspects and diverse prospects for enhancing biogas production in sustainable means. Renew. Sust. Energ. Rev. 42, 690–699.
Dutta, S., Yu, I.K.M., Tsang, D.C.W., Su, Z., Hu, C., Wu, K.C.W., Yip, A.C.K., Ok, Y.S., Poon, C.S., 2020. Influence of green solvent on levulinic acid production from lignocellulosic paper waste. Bioresour. Technol. 298, 122544.
Eastman, J.A., Ferguson, J.F., 1981. Solubilization of particulate organic carbon during the acid phase of anaerobic digestion. J. Water Pollut. Control Fed., 352–366.
Eiteman, M.A., Lee, S.A., Altman, E., 2008. A co-fermentation strategy to consume sugar mixtures effectively. J. Biol. Eng. 2 (1), 3–10.
Elisashvili, V., Penninckx, M., Kachlishvili, E., Tsiklauri, N., Metreveli, E., Kharziani, T., Kvesitadze, G., 2008. *Lentinus edodes* and *Pleurotus* species lignocellulolytic enzymes activity in submerged and solid-state fermentation of lignocellulosic wastes of different composition. Bioresour. Technol. 99 (3), 457–462.
Eriksson, E., Auffarth, K., Henze, M., Ledin, A., 2002. Characteristics of grey wastewater. Urban water 4 (1), 85–104.
Evans, J.M., Wilkie, A.C., 2010. Life cycle assessment of nutrient remediation and bioenergy production potential from the harvest of hydrilla (*Hydrilla verticillata*). J. Environ. Manag. 91, 2626–2631.
Farrell, A.E., Plevin, R.J., Turner, B.T., Jones, A.D., O'Hare, M., Kammen, D.M., 2006. Ethanol can contribute to energy and environmental goals. Science 311 (5760), 506–508.
FitzPatrick, M., Champagne, P., Cunningham, M.F., Whitney, R.A., 2010. A biorefinery processing perspective: treatment of lignocellulosic materials for the production of value-added products. Bioresour. Technol. 101, 8915–8922.
Gamage, N.P., Asaeda, T., 2005. Decomposition and mineralisation of *Eichhornia crassipes* litter under aerobic conditions with and without bacteria. Hydrobiologia 541 (1), 13–27.
Gaur, S., Singhal, P.K., Hasija, S.K., 1992. Relative contributions of bacteria and fungi to water hyacinth decomposition. Aquat. Bot. 43, 1–15.
Gujer, W., Zehnder, A.J., 1983. Conversion processes in anaerobic digestion. Water Sci. Technol. 15 (8–9), 127–167.
Hammel, K.E., 1997. Fungal degradation of lignin. In: Driven by Nature: Plant Litter Quality and Decomposition. vol. 33, p. 45.

Harmon, M.E., Franklin, J.F., Swanson, F.J., Sollins, P., Gregory, S.V., Lattin, J.D., Anderson, N.H., Cline, S.P., Aumen, N.G., Sedell, J.R., Lienkaemper, G.W., 1986. Ecology of coarse woody debris in temperate ecosystems. Adv. Ecol. Res. 15, 133–302 (Academic Press).

Havilah, P.R., Sharma, P.K., Gopinath, M., 2019. Combustion characteristics and kinetic parameter estimation of *Lantana camara* by thermogravimetric analysis. Biofuels 10 (3), 365–372.

Herrera, S., 2006. Bonkers about biofuels. Nat. Biotechnol. 24 (7), 755–760.

Hooda, R., Bhardwaj, N.K., Singh, P., 2018. Brevibacillus parabrevis MTCC 12105: a potential bacterium for pulp and paper effluent degradation. World J. Microbiol. Biotechnol. 34 (2), 31.

Horn, S.J., Vaaje-Kolstad, G., Westereng, B., Eijsink, V., 2012. Novel enzymes for the degradation of cellulose. Biotechnol. Biofuels 5 (1), 45.

Howard, R.L., Abotsi, E.L.J.R., Van Rensburg, E.J., Howard, S., 2003. Lignocellulose biotechnology: issues of bioconversion and enzyme production. Afr. J. Biotechnol. 2 (12), 602–619.

Iakovou, E., Karagiannidis, A., Vlachos, D., Toka, A., Malamakis, A., 2010. Waste biomass-to-energy supply chain management: a critical synthesis. Waste Manag. 30 (10), 1860–1870.

Javier, P.F., Óscar, G., Sanz-Aparicio, J., Díaz, P., 2007. Xylanases: Molecular Properties and Applications. Industrial Enzymes, Springer, Dordrecht, pp. 65–82.

Kainthola, J., Kalamdhad, A.S., Goud, V.V., 2019. A review on enhanced biogas production from anaerobic digestion of lignocellulosic biomass by different enhancement techniques. Process Biochem. 84, 81–90.

Kalle, G.P., Nayak, K.K., De Sa, C., 1985. An approach to improve methanogenesis through the use of mixed cultures isolated from biogas digester. J. Biosci. 9 (3–4), 137–144.

Kawaguchi, H., Hasunuma, T., Ogino, C., Kondo, A., 2016. Bioprocessing of bio-based chemicals produced from lignocellulosic feedstocks. Curr. Opin. Biotechnol. 42, 30–39.

Keller, M.B., Badino, S.F., Blossom, B.M., McBrayer, B., Borch, K., Westh, P., 2020. Promoting and impeding effects of lytic polysaccharide monooxygenases on glycoside hydrolase activity. ACS Sustain. Chem. Eng. 8 (37), 14117–14126.

Kersten, P., Cullen, D., 2007. Extracellular oxidative systems of the lignin-degrading basidiomycete *Phanerochaete chrysosporium*. Fungal Genet. Biol. 44 (2), 77–87.

Kim, T.H., Kim, J.S., Sunwoo, C., Lee, Y.Y., 2003. Pretreatment of corn Stover by aqueous ammonia. Bioresour. Technol. 90 (1), 39–47.

Kirk, T.K., Connors, W.J., Zeikus, J.G., 1976. Requirement for a growth substrate during lignin decomposition by two wood-rotting fungi. Appl. Environ. Microbiol. 32 (1), 192–194.

Kumar, A., Singh, A.K., Ghosh, S., 2009. Bioconversion of lignocellulosic fraction of water-hyacinth (*Eichhornia crassipes*) hemicellulose acid hydrolysate to ethanol by *Pichia stipites*. Bioresour. Technol. 100, 3293–3297.

Lalov, I.G., Krysteva, M.A., Phelouzat, J.L., 2001. Improvement of biogas production from vinasse via covalently immobilised methanogens. Bioresour. Technol. 79 (1), 83–85.

Leonowicz, A., Matuszewska, A., Luterek, J., Ziegenhagen, D., Wojtaś-Wasilewska, M., Cho, N.S., Hofrichter, M., Rogalski, J., 1999. Biodegradation of lignin by white rot fungi. Fungal Genet. Biol. 27 (2–3), 175–185.

Li, W., Loh, K.C., Zhang, J., Tong, Y.W., Dai, Y., 2018a. Two-stage anaerobic digestion of food waste and horticultural waste in high-solid system. Appl. Energy 209, 400e408.

Li, X., Guo, S., Peng, Y., He, Y., Wang, S., Li, L., Zhao, M., 2018b. Anaerobic digestion using ultrasound as pretreatment approach: changes in waste activated sludge, anaerobic digestion performances and digestive microbial populations. Biochem. Eng. J. 139, 139–145.

Liu, X., Hiligsmann, S., Gourdon, R., Bayard, R., 2017. Anaerobic digestion of lignocellulosic biomasses pretreated with *Ceriporiopsis subvermispora*. J. Environ. Manag. 193, 154–162.

Loder, A.J., Zeldes, B.M., Conway, J.M., Counts, J.A., Straub, C.T., Khatibi, P.A., Rubinstein, G.M., 2017. Extreme thermophiles as metabolic engineering platforms: strategies and current perspective. Industrial Biotechnology: Microorganisms 2, 505–580.

Ma, J., Frear, C., Wang, Z.W., Yu, L., Zhao, Q., Li, X., Chen, S., 2013. A simple methodology for rate-limiting step determination for anaerobic digestion of complex substrates and effect of microbial community ratio. Bioresour. Technol. 134, 391–395.

Mah, R.A., 1982. Methanogenesis and methanogenic partnerships. Philos. Trans. R. Soc. B, Biological Sciences 297 (1088), 599–616.

Majumdar, S., Priyadarshinee, R., Kumar, A., Mandal, T., Mandal, D.D., 2019. Exploring *Planococcus* sp. TRC1, a bacterial isolate, for carotenoid pigment production and detoxification of paper mill effluent in immobilized fluidized bed reactor. J. Clean. Prod. 211, 1389–1402.

Malik, A., 2007. Environmental challenge Vis a Vis opportunity: the case of water hyacinth. Environ. Int. 33, 122–138.

Mansfield, S.D., Mooney, C., Saddler, J.N., 1999. Substrate and enzyme characteristics that limit cellulose hydrolysis. Biotechnol. Prog. 15 (5), 804–816.

Martinez, D., Larrondo, L.F., Putnam, N., Gelpke, M.D.S., Huang, K., Chapman, J., Helfenbein, K.G., Ramaiya, P., Detter, J.C., Larimer, F., Coutinho, P.M., 2004. Genome sequence of the lignocellulose degrading fungus *Phanerochaete chrysosporium* strain RP78. Nat. Biotechnol. 22 (6), 695–700.

Masai, E., Katayama, Y., Fukuda, M., 2007. Genetic and biochemical investigations on bacterial catabolic pathways for lignin-derived aromatic compounds. Biosci. Biotechnol. Biochem. 71, 1–15.

Mata-Alvarez, J., Macé, S., Llabres, P., 2000. Anaerobic digestion of organic solid wastes. An overview of research achievements and perspectives. Bioresour. Technol. 74 (1), 3–16.

Mathew, A.K., Bhui, I., Banerjee, S.N., Goswami, R., Chakraborty, A.K., Shome, A., Balachandran, S., Chaudhury, S., 2015. Biogas production from locally available aquatic weeds of Santiniketan through anaerobic digestion. Clean Techn. Environ. Policy 17 (6), 1681–1688.

McInerney, M.J., Bryant, M.P., Stafford, D.A., 1980. Metabolic stages and energetics of microbial anaerobic digestion. In: Stafford, D.A., Wheatley, B.I., Hughes, D.E. (Eds.), Anaerobic Digestion: [Proceedings of the first International Symposium on Anaerobic Digestion, held at University College, Cardiff, Wales, September 1979]. Applied Science Publishers, London.

Medie, F.M., Davies, G.J., Drancourt, M., Henrissat, B., 2012. Genome analyses highlight the different biological roles of cellulases. Nat. Rev. Microbiol. 10 (3), 227–234.

Melero, J.A., Iglesias, J., Garcia, A., 2012. Biomass as renewable feedstock in standard refinery units. Feasibility, opportunities and challenges. Energy Environ. Sci. 5 (6), 7393–7420.

Menardo, S., Airoldi, G., Balsari, P., 2012. The effect of particle size and thermal pretreatment on the methane yield of four agricultural by-products. Bioresour. Technol. 104, 708–714.

Moreno, A.D., Ibarra, D., Alvira, P., Tomás-Pejó, E., Ballesteros, M., 2015. A review of biological delignification and detoxification methods for lignocellulosic bioethanol production. Crit. Rev. Biotechnol. 35 (3), 342–354.

Mosier, N., Wyman, C., Dale, B., Elander, R., Lee, Y.Y., Holtzapple, M., Ladisch, M., 2005. Features of promising technologies for pretreatment of lignocellulosic biomass. Bioresour. Technol. 96 (6), 673–686.

Moyson, E., Verachtert, H., 1991. Growth of higher fungi on wheat straw and their impact on the digestibility on the substrate. Appl. Microbiol. Biotechnol. 36, 421–423.

Mthethwa, N.P., Nasr, M., Kiambi, S.L., Bux, F., Kumari, S., 2019. Biohydrogen fermentation from Pistia stratiotes (aquatic weed) using mixed and pure bacterial cultures. Int. J. Hydrog. Energy 44 (33), 17720–17731.

Mukherjee, R., Ghosh, M., Nandi, B., 2004. Improvement of dry matter digestibility of water hyacinth by solid state fermentation using white rot fungi. Indian J. Exp. Biol. 42 (8), 837–843.

Mukhopadhyay, S.B., Chatterjee, N.C., 2010. Bioconversion. Bioresources 5 (2), 1301–1310.

Mustafa, A.M., Poulsen, T.G., Xia, Y., Sheng, K., 2017. Combinations of fungal and milling pretreatments for enhancing rice straw biogas production during solid-state anaerobic digestion. Bioresour. Technol. 224, 174–182.

Neshat, S.A., Mohammadi, M., Najafpour, G.D., Lahijani, P., 2017. Anaerobic co-digestion of animal manures and lignocellulosic residues as a potent approach for sustainable biogas production. Renew. Sust. Energ. Rev. 79, 308–322.

Nigam, J.N., 2002. Bioconversion of water-hyacinth (*Eichhornia crassipes*) hemicellulose acid hydrolysate to motor fuel ethanol by xylose-fermenting yeast. J. Biotechnol. 97, 107–111.

Pérez, J., Muñoz-dorado, J., de la Rubia, T., Martínez, J., 2002. Biodegradation and biological treatments of cellulose, hemicellulose. Int. Microbiol. 5, 53–63.

Pester, M., Brune, A., 2007. Hydrogen is the central free intermediate during lignocellulose degradation by termite gut symbionts. ISME J. 1 (6), 551–565.

Pilli, S., Bhunia, P., Yan, S., LeBlanc, R.J., Tyagi, R.D., Surampalli, R.Y., 2011. Ultrasonic pretreatment of sludge: a review. Ultrason. Sonochem. 18 (1), 1–18.

Pold, G., Melillo, J.M., DeAngelis, K.M., 2015. Two decades of warming increases diversity of a potentially lignolytic bacterial community. Front. Microbiol. 6, 480–492.

Principi, P., König, R., Cuomo, M., 2019. Anaerobic digestion of lignocellulosic substrates: benefits of pretreatments. Curr. Sustain./Renew. Energy Rep. 6 (3), 61–70.

Priya, P., Nikhitha, S.O., Anand, C., Nath, R.D., Krishnakumar, B., 2018. Biomethanation of water hyacinth biomass. Bioresour. Technol. 255, 288–292.

Psomopoulos, C.S., Bourka, A., Themelis, N.J., 2009. Waste-to-energy: a review of the status and benefits in USA. Waste Manag. 29 (5), 1718–1724.

Putro, J.N., Soetaredjo, F.E., Lin, S.Y., Ju, Y.H., Ismadji, S., 2016. Pretreatment and conversion of lignocellulose biomass into valuable chemicals. RSC Adv. 6 (52), 46834–46852.

Quintero, J.C., Gumersindo, F.C., Lema, J.M., 2006. Production of ligninolytic enzymes from basidiomycete fungi on lignocellulosic materials. Vitae 13 (2), 61–67.

Rabinovich, M.L., Bolobova, A.V., Vasil'chenko, 2004. Fungal decomposition of natural aromatic structures and xenobiotics: a review. Appl. Biochem. Microbiol. *40*, 1–17.

Raji, K.P., Natarajan, P., Kurup, G.M., 2008. Bioconversion of lignocellulosic residues of water hyacinth to commercial products. Bioconversion of Lignocellulosic Residues 2 (2), 261–264.

Ramachandra, M., Crawford, D.L., Hertel, G., 1988. Characterisation of an extracellular lignin peroxidase of the lignocellulolytic actinomycete *Streptomyces viridosporus*. Appl. Environ. Microbiol. 54 (12), 3057–3063.

Rodríguez, J., Ferraz, A., Nogueira, R.F., Ferrer, I., Esposito, E., Durán, N., 1997. Lignin biodegradation by the ascomycete *Chrysonilia sitophila*. Appl. Biochem. Biotechnol. 62 (2–3), 233.

Rüttimann, C., Vicuña, R., Mozuch, M.D., Kirk, T.K., 1991. Limited bacterial mineralization of fungal degradation intermediates from synthetic lignin. Appl. Environ. Microbiol. 57 (12), 3652–3655.

Saito, Y., Tsuchida, H., Matsumoto, T., Makita, Y., Kawashima, M., Kikuchi, J., Matsui, M., 2018. Screening of fungi for decomposition of lignin-derived products from Japanese cedar. J. Biosci. Bioeng. 126 (5), 573–579.

Salihu, A., Alam, M.Z., 2016. Pretreatment methods of organic wastes for biogas production. J. Appl. Sci. 16 (3), 124–137.

Sánchez, C., 2009. Lignocellulosic residues: biodegradation and bioconversion by fungi. Biotechnol. Adv. 27 (2), 185–194.

Sarkar, P., Meghvanshi, M., Singh, R., 2011. Microbial consortium: a new approach in effective degradation of organic kitchen wastes. Int. J. Environ. Sci. Develop 2 (3), 170–174.

Schink, B., 1997. Energetics of syntrophic cooperation in methanogenic degradation. Microbiol. Mol. Biol. Rev. 61 (2), 262–280.

Schink, B., Montag, D., Keller, A., Müller, N., 2017. Hydrogen or formate: alternative key players in methanogenic degradation. Environ. Microbiol. Rep. 9 (3), 189–202.

Sedlak, M., Edenberg, H.J., Ho, N.W., 2003. DNA microarray analysis of the expression of the genes encoding the major enzymes in ethanol production during glucose and xylose co-fermentation by metabolically engineered *Saccharomyces* yeast. Enzym. Microb. Technol. 33 (1), 19–28.

Sharma, A., Unni, B.G., Singh, H.D., 1999. A novel fed batch digestion system for biomethanation of plant biomasses. J. Biosci. Bioeng. 87, 678–682.

Sharma, H.K., Xu, C., Qin, W., 2019. Biological pretreatment of lignocellulosic biomass for biofuels and bioproducts: an overview. Waste Biomass Valoriz. 10 (2), 235–251.

Shin, H.S., Youn, J.H., 2005. Conversion of food waste into hydrogen by thermophilic acidogenesis. Biodegradation 16, 33–44.

Sindhu, R., Binod, P., Pandey, A., 2016. Biological pretreatment of lignocellulosic biomass–an overview. Bioresour. Technol. 199, 76–82.

Sindhu, R., Binod, P., Pandey, A., Madhavan, A., Alphonsa, J.A., Vivek, N., Gnansounou, E., Castro, E., Faraco, V., 2017. Water hyacinth a potential source for value addition: an overview. Bioresour. Technol. 230, 152–162.

Singh, R., Roy, S., Singh, S., 2014. Energy recovery from waste a review. In: Impending–Power Demand and Innovative Energy Paths. Vol. 1, pp. 298–303.

Singh, R., Krishna, B.B., Kumar, J., Bhaskar, T., 2016. Opportunities for utilization of non-conventional energy sources for biomass pretreatment. Bioresour. Technol. 199, 398–407.

Singhal, P.K., Hasija, S.K., Agarwal, G.P., 1991. Microbial degradation of lignocellulosic biomass in aquatic environments; implications to water quality and resource use. Curr. Trends Limnol. Res. 1, 37–46.

Sokolov, A.P., Stone, P.H., Forest, C.E., Prinn, R.G., Sarofim, M.C., Webster, M., Paltsev, S., Schlosser, C.A., Kicklighter, D., Dutkiewicz, S., Reilly, J.M., 2009. Probabilistic forecast for twenty-first-century climate based on uncertainties in emissions (without policy) and climate parameters. J. Clim. 22 (19), 5175–5204.

Stanislaus, M.S., Zhang, N., Zhao, C., Zhu, Q., Li, D., Yang, Y., 2017. Ipomoea aquatica as a new substrate for enhanced biohydrogen production by using digested sludge as inoculum. Energy 118, 264–271.

Strickland, M.S., Rousk, J., 2010. Considering fungal: bacterial dominance in soils–methods, controls, and ecosystem implications. Soil Biol. Biochem. 42 (9), 1385–1395.

Syaichurrozi, I., 2018. Biogas production from co-digestion *Salvinia molesta* and rice straw and kinetics. Renew. Energy 115, 76–86.

Ten Have, R., Teunissen, P.J., 2001. Oxidative mechanisms involved in lignin degradation by white-rot fungi. Chem. Rev. 101 (11), 3397–3414.

Uhlenhut, F., Schlüter, K., Gallert, C., 2018. Wet biowaste digestion: ADM1 model improvement by implementation of known genera and activity of propionate oxidizing bacteria. Water Res. 129, 384–393.

Van Maris, A.J., Abbott, D.A., Bellissimi, E., van den Brink, J., Kuyper, M., Luttik, M.A., Wisselink, H.W., Scheffers, W.A., van Dijken, J.P., Pronk, J.T., 2006. Alcoholic fermentation of carbon sources in biomass hydrolysates by *Saccharomyces cerevisiae*: current status. Antonie Van Leeuwenhoek 90 (4), 391–418.

Veeken, A.H.M., Hamelers, B.V.M., 2000. Effect of substrate-seed mixing and leachate recirculation on solid state digestion of biowaste. Water Sci. Technol. 41 (3), 255–262.

Větrovský, T., Steffen, K.T., Baldrian, P., 2014. Potential of cometabolic transformation of polysaccharides and lignin in lignocellulose by soil actinobacteria. PLoS One 9, e89108.

Vicuña, R., 1988. Bacterial degradation of lignin. Enzym. Microb. Technol. 10 (11), 646–655.

Vishan, I., Sivaprakasam, S., Kalamdhad, A., 2017. Isolation and identification of bacteria from rotary drum compost of water hyacinth. Int. J. Recycl. Organic Waste Agric. 6 (3), 245–253.

Wazeri, A., Elsamadony, M., Le Roux, S., Peu, P., Tawfik, A., 2018. Potentials of using mixed culture bacteria incorporated with sodium bicarbonate for hydrogen production from water hyacinth. Bioresour. Technol. 263, 365–374.

Weib, S., Tauber, M., Somitsch, W., Meincke, R., Müller, H., Berg, G., Guebitz, G.M., 2010. Enhancement of biogas production by addition of hemicellulolytic bacteria immobilised on activated zeolite. Water Res. 44 (6), 1970–1980.

Wilhelm, R.C., Singh, R., Eltis, L.D., Mohn, W.W., 2019. Bacterial contributions to delignification and lignocellulose degradation in forest soils with metagenomic and quantitative stable isotope probing. The ISME J. 13 (2), 413–429.

Wirth, R., Kovács, E., Maróti, G., Bagi, Z., Rákhely, G., Kovács, K.L., 2012. Characterisation of a biogas-producing microbial community by short-read next generation DNA sequencing. Biotechnol. Biofuels 5 (1), 41–56.

Wong, D.W., 2009. Structure and action mechanism of ligninolytic enzymes. Appl. Biochem. Biotechnol. 157 (2), 174–209.

Xu, Z., Qin, L., Cai, M., Hua, W., Jin, M., 2018. Biodegradation of Kraft lignin by newly isolated *Klebsiella pneumoniae*, *Pseudomonas putida*, and *Ochrobactrum tritici* strains. Environ. Sci. Pollut. Res. 25 (14), 14171–14181.

Yu, T., Wu, W., Liang, W., Lever, M.A., Hinrichs, K.U., Wang, F., 2018. Growth of sedimentary *Bathyarchaeota* on lignin as an energy source. Proc. Natl. Acad. Sci. 115 (23), 6022–6027.

Zeng, Y., Zhao, S., Yang, S., Ding, S.Y., 2014. Lignin plays a negative role in the biochemical process for producing lignocellulosic biofuels. Curr. Opin. Biotechnol. 27, 38–45.

Zhao, C., Zou, Z., Li, J., Jia, H., Liesche, J., Chen, S., Fang, H., 2018. Efficient bioethanol production from sodium hydroxide pretreated corn Stover and rice straw in the context of on-site cellulase production. Renew. Energy 118, 14–24.

Zimmermann, W., 1990. Degradation of lignin by bacteria. J. Biotechnol. 13 (2–3), 119–130.

CHAPTER

11

Recent trends in bioremediation of pollutants by enzymatic approaches

M. Srinivasulu[a], *M. Subhosh Chandra*[b], *Naga Raju Maddela*[c], *Narasimha Golla*[d], *and Bellamkonda Ramesh*[e]

[a]Department of Biotechnology, Yogi Vemana University, Kadapa, Andhra Pradesh, India [b]Department of Microbiology, Yogi Vemana University, Kadapa, Andhra Pradesh, India [c]Departamento de Ciencias Biológicas, Facultad de Ciencias de la Salud, Universidad Técnica de Manabí, Portoviejo, Manabí, Ecuador [d]Department of Virology, Sri Venkateswara University, Tirupati, Andhra Pradesh, India [e]Department of Food Technology, Vikrama Simhapuri University, Nellore, Andhra Pradesh, India

OUTLINE

11.1 Introduction

Bioremediation is the remediation of unwanted materials from the polluted site of interest by using the biological sources such as plants and microorganisms or the substances released from them. Although it is a natural process, many interventions can be introduced to make it rapid, efficient, and specific for the degradation of the pollutants at the polluted sites (Hlihor et al., 2017; Iwamoto and Nasu, 2001; Shankar et al., 2011; Vidali, 2001). With increase in scientific knowledge, socioeconomic perception, human health problems, and ecological apprehensions, people are

https://doi.org/10.1016/B978-0-12-822933-0.00018-8

more concerned about the widespread environmental contaminants. Therefore, the occurrence of newly identified contaminants and emerging pollutants (EPs) or emerging contaminants (ECs) in our major waterbodies is of continued and burning concern globally. The undesirable EPs or ECs are discharged intentionally/unintentionally with/without partial treatments into aquatic environments that cause serious health problems and affect the entire living ecosystem (Ishtiaq et al., 2017).

The worldwide population explosion has obligated the production of more and more grains, fibers, and medicines and the construction of additional houses. The revolution in the industrial sector tied with information technology has brought a remarkable load for the environment to carry. The capacity to supply food to the current population of the world has been exceeded by the demand. We must apply pesticides, insecticides, herbicides and other plant growth promoters (PGPs), and other protecting agents to obtain high yields, and all of these have harmful effects on the environmental health. Domestic, industrial, and agricultural processes irregularly add a variety of polluting agents to the environment, such as polychlorinated biphenyls, chlorinated solvents, chlorinated phenols, polyaromatic hydrocarbons, explosives, heavy metals, radionuclides, biocides, oxidized metals, plastics, fertilizers, and dyes (Damlas and Eleftherohorinos, 2011; Lew et al., 2013; Nicolopoulou-Stamati et al., 2016; Wasim et al., 2009). For instance, the compounds like dieldrin and aldrin; organophosphates such as parathion and malathion; 1,2,3-trichloropropane; lindane; dichloro diphenyl trichloro ethane (DDT); and hexachlorohexane cause contamination in soil, water, and air and can result serious diseases in humans and animals, or impact the biodiversity as entire (Jayaraj et al., 2016; Loredana et al., 2017; Saadati et al., 2012; Setyo et al., 2017).

Bioremediation of pollutants by microorganisms, due to the smaller size, enables them to contact contaminants very easily, facilitating rapid and efficient degradation or reduction to an acceptable or less hazardous state (Hong et al., 2018). A variety of factors, such as (1) type of organisms, (2) type of contaminants, and (3) geological and chemical conditions at the polluted sites, influence the accomplishment of bioremediation (Dixit et al., 2015). The compounds like radionuclides and heavy metals cannot be biodegraded but can be reduced to a less hazardous state, as a result posing a limitation on the appliance of bioremediation as it is limited to the biodegradable materials (Ayangbenro and Babalola, 2017). The sources and the schematic mechanism of heavy metals' bioremediation are presented in Figs. 11.1 and 11.2, respectively. The enzymatic bioremediation refers to the application of naturally available enzymes in the microbes or plants to degrade or reduce hazardous, unwanted, and recalcitrant environmental contaminants in order to clean the polluted sites (Chandrakant and Shwetha, 2011).

The enzymes are biocatalysts which lower the activation energy and facilitate rapid and total breakdown of the substrates (Fan and Krishnamurthy, 1995). The enzyme-mediated bioremediation method has evolved over time to deal with slow growth-dependent microbial bioremediation and to minimize the toxic waste produced by the microorganisms themselves. It is a more eco-friendly method with the advantage of specific degradation abilities

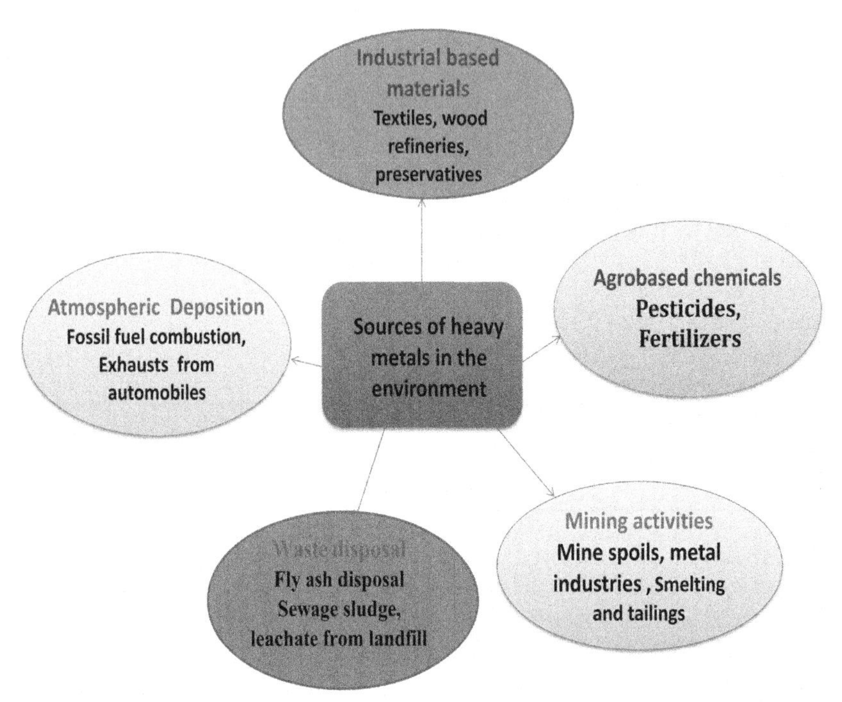

FIG. 11.1 Sources of heavy metals in the environment.

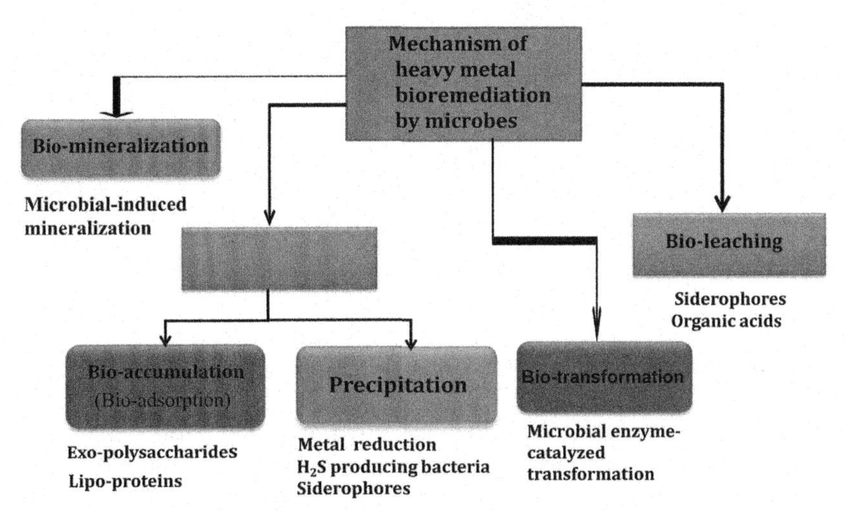

FIG. 11.2 Bioremediation of heavy metals in the contaminated soil.

with improved action on the compounds of interest. The smaller size of enzymes when compared to microbial cells enables them to contact pollutants easily, aiding quicker mobility, increased contact with pollutants, and facilitating more targeted quick and efficient degradation or reduction to an admissible or less hazardous state (Rao et al., 2010). Sharma et al. (2018) observed the function of different enzymes (oxygenases, peroxidases, laccases, haloalkane carboxylesterases, dehalogenases, lipases, phosphotriesterases, and cellulases) in the degradation pathways of various contaminants.

Enzymes are the biological catalysts that play a key role in the conversion of substrates into the products. Enzymes may be made with a protein or glycoprotein and contain at least one polypeptide moiety. The part of enzyme that is directly involved in the catalytic process is known as the active site. Enzymes may be having one or many groups that are necessary for the catalytic activity connected with active sites by either covalent or non-covalent bonds, and the protein or glycoprotein moiety in such enzyme is known as the apoenzyme, whereas the non-protein part is known as the prosthetic group. The apoenzyme with the prosthetic group is called the holoenzyme. A number of enzymes produced from the bacteria, fungi, and plants play a major role in the biodegradation of lethal organic contaminants. The bioremediation is a cost-effective and environment-friendly biotechnology that is driven by the microbial enzymes. Research findings in this particular area can contribute to develop the superior bioprocess technology for reducing the toxicity of contaminants and also to get valuable substances (Chandrakant and Shwetha, 2011).

11.2 Microbial oxidoreductases

Detoxification of the toxic organic compounds by different bacteria, fungi (Gianfreda et al., 1999), and higher plants (Bollag and Dec, 1998) through oxidative coupling is mediated with oxidoreductases. The microorganisms get energy from the biochemical reactions, which are mediated via these enzymes to cleave the chemical bonds and to help the carry of electrons from the reduced organic substrate (i.e., donor) to other chemical compound (acceptor). Oxidoreductases are involved in the humification of various phenolic compounds that are formed by the decomposition of lignin in the soil environment. Likewise, the oxidoreductases can also detoxify the toxic pollutants, like phenolic/anilinic compounds, by the polymerization and copolymerization with other substrates or binding to the humic materials (Park et al., 2006). Microbial enzymes have been exploited in the decolorization and degradation of azo dyes (Vidali, 2001; Williams, 1977; Husain, 2006). Several bacteria reduce the radioactive metals from the oxidized soluble state to a reduced insoluble form. During the course of energy production, bacterium takes up electrons from organic compounds and utilizes radioactive metal as the final electron acceptor. Few bacterial species reduces the radioactive compounds indirectly by the intermediate electron donor. Ultimately, precipitant is observed as a result of the redox reactions in the metal-reducing bacteria (Leung, 2004). The chlorinated phenolic substances are the largely abundant recalcitrant wastes, which are present in the effluents released by the pulp and paper industries. These compounds are produced by the incomplete degradation of lignin during the pulp bleaching process. The numerous fungal species are the most appropriate for the elimination of chlorinated phenolic compounds from the polluted sites in the environment. The fungal activity is mainly because of action of the extracellular oxidoreductase enzymes, such as lignin peroxidase, laccase, and manganese peroxidase that are produced from the fungal mycelium and released into their surrounding environment. Being filamentous, fungi can reach soil pollutants more efficiently than bacteria (Rubilar et al., 2008). The water polluted with phenolic compounds is decontaminated by enzymes exuded by plant roots. Plants related to *Fabaceae, Solanaceae,* and *Gramineae* families release oxidoreductases which take part in the oxidative degradation of certain soil components. The phytoremediation of organic pollutants has usually focused on three classes of compounds, explosives, chlorinated solvents, and petroleum hydrocarbons (Duran and Esposito, 2000; Newman et al., 1998). The overview of bioremediation of various contaminants in the environment is presented in Fig. 11.3.

11.3 Microbial oxygenases

Oxygenases are part of the oxidoreductases family. Hence they bolonging to the oxidoreductases. These enzymes participate in the oxidation of reduced substrates by transferring oxygen from the molecular oxygen (O_2) utilizing FAD/NADH/NADPH as co-substrate. The oxygenases are grouped into two categories: monooxygenases and dioxygenases based on the number of oxygen atoms utilized for oxygenation. They play very important role in the metabolism of organic compounds with increasing their reactivity/water solubility or bring about the cleavage of aromatic ring. The oxygenases have a broad substrate range and are active against wide range of compounds, including chlorinated aliphatic compounds. Frequently, the introduction of oxygen atoms into the organic molecule by the oxygenase enzyme results in the aromatic ring cleavage. Traditionally, the major studied enzymes in bioremediation are bacterial monooxygenases or dioxygenases. The detailed study of the role of oxygenases in biodegradation process is reported (Arora et al., 2009; Fetzner and Lingens, 1994; Fetzner, 2003).

Halogenated organic compounds contain the major groups of pollutants as a result of their extensive application as fungicides, insecticides, herbicides, plasticizers, and intermediates for the chemical synthesis. The degradation of these contaminants is achieved by specific oxygenases. The oxygenase enzymes can also mediate the dehalogenation reactions of halogenated compounds such as ethanes, methanes, and also ethylenes in connection with multifunctional enzymes (Fetzner and Lingens, 1994).

11.3.1 Monooxygenases

The monooxygenases integrate one atom of oxygen molecule into the substrate. The monooxygenases are grouped into two subclasses on the basis of presence of cofactor: flavin-dependent monooxygenase and P450 monooxygenase. The flavin-dependent monooxygenases contain flavin as prosthetic group and require NADPH or NADP as coenzyme. The P450 monooxygenases are heme-containing oxygenases that are present in both prokaryotic and

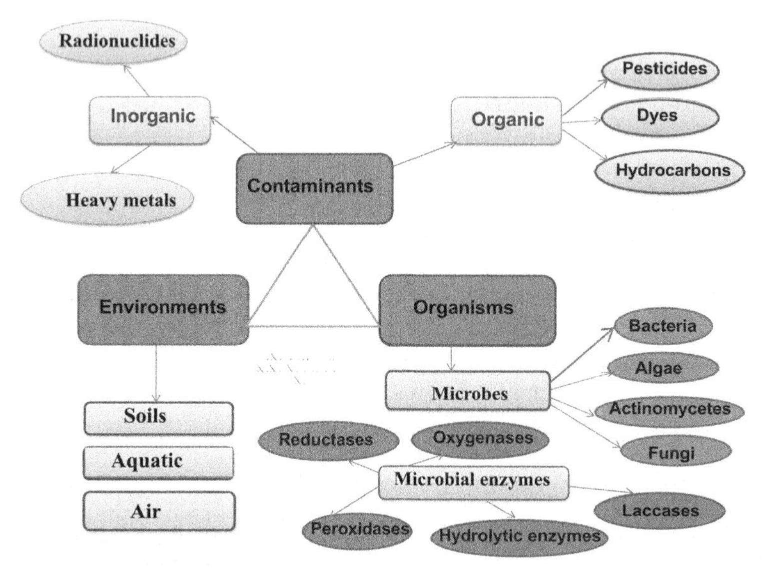

FIG. 11.3 Bioremediation of various contaminants in the environment.

eukaryotic organisms. Monooxygenase enzymes contain versatile super family of enzymes that catalyze the oxidative reactions of the substrates ranging from the alkanes to the complex endogenous molecules such as steroids and fatty acids.

The monooxygenases work as biocatalysts in the process of bioremediation and synthetic chemistry due to their high region selectivity and stereoselectivity on broad range of substrates. Most of the monooxygenase enzymes studied earlier are having cofactor; however, there are certain monooxygenase enzymes that function as independent cofactor. Monooxygenases need only the molecular oxygen for their activity and utilize the substrate as a reducing agent (Arora et al., 2010; Cirino and Arnold, 2002).

11.3.2 Dioxygenases

The dioxygenases are multicomponent enzyme systems that can introduce molecular oxygen into their substrate. An aromatic hydrocarbon dioxygenase enzyme belongs to the large family of the Rieske nonheme iron oxygenases. These enzymes catalyze enantio-specifically the oxygenation of wide range of substrates. Dioxygenases are primarily oxidized aromatic compounds and, hence, have the advantages in the environmental bioremediation. All the dioxygenase families have one or two electron-transport proteins prior to their oxygenase components. Crystal structure of the naphthalene dioxygenase showed the presence of a Rieske (2Fe-2S) cluster and mono-nuclear iron in each alpha subunit (Dua et al., 2002).

11.4 Microbial peroxidases

Peroxidases (donor: hydrogen peroxide oxidoreductases) are universal enzymes that catalyzes the oxidation of lignin and other phenolic compounds at the expenditure of hydrogen peroxide (H_2O_2) in the presence of a mediator. These peroxidases are the heme and nonheme proteins. In the mammals, they involved in the biological processes such as immune system or regulation of hormone. In plants, they involved in the auxin metabolism, lignin and suberin formation, cross-linking of cell wall components, cell elongation, or defense against pathogens (Hiner et al., 2002; Koua et al., 2009).

11.4.1 Classification of peroxidase enzymes

The peroxidases have been classified into many types based on its source and activity (http://peroxibase.toulouse.inra.fr/). Among the peroxidases, lignin peroxidase (LiP), manganese-dependent peroxidase (MnP), and versatile peroxidase (VP) have been studied the most because of their high potential to degrade toxic compounds in the environment.

11.4.1.1 Microbial lignin peroxidases

The lignin peroxidases are the heme proteins secreted mostly by white rot fungus during the secondary metabolism. In the presence of the co-substrate H_2O_2 and intermediary such as veratryl alcohol, LiP degrades the lignin and also other phenolic compounds. In the reaction, H_2O_2 gets reduced to H_2O with the acquisition of electron from LiP (which itself gets oxidized). Lignin peroxidase (oxidized) with acquiring an electron from the veratryl alcohol returns to its native state (reduced), and a veratryl aldehyde is produced. The veratryl aldehyde again gets reduced and back to veratryl alcohol by acquiring the electron from substrate. Consequently, the oxidation of halogenated phenolic compounds, polycyclic aromatic compounds, and other similar compounds is followed by a sequence of non-enzymatic reaction (Yoshida, 1998; Ten Have and Teunissen, 2001). The lignin peroxidases (LiP) play a key role in the biodegradation lignin present in the plant cell wall (Piontek et al., 2001).

11.4.1.2 Microbial manganese peroxidases

The manganese peroxidase (MnP) is an extracellular enzyme formed from the basidiomycetes fungi, which oxidizes Mn^{2+} to oxidant Mn^{3+} in the multistep reaction. The Mn^{2+} enhances the MnP production and functions as a substrate for the MnP. The Mn^{3+}, generated by MnP, acts as a mediator for oxidation of the different phenolic compounds. As a result, Mn^{3+} chelate oxalate is little and sufficient to diffuse into the areas unreachable even to enzyme, because in the case of lignin or analogous structures like recalcitrance/xenobiotic contaminants buried deeply in the soil, that are not essentially available to the enzymes (Ten Have and Teunissen, 2001).

11.4.1.3 Microbial versatile peroxidases

The versatile peroxidase enzymes are capable of directly oxidizing Mn^{2+}, methoxybenzenes, and phenolic aromatic substrates like MnP, LiP, and horseradish peroxidase. The versatile peroxidase has unusual broad substrate specificity and also has tendency to oxidize substrates in the absence of manganese in comparison with other peroxidases. It has demonstrated that versatile peroxidase is also able to oxidize both phenolic and non-phenolic lignin model dimers (Ruiz-Duenas et al., 2007). Thus, a highly efficient versatile peroxidase overproduction system is required for the biotechnological use in the industrial processes and bioremediation of recalcitrant contaminants (Wong, 2009; Tsukihara et al., 2006).

11.5 Biosensors

Biosensors (electrochemical, optical, piezoelectric, and thermal) are versatile, rapid, portable, and cost-effective analytical devices used for the detection of the concentration of a variety of analytes at the site of interest by the application of various biological agents (microbes, enzymes, nucleic acids, antibody, etc.) and transducers (electrochemical, gravimetric, piezoelectric, optical, calorimetric, etc.). Transducers convert biological signals into electrical signals which are recorded and displayed digitally to the consumer (Kashem et al., 2015). Biosensors generate either constant or distinct electronic signals that are directly proportional to the concentrations of analytes of interest. The examples of biosensors include enzyme-based biosensors, immunosensors, organism-, and whole cell-based biosensors. Nano-bioelectronics, the integration of several modern technologies such as nanotechnology, microfluidics, electronics, and biotechnology, has paved the way for the production of a miniaturized form of biosensors with strong recognition and determination of the analytes at the specific site of interest in an easy and rapid manner (Wang et al., 2014). The biosensors can be applied in various sectors, such as medicine, drug discovery, biomedical research, the environment, process industries, food, security, and defense (Rotariu et al., 2016). These are cost-effective and elegant (miniaturized, portable, sample volume requirement) substitutes for the usual analytical techniques that are able to provide real-time online observation of the desired site. The latest biosensors are now using various polymers, plastics, papers, and composites for the fabrication of biosensors, therefore making them eco-friendly and economically sustainable (Silveira et al., 2016; Khan et al., 2017).

11.6 Nanozymes

The nanozymes are nano-materials of size range 0–100 nm that shows enzyme-like characteristics. These are the very significant way to connect nano-materials like clusters, nanorods, crystals, nanobelts, and nanowires to the biological systems. They show exceptional advantages like excellent robustness, stability, and low-cost production with easy scale-up, which are the important qualities crucially needed as an alternative to the natural enzymes. Some of the examples are metal oxide-based nanozymes, metal-based nanozymes (Au NPs) and platinum nanoparticles (Pt NPs), and carbon-based nanozymes (carbon nanotubes, graphene oxide etc.) (Maduraiveeran and Jin, 2017). Over the past few decades, researchers have developed several important artificial enzymes by various materials to mimic structures and functions of the natural enzymes. In recent times, nanozymes, nano-materials with enzyme-like characteristics, have been emerged as novel artificial enzymes and attracted vast interest from many researchers. The amazing progress made in the field of nanozymes is because of their special properties when compared with natural and usual artificial enzymes. The nanozyme-based methods are the powerful, cost-effective, and simple techniques for the degradation and mineralization of the organic dyes from industrial practices (Yu et al., 2014). Most significantly, metal nanoparticles (MNPs) like peroxidase were investigated for the degradation of the organic contaminants, like phenol, rhodamine B, and methylene blue. The nanozymes have many advantages in terms of their low cost, robustness, high stability, long-term storability, and ease of mass production, when compared with normal enzymes. Nanozymes also have disadvantages, which include their low activity when compared with natural enzymes. The nanozymes also have low selectivity to targets, because of the absence of active sites where a substrate molecule binds and undergoes a chemical reaction in case of natural enzyme. The toxicity of nanozymes to the ecosystem and humans is also an important problem to be solved in view point of environmental and therapeutic applications (Hui and Erkang, 2013; Yang et al., 2008; Zeng et al., 2017).

11.7 Bioremediation of toxic compounds by enzymatic methods

The bioremediation is an eco-friendly process of the degradation of toxic contaminants into the nontoxic or less harmful compounds mediated by the novel microbial enzymes. The bioremediation mainly involves the enzymatic attack of microorganisms over the pollutant, therefore converting it into a nontoxic product. For the effective bioremediation of pollutants, the environmental conditions must favor the growth of microorganisms and their activity, and thus, the manipulation of environmental factors is required for the growth of microorganisms in order to carry out the degradation process at a faster rate. The enzymes are biological catalysts that usually speed up the conversion of substrates into products by providing the suitable conditions that lower the activation energy of the reaction. Almost all the enzymes belong to the six classes: transferases, hydrolases, oxidoreductases, isomerases, ligases, and lyases (Veni et al., 2019).

11.7.1 Enzymes used in decolorization and degradation of dyes

11.7.1.1 Enzymes used in bacterial systems

The alternative approach to general physicochemical methods, an enzymatic treatment has gained great attention in relation to decolorization and degradation of azo dyes. Different oxidoreductive enzymes are able to undergo a complex series of spontaneous cleavage reactions by generating very reactive free radicals. The decolorization and degradation of azo dyes by bacteria utilize various oxidoreductive enzymes as shown below.

11.7.1.2 Reductive enzyme

The azoreductase enzymes generate colorless aromatic amine products by catalyzing the reductive cleavage of azo bonds (Chang et al., 2001). They have been observed in many organisms like rat liver enzyme, rabbit liver aldehyde oxidase, and intestinal microbiota (Chen et al., 2005a,b). Based on their functions, they can be divided into two types: flavin-dependent azoreductases and flavin-independent azoreductases (Blumel and Stolz, 2003). Flavin-dependent azoreductase enzymes need nicotinamide adenine dinucleotide hydrogen (NADH), nicotinamide adenine dinucleotide phosphate hydrogen (NADPH), or flavin adenine dinucleotide (FADH) cofactor as electron donor for the reduction of azo bond (Russ et al., 2000). The functional group, which is present close to the azo bond, controls the substrate specificity of the azoreductases. The specificity of these enzymes has been studied through the synthetic model substrates based on the disodium-(R)-benzyl-azo-2,7 dihydroxy-3,6-disulfonyl naphthalene (Maier et al., 2004). The purification and characterization of the azoreductase enzymes have been systematically studied from different bacteria such as *Pigmentiphaga kullae* K24, *Staphylococcus aureus*, *Xenophilus azovorans* KF46F, *Escherichia coli*, *Rhodobacter sphaeroides*, and *Bacillus* sp. OY1-2 (Blumel and Stolz, 2003; Chen et al., 2005a,b; Suzuki et al., 2001). Azoreductases are restrained to extracellular or intracellular sites of cell membrane of bacteria; however in latest years, the function of intracellular azoreductase in the bacterial decolorization is unknown as azo dyes are hard to pass via the cell membrane due to the highest polarities and the complexity in their structure, but in many bacteria intracellular azoreductase enzyme activity has been determined (Saratale, 2009; Telke et al., 2008). It was demonstrated that an aerobic FMN-dependent azoreductase from *Enterococcus faecalis* has the capacity to degrade a broad range of azo dyes and the human intestinal bacterium *Pseudomonas aeruginosa* also exhibits the oxygen-insensitive azoreductase activity toward the several azo dyes (Chen et al., 2005a,b).

11.7.1.3 Nicotinamide adenine dinucleotide hydrogen (NADH) and 2,6-dichlorophenolindophenol (DCIP) reductases

There are many important enzymes of bacterial and fungal mixed function oxidase systems, which are mainly involved in the degradation of the xenobiotic compounds (Bhosale et al., 2006; Saratale et al., 2007a). The 2,6-dichlorophenolindophenol (DCIP) is reduced by accepting the electron from NADH in the presence of the enzyme. In the oxidized state, DCIP is in blue color and it becomes colorless after the reduction. The function of NADH-DCIP was reported during the degradation of reactive orange 16 by *Bacillus* sp. ADR, methyl red by *Brevibacillus laterosporus*, direct brown MR by *Acinetobacter calcoaceticus*, and red BL1 by *Pseudomonas* sp. SUK1.

11.7.1.4 Riboflavin reductases

The organisms usually need a system to catalyze the non-enzymatic reduction of the free flavins by NADPH/NADH because it is a relatively slow reaction. Subsequently, the enzyme NAD(P)H: flavin oxidoreductase or flavin reductase can be developed. The reduction of many flavins, such as FAD, FMN, and riboflavin, and the reduced pyridine nucleotides are catalyzed by an enzyme flavin reductase (Ingelman et al., 1999). The function of riboflavin reductase has been determined during the degradation of the dye, mordant yellow 10 by using the brilliant blue G and the anaerobic granular sludge by consortium GB, which is *Galactomyces geotrichum* MTCC 1360 and *Bacillus* sp. VUS (Field and Brady, 2003; Jadhav et al., 2008a,b). Moreover, the use of both enzymes in the degradation of scarlet R by consortium GR, which is, *Proteus vulgaris* and *Micrococcus glutamicus* and during the degradation of navy blue GL and reactive green 19A by *Bacillus* sp. VUS and *Micrococcus glutamicus* NCIM 2168 respectively have also studied (Dawkar et al., 2009; Saratale et al., 2009).

11.7.1.5 Lignin peroxidase

LiP is a glycoprotein with approximate molecular weights of 38–46 kDa. They belong to the oxidoreductase family, which has been used in the breakdown of different aromatic compounds like synthetic dyes, three- and four-ring polyaromatic hydrocarbons (PAHs), and also polychlorinated biphenyls (Duran and Esposito, 2000; Dawkar et al., 2009). The oxidation mechanism of azo dye by peroxidases takes place through the oxidation of phenolic group, which produces a radical at carbon with azo linkages. Phenyldiazene is formed by cleaving of the molecule by the attack of water to phenolic carbon, and this phenyldiazene with one electron reaction can be oxidized and generates N_2 (Chivukula and Renganathan, 1995). The different sulfonated azo dyes have been decolorized efficiently by the LiP isolated from *B. laterosporus* MTCC 2298 and *A. calcoaceticus* NCIM 2890 (Ghodake et al., 2009)

11.7.1.6 Laccases

The laccases (*p*-diphenol:dioxygen oxidoreductase) comprise a family of multicopper oxidases formed by certain insects, plants, fungi, and bacteria, which catalyzes the oxidation of a wide variety of reduced phenolic and aromatic compounds with simultaneous reduction of molecular oxygen to water (Gianfreda et al., 1999, Mai et al., 2000). The laccases are well-known to occur in several isoenzyme forms each of which is encoded by a separate gene (Giardina et al., 1995), and in few cases, the genes were expressed differently depending upon the nature of the inducer (Rezende et al., 2005). Laccase enzymes belong to the family of multicopper oxidase protein and are also known as the phenol oxidases. These enzymes catalyze the oxidation of substituted phenolic and non-phenolic compounds in the presence of oxygen as an electron donor (Sharma et al., 2007). Molecular weight of the laccase was about 60–390 kDa (Kalme et al., 2009). The extensive reaction capabilities, large substrate specificity, and no dependence on the cofactors are some of the factors, which make laccases very much efficient for the degradation of dye. These enzymes also prevent the formation of harmful aromatic amines during the decolorization of a few azo dyes because they decolorize them through a highly nonspecific free radical mechanism, without the direct cleavage of the azo bond (Kalme et al., 2009). The first prokaryotic laccase was reported from a rhizospheric bacterium, *Azospirillum lipoferum* (Givaudan et al., 1993). Decolorization of the various synthetic dyes through laccase activity was mostly reported in the lignolytic fungi (Abadulla et al., 2000). In spite of the studies on the biochemical characterization from the sources of bacteria (Diamantidis et al., 2000; McMahon et al., 2007), the information available on the substrate specificities for dye decolorization is limited. The 100% decolorization of many dyes, such as red HE7B, green HE4B, and direct blue 6, has been demonstrated by the laccase, which was isolated from *Pseudomonas desmolyticum* NCIM 2112 (Kalme et al., 2009).

11.7.1.7 Tyrosinases

The tyrosinases are copper-containing enzymes, which catalyze the oxidation of phenol, and thus utilized in the removal of phenol for that molecular oxygen is an oxidant. In the enzymatic method, initially the formation of o-diphenols occurs by the hydroxylation of monophenols, and in the second step, o-diphenols are oxidized to o-quinones. This enzyme is mainly involved in the degradation of azo dyes as a marker of the oxidative enzymes. In order to decontaminate the water polluted by phenols, the use of tyrosinase enzyme from *Bacillus thuringiensis* has been reported (El-Shora and Metwally, 2008). The function of the tyrosinase has been reported in the degradation of direct blue 6 by *Pseudomonas desmolyticum* NCIM 2112 (Kalme et al., 2007). The microbial consortia, which consists of *Galactomyces geotrichum* MTCC 1360 and *Bacillus* sp. VUS, also utilize the application of tyrosinase in the effective degradation of dispersed dye brown 3REL (Jadhav et al., 2008a,b). The tyrosinase-based degradation of azo dyes from *Alcaligenes faecalis* PMS-1 and *B. laterosporus* MTCC 2298 has also been reported (Saratale et al., 2011; Shah et al., 2012).

11.7.2 Enzymes involved in fungal systems

In order to survive the fungi, their ability to adapt their metabolism to various carbon and nitrogen sources is carried out by the production of various intra- and extracellular enzymes that are able to degrade many complex organic pollutants like PAHs, organic waste, and dye effluents (Gadd and Gadd, 2001; Humnabadkar et al., 2008; McMullan et al., 2001; Saratale et al., 2007a,b). The application of the fungal systems is one of the most appropriate approaches for the elimination of toxicity from the various colored and metallic discharges (Ezeronye and Okerentugba, 1999). Actually, the complex structures challenging for the bacteria are likely to be degraded by the fungi with their extracellular enzymatic secretion (Forss and Welander, 2009). The lignolytic fungi are the most widely studied fungi in relation to dye degradation. The nonspecific nature of enzymes such as LiP, manganese peroxidase, and laccase formed by white rot fungi permits the breakdown of a wide variety of aromatic compounds (Ehlers and Rose, 2005; Forgacs et al., 2004; Harazono and Nakamura, 2005; Srebotnik and Boisson, 2005; Toh et al., 2003).

Among the white rot fungi, *Phanerochaete chrysosporium* is the most widely researched. Furthermore, *Bjerkandera adusta*, species of *Pleurotus* and *Phlebia*, *Trametes* (Coriolus) *versicolor*, *Aspergillus ochraceus* etc. also gained significant consideration. According to some reports, *P. chrysosporium* is principally concerned with a manganese peroxidase and *B. adusta* with a LiP as the chief dye-decolorizing enzyme (Chagas and Durrant, 2001; Robinson et al., 2001). The biodegradation of 3 azo dyes (orange II, Congo red, and tropaeolin O) has been studied with fungus *P. chrysosporium* (Cripps et al., 1990). Palmieri et al. (2005) reported the participation of the fungus *basidiomycete Pleurotus ostreatus* in the decolorization of Remazol brilliant blue R, for that laccases appear to be principally involved, although *P. ostreatus* produces other enzymes such as veratryl alcohol oxidase and peroxidase. But, white rot fungi, which are used for the elimination of textile wastewater dyes also, have some drawbacks like their long growth cycle and they require nitrogen-limiting conditions. White rot fungi do not naturally occur in the wastewater; for this reason, the enzyme production may not be consistent (Robinson et al., 2001). Moreover, the fungal decolorization system performance is also restricted by the long hydraulic retention time necessary for the total decolorization (Chang et al., 2004) and the maintenance of the bioreactor fungi is also a matter of concern (Stolz, 2001).

11.8 Petroleum hydrocarbons (PHCs)

The industrial revolution has been increasingly enabled humans to exploit the natural resources such as petroleum hydrocarbons (PHCs) to be used as the principal source of energy. However, this has generated unparalleled disturbances in overall elemental cycles. The wide-scale production, transport, use, and disposal of petroleum worldwide have made it a main pollutant in both occurrence and quantity in the environment. Thus, agricultural soil in the surrounding area of refineries and industries has been polluted by hydrocarbons. Petroleum/hydrocarbon-polluted soil causes contamination of local groundwater by organics, makes threats to the safety of potable water, restricts the use of groundwater, and causes huge economic losses and also ecological calamity (Wang et al., 2007; Xu et al., 2006).

11.8.1 Petroleum-degrading enzymes and their roles

The variety of microorganisms, like bacteria, fungi, actinomycetes, green algae, and cyanobacteria, are capable of degrading various petroleum products under diverse environmental conditions, such as alkaline, acidic, aerobic, and anaerobic conditions. Different enzymes provide such capabilities to the microorganisms. The petroleum degradation occurs through enzyme-mediated metabolism of its compounds. The genes, which are responsible for the degradation of petroleum hydrocarbons, may be found in plasmid or chromosomal DNA (Broderick, 1999). The hydrocarbon (aliphatic/aromatic) biodegradation may take place under aerobic or anaerobic conditions in the environment (Hamme van et al., 2003). Under the aerobic conditions, oxygenases introduce oxygen atoms into the hydrocarbons. Under anaerobic conditions, anaerobes, such as sulfate-reducing bacteria, utilize different terminal electron acceptors in the place of oxygen (Hamme van et al., 2003). The degradation of hydrocarbons in the aerobic conditions occurs faster because aerobic microbes use oxygen as the terminal electron acceptor (Cao et al., 2009). The acetyl coenzyme A (CoA) is the final product after the oxidation of saturated aliphatic hydrocarbons that are metabolized during the citric acid cycle, besides the production of electrons in the electron transport chain. This cycle can be repeated, and the hydrocarbons are oxidized to CO_2 (Madigan et al., 2010). Toluene, benzene, naphthalene, xylene, and other aromatic hydrocarbons are also degraded under the aerobic conditions. The catechol formation occurs during initial step in the biodegradation. In addition, catechol is degraded into the alcohol, aldehydes, fatty acid, and acetyl-CoA, and then, it can be entered into citric acid cycle and catabolized into CO_2 (Cao et al., 2009; Madigan et al., 2010). A variety of bacteria, fungi, algae, and yeast have alkane hydroxylases that are alkane-degrading enzymes, which are distributed in the different species (Beilen van & Funhoff, 2007). Various alkane-degrading enzymes, like methane monooxygenase oxidized compounds ranges from the methane (C1) to butane (C4), integral membrane nonheme iron or cytochrome P450 enzyme oxidized compounds ranges from the pentane (C5) to hexadecane (C16), and C171 or longer alkanes are oxidized by unknown enzyme systems (Beilen van & Funhoff, 2007). The alkane degradation was studied in the *Pseudomonas putida* Gpo1, which is encoded by the octane plasmid (Beilen van et al., 2001; Beilen van, Wubbolts and Witholt, 1994). Here, the membrane monooxygenase enzyme transfers an alkane into alcohol. In the gram-negative bacteria, the location and function of the alkane (ALK) gene product were discovered by Hamme van et al. (2003). The degradation of aromatic hydrocarbon has been carried out by the enzyme catechol dioxygenase under the aerobic conditions. These enzymes are able to catalyze by the incorporation of molecular oxygen atoms into the 1,2-dihydroxybenzene (catechol), with subsequent opening of an aromatic ring and its cleavage (Cao et al., 2009; Hamme van et al., 2003; Madigan et al., 2010). The most important enzymes like catechol dioxygenases are responsible for the degradation of aromatic compounds (Broderick, 1999). The PHCs are able to degrade faster under the aerobic conditions; however, anaerobic degradation is also necessary for the process of bioremediation because of the several environmental conditions like aquifers, mangroves, and sludge digesters where the oxygen availability is restricted (Santos et al., 2011). Under the anaerobic conditions, aromatic HC degradation benzoyl-CoA reductase transfers aromatic compounds into benzoyl-CoA (Hosoda et al., 2005). Various terminal electron acceptors, like sulfate, nitrate, and Fe (III), are used in the degradation pathway (Cao et al., 2009).

11.9 Hydrolytic enzymes for bioremediation

In addition to the abovementioned enzymes, few hydrolytic enzymes play an important role in the bioremediation of the soil contaminants. The water polluted by the industrial and petroleum waste has been degraded by the hydrolytic enzymes like amylases, lipases, proteases, cellulases, pullulanases, and xylanases. These hydrolytic enzymes have diverse function in different areas including various industries and also in the biochemical sciences (Narendra et al., 2020).

11.9.1 Amylases

The alpha amylases are extracellular enzymes which break the α-1,4-glucosidic linkage in starch and produce the oligosaccharides. The β-amylase also breaks the second α-1,4-glycosidic bond of maltose and is synthesized in plants and also in bacteria. The amylases are significant enzymes for their specific application in the industrial starch conversion process. These enzymes work particularly on the disaccharides (sucrose) and polysaccharides (starch) and categorized into the group of glycoside hydrolases. The application of these enzymes has been established in the starch liquefaction, food, paper, sugar, and pharmaceutical industries (Gopinath et al., 2017). The pollutants found in wastewater decrease the level of dissolved oxygen (DO). Numerous pharmaceutical, textile, and food industry effluents are rich in fiber, starch, and other organic pollutants which are lethal to the environment. Priya and Renu (2018) reported that both commercial amylases and amylases from *Penicillium* sp. SP2 increased the DO content in wastewater. Wastewater contains effluents, ions, inhibitors, surfactants, and other insoluble matters. Therefore, the activities of amylase in wastewater treatment mainly depend on the biochemical properties of enzyme (Ali et al., 2014). Increased DO content in the amylase-treated wastewater revealed that the enzyme breakdowns organic matters in wastewater. The enzyme amylase from *Penicillium* sp. SP1 was effective for the treatment of wastewater (Priya and Renu, 2018). Uwadiae (2011) reported the treatment of pharmaceutical effluent by amylase and noticed the increase in the level of DO after the application of amylase. *Penicillium* sp. SP2 has the potential to utilize jack fruit seed for the growth and production of amylase. The addition of carbon and nitrogen source and Ca ions improved amylase production in solid substrate fermentation (SSF). Considering its nutritive value and cheap cost, jack fruit seeds could be useful for the production of amylase in industrial scale. This organism hydrolyzed carbohydrates, and as a result, DO was increased. Therefore, *Penicillium* sp. SP2 could be useful for the treatment of sewage water (Priya and Renu, 2018)

11.9.2 Lipases

The lipases are present in the plants and animals and also in the microorganisms and play a key role in the degradation of lipids. This enzyme acts on the soil organic pollutants and helps in the reduction of contaminants from the polluted soil; the reactions include esterification, hydrolysis, and aminolysis. It is an indicator to degrade hydrocarbon pollutants in the soil environment. Lipase has different applications in the food, paper, pulp, cosmetic, and chemical industries. However, the production cost is restrictive (Nigam, 2013; Okino-Delgado et al., 2017). The lipase enzymes degrade lipids derived from the variety of microbes in plants and animals. The recent reports showed that lipase is closely related to the organic contaminants present in the soil. The activity of lipase was responsible for the drastic reduction of total hydrocarbon from the polluted soil. The research has been undertaken in this area is expected to progress the awareness in bioremediation of the oil spills (Margesin et al., 1999; Riffaldi et al., 2006). The lipases are extracted from bacteria, actinomycetes, plant, and animal cells. Among these, microbial lipases are more versatile due to their potential use in industries. These enzymes catalyze different reactions such as hydrolysis, interesterification, esterification, alcoholysis, and aminolysis (Prasad and Manjunath, 2011).

11.9.2.1 Proteases

These are the group of hydrolytic enzymes, which break peptide bonds that are present in the primary structure of polypeptides and also other proteins. Proteases are the well-known industrial enzymes, with an important role in the different industries, such as pharmaceutical, feed food, and textile, and can be extracted from the plants, animals, and microbes. Protease enzyme provides potential applications for the waste management from some processing industries and also household activities. Besides, it is also useful in the poultry processing industry (Bhunia and Basak, 2014).

The proteases hydrolyze the breakdown of proteinaceous substances which can enter into environment as a result of shedding and moulting of appendages, death of the animals, and also as byproduct of the few industries such as fishery, poultry, and leather. The proteases are a class of enzymes, which hydrolyze the peptide bonds in the aqueous environment and produce them in the nonaqueous environment. Proteases have a broad range of uses in food, detergent, leather, and pharmaceutical industries (Singh, 2003; Beena and Geevarghese, 2010).

11.9.3 Cellulases

The cellulases assure the potential of converting waste cellulosic material into foods to meet growing population and have been the subject of intense research (Bennet et al., 2002). Some of the organisms synthesize cell-bound and envelope-associated, and few extracellular, cellulase enzymes. The extracellular cellulases, hemicellulases, and pectinases have been shown to be constitutively expressed at very low levels by a few bacteria and fungi (Rixon et al., 1992; Adriano-Anaya et al., 2005). The biodegradation of cellulosic biomass by enzymatic hydrolysis requires low volumes of chemicals and is conducted under mild conditions compared with chemical hydrolysis. Furthermore, chemical hydrolyzates required to be detoxified before fermentation. Hence, the enzymatic hydrolysis of lignocellulosic substrates is an effective process (Rodhe et al., 2011).

Various types of the microbes are involved in the enzymatic hydrolysis of cellulose with the help of multienzymatic system. Cellulases are the inducible enzymes synthesized by a variety of microbes which include bacteria, fungi, and actinomycetes during their growth on the cellulosic substances. These microbes include aerobic, anaerobic, mesophilic, and thermophilic (Koo, 2001; Kubicek, 1993). However, relatively few fungi and bacteria generate high levels of extracellular cellulases capable of solubilizing crystalline celluloses (Johnson et al., 1982; Wood, 1989).

The municipal solid waste (MSW) contains elevated levels of cellulosic materials, which is an ideal organic waste for growth of the majority microorganisms as well as composting by potential microorganisms. The MSW is mainly composed of 40%–50% cellulose, 9%–12% hemicelluloses, and 10%–15% lignin on the basis of dry weight. Unscientific disposal of waste led to an adverse impact on all the organisms of the environment and human health. A variety of microbes are found in MSW. The MSW is appropriate for composting because of the presence of high amount of the organic matter (Rani and Nand, 2000; Gautam et al., 2010a).

The extracellular cellulases are highly activated in depolymerizing the cellulosic substrates. Several cellulolytic microorganisms including *Penicillium, Aspergillus, Trichoderma,* and *Humicola* (Gautam et al., 2009a) are able to degrade cellulosic materials and produce huge quantities of extracellular cellulases. The cellulolytic enzymes were used in the organic waste degradation process, and the studies were carried out with *Trichoderma* sp., *Penicillium* sp., and *Aspergillus* spp. (Brown et al., 1987; Gautam et al., 2009b). The bacteria such as *Bacillus subtilis, Bacillus* spp., *Clostridium thermocellum, Ruminococcus* spp., *Streptomyces* spp., *Cellulomonas* spp., *Pseudomonas* spp., *Staphylococcus* spp., *Proteus,* and *Serratia* produce numerous cellulases, which mainly

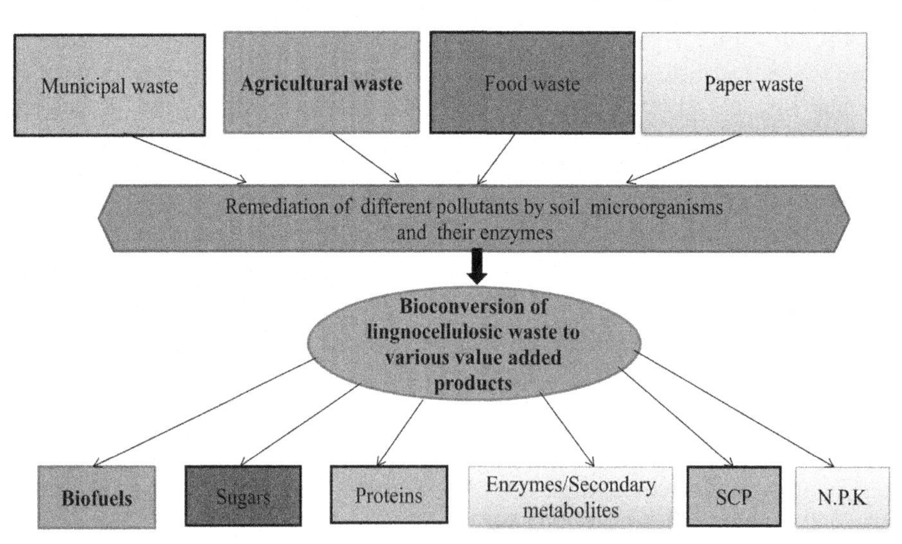

FIG. 11.4 Bioremediation of agricultural and food waste.

bound to their cell wall and involve in hydrolyzing the native lignocellulose preparations (Wood and Bhat, 1988; Gautam et al., 2010c). Bioconversion of different types of wastes/pollutants into value-added products is presented in Fig. 11.4.

11.10 Advantages of enzymes over microorganisms and plants

The use of enzymes in preference to microorganisms and plants provides more flexibility in terms of operational conditions. They can be used under more cruel environmental conditions such as concentration of pollutant, pH, temperature, and salinity which are not suitable for population of microorganisms (Choi et al., 2015). The enzymes do not require any nutrient supply or necessitate acclimatization as do population of microorganisms, and also, the production of metabolic products does not occur here. The mass transfer limitation is very much reduced under enzymatic treatment when compared to microbial or whole cell treatment. Furthermore, it is easier to manage an enzymatic process than microbial remediation. The application of enzymes when compared to plants and microbes for bioremediation is cost-effective, rapid, accessible, and also very specific (Demarche et al., 2012). The microbial whole cells require the addition of nutrients and air to maintain the optimum growth rate, and solvents and/or surfactants to increase the bioavailability and immobilization. It is also more feasible from an enzymatic perspective than using whole cells. Moreover, enzyme-mediated biotransformation decreases the generation of toxic by-products drastically when compared to the chemical and few types of microbial bioremediation. The enzyme-based biosensors have further contributed to the miniaturization of analytical instruments with greater qualitative and quantitative assessment of the compounds of interest at the polluted site (Patel, 2018). The enzyme technology with enhanced sensitivity and performance provides an absolute solution to all the issues linked with microbial remediation. The recent advancements in the nanotechnology coupled with enzyme technology have introduced the new concept of nanozymes to treat the contaminants or pollutants, which is more selective, productive, and faster than enzymatic treatment alone (Raman and Henning, 2013). Various microorganisms (microbial whole cells) and their enzymes used for the bioremediation of contaminants are presented in Tables 11.1 and 11.2, respectively.

11.11 Limitations to enzyme-mediated bioremediation strategies

Limitations to enzyme-mediated bioremediation strategies are as follows:

1. Isolation and identification of source microbes and plants. 2. The extraction, purification, characterization, and industrial production of enzymes are time- and capital-intensive processes. 3. Scale-up. 4. The stability and activity of enzymes across a wide range of pollutants and environmental conditions. 5. Performance or efficiency of enzymatic treatments varies from lab to pilot to field conditions (Kumar and Bharadvaja, 2019).

TABLE 11.1 Microorganisms (microbial whole cells) used for the remediation of pollutants.

S. no	Microorganisms	Pollutants	Remediation and result	Reference
1.	*Pantoea agglomerans*	Heavy metals	66 > 99% metal adsorption	Audu et al. (2020)
2.	Bacterial consortium[a]	Pesticides	> 90% of degradation of atrazine, carbofuran, and glyphosate	Góngora-Echeverría et al. (2020)
3.	*Cupriavidus* sp.	Heavy metals	Removed chromium (60%) and cadmium (30%) from the culture medium	Minari et al. (2020)
4.	*Acinetobacter* sp. Y2	Hydraulic fracturing flow back and produced water	In 7 days, COD, n-alkanes, and PAHs were decreased by 77%, 94%, and 77%, respectively	Zhou et al. (2020)
5.	Mixotrophic sulfide-oxidizing bacteria[b] (SOB)	Odorous surface water	Removal efficiencies of sulfide were > 84%	Sun et al. (2019)
6.	Mixed culture of bacteria and fungi[c]	Crude oil-polluted soil	TPHs removed (30 d) in solid phase and slurry phase 79 and 88%, respectively (*in vitro*)	Maddela et al. (2016a)
7.	Mixed microbial inoculum[d]	Crude oil-treated soil	Percent removal of TPHs in 90 days was 87% (*in vivo*)	Maddela et al. (2016b)
8.	*Geomyces* sp. strain UEAFg	Metal	100% metal (Cu(II)) removal in 7 days from the medium	Maddela et al. (2015a)
9.	Mixed fungi culture (*Geomyces* spp.)	Heavy crude oil	80% TPHs removed in 30 d (*in vitro*)	Maddela et al. (2015b)
10.	Mixed bacteria culture (*Bacillus* spp.)	Crude oil and spent lubricant oil	Percent removal of TPHs in crude oil- and spent lubricant oil-polluted soil was 84% and 28%, respectively (*in vitro*)	Raju et al. (2017)
11.	*Rhodococcus ruber* strain MCP-2	Pesticide	Removal of 45% degraded monocrotophos in 4 days from the medium	Srinivasulu et al. (2017)
12.	Soil bacteria	Pesticide	Utilization of acephate in the medium. Growth increment by 220% in 6 hours	Maddela and Venkateswarlu (2018)
13.	*Planococcus* sp. S5	Drug	Removed 30% of naproxen after 35 days as sole carbon source. Under co-metabolic conditions, 75.14% and 86.27% of naproxen removed in the presence of glucose and phenol, respectively	Domaradzka et al. (2015)
14.	*Phanerochaete sordida*	Drug	Removed 90% of diclofenac after 6 days	Hata et al. (2010)
15.	*Trametes trogii*	Drug	Rate of diclofenac degradation was 100%	Doruk et al. (2018)

[a] *Twenty one identified bacteria species and the most abundant (52%) species was Pseudomonas nitroreducens.*
[b] *Paracoccus sp. N1, Pseudomonas sp. N2, and Pseudomonas sp. S4.*
[c] *Bacillus cereus, Bacillus thuringiensis, Geomyces pannorum, and Geomyces sp.*
[d] *Bacillus thuringiensis B3, Bacillus cereus B6, Geomyces pannorum HR, and Geomyces sp. strain HV.*

11.12 Conclusions

Microbial enzymes play a tremendous role in the bioremediation of various types of pollutants from the environment. For the successful bioremediation of contaminants, the environmental conditions have to favor the growth of microorganisms and their activity, and therefore, the manipulation of environmental factors is crucial for the growth of microbes and in turn to accelerate the degradation at a faster rate. The oxidoreductases such as lignin peroxidase, laccase, and manganese peroxidase produced from the fungal mycelium are released into their surrounding environment. Being filamentous, fungi are able to reach the soil pollutants more effectively than bacteria.

TABLE 11.2 Microbial enzyme-mediated remediation of environmental pollutants.

S. no	Enzymes	Pollutants	Remediation and result	Reference
1.	Laccase (Bacterial consortium[a])	Heavy oil	~ 80% degradation of saturated hydrocarbons and aromatic hydrocarbons	Dai et al. (2020)
2.	*Cb*- FDH (*Candida boidinii*)	Oil	Percent degradation was 36 in 12h	Ji et al. (2019)
3.	MNPs-(Immobilized peroxidase)	Azo dyes	Dye disappeared after 4h of incubation	Darwesh et al. (2019)
4.	*Sso* Pox (*Sulfolobus solfataricus*)	Pesticides	95% of pesticides (four types) were removed within 5–43min	Poirier et al. (2019)
5.	Organophosphate-degrading enzyme A (OpdA) (Immobilized)	Pesticide	Degraded methyl parathion in 2h	Gao et al. (2014)
6.	Lignin peroxidase (LiP) (*Phanerochaete chrysosporium*)	Azo dyes	Dye degradation in 8h	Ilić Đurđić et al. (2021)
7.	Peroxidase (Immobilized)	Phenolic compounds	Removed 92% of phenol and 98% of p-cresol	Sellami et al. (2021)
8.	Enzyme mimics-based membrane reactor	Plasticizers	Di (2-ethylhexyl) phthalate was degraded in a continuous cycling mode	Li et al. (2021)
9.	Laccase (*Rigidoporus* sp. FMD21 crude extracellular enzyme extract)	Dioxins	77% of 2,3,7,8-TCDD was degraded within 36 days	Dao et al. (2021)
10.	Peroxidase (*Brassica rapa*)	Organic compounds	90% of phenol degradation after 40min. Conditions were 90mg/L of initial phenol amount, 2mM of H_2O_2, and 2.5UI/mL of BRP-CLEAs	Tandjaoui et al. (2019)
11	Acrylamidase (*Cupriavidusoxa laticus* ICTDB921)	Organic compound	Acrylamide (~ 1.75g/L) in industrial wastewater was degraded after 60min	Kulkarni et al. (2020)
12	Laccase (cross-linked with enzyme aggregates)	Phenol	Phenol was efficiently removed by E-CLEA	Fathali et al. (2019)
13	Protease (cross-linked with enzyme aggregates)	Silver	Enzyme showed best activity with minimum time in the silver removal	Asgher et al. (2018)
14.	Crude laccase from *Trametes* trogii	Drug	97% removal of diclofenac in 48h	Doruk et al. (2018)

[a] *Brevibacillus sp. DL-1, Bacillus sp. DL-13, and Acinetobacter sp. DL-34.*

The oxygenases play a key role in the metabolism of organic compounds with increasing their reactivity and involve in the cleavage of aromatic ring. The oxygenases have a broad substrate range and active against broad range of compounds include chlorinated aliphatic compounds. The major studied enzymes in the bioremediation are the bacterial monooxygenases or dioxygenases. The lignin peroxidases (LiP) play an important role in the degradation of lignin present in the plant cell wall. In the presence of co-substrate H_2O_2 and intermediary such as veratryl alcohol, LiP degrades the lignin and also other phenolic compounds. The potential versatile peroxidase overproduction system could be useful in the industrial processes and bioremediation of recalcitrant pollutants. Biosensors are versatile, rapid, portable, and cost-effective analytical devices used for the detection of the concentration of a variety of analytes at the site of interest by the application of various biological agents such as microbes, enzymes, and nucleic acids, and transducers include electrochemical, gravimetric, and piezoelectric. The nanozymes are potential and cost-effective remedial agents for the degradation and mineralization of organic dyes in the industrial practices when compared with usual enzymes. Metal nanoparticles (MNPs) such as peroxidases were investigated for the degradation of organic contaminants, like phenol, rhodamine B, and methylene blue. The biodegradation of hydrocarbon may occur in aerobic/anaerobic conditions in the environment. The genes, which are responsible for the degradation of petroleum hydrocarbons, might be found in plasmid or chromosomal DNA. Toluene, benzene, naphthalene, xylene, and other aromatic hydrocarbons are also degraded under the aerobic conditions. Catechol is degraded into the alcohol, aldehydes, fatty acid, and acetyl-CoA, and it can be entered into citric acid cycle and finally catabolized into CO_2.

The wastewater consists of effluents, ions, inhibitors, surfactants, and other insoluble matters. Amylase from *Penicillium* sp. SP1 was effective for the treatment of wastewater. The lipases are extracted from bacteria, actinomycetes, plant, and animal cells. Among these, microbial lipases are more flexible because of their potential applications in industries. Proteases provide potential applications for the waste management from processing industries to household activities. Various types of the microbes are involved in the enzymatic hydrolysis of cellulose with the help of multienzymatic system. Cellulases are synthesized by a variety of microbes which include bacteria, fungi, and actinomycetes during their growth on the cellulosic wastes. Modern advancements in nanotechnology linked with enzyme technology have introduced the new concept of nanozymes for the treatment of pollutants. Enzyme-mediated biotransformation reduces the formation of hazardous by-products significantly, when compared with chemical and microbial remediation. The application of enzymes for the remediation of pollutants is cost-effective, rapid, and more accessible in comparison with plants and microorganisms. During the bioremediation of municipal, paper, and agricultural wastes by microbial enzymes, it could be possible to produce value-added products such as bioethanol, biogas, and SCP.

References

Abadulla, E., Tzanov, T., Costa, S., Robra, K.H., Cavaco-Paulo, A., Gubitz, G.M., 2000. Decolorization and detoxification of textile dyes with a laccase from Trametes hirsuta. Appl. Environ. Microbiol 66 (8), 3357–3362.

Adriano-Anaya, M., Salvador-Figueroa, M., Ocampo, J.A., Garcıa-Romera, I., 2005. Plant cell wall degrading hydrolytic enzymes of *Gluconacetobacter diazotrophicus*. Symbiosis 40, 151–156.

Ali, I., Akbar, A., Yanwisetpakdee, B., Prasongsuk, S., Lotrakul, P., Punnapayak, H., 2014. Purification, characterization, and potential of saline waste water remediation of a polyextremophilic-amylase from an obligate halophilic *Aspergillus gracilis*. BioMed. Res. Int. https://doi.org/10.1155/2014/106937, 106937.

Arora, P.K., Kumar, M., Chauhan, A., Raghava, G.P., Jain, R.K., 2009. OxDBase: a database of oxygenases involved in biodegradation. BMC Res. Notes 2, 67.

Arora, P.K., Srivastava, A., Singh, V.P., 2010. Application of monooxygenases in dehalogenation, desulphurization, denitrification and hydroxylation of aromatic compounds. J. Bioreme. Biodegr. 1, 1–8.

Asgher, M., Bashir, F., Iqbal, H.M.N., 2018. Protease-based cross-linked enzyme aggregates with improved catalytic stability, silver removal, and dehairing potentials. Int. J. Biol. Macromol. 118, 1247–1256.

Audu, K.E., Adeniji, S.E., Obidah, J.S., 2020. Bioremediation of toxic metals in mining site of Zamfara metropolis using resident bacteria (*Pantoea agglomerans*): a optimization approach. Heliyon 6 (8). https://doi.org/10.1016/j.heliyon.2020.e04704.

Ayangbenro, A.S., Babalola, O.O., 2017. A new strategy for heavy metal polluted environments: a review of microbial biosorbents. Environ. Res. Public Heal. 14, 1–16.

Beena, A.K., Geevarghese, P.I., 2010. A solvent tolerant thermostable protease from a psychrotrophic isolate obtained from pasteurized milk. Develop. Microbiol. Mol. Biol. 1, 113–119.

Beilen van, J.B., Funhoff, E.G., 2007. Alkane hydroxylases involved in microbial alkane degradation. Appl. Microbiol. Biotechnol. 74 (1), 13–21.

Beilen van, J.B., Wubbolts, M.G., Witholt, B., 1994. Genetics of alkane oxidation by Pseudomonas oleovorans. Biodegradation 5 (3-4), 161–174.

Beilen van, J.B., Panke, S., Lucchini, S.A., Franchini, G., Rothlisberger, M., Witholt, B., 2001. Analysis of Pseudomonas putida alkane-degradation gene clusters and flanking insertion sequences: Evolution and regulation of the ALK genes. Microbiol 147 (6), 1621–1630.

Bennet, J.W., Wunch, K.G., Faison, B.D., 2002. Use of Fungi Biodegradation. ASM Press, Washington, DC, USA.

Bhosale, S., Saratale, G., Govindwar, S., 2006. Biotransformation enzymes in Cunninghamella blakesleeana (NCIM-687). J. Basic Microbiol. 46 (6), 444–448.

Bhunia, B., Basak, B., 2014. A Review on Application of Microbial Protease in Bioremediation. Industrial and Environmental Biotechnology. Studium Press, India, pp. 217–228.

Blumel, S., Stolz, A., 2003. Cloning and characterization of the gene coding for the aerobic azoreductase from Pigmentiphaga kullae K24. Appl. Microbiol. Biotechnol 62 (2-3), 186–190.

Bollag, J.M., Dec, J., 1998. Use of Plant Material for the Removal of Pollutants by Polymerization and Binding to Humic Substances. Tech. Rep. R-82092, Center for Bioremediation and Detoxification Environmental Resources Research Institute, The Pennsylvania State University, University Park, PA, USA.

Broderick, J.B., 1999. Catechol dioxygenases. Essays Biochem. 34 (11), 173–189.

Brown, J.A., Collin, S.A., Wood, T.M., 1987. Development of a medium for high cellulase, xylanase and b-glucosidase production by a mutant strain (NTG III/6) of the cellulolytic fungus *Penicillium pinophilum*. Enzyme Microb. Technol. 9, 355–360.

Cao, B., Nagarajan, K., Loh, K.C., 2009. Biodegradation of aromatic compounds: current status and opportunities for biomolecular approaches. Appl. Microbiol. Biotechnol. 85 (2), 207–228.

Chagas, E., Durrant, L., 2001. Decolorization of azo dyes by Phanerochaete chrysosporium and Pleurotus sajorcaju. Enzyme Microb. Technol. 29, 473–477.

Chandrakant, S.K., Shwetha, S.R., 2011. Role of microbial enzymes in the bioremediation of pollutants: a review. Enzyme Res. 805187, 1–11. https://doi.org/10.4061/2011/805187.

Chang, J.S., Chou, C., Lin, Y.C., Lin, P.J., Ho, J.Y., Hu, T.L., 2001. Kinetic characteristics of bacterial azo-dye decolorization by *Pseudomonas luteola*. Water Res. 35 (12), 2841–2850.

Chang, J.S., Chen, B.Y., Lin, Y.S., 2004. Stimulation of bacterial decolorization of an azo dye by extracellular metabolites from Escherichia coli strain NO_3. Biores. Technol. 91 (3), 243–248.

Chen, B.-Y., Chen, S.-Y., Chang, J.-S., 2005a. Immobilized cell fixed-bed bioreactor for wastewater decolorization. Process Biochem. 40 (11), 3434–3440.

Chen, H., Hopper, S.L., Cerniglia, C.E., 2005b. Biochemical and molecular characterization an azoreductase from Staphylococcus aureus, a tetrameric NADPH-dependent flavoprotein. Microbiol 151, 1433–1441.

Chivukula, M., Renganathan, V., 1995. Phenolic azo dye oxidation by laccase from Pyricularia oryzae. Appl. Environ. Microbiol. 61 (12), 4374–4377.

Choi, J., Han, S., Kim, H., 2015. Industrial applications of enzyme biocatalysis: current status and future aspects. Biotechnol. Adv. 33, 1443–1454.

Cirino, P.C., Arnold, F.H., 2002. Protein engineering of oxygenases for biocatalysis. Curr. Opi. Chem. Biol. 6, 130–135.

Cripps, C., Bumpus, J.A., Aust, S.D., 1990. Biodegradation of azo and heterocyclic dyes by Phanerochaete chrysosporium. Appl. Environ. Microbiol 56 (4), 1114–1118.

Dai, X., Lv, J., Yan, G., Chen, C., Guo, S., Fu, P., 2020. Bioremediation of intertidal zones polluted by heavy oil spilling using immobilized laccase-bacteria consortium. Biores. Technol. 309. https://doi.org/10.1016/j.biortech.2020.123305.

Damlas, C.A., Eleftherohorinos, I.G., 2011. Pesticide exposure, safety issues, and risk assessment indicators. Int. J. Environ. Res. Public Heal. 8, 1402–1419.

Dao, A.T.N., Loenen, S.J., Swart, K., Dang, H.T.C., Brouwer, A., de Boer, T.E., 2021. Characterization of 2,3,7,8-tetrachlorodibenzo-p-dioxin biodegradation by extracellular lignin-modifying enzymes from ligninolytic fungus. Chemosphere 263. https://doi.org/10.1016/j.chemosphere.2020.128280.

Darwesh, O.M., Matter, I.A., Eida, M.F., 2019. Development of peroxidase enzyme immobilized magnetic nanoparticles for bioremediation of textile wastewater dye. J. Environ. Chem. Eng., 7. https://doi.org/10.1016/j.jece.2018.11.049.

Dawkar, V.V., Jadhav, U.U., Ghodake, G.S., Govindwar, S.P., 2009. Effect of inducers on the decolorization and biodegradation of textile azo dye navy blue 2GL by *Bacillus* sp. VUS. Biodegradation 20 (6), 777–787.

Demarche, P., Junghanns, C., Nair, R.R., Agathos, S.N., 2012. Harnessing the power of enzymes for environmental stewardship. Biotechnol. Adv. 30, 933–953.

Diamantidis, G., Effosse, A., Potier, P., Bally, R., 2000. Purification and characterization of the first bacterial laccase in the rhizospheric bacterium *Azospirillum lipoferum*. Soil Biol. Biochem 32 (7), 919–927.

Dixit, R., Malaviya, D., Pandiyan, K., Singh, U.B., Sahu, A., Shukla, R., 2015. Bioremediation of heavy metals from soil and aquatic environment: an overview of principles and criteria of fundamental processes. Sustainability 7, 2189–2212.

Domaradzka, D., Guzik, U., Hupert-Kocurek, K., Wojcieszyńska, D., 2015. Co-metabolic degradation of naproxen by *Planococcus* sp. strain S5. Water Air Soil Pollut. 226, 297. https://doi.org/10.1007/s11270-015-2564-6.

Doruk, Y.A., Hakan, G., Cihangir, N., 2018. Biodegradation of diclofenac with fungal strains. Arch. Environ. Protec. 44, 55–62.

Dua, M., Singh, A., Sethunathan, N., Johri, A., 2002. Biotechnology and bioremediation: successes and limitations. Appl. Microbiol. Biotechnol. 59, 143–152.

Duran, N., Esposito, E., 2000. Potential applications of oxidative enzymes and phenoloxidase-like compounds in wastewater and soil treatment: a review. Appl. Catal. B 28, 83–99.

Ehlers, G.A., Rose, P.D., 2005. Immobilized white-rot fungal biodegradation of phenol and chlorinated phenol in trickling packed-bed reactors by employing sequencing batch operation. Biores. Technol. 96 (11), 1264–1275.

El-Shora, H.M., Metwally, M., 2008. Use of tyrosinase enzyme from *Bacillus thuringiensis* for the decontamination of water polluted with phenols. Biotechnology 7 (2), 305–310.

Ezeronye, O.U., Okerentugba, P.O., 1999. Performance and efficiency of a yeast biofilter for the treatment of a Nigerian fertilizer plant effluent. World J. Microbiol. Biotechnol. 15, 515–516.

Fan, C., Krishnamurthy, S., 1995. Enzymes for enhancing bioremediation of petroleum-contaminated soils: a brief review. Air Waste Manag. Assoc. 45, 453–460.

Fathali, Z., Rezaei, S., Faramarzi, M.A., Habibi-Rezaei, M., 2019. Catalytic phenol removal using entrapped cross-linked laccase aggregates. Int. J. Biol. Macromol. 122, 359–366.

Fetzner, S., 2003. Oxygenases without requirement for cofactors or metal ions. Appl. Microbiol. Biotech. 60, 243–257.

Fetzner, S., Lingens, F., 1994. Bacterial dehalogenases: biochemistry, genetics, and biotechnological applications. Microbiol. Rev. 58, 641–685.

Field, J.A., Brady, J., 2003. Riboflavin as a redox mediator accelerating the reduction of the azo dye mordant yellow 10 by anaerobic granular sludge. Water Sci. Technol. 48 (6), 187–193.

Forgacs, E., Cserhati, T., Oros, G., 2004. Removal of synthetic dyes from wastewaters: a review. Environ. Int. 30 (7), 953–971.

Forss, J., Welander, U., 2009. Decolourization of reactive azo dyes with microorganisms growing on soft wood chips. Int. Biodeter. Biodegr. 63 (3), 752–758.

Gadd, G.M., Gadd, G.M., 2001. Fungi in Bioremediation. vol. 23 Cambridge University Press, p. 481.

Gao, Y., Truong, Y.B., Cacioli, P., Butler, P., Kyratzis, I.L., 2014. Bioremediation of pesticide contaminated water using an organophosphate degrading enzyme immobilized on nonwoven polyester textiles. Enzyme Microb. Technol. 54, 38–44.

Gautam, S., Bundela, P., Pandey, A., Awasthi, M., Sarsaiya, S., 2009a. Prevalence of fungi in municipal solid waste of Jabalpur city (MP). J. Basic Appl. Mycol. 8, 80–81.

Gautam, S., Bundela, P., Pandey, A., Jain, R., Deo, P., Khare, S., Awasthi, M., Sarsaiya, S., 2009b. Biodegradation and recycling of urban solid waste. Am. J. Environ. Sci. 5, 653.

Gautam, S., Bundela, P., Pandey, A., Awasthi, M., Sarsaiya, S., 2010a. Composting of municipal solid waste of Jabalpur City. Global J. Environ. Res. 4, 43–46.

Gautam, S., Bundela, P., Pandey, A., Awasthi, M., Sarsaiya, S., 2010c. Cellulase production by *Pseudomonas* sp. isolated from municipal solid waste compost. Int. J. Acad. Res. 2.

Ghodake, G., Jadhav, S., Dawkar, V., Govindwar, S., 2009. Biodegradation of diazo dye direct brown MR by *Acinetobacter calcoaceticus* NCIM 2890. Int. Biodeter. Biodegrad. 63 (4), 433–439.

Gianfreda, L., Xu, F., Bollag, J.M., 1999. Laccases: a useful group of oxidoreductive enzymes. Bioreme. J. 3, 1–25.

Giardina, P., Cannio, R., Martirani, L., Marzullo, L., Palmieri, G., Sannia, G., 1995. Cloning and sequencing of a laccase gene from the lignin-degrading basidiomycete Pleurotus ostreatus. Appl. Env. Microbiol. 61, 2408–2413.

Givaudan, A., Effosse, A., Faure, D., Potier, P., Bouillant, M.L., Bally, R., 1993. Polyphenol oxidase in *Azospirillum lipoferum* isolated from rice rhizosphere: Evidence for laccase activity in non-motile strains of *Azospirillum lipoferum*. FEMS Microbiol. Lett. 108 (2), 205–210.

Góngora-Echeverría, V.R., García-Escalante, R., Rojas-Herrera, R., Giácoman-Vallejos, G., Ponce-Caballero, C., 2020. Pesticide bioremediation in liquid media using a microbial consortium and bacteria-pure strains isolated from a biomixture used in agricultural areas. Ecotoxicol. Environ. Saf. 200. https://doi.org/10.1016/j.ecoenv.2020.110734.

Gopinath, S.C.B., Anbu, P., Arshad, M.K.M., Lakshmipriya, T., Voon, C.H., Hashim, U., et al., 2017. Biotechnological processes in microbial amylase production. BioMed Res. Int., 1272193. https://doi.org/10.1155/2017/1272193.

Hamme van, J.D., Singh, A., Ward, O.P., 2003. Recent advances in petroleum microbiology. Microbiol. Mol. Biol. Rev. 67 (4), 503–549.

Harazono, K., Nakamura, K., 2005. Decolorization of mixtures of different reactive textile dyes by the white-rot basidiomycete Phanerochaete sordida and inhibitory effect of polyvinyl alcohol. Chemosphere 59 (1), 63–68.

Hata, T., Kawai, S., Okamura, H., Nishida, T., 2010. Removal of diclofenac and mefenamic acid by the white rot fungus *Phanerochaete sordida* YK-624 and identifi cation of their metabolites after fungal transformation. Biodegradation 21, 681–689.

Hiner, A.N.P., Ruiz, J.H., Rodri, J.N., García-Cánovas, F., Nigel, C.B., Andrew, T.S., et al., 2002. Reactions of the class II peroxidases, lignin peroxidase and *Arthromyces ramosus* peroxidase, with hydrogen peroxide: catalase-like activity, compound III formation, and enzyme inactivation. J. Biol. Chem. 277, 26879–26885.

Hlihor, R.M., Gavrilescu, M., Tavares, T., Favier, L., Olivieri, G., 2017. Bioremediation: An Overview on Current Practices, Advances, and New Perspectives in Environmental Pollution Treatment. Hindawi, pp. 3–4.

Hong, G., Yuebo, X., Sarfaraz, H., Alamgir, K.K., Xiaolin, W., Huiyong, X., 2018. Application of microbial technology used in bioremediation of urban polluted river: study of Chengnan river, China. Water 10, 643–664.

Hosoda, A., Kasai, Y., Hamamura, N., Takahata, Y., Watanabe, K., 2005. Development of a PCR method for the detection and quantification of benzoyl-CoA. reductase genes and its application to monitored natural attenuation. Biodegradation 16 (6), 591–601.

Hui, W., Erkang, W., 2013. Nanomaterials with enzyme-like characteristics (nanozymes): next-generation artificial enzymes. Chem. Soc. Rev. 42, 6060–6093.

Humnabadkar, R., Saratale, G., Govindwar, S., 2008. Decolorization of purple 2R by *Aspergillus ochraceus* (NCIM-1146). Asian J. Microbiol. Biotechnol. Sci. 10, 693–697.

Husain, Q., 2006. Potential applications of the oxidoreductive enzymes in the decolorization and detoxification of textile and other synthetic dyes from polluted water: a review. Crit. Rev. Biotech. 26, 201–221.

Ilić Đurđić, K., Ostafe, R., Prodanović, O., Đurđević Đelmaš, A., Popović, N., Fischer, R., et al., 2021. Improved degradation of azo dyes by lignin peroxidase following mutagenesis at two sites near the catalytic pocket and the application of peroxidase-coated yeast cell walls. Front. Environ. Sci. Eng. 15. https://doi.org/10.1007/s11783-020-1311-4.

Ingelman, M., Ramaswamy, S., Niviere, V., Fontecave, M., Eklund, H., 1999. Crystal structure of NAD(P)H: flavin oxidoreductase from *Escherichia coli*. Biochemistry 38 (22), 7040–7049.

Ishtiaq, A., Hafiz, M.N.I., Kuldeep, D., 2017. Enzyme-based biodegradation of hazardous pollutants–an overview. J. Exp. Biol. Agric. Sci. 5, 402–411.

Iwamoto, T., Nasu, M., 2001. Current bioremediation practice and perspective. Biosci. Bioeng. 92 (1), 1–8.

Jadhav, S., Jadhav, M., Kagalkar, K., Govindwar, S., 2008a. Decolorization of brilliant blue G dye mediated by degradation of microbial consortium of *Galactomyces geotrichum* and Bacillus sp. J. Chin. Inst. Chem. Eng. 39, 563–570.

Jadhav, S., Jadhav, U., Dawkar, V., Govindwar, S., 2008b. Biodegradation of disperse dye brown 3REL by microbial consortium of Galactomyces geotrichum MTCC 1360 and Bacillus sp. VUS. Biotechnol. Bioproc. Eng. 13 (2), 232–239.

Jayaraj, R., Megha, P., Sreedev, P., 2016. Organochlorine pesticides, their toxic effects on living organisms and their fate in the environment. Interdiscip. Toxicol. 9, 90–100.

Ji, L., Fu, X., Wang, M., Xu, C., Chen, G., Song, F., Guo, S., Zhang, Q., 2019. Enzyme cocktail containing NADH regeneration system for efficient bioremediation of oil sludge contamination. Chemosphere 233, 132–139.

Johnson, E.A., Sakajoh, M., Halliwell, G., Madia, A., Demain, A.L., 1982. Saccharification of complex cellulosic substrates by the cellulase system from *Clostridium thermocellum*. Appl. Environ. Microbiol. 43, 1125–1132.

Kalme, S.D., Parshetti, G.K., Jadhav, S.U., Govindwar, S.P., 2007. Biodegradation of benzidine based dye direct blue-6 by *Pseudomonas desmolyticum* NCIM 2112. Biores. Technol. 98, 1405–1410.

Kalme, S., Jadhav, S., Jadhav, M., Govindwar, S., 2009. Textile dye degrading laccase from *Pseudomonas desmolyticum* NCIM 2112. Enzyme Microb. Technol. 44 (2), 65–71.

Kashem, A., Suzuki, M., Kimoto, K., Iribe, Y., 2015. An optical biochemical oxygen demand biosensor chip for environmental monitoring. Sens. Act. B Chem. 221, 1594–1600.

Khan, M.S., Misra, S.K., Wang, Z., Daza, E., Schwartz-duval, A.S., Kus, J.M., et al., 2017. Paper-based analytical biosensor chip designed from graphene-nanoplatelet amphiphilic- diblock-co-polymer composite for cortisol detection in human saliva. Anal. Chem. 1-9.

Koo, Y.-M., 2001. Pilot-scale production of cellulase using *Trichoderma reesei* Rut C-30 in fed-batch mode. J. Microbiol. Biotechnol. 11, 229–233.

Koua, D., Cerutti, L., Falquet, L., Christian, J.A.S., Gregory, T., Nicolas, H., et al., 2009. Peroxi Base: a database with new tools for peroxidase family classification. Nucleic Acids Res. 37, 261–266.

Kubicek, C., 1993. In: From Cellulose to Cellulase Inducers: Facts and Fiction. Proceedings of the Second Tricel Symposium on *Trichoderma reesei* Cellulases and Other Hydrolases, Espoo, Finland. Foundation for Biotechnical and Industrial Fermentation Research, Helsinki, pp. 181–188.

Kulkarni, N.H., Muley, A.B., Bedade, D.K., Singhal, R.S., 2020. Cross-linked enzyme aggregates of arylamidase from Cupriavidus oxalaticus ICTDB921: process optimization, characterization, and application for mitigation of acrylamide in industrial wastewater. Bioproc. Biosyst. Eng. 43, 457–471.

Kumar, L., Bharadvaja, N., 2019. Enzymatic bioremediation: a smart tool to fight environmental pollutants, Chapter 6. In: Bhatt, P. (Ed.), Smart Bioremediation Technologies; Microbial Enzymes. Academic Press/Elsevier.

Leung, M., 2004. Bioremediation: techniques for cleaning up a mess. J. Biotech. 2, 18–22.

Lew, S., Lew, M., Biedunkiewicz, M., Szarek, J., 2013. Impact of pesticide contamination on aquatic microorganism populations in the Littoral zone. Arch. Environ. Contam. Toxicol 64, 399–409.

Li, X., Li, J., Hao, S., Han, A., Yang, Y., Fang, G., Liu, J., Wang, S., 2021. Enzyme mimics based membrane reactor for di(2-ethylhexyl) phthalate degradation. J. Hazar. Mater. 403. https://doi.org/10.1016/j.jhazmat.2020.123873.

Loredana, S., Graziano, P., Antonio, M., Carlotta, N.M., Caterina, L., Maria, A.A., 2017. Lindane bioremediation capability of bacteria associated with the demosponge Hymeniacidon perlevis. Mar. Drugs, 1–15.

Maddela, N.R., Venkateswarlu, K., 2018. Bacterial Utilization of Acephate and Buprofezin. Insecticides–Soil Microbiota Interactions. Springer, pp. 87–101.

Maddela, N.R., Reyes, J.J.M., Viafara, D., Gooty, J.M., 2015a. Biosorption of copper (II) by the microorganisms isolated from the crude-oil-contaminated soil. Soil Sediment Contam. Int. J. 24, 898–908.

Maddela, N.R., Scalvenzi, L., Pérez, M., Montero, C., Gooty, J.M., 2015b. Efficiency of indigenous filamentous fungi for biodegradation of petroleum hydrocarbons in medium and soil: laboratory study from Ecuador. Bull. Environ. Contam. Toxicol. 95, 385–394.

Maddela, N.R., Burgos, R., Kadiyala, V., Carrion, A.R., Bangeppagari, M., 2016a. Removal of petroleum hydrocarbons from crude oil in solid and slurry phase by mixed soil microorganisms isolated from Ecuadorian oil fields. Int. Biodeter. Biodegrad. 108, 85–90.

Maddela, N.R., Scalvenzi, L., Kadiyala, V., 2016b. Microbial degradation of total petroleum hydrocarbons in crude oil: a field-scale study at the low-land rainforest of Ecuador. Environ. Technol, 1–24.

Madigan, M.T., Martinko, J.M., Dunlap, P.V., Clark, D.P., 2010. Brock Biology of Microorganisms, twelfth ed. Benjamin Cummings.

Maduraiveeran, G., Jin, W., 2017. Nanomaterials based electrochemical sensor and biosensor platforms for environmental applications. Trends Environ. Anal. Chem. 13, 10–23.

Mai, C., Schormann, W., Milstein, O., Huttermann, A., 2000. Enhanced stability of laccase in the presence of phenolic compounds. Appl. Microbiol. Biotech. 54, 510–514.

Maier, J., Kandelbauer, A., Erlacher, A., Cavaco-Paulo, A., Gubitz, G.M., 2004. A new alkali-thermostable azoreductase from *Bacillus* sp. strain SF. Appl. Environ. Microbiol 70 (2), 837–844.

Margesin, R., Zimmerbauer, A., Schinner, F., 1999. Soil lipase activity-A useful indicator of oil biodegradation. Biotechnol. Tech. 13, 859–863.

McMahon, A.M., Doyle, E.M., Brooks, S., O'Connor, K.E., 2007. Biochemical characterisation of the coexisting tyrosinase and laccase in the soil bacterium *Pseudomonas putida* F6. Enzyme Microb. Technol. 40 (5), 1435–1441.

McMullan, G., Meehan, C., Conneely, A., Kirby, N., Robinson, T., Nigam, P., et al., 2001. Microbial decolourisation and degradation of textile dyes. Appl. Microbiol. Biotechnol. 56 (1-2), 81–87.

Minari, G.D., Saran, L.M., Lima Constancio, M.T., Correia da Silva, R., Rosalen, D.L., José de Melo, W., Carareto Alves, L.M., 2020. Bioremediation potential of new cadmium, chromium, and nickel-resistant bacteria isolated from tropical agricultural soil. Ecotoxicol. Environ. Saf. 204. https://doi.org/10.1016/j.ecoenv.2020.111038.

Narendra, K., Neha, J., Narendra, K., Saurabh, G., Hukum, S., 2020. Phytoremediation facilitating enzymes: an enzymatic approach for enhancing remediation process. Chapter 15. In: Bhatt, P. (Ed.), Smart Bioremediation Technologies. vol. 3. Academic Press; Elsevier, pp. 289–306.

Newman, L.A., Sharon, L.D., Katrina, L.G., Paul, E.H., Induluis, M., Tanya, Q.S., et al., 1998. Phytoremediation of organic contaminants: a review of phytoremediation research at the University of Washington. Soil Sedim. Contam. 7, 531–542.

Nicolopoulou-Stamati, P., Maipas, S., Kotampasi, C., Stamatis, P., Hens, L., 2016. Chemical pesticides and human health: the urgent need for a new concept in agriculture. Front. Public Heal. 4 (148), 1–8.

Nigam, P.S., 2013. Microbial enzymes with special characteristics for biotechnological applications. Biomolecules 3, 597–611. https://doi.org/10.3390/biom3030597.

Okino-Delgado, C.H., Prado, D.Z., Facanali, R., Marques, M.M.O., Nascimento, A.S., Fernandes, J.C., 2017. Bioremediation of cooking oil waste using lipases from wastes. PLoS One 12 (10), e0186246. https://doi.org/10.1371/journal.pone.018624.

Palmieri, G., Cennamo, G., Sannia, G., 2005. Remazol brilliant blue R decolourisation by the fungus Pleurotus ostreatus and its oxidative enzymatic system. Enzyme Microb. Technol. 36 (1), 205–210.

Park, J.W., Park, B.K., Kim, J.E., 2006. Remediation of soil contaminated with 2,4-dichlorophenol by treatment of minced shepherd's purse roots. Arch. Environ. Contam. Toxicol. 50, 191–195.

Patel, R.N., 2018. Biocatalysis for synthesis of pharmaceuticals. Bioorg. Med. Chem. 26, 1252–1274.

Piontek, K., Smith, A.T., Blodig, W., 2001. Lignin peroxidase structure and function. Biochem. Soc. Trans. 29, 111–116.

Poirier, L., Pinault, L., Armstrong, N., Ghigo, E., Daudé, D., Chabrière, E., 2019. Evaluation of a robust engineered enzyme towards organophosphorus insecticide bioremediation using planarians as biosensors. Chem. Biolog. Interact. 306, 96–103.

Prasad, M.P., Manjunath, K., 2011. Comparative study on biodegradation of lipid-rich wastewater using lipase producing bacterial species. Ind. J. Biotechnol. 10, 121–124.

Priya, F.S., Renu, A., 2018. Efficacy of amylase for wastewater treatment from *Penicillium* sp. SP2 isolated from stagnant water. J. Environ. Biol. 39, 189–195.

Raju, M.N., Leo, R., Herminia, S.S., Moran, R.E., Venkateswarlu, K., Laura, S., 2017. Biodegradation of diesel, crude oil and spent lubricating oil by soil isolates of *Bacillus* spp. Bull. Environ. Contam. Toxicol. 98, 698–705.

Raman, K., Henning, P., 2013. Environmental assessment of enzyme use in industrial production: a literature review. Clean. Prod. 42, 228–240.

Rani, D.S., Nand, K., 2000. Production of thermostable cellulase-free xylanase by *Clostridium absonum* CFR-702. Process Biochem. 36, 355–362.

Rao, M.A., Scelza, R., Scotti, R., Gianfreda, L., 2010. Role of enzymes in the remediation of polluted environments. J. Soil Sci. Plant Nutr. 10 (3), 333–353.

Rezende, M.I., Barbosa, A.M., Vasconcelos, A.F.D., Haddad, R., Dekker, R.F.H., 2005. Growth and production of laccases by the ligninolytic fungi, *Pleurotus ostreatus* and *Botryosphaeria rhodina*, cultured on basal medium containing the herbicide, Scepter- (imazaquin). J. Basic. Microbiol. 45, 460–469.

Riffaldi, R., Levi-Minzi, R., Cardelli, R., Palumbo, S., Saviozzi, A., 2006. Soil biological activities in monitoring the bioremediation of diesel oil-contaminated soil. Water, Air, Soil Pollut. 170, 3–15.

Rixon, J.E., Ferreira, L.M.A., Durrant, A.J., Laurie, J.I., Hazlewood, G.P., Gilbert, H.J., 1992. Characterization of the gene Cel D and its encoded product 1,4-β-D-glucan glucohydrolase D from *Pseudomonas fluorescens* subsp. cellulosa. Biochem. J. 285 (3), 947–955.

Robinson, T., McMullan, G., Marchant, R., Nigam, P., 2001. Remediation of dyes in textile effluent: a critical review on current treatment technologies with a proposed alternative. Biores. Technol. 77 (3), 247–255.

Rodhe, A.V., Sateesh, L., Sridevi, J., Venkateswarlu, B., Rao, L.V., 2011. Enzymatic hydrolysis of sorghum straw using native cellulase produced by *T. reesei* NCIM 992 under solid state fermentation using rice straw. 3 Biotech. 1, 207–215.

Rotariu, L., Lagarde, F., Jaffrezic-Renault, N., Bala, C., 2016. Electrochemical biosensors for fast detection of food contaminants-trends and perspective. Trends Anal. Chem. 79, 80–87.

Rubilar, O., Diez, M.C., Gianfreda, L., 2008. Transformation of chlorinated phenolic compounds by white rot fungi. Critic. Rev. Env. Sci. Tech. 38, 227–268.

Ruiz-Duenas, F.J., Morales, M., Perez-Boada, M., Thomas, C., María, J.M., Klaus, P., et al., 2007. Manganese oxidation site in *Pleurotus eryngii* versatile peroxidase: a site-directedmutagenesis, kinetic, and crystallographic study. Biochemistry 46, 66–77.

Russ, R., Rau, J., Stolz, A., 2000. The function of cytoplasmic flavin reductases in the reduction of azo dyes by bacteria. Appl. Environ. Microbiol 66 (4), 1429–1434.

Saadati, N., Abdullah, P., Zakaria, Z., Rezayi, M., Hosseinizare, N., 2012. Distribution and fate of HCH isomers and DDT metabolites in a tropical environment-Case study Cameron highlands-Malaysia. Chem. Central J. 6, 130–145.

Santos, H.F., Carmo, F.L., Paes, J.E., Rosado, A.S., Peixoto, R.S., 2011. Bioremediation of mangroves impacted by petroleum. Water, Air Soil Pollut. 216 (1-4), 329–350.

Saratale, R., 2009. Development of Efficient Microbial Consortium for Biodegradation of Azo Dyes. Shivaji University, Kolhapur, India.

Saratale, G., Kalme, S., Bhosale, S., Govindwar, S., 2007a. Biodegradation of kerosene by *Aspergillus ochraceus* NCIM 1146. J. Basic Microbiol. 47 (5), 400–405.

Saratale, G., Humnabadkar, R., Govindwar, S., 2007b. Study of mixed function oxidase system in *Aspergillus ochraceus* (NCIM 1146). Ind. J. Microbiol. 47 (4), 304–309.

Saratale, R.G., Saratale, G.D., Chang, J.S., Govindwar, S.P., 2009. Ecofriendly degradation of sulfonated diazo dye C.I. reactive green 19A using *Micrococcus glutamicus* NCIM-2168. Biores. Technol. 100 (17), 3897–3905.

Saratale, R., Saratale, G., Chang, J., Govindwar, S., 2011. Bacterial decolorization and degradation of azo dyes: a review. J. Taiwan Inst. Chem. Eng. 42, 138–157.

Sellami, K., Couvert, A., Nasrallah, N., Maachi, R., Tandjaoui, N., Abouseoud, M., Amrane, A., 2021. Bio-based and cost effective method for phenolic compounds removal using cross-linked enzyme aggregates. J. Hazar. Mater. 403, 124021.

Setyo, A., Refdinal, P., Fahimah, N., Kuniyoshi, M., 2017. Biodegradation of aldrin and dieldrin by the white-rot fungus Pleurotus ostreatus. Curr. Microbiol. 74, 320–324.

Shah, P., Dave, S., Rao, M., 2012. Enzymatic degradation of textile dye reactiveorange 13 by newly isolated bacterial strain *Alcaligenes faecalis* PMS-1. Int. Biodeter. Biodegrad. 69, 41–50.

Shankar, J., Abhilash, P.C., Singh, H.B., Singh, R.P., Singh, D.P., 2011. Genetically engineered bacteria: an emerging tool for environmental remediation and future research perspectives. Gene 480, 1–9.

Sharma, P., Goel, R., Capalash, N., 2007. Bacterial laccases. World J. Microbiol. Biotechnol. 23 (6), 823–832.

Sharma, B., Dangi, A.K., Shukla, P., 2018. Contemporary enzyme based technologies for bioremediation: a review. Environ. Manag. 210, 10–22.

Silveira, C.M., Monteiro, T., Almeida, M.G., 2016. Biosensing with paper-based miniaturized printed electrodes-a modern trend. Biosensors 51 (6), 1–17.

Singh, C.J., 2003. Optimization of an extracellular protease of *Chrysosporium keratinophilum* and its potential in bioremediation of keratinic wastes. Mycopathologia 156, 151–156.

Srebotnik, E., Boisson, J., 2005. Peroxidation of linoleic acid during the oxidation of phenols by fungal laccase. Enzyme Microb. Technol 36, 785–789.

Srinivasulu, M., Nilanjan, P.C., Chakravarthi, B., Jayabaskaran, C., Jaffer, M.G., Naga, R.M., Manjunatha, B., Darwin, R.O., Juan, O.T., Rangaswamy, V., 2017. Biodegradation of monocrotophos by bacteria isolated from soil. Afr. J. Biotechnol. 16, 408–417.

Stolz, A., 2001. Basic and applied aspects in the microbial degradation of azo dyes. Appl. Microbiol. Biotechnol. 56 (1-2), 69–80.

Sun, Z., Pang, B., Xi, J., Hu, H.-Y., 2019. Screening and characterization of mixotrophic sulfide oxidizing bacteria for odorous surface water bioremediation. Biores. Technol. 290. https://doi.org/10.1016/j.biortech.2019.121721.

Suzuki, Y., Yoda, T., Ruhul, A., Sugiura, W., 2001. Molecular cloning and characterization of the gene coding for azoreductase from Bacillus sp. OY1-2 isolated from soil. J. Biolog. Chem. 276 (12), 9059–9065.

Tandjaoui, N., Abouseoud, M., Couvert, A., Amrane, A., Tassist, A., 2019. A combination of absorption and enzymatic biodegradation: phenol elimination from aqueous and organic phase. Environ. Technol. (UK) 40, 625–632.

Telke, A., Kalyani, D., Jadhav, J., Govindwar, S., 2008. Kinetics and mechanism of reactive red 141 degradation by a bacterial isolate *Rhizobium radiobacter* MTCC 8161. Acta Chim. Slov. 55, 320–329.

Ten Have, R., Teunissen, P.J.M., 2001. Oxidative mechanisms involved in lignin degradation by white-rot fungi. Chem. Rev. 101, 3397–3413.

Toh, Y., Yen, J.J.L., Obbard, J.P., Ting, Y., 2003. Decolourisation of azo dyes by white-rot fungi (WRF) isolated in Singapore. Enzyme Microb. Technol. 33, 569–575.

Tsukihara, T., Honda, Y., Sakai, R., Watanabe, T., Watanabe, T., 2006. Exclusive overproduction of recombinant versatile peroxidase MnP2 by genetically modified white rot fungus, *Pleurotus ostreatus*. J. Biotechnol. 126, 431–439.

Uwadiae, S.E., Yerima, Y., Azike, R.U., 2011. Enzymatic biodegradation of pharmaceutical wastewater. Int. J. Ener. Environ. 2, 683–690.

Veni, P., Satish, C.P., Tushar, J., Diksha, S., Saurabh, G., Saurabh, K., Mukesh, S., 2019. Biodegradation of toxic dyes: a comparative study of enzyme action in a microbial system. Chapter 14. In: Bhatt, P. (Ed.), Smart Bioremediation Technologies. Academic Press; Elsevier, pp. 255–287.

Vidali, M., 2001. Bioremediation. an overview. Pure Appl. Chem. 73, 1163–1172.

Wang, J., Xu, H.K., Guo, S.H., 2007. Isolation and characteristics of a microbial consortium for effectively degrading phenanthrene. Petrol. Sci. 4 (3), 68–75.

Wang, X., Lu, X., Chen, J., 2014. Development of biosensor technologies for analysis of environmental contaminants. Trends Environ. Anal. Chem. 2, 25–32.

Wasim, A., Sengupta, D., Chowdhury, A., 2009. Impact of pesticides use in agriculture: their benefits and hazards. Interdiscip. Toxicol. 2 (1), 1–12.

Williams, P.P., 1977. Metabolism of synthetic organic pesticides by anaerobic microorganisms. Residue Rev. 66, 63–135.

Wong, D.W.S., 2009. Structure and action mechanism of ligninolytic enzymes. Appl. Biochem. Biotech. 157, 174–209.

Wood, T., 1989. Mechanisms of cellulose degradation by enzymes from aerobic and anaerobic fungi. Enzyme Syst. Lignocellul. Degrad., 17–35.

Wood, T.M., Bhat, K.M., 1988. Methods for measuring cellulase activities. Methods Enzymol. 160, 87–112.

Xu, S.Y., Chen, Y.X., Wu, W.X., 2006. Enhanced dissipation of phenathrene and pyrene in spiked soils by combined plants cultivation. Sci. Tot. Environ. 363, 206–215.

Yang, Z., Si, S., Zhang, C., 2008. Magnetic single-enzyme nanoparticles with high activity and stability. Biochem. Biophys. Res. Commun. 367, 169–175.

Yoshida, S., 1998. Reaction of manganese peroxidase of *Bjerkandera adusta* with synthetic lignin in acetone solution. J. Wood Sci. 44, 486–490.

Yu, X., Liu, Z., Huang, X., 2014. Nanostructured metal oxides/hydroxides-based electrochemical sensor for monitoring environmental micropollutants. Trends Environ. Anal. Chem. 4, 28–35.

Zeng, Y., Miao, F., Zhao, Z., Zhu, Y., Liu, T., Chen, R., et al., 2017. Low-cost nanocarbon- based peroxidases from graphite and carbon fibers. Appl. Sci. 7, 924–935.

Zhou, H., Huang, X., Liang, Y., Li, Y., Xie, Q., Zhang, C., You, S., 2020. Enhanced bioremediation of hydraulic fracturing flowback and produced water using an indigenous biosurfactant-producing bacteria *Acinetobacter* sp. Y2. Chem. Eng. J. 397, 125348.

CHAPTER

12

Phytoremediation of heavy metals and petroleum hydrocarbons using *Cynodon dactylon (L.) Pers.*

Srujana Kathi

ICSSR Postdoctoral Research Fellow, Department of Applied Psychology, Pondicherry University, Puducherry, India

OUTLINE

12.1 Introduction

Phytoremediation technology has been receiving more attention largely, and the result from trials indicated a cost-saving treatment compared to existing treatment (Prathap et al., 2019). Decontamination of heavy metal-contaminated soil is critical for the maintenance of environmental health and ecological restoration (Ali et al., 2012). Restoring the soil contaminated by petroleum hydrocarbons is a challenging task because of their complexity and persistence in the environment (Baoune et al., 2019). Phytoremediation process employs plants to remove organic and inorganic contaminants from the environment (Sanusi et al., 2012). Co-contamination of metals and crude oil is a common feature in nature (Atagana, 2011). Plants contribute to contaminant removal from the soil through phytoextraction, phytodegradation, rhizosphere degradation, rhizofiltration, phytostabilization, phytovolatilization, and phytorestoration mechanisms (Peer et al., 2005; Tangahu et al., 2011). Several plants have also been described to have efficiency to clean up petroleum hydrocarbon-contaminated soils (Njoku et al., 2009). Composting is additionally applied in the process of bioremediation as a means of degrading toxic organic compounds and for decreasing the toxicity of metallic contaminants in organic residues, wastes, and byproducts (Barker and Bryson, 2002).

Sinha et al. (2009) stated that soil microorganisms bio-accumulate the metals in concentrations up to 50 times higher than the surrounding soil. Microorganisms resist toxic metals by oxidation and reduction of metals, methylation and demethylation, enzymatic reduction, formation of metallorganic complexes, metal ligand degradation,

https://doi.org/10.1016/B978-0-12-822933-0.00005-X

metal sequestration, metal efflux pumps, exclusion by permeability barrier, and production of metal chelators such as metallothioneins and biosurfactants. Rhizosphere, an interface of soil and plant, plays an important role in phytoremediation of soil contaminated by heavy metals, where microbial populations were known to affect heavy metal mobility and availability to the plant by releasing chelating agents, acidification, phosphate solubilization, and redox changes, thus possessing the potential to enhance phytoremediation processes (Jing et al., 2007).

Plant systems and degrading microorganisms support each other in the remediation process, and different plants could enhance degradation of PAHs in their root system to variable extent (Rezek et al., 2008). Some of the bacteria capable of destroying PAHs belong to the genera *Pseudomonas, Alcaligenes, Rhodococcus, Sphingomonas,* and *Mycobacterium* (Van Gestel et al., 2003; Sinha et al., 2009). Hydrocarbonoclastic strains like *Gordonia, Mycobacterium,* and *Rhodococcus* could be successfully applied to bioremediation for bioaugmentation of soil co-contaminated with hydrocarbons, metals, and even pesticides (Garcia-Diaz et al., 2013).

Plants from Poaceae family were found to reduce total petroleum hydrocarbons to a greater extent (Sanusi et al., 2012) with grasses having dense fibrous root system and their root biomass being larger than the dicots. Larger root surface area is expected to enhance microbial density by providing surface area for adhesion of metabolic substances (Olson et al., 2007). Plants and their associated bacteria interact with each other whereby plant supplies the bacteria with a special carbon source that stimulates the bacteria to degrade organic contaminants in the soil. In return, plant-associated bacteria extend support to their host plant to overcome stress responses induced by the contaminants and improve plant growth and development. Further, plants get benefitted from associated bacteria that possess hydrocarbon degradation potential, leading to enhanced hydrocarbon mineralization and lowering of both phytotoxicity and evapotranspiration of volatile hydrocarbons (Sumia et al., 2013). Yoon et al. (2006) showed that native plant species growing on contaminated sites might possess the potential for phytoremediation. Sudan grass (*Sorghum sudanense* (Piper) Stapf.) with developed root system and strong PAHs and heavy metal tolerance is also a well-established choice for phytoremediation (Liu et al., 2020).

Cynodon dactylon is a major turf and forage grass which propagates vegetatively with fibrous root system with deep rhizomes. It grows well on a variety of soil, tolerating both acidic and alkaline soil conditions, and is also highly tolerant to salt, drought, anoxia, cold, and soil compaction. It is a promising species for phytoremediation of oil-contaminated soil (Razmjoo and Adavi, 2012) and was reported to enhance biodegradation of PAHs in the soil (White et al., 2003; Tang et al., 2010). The grass releases organic compounds in the form of root exudates capable of binding metals (Thomas et al., 2014). This species demonstrated adaptability to a wide range of soil Cd concentrations and accumulated relatively higher concentrations of Cu and Pb than other plant species and was found growing in a wide range of soil with Cu and Pb concentrations. *C. dactylon* was able to accumulate Zn also at higher leaf tissue content (Archer and Caldwell, 2004).

The present study aims at evaluating the phytoremediation efficiency of *Cynodon dactylon* for extraction of crude oil and removal of Pb, Zn, Cu, Mn, and Cd in both composted and *Rhodococcus ruber* inoculated treatments.

12.2 Materials and methods

12.2.1 Study area

The uncontaminated soil (crude oil concentration of 4.55% w/w) used for the pot studies was the soil collected from the Pondicherry University garden (79°85'14.74"E, 12°01'98.24"N), Pondicherry, India.

12.2.2 Experimental design

12.2.2.1 Preparation of crude oil-contaminated microcosms

The experiment was conducted in a net house. The lighting for plant growth was supplied from the natural sunlight, and the temperature was found to be in the range of 24–35°C. Soil moisture was maintained at 60% of the field water-holding capacity by daily watering during the experiment. The pots were watered twice a day, and on sunny days when the rate of evapotranspiration is very high, water was sprinkled on top of plants.

Garden soil was sieved with a 2mm sieve, air-dried for two weeks, and then added with 10mg/kg, 30mg/kg, 50mg/kg, and 70mg/kg of No. 2 fuel oil (Carman et al., 1998) concentration (w/w) in acetone (1:1) and then placed overnight under the fume hood (Adam and Duncan, 2002; Sanusi et al., 2012). Care was taken to ensure uniform mixing of soil samples with crude oil. Uncontaminated soil was used as experimental control. Contaminated experimental soil was allowed to attenuate for one month. The crude oil-contaminated soil was then placed into pots and planted with a single grass species.

12.2.2.2 Analytical methods

Soil pH and moisture were measured with portable handheld soil moisture and pH meter. Total nitrogen in the soil was measured by the Kjeldahl method using concentrated H_2SO_4, K_2SO_4, and HgO to digest the sample (Bremner and Sparks, 1996). Soil particle distribution was done by the hydrometer method, and soil texture was determined by the Bouyoucos hydrometer method (Gee et al., 1986).

12.2.2.3 Preparation of plant

Prior to planting in the pots, *Cynodon dactylon* (L.) Pers. plants were collected from hydrocarbon-contaminated areas and seeds were germinated in the pots. Plants survived well up to 50% crude oil concentration during screening. After 2 weeks of germination, plants of 1cm shoot and root length were selected and then planted in different pots (Basumatary et al., 2013).

Fifty seedlings per pot were dispersed with three replicates in each treatment (Merkl et al., 2005). Each pot was filled with 5 kgs of soil, and each seedling of 1 cm shoot and root length, respectively, was planted in each pot. In the first set of treatments (T1 to T4), each pot was filled with 5 kgs of contaminated soil and topped with 1 kg of compost. Compost used in our experiments showed the following chemical characteristics: pH 6.8, 11.5% total carbon, 1.13% total nitrogen, 0.47% available phosphorous, 352.38 mg/100 g of sodium, 16.19 mg/100 g of zinc, 53.86 mg/100 g of manganese, and 5.16 mg/100 g of copper. In the second set of treatments (T1–T4) where soil was inoculated with *R. ruber*, each pot was filled with 5 kg of contaminated soil and inoculated with (1 mL) *R. ruber*. Pots were placed under experimental sheds for 90 days. Sampling was performed on every 10th, 50th, and 90th day.

12.2.2.4 Isolation and inoculation of crude oil-degrading bacteria

Soil samples for isolation of bacterial species were collected from 10 automobile workshops contaminated with petroleum hydrocarbons located in Pondicherry, India. Enrichment cultures using BHB minimal medium were carried out with representative soil samples. Enrichment was carried out in 250-mL Erlenmeyer flasks containing 100 mL of the BHB medium with 10 g of soil. 50 mg/mL of phenanthrene (C14H10), fluorene (C13H10), fluoranthene (C16H10), chrysene (C18H12), and benzo[*a*]anthracene (C18H12) were prepared in acetone, while naphthalene (C10H8) was dissolved in methanol at a final concentration of 100 mg/mL. Phenanthrene, fluoranthene, and fluorene were sprayed onto a solid agar medium after streak plating of bacteria (Hilyard et al., 2008; Kiyohara et al., 1982). Naphthalene dissolved in methanol was applied to a petri dish lid; the methanol was allowed to evaporate leaving volatile naphthalene as the sole carbon and energy source for microbes on BH agar plates. Tests of PAH degradability were visualized by a distinct phenanthrene clearing zone surrounding individual colonies. These representative colonies were aseptically selected, removed, and subcultured in liquid BHB medium containing phenanthrene. Flasks were incubated at 30°C in an incubator shaker at 200 rpm agitation. The enrichment was maintained by consecutive transfers (10% of cultured medium) to fresh medium at 30 days interval until pure colonies were obtained.

The culture filtrate was diluted in 1:1 proportion with distilled water. The presence of the microbe in the inoculated soil was confirmed on 10th, 50th, and 90th day by serially diluting the soil sample followed by isolation in BH medium.

12.2.2.5 Treatment details

Set 1
T1: Soil 5 kg + crude oil (10 mg/kg) + 1 kg compost (top layer) + *C. dactylon* (L.) Pers.
T2: Soil 5 kg + crude oil (30 mg/kg) + 1 kg compost (top layer) + *C. dactylon* (L.) Pers.
T3: Soil 5 kg + crude oil (50 mg/kg) + 1 kg compost (top layer) + *C. dactylon* (L.) Pers.
T4: Soil 5 kg + crude oil (70 mg/kg) + 1 kg compost (top layer) + *C. dactylon* (L.) Pers.
Control: Soil 5 kg + *C. dactylon* (L.) Pers. + 1 kg compost (top layer)

Set 2
T1: Soil 5 kg + crude oil (10 mg/kg) + *R. ruber* inoculum (1 mL) + *C. dactylon* (L.) Pers.
T2: Soil 5 kg + crude oil (30 mg/kg) + *R. ruber* inoculum (1 mL) + *C. dactylon* (L.) Pers.
T3: Soil 5 kg + crude oil (50 mg/kg) + *R. ruber* inoculum (1 mL) + *C. dactylon* (L.) Pers.
T4: Soil 5 kg + crude oil (70 mg/kg) + *R. ruber* inoculum (1 mL) + *C. dactylon* (L.) Pers.
Control: Soil 5 kg + *C. dactylon* (L.) Pers. + *R. ruber* inoculum (1 mL)

12.2.2.6 Estimation of heavy metals

To determine the total heavy metal content in soils, air-dried 0.5 g of dry soil was sieved through 250 μm nylon sieve (Addo et al., 2012) and digested in aqua regia using microwave-assisted digester (Anton Paar, Austria) in accordance with the USEPA method SW 3051 (USEPA, 2000). Metal concentrations in the final solution were determined using Atomic Absorption Spectrometer (AAS, GBC Scientific Equipment, Australia) that consists of a hollow cathode lamp, slit width of 0.7 nm, and an acetylene flame. Zn was analyzed at a wavelength of 213.86 nm, Mn at 232.00 nm, Cd at 228.80 nm, Pb at 283.31 nm, and Cu at 324.75 nm. Standard stock solutions for all the elements were suitably diluted following the procedures (APHA, 2005), and certified reference material was employed for quality assurance and quality control purposes. The recoveries from the certified reference material were within 85%–115% of the values given in the standard certificate. Deionized water and analytical-grade reagents were used throughout this work. Reagent blank was used to minimize the impurity of metal of interest.

12.2.3 Determination of TPHs

Total petroleum hydrocarbon (TPH) was measured with an InfraCal cuvette holder model Total organic grease (TOG)/TPH Analyzer (Wilks Enterprise Co.) according to the USEPA Methods 418.1 that uses S-316 as extraction solvent in replacement to Freon (ASTM, 2004). The common method for estimating TPH in soil was USEPA Methods 418.1 (Schwartz et al., 2012). This is quick and easy field and laboratory analysis method for determining TPH and oil and grease concentration levels in soil and water. Total petroleum hydrocarbons were calculated from the sample using the formula (USEPA, 1978):

$$\text{Concentration of TPHs} = \frac{R * D}{V}$$

where,
R = mg of petroleum hydrocarbons as determined from the calibration plot
D = extract dilution factor
V = volume of sample, in liters

12.2.4 Statistical analyses

Statistical analysis was performed using SPSS version 16.0 for windows (SPSS Inc., USA). In the experiments concerning interactions of each metal and each treatment with composted and *R. ruber* inoculated treatments, means were compared by a one-way ANOVA followed by Tukey test. Substantial variations in biomass production and total petroleum hydrocarbon concentrations of metals among treatments were also measured using one-way ANOVA.

12.3 Results and discussion

The effectiveness of phytoremediation depends on the selection of plants native to the target since they adapt to local climate, insects, and diseases (Mahdavi et al., 2014). The following plants were found to exist in the localities where automobile wastes were discarded: *Tridax procumbens, Vernonia cinerea, Commelina diffusa, Cyperus rotundus* (Basumatary et al., 2013), *Cyperus tuberosus, Euphorbia hyssopifolia, Leucas aspera, Euphorbia hirta, Sida rhombifolia, Ricinus communis, Phyllanthus maderaspatensis, Peperomia pellucida,* Bermuda grass (*Cynodon dactylon* (L.) Pers.), *Solanum nigrum, Oldenlandia corymbosa, Mimosa pudica,* and *Commelina diffusa* (White et al., 2003; Olson et al., 2007; Tang et al., 2010).

Among them, *Cyperus rotundus* and *Cynodon dactylon* were the only perennial grasses. All other weeds are not only specific to the oil-contaminated sites as they are commonly found in other uncontaminated waste lands and roads in Pondicherry. *Cyperus rotundus* is an invasive weed with a 20 m long root and a tuber which is a tough competitor for ground resources. So, *C. dactylon* was selected as it is the only aggressive turf grass spread by both above-ground (stolons) and below-ground (rhizomes) stems with potential for soil remediation and stabilization. It is a widely distributed and drought-resistant grass species. Using this grass, it is most cost-efficient to control water and wind erosion combined with the groundwater recharge, organic chemical decomposition, and inorganic chemical stabilization. *C. dactylon* is quite effective in soil and water preservation. It can be used in economic phytoextraction technique as they do not require replanting every year (Koduah Owusu Ansah, 2012). The physicochemical properties of experimental soil are presented in Table 12.1.

TABLE 12.1 Physicochemical analysis of experimental soil.

Parameter	Value
Texture	Sandy clay
Electrical conductivity (dS/m)	0.21
pH	5.65–7.5
Soil moisture	7.08–50.43
Sodium (ppm)	148–283
Potassium (ppm)	168.00–271.04
Nitrogen (ppm)	87
Phosphorous (ppm)	27
Potassium (ppm)	80

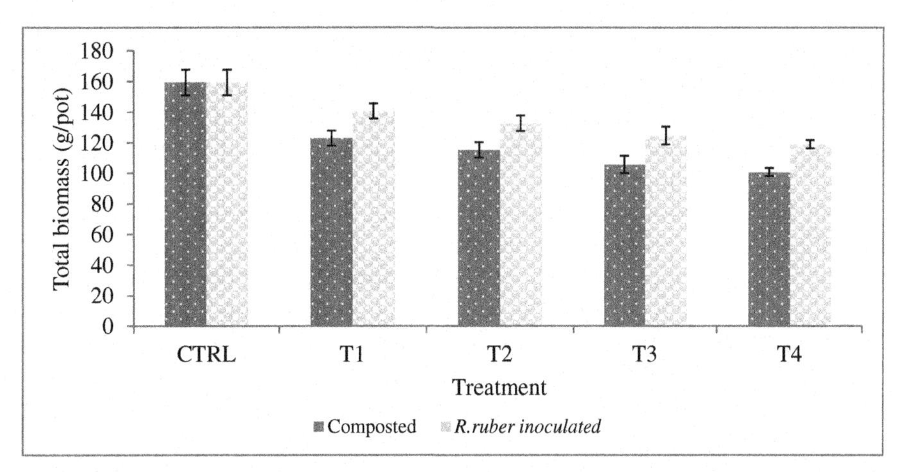

FIG. 12.1 Comparison of biomass in composted and *R. ruber* inoculated phytoremediation treatments after 90 days.

12.3.1 Estimation of biomass of Bermuda grass (*C. dactylon*) in composted and inoculated treatments

The plant species employed in the current study Bermuda grass (*C. dactylon*) showed a promising behavior in petroleum hydrocarbon-contaminated soil. Plant growth was significantly reduced from T1 to T4 as crude oil concentration increased in the soil assessed, by measuring the biomass toward the end of the experiment (Fig. 12.1). Many plant species were sensitive to petroleum hydrocarbon contamination as indicated by their slow growth rate (Diab and Sandouka, 2010). The variation in biomass between the composted and microbe-inoculated treatments could be clearly discernable where *R. ruber* inoculated treatment showed better growth in all the treatments up to 90 days of cultivation. Significant effect of crude oil contamination on plant biomass was observed ($P < 0.05$) as compared to control. Crude oil contamination in soil reduced the biomass along composted treatments (T1–T4) by 68.13% compared to control.

The percentage reduction in biomass of *R. ruber* inoculated treatment was 74.68 when compared to control. Species such as *C. dactylon* produce adventitious roots at leaf nodes resulting in the rapid expansion of the rhizosphere. The profuse growth of the plants and the liability to grow in the presence of the metal and contaminant oil suggested that the plants would survive in a pilot trial (Atagana, 2011). The biomass of the plant was reduced across the treatments with increase in concentration of crude oil but there was an increase in biomass toward the end of experiment in all the treatments. Rajkumar et al. (2012) while investigating the role of plant-associated microbes in protection against heavy metal stress have demonstrated that the bacterial colonization resulted in increased nutrient uptake and an increase in plant biomass.

Reduction in biomass across the treatments was reported in maize by Zand et al. (2009) in highly contaminated aged soil from the oil refinery in Tehran. The tested plant species *Gaillardia aristata*, *Echinacea purpurea*, Fawn (*Festuca arundinacea* Schreb), Fire Phoenix (a combined *F. arundinacea*), and *Medicago sativa* L. grown in the contaminated soil yielded less dry weight with 50% decrease than the control (Liu et al., 2012). Revathi et al. (2011) reported a significant reduction in biomass in phytoremediation of metal (Cd)-contaminated soil of Ranipet Tanneries utilizing Sorghum plant.

12.3.2 Estimation of total petroleum hydrocarbons

After treatment with different concentrations of crude oil, reduction in TPH was observed toward the end of the experiments in both composted and *R. ruber* inoculated treatments. TPH of the soil at 10 days, 50 days, and 90 days is presented in Table 12.2. The concentration of total petroleum hydrocarbons varied significantly along treatments on 10th day ($F=2437.064$, $P<0.05$), 50th day ($F=550.652$, $P<0.05$), and 90th day ($F=582.513$, $P<0.05$).

Our results corroborate with the findings of Beaudin et al. (1996) who studied the composting of soil contaminated by hydrocarbons with alfalfa and maple leaves in a laboratory-scale reactor. In soils contaminated with mineral oils and grease at 17,000 mg/kg total hydrocarbons, 60% of the aliphatic, 54% of the aromatic, and 83% of the polar compounds were degraded during 180 days of composting. After 287 days of composting, 73% of all mineral oils and grease were degraded.

The TPH concentration (ppm) in *R. ruber* inoculated treatment of the soil at 10th, 50th, and 90th day is presented in Table 12.2. The concentration of total petroleum hydrocarbons varied significantly along treatments on 10th day ($F=1158.66$, $P<0.05$), 50th day ($F=617.28$, $P<0.05$), and 90th day ($F=303.5$, $P<0.05$).

Natural attenuation in comparison with Tall fescue plant's presence reduced TPH level by 57.4% (Zand et al., 2009). Techer et al. (2012) reported that slight decrease in total four-ring PAH concentration would suggest a positive influence of growing plants in PAH-contaminated soil in the long term. Muratova et al. (2008) reported oil sludge degradation in planted soil during a two-and-a-half-month pot experiment. Oil sludge as a whole was degraded successfully in the soil planted with the prairie grasses, in the order of: perennial ryegrass (46%) > crested wheatgrass (45%)> couch grass (44%). The most effective phytoremediation was reported in the soil planted with rye (52%). Among the legumes, the best reductions were obtained with alfalfa (41%), where the TPH reduction in the unplanted control was 34%. Reduction of 35% diesel oil content was reached in the soil which was spiked with diesel oil and mixed with biowaste at room temperature in the composting bin system for 12 weeks (Van Gestel et al., 2003). By multiple inoculation of soil with indigenous microorganisms followed by their application on to the soil contaminated with double high concentration of diesel oil and aircraft fuel, the removal efficiency of 80% for diesel oil and 98% for aircraft fuel, respectively, was reported (Lebkowska et al., 2011).

During the first 10 days, the average removal efficiency of the TPHs in the composted soil was less than 30%. At 50 days, maximum removal rate of TPHs at 57.02% was recorded in T4 of *R. ruber* inoculated treatment. The reduction was not uniform along treatments but significantly higher than that in the control. At the end of the experiment, higher TPH reduction was observed in all the *R. ruber* inoculated treatments when compared to composted soil. Maximum percentage reduction of 79.7 was recorded in T1 in *R. ruber* inoculated treatment at the end of the experiment (Fig. 12.2). Perennial ryegrass cultivation showed a 50% reduction in PAHs. The role of plants in phytoremediation could support diverse microbial communities that contain more efficient strains of microorganisms (Rezek et al., 2008).

TABLE 12.2 Total petroleum hydrocarbons present in composted and inoculated soil after remediation by *C. dactylon* grass.

	Day 10		Day 50		Day 90	
Treatment	**Composted**	***R. ruber* inoculated**	**Composted**	***R. ruber* inoculated**	**Composted**	***R. ruber* inoculated**
Control	0.33 ± 0.58^a	1.00 ± 1.00^a	0.50 ± 1.87^a	1.04 ± 0.06^a	0.60 ± 0.57^a	1.12 ± 0.38^a
T1	8.03 ± 1.20^a	7.00 ± 1.33^a	4.93 ± 0.57^a	4.99 ± 0.68^a	3.00 ± 0.26^a	2.03 ± 0.27^a
T2	27.14 ± 0.3^a	27.23 ± 0.86^a	16.55 ± 0.93^a	15.05 ± 0.82^a	11.44 ± 0.32^a	10.68 ± 1.13^a
T3	42.97 ± 0.73^a	38.24 ± 0.69^a	32.19 ± 1.87^a	32.99 ± 0.58^a	24.99 ± 0.71^a	15.01 ± 0.38^a
T4	67.01 ± 1.44^a	66.95 ± 2.25^a	44.23 ± 1.95^a	30.08 ± 1.89^a	23.95 ± 1.54^a	20.2 ± 1.31^a

All values are represented by the unit ppm.

[a] *The mean difference is significantly different to control at P<0.05.*

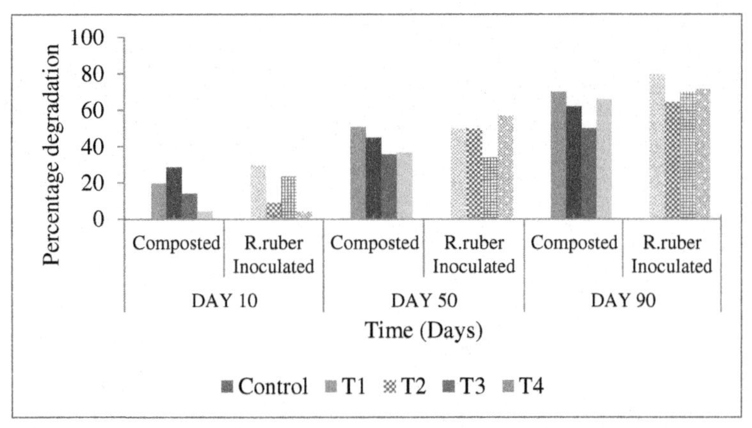

FIG. 12.2 Comparison of TPH in composted and *R. ruber* inoculated soil.

Razmjoo and Adavi (2012) found up to 80,000 mg/kg TPH tolerance among some cultivars of *C. dactylon*. Liu et al. (2012) reported that after a 30-day pot culture with 14 ornamental plants, PAH removal rates were 37.16%, 46.74%, 49.42%, 41.00%, and 37.93%, respectively, significantly higher than that of the control (only 12.93%). Njoku et al. (2009) reported that the growth of *Glycine max* in crude oil-contaminated soils reduced the toxicity of crude oil in the soil. Basumatary et al. (2013) reported TPH degradation by both the plants *Cyperus rotundus* and *C. brevifolius*. However, in their field experiments *C. rotundus* showed comparatively more TPH degradation (75%) than *C. brevifolius* (64%). They attributed significant differences of TPH degradation in the vegetated and unvegetated plots in the 360-day period to the difference in the existing field conditions.

Wang et al. (2008) through pot experiments showed that phytoremediation of petroleum-contaminated soil with three species of grasses, namely *Pannicum, Eleusine indica* (L.) Gaerth, and Tall fescue, for 150 days could degrade petroleum hydrocarbons by 30% to 40% in the planted soil. The results of the biodegradation of oil and its fractions showed that TPHs were reduced by 30% in the rhizosphere soil of *Vicia faba* as compared to 16.8% and 13.7% reduction in rhizosphere soil of *Zea mays* and *Triticum aestivum,* respectively (Diab, 2008).

Tang et al. (2010) worked on soil petroleum remediation using ryegrass (*Lolium perenne*) and the microbial strains of *Bacillus subtilis, Sphingobacterium multivolume, Acinetobacter radioresistens, Rhodococcus erythropolis,* and *Pseudomonas fluorescens* both as a single agent and a mixed agent. The combination of ryegrass with mixed microbial strains resulted in a TPH degradation rate of 58% after 162 days. These results support our experimental findings where higher percentage degradation of TPH in *R. ruber* inoculated soil could be obtained compared to *C. dactylon* in composted soil.

12.3.3 Estimation of heavy metals

From the remediation experiments for all the metals, it was evident that *R. ruber* inoculated treatment showed comparatively higher degradation than composted treatments (Tables 12.3–12.6).

TABLE 12.3 Variation in the concentration of lead during phytoremediation with C. *dactylon* in composted and inoculated treatments.

	Day 10		Day 50		Day 90	
Treatment	**Composted**	***R. ruber* inoculated**	**Composted**	***R. ruber* inoculated**	**Composted**	***R. ruber* inoculated**
Control	91.48 ± 2.40	91.48 ± 2.40	89.64 ± 1.18[a]	89.64 ± 1.18[a]	88.15 ± 2.13[a]	88.15 ± 2.13[a]
T1	97.98 ± 7.99	86.63 ± 3.26	69.72 ± 3.26[a]	67.54 ± 5.83[a]	57.21 ± 2.26[a]	51.04 ± 2.58[a]
T2	98.34 ± 3.10	88.93 ± 3.92	69.86 ± 3.92[a]	61.23 ± 3.62[a]	56.43 ± 1.55[a]	50.99 ± 0.25[a]
T3	98.63 ± 5.16	90.60 ± 7.05	64.60 ± 5.07[a]	59.89 ± 3.71[a]	53.89 ± 3.27[a]	47.07 ± 2.46[a]
T4	99.26 ± 8.14[a]	93.53 ± 7.05[a]	65.98 ± 4.26[a]	58.29 ± 3.16[a]	56.67 ± 3.64[a]	46.48 ± 3.17[a]

All values are represented in mg/kg.

[a] *The mean difference is significantly different to control at P = 0.05.*

TABLE 12.4 Variation in the concentration of zinc during phytoremediation with C. *dactylon* in composted and inoculated treatments.

	Day 10		Day 50		Day 90	
Treatment	**Composted**	***R. ruber* inoculated**	**Composted**	***R. ruber* inoculated**	**Composted**	***R. ruber* inoculated**
Control	25.74 ± 0.62	26.19 ± 1.20	21.79 ± 0.59	23.04 ± 0.76	19.50 ± 0.83	22.25 ± 0.18
T1	38.49 ± 0.81	29.24 ± 2.00	33.71 ± 1.51[a]	24.72 ± 0.60	24.96 ± 1.99	22.96 ± 0.78
T2	30.89 ± 2.96	25.29 ± 0.52	23.59 ± 2.83[a]	22.42 ± 0.61	29.76 ± 3.94	19.80 ± 0.04[a]
T3	27.76 ± 1.61	25.33 ± 1.23	23.91 ± 2.22	23.20 ± 2.20	24.26 ± 1.11	18.70 ± 0.24[a]
T4	26.40 ± 6.13	26.69 ± 2.43	25.84 ± 0.31[a]	23.48 ± 0.59	21.00 ± 2.69	18.64 ± 1.13[a]

All values are represented in mg/kg.
[a] *The mean difference is significantly different to control at P = 0.05.*

TABLE 12.5 Variation in the concentration of copper during phytoremediation with C. *dactylon* in composted and inoculated treatments.

	Day 10		Day 50		Day 90	
Treatment	**Composted**	***R. ruber* inoculated**	**Composted**	***R. ruber* inoculated**	**Composted**	***R. ruber* inoculated**
Control	25.39 ± 0.39	24.82 ± 0.28	23.18 ± 0.76	23.28 ± 1.00	22.90 ± 0.52	22.14 ± 0.70
T1	22.94 ± 1.62	24.79 ± 1.01	22.08 ± 0.69	22.27 ± 0.43	20.01 ± 0.35[a]	20.78 ± 0.16
T2	22.66 ± 0.77	24.83 ± 0.81	20.36 ± 0.21[a]	22.26 ± 0.49	19.25 ± 0.70[a]	19.27 ± 0.62[a]
T3	22.53 ± 1.40	24.35 ± 0.83	20.19 ± 0.12[a]	22.09 ± 0.10	18.52 ± 1.21[a]	18.02 ± 0.83[a]
T4	22.38 ± 0.58	24.29 ± 0.60	20.35 ± 0.68[a]	21.88 ± 0.84	17.77 ± 0.86[a]	17.86 ± 0.48[a]

All values are represented in mg/kg.
[a] *The mean difference is significantly different to control at P = 0.05.*

TABLE 12.6 Variation in the concentration of cadmium during phytoremediation with C. *dactylon* in composted and inoculated treatments.

	Day 10		Day 50		Day 90	
Treatment	**Composted**	***R. ruber* inoculated**	**Composted**	***R. ruber* inoculated**	**Composted**	***R. ruber* inoculated**
Control	1.45 ± 0.02	1.44 ± 0.02	1.31 ± 0.01	1.35 ± 0.02	1.29 ± 0.01[a]	1.27 ± 0.01
T1	1.41 ± 0.03	1.42 ± 0.01	1.32 ± 0.02	1.34 ± 0.01	1.25 ± 0.01[a]	1.24 ± 0.01
T2	1.39 ± 0.03[a]	1.39 ± 0.03	1.36 ± 0.03[a]	1.33 ± 0.02	1.23 ± 0.01[a]	1.16 ± 0.01[a]
T3	1.44 ± 0.02	1.38 ± 0.01[a]	1.36 ± 0.01[a]	1.27 ± 0.02[a]	1.19 ± 0.03[a]	1.10 ± 0.03[a]
T4	1.42 ± 0.02	1.41 ± 0.02	1.35 ± 0.02	1.26 ± 0.01[a]	1.17 ± 0.00[a]	1.09 ± 0.03[a]

All values are represented in mg/kg.
[a] *The mean difference is significantly different to control at P = 0.05.*

12.3.3.1 *Reduction in the concentration of lead (Pb)*

Reduction in the concentration of lead along the treatments in composted and inoculated soil is presented in Table 12.3. In composted treatment, the percentage degradation at the end of the experiment was 40.58% in T1, 42.28% in T2, 43.33% in T3, and 47.95% in T4, whereas in uncontaminated control it was 3.64%; in *R. ruber* inoculated treatment, it was 41.09% in T1, 42.66% in T2, 48.05% in T3, and 50.31% in T4, whereas it was 3.64% in uncontaminated control, and these results corroborate with the findings of Yoon et al. (2006).

Peer et al. (2005) reported that several plant species can hyperaccumulate Pb. *Piptathertan miliacetall*, a grass, accumulated Pb directly correlating to soil concentrations without showing any symptoms of toxicity in the experiments conducted for 3 weeks. *Sesbania drummondii* tolerated Pb levels up to 1500 mg/L while *Brassica juncea* showed reduced growth at 645 µg/g Pb in the soil substrate. The phytoextraction efficiency of lead in vetiver grass was reported by Wilde et al. (2005). Vetiver acted as a potential phytostabilizer of metals in red mud-treated soil (Gautam and Agrawal, 2017). By hyperaccumulating and phytoextraction mechanisms, plants were able to reduce the contaminant levels of the soil. Thomas et al. (2014) indicated compositional changes in Bermuda grass root exudates as organic carbon concentrations increased overtime demonstrating the ability of Bermuda grass root exudates to complex with lead in aqueous media.

12.3.3.2 Reduction in the concentration of zinc (Zn)

Though Zn is an essential microelement, it is toxic to animals and plants at high concentrations. During the 90 days of experimental period (Table 12.4), the reduction in zinc content in composted treatment was 21.58% in T1, 22.01% in T2, 28.24% in T3, and 36.57% in T4, whereas in uncontaminated control it was 15.15%; in *R.ruber* inoculated treatment, the reduction was 21.48% in T1, 21.7% in T2, 26.18% in T3, and 30.16% in T4, whereas in uncontaminated control it was 17.71%. Peer et al. (2005) reported that from sewer sludge contaminated agricultural soil in UK and a site contaminated by mine tailings in Silver Bow Creek, *Thlaspi caerulescens* reduced Cd and Zn from the soil. Bennett et al. (2003) observed that *Brassica juncea* plants genetically modified with bacterial genes were found to accumulate 1.5 times more Cd and Zn compared to wild-type *B. juncea* growing on metal-contaminated soil from a USEPA Superfund site near Leadville Company. Similar results were reported in a laboratory-scale experiment that aimed to verify the possibility of using the vegetable species *Mirabilis jalapa* for the decontamination of soils which were artificially contaminated with different zinc and lead contents for one year. Montevecchio soils were diluted with vegetative soil and compost with a dilution factor 4 and 8 (600–1700 ppm Zn; 5700–19,000 ppm Pb). Reduction in bioavailable metal fraction revealed that *Mirabilis jalapa* can be used for phytostabilization (Carucci et al., 2005).

12.3.3.3 Reduction in the concentration of copper (Cu)

The reduction in copper content in composted treatment was 12.82% in T1, 15.05% in T2, 17.8% in T3, and 20.6% in T4, whereas in uncontaminated control it was 9.8%; in *R. ruber* inoculated treatment, it was 16.14% in T1, 22.39% in T2, 25.99% in T3, and 26.47% in T4, whereas in uncontaminated control it was 10.8% by the end of the experiment (Table 12.5). Chigbo and Batty (2013) observed that the total removal of Cu from pyrene and copper metal co-contaminated soil was always higher by two- to threefold in aged soil than in freshly spiked soil.

12.3.3.4 Reduction in the concentration of cadmium (Cd)

Cd is a toxic heavy metal pollutant whose presence is of great concern in agricultural soils and crops. Soil bioaugmentation promoted cadmium phytoremediation in Napier grass (Wiangkham and Prapagdee, 2018). Naturally occurring Cd concentration in soils ranged from 0.01 to 7 ppm (Ramachandran and D'Souza, 1998). As presented in Table 12.6, the reduction in cadmium content toward the end of the experiment in composted treatment was 11.35% in T1, 11.51% in T2, 17.36% in T3, and 17.61% in T4, whereas in uncontaminated control it was 11.03%; in *R. ruber* inoculated treatment, it was 12.68% in T1, 16.55% in T2, 20.07% in T3, and 22.7% in T4, whereas in uncontaminated control it was 11.89%.

Shi and Cai (2009) reported that crops namely hemp (*Cannabis sativa*), flax (*Linum usitatissimum*), castor (*Ricinus communis*), and peanut (*Arachis hypogea*) were more tolerant to cadmium. Three of these crops (hemp, flax, and peanut) were excellent candidates for phytoremediation in soils contaminated with cadmium. Plant's capability to mediate the removal of crude oil in contaminated soil was not significantly affected by the concentrations of metals in the soil (Atagana, 2011). Cao et al. (2008) reported that when *Trichoderma atroviride* F6, a fungus isolated from decaying feather and resistant to 100 mg/L Cd^{2+} and 250 mg/L Ni^{2+}, was applied for rhizoremediation of Cd, Ni, and Cd–Ni combination contaminated soils through association with *Brassica juncea* (L.) Coss.var. foliosa, the efficiency of phytoextraction for *B. juncea* (L.) Coss.var. foliosa got enhanced after inoculating with *T. atroviride* F6.

By the end of the experiment (90 days), higher reduction in the levels of lead contaminants followed by Zn, Cu, and Cd was observed. Based on the percentage degradation of heavy metal content in soil, the metals studied can be classified, from the higher degradability to the lower degradability, as follows: Pb > Zn > Cu > Cd. The concentration of chromium was found to be below detectable levels. Yoon et al. (2006) reported that the experimental soil mainly contaminated with Pb showed elevated levels of Cu and Zn supposed to be derived from a single source. Substantial amount of heavy metal(loid)s was translocated into *Echinochloa crus-galli* (cockspur grass), and their removal efficiency after 3 weeks was as follows: Ni ≥ Cu > Co > Zn > Mn > Fe > As (Jung et al., 2020).

12.4 Conclusions

The plant species (*C. dactylon*) employed in the current phytoremediation study showed a promising behavior, surviving, and phytoaccumulating heavy metals in crude oil-contaminated soil. Composted and *R. ruber*-amended treatments were maintained separately and analyzed for degradation of hydrocarbon contamination and assessed using biomass, percentage degradation of total petroleum hydrocarbons, and concentration of heavy metals during 10 days, 50 days, and 90 days of sampling periods. It was understood from our studies on biomass that crude oil contamination reduced the amount of plant growth in all the treatments. The decrease in both heavy metals and TPH contaminants in the *R. ruber* inoculated treatments indicated the effectiveness of the microbe-inoculated phytoremediation system. Higher percentage degradation in case of all the metals in *R. ruber* inoculated treatments was observed compared to composted treatments. The order of degradation of metals was found to be Pb > Zn > Cu > Cd. Combining *C. dactylon* with *R. ruber* offered higher potential to efficiently remediate crude oil-contaminated sites.

References

Adam, G., Duncan, H., 2002. Influence of diesel on seed germination. Environ. Pollut. 120, 363–370.

Addo, M.A., Darko, E.O., Gordon, C., Nyarko, B.J.B., Gbadago, J.K., 2012. Heavy metal concentrations in road deposited dust at Ketu-south district, Ghana. Intr. J. Sci. Tech. 2 (1), 28–39.

Ali, H., Naseer, M., Sajad, M.A., 2012. Phytoremediation of heavy metals by *Trifolium alexandrinum*. Intr. J. Environ. Sci. 2 (3), 1459–1469.

American Public Health Association (APHA), 2005. Standard Methods for the Examination of Water and Wastewater, twenty-first ed. APHA, AWWA, WPCF, Washington, DC, USA, pp. 1–1207.

Archer, M.J.G., Caldwell, R.A., 2004. Response of six Australian plant species to heavy metal contamination at an abandoned mined site. Water Air Soil Pollut. 157, 257–267.

ASTM, 2004. Standard Test Method for Dimer/Trimer of Chlorotrifluoroethylene (S-316) Recoverable Oil and Grease and Nonpolar Material by Infrared Determination, American Society for Testing Materials, Method Designation D 7066-04. ASTM, West Conshohocken, PA, USA, pp. 1–9. www.astm.org.

Atagana, H.I., 2011. Bioremediation of co-contamination of crude oil and heavy metals in soil by phytoremediation using *Chromolaena odorata* (L.) King, Robinson HE. Water Air Soil Pollut. 215, 261–271.

Baoune, H., Aparicio, J.D., Acuña, A., El Hadj-khelil, A.O., Sanchez, L., Polti, M.A., Alvarez, A., 2019. Effectiveness of the *Zea mays— Streptomyces* association for the phytoremediation of petroleum hydrocarbons impacted soils. Ecotoxicol. Environ. Saf. 184, 109591.

Barker, A.V., Bryson, G.V., 2002. Bioremediation of heavy metals and organic toxicants by composting. Sci. World J. 2, 407–420.

Basumatary, B., Saikia, R., Das, H.C., Bordoloi, S., 2013. Field note: Phytoremediation of petroleum sludge contaminated field using sedge species, *Cyperus rotundus* (Linn.) and *Cyperus brevifolius* (Rottb.) Hassk. Intr. J. Phytorem. 15 (9), 877–888.

Beaudin, N., Caron, R.F., Legros, R., Ramsay, J., Lawlor, L., Ramsay, B., 1996. Co-composting of weathered hydrocarbon contaminated soil. Compost. Sci. Util. 4 (2), 37–45.

Bennett, L.E., Burkhead, J.L., Hale, K.L., Terry, N., Pilon, M., Pilon-Smits, E.A., 2003. Analysis of transgenic Indian mustard plants for phytoremediation of metal contaminated mine tailings. J. Environ. Qual. 32, 432–440.

Bremner, J.M., Sparks, D.L. (Eds.), 1996. Total nitrogen. In: Methods of Soil Analysis Chemical Methods Part 3. Soil Science Society of America Inc., American Society of Agronomy, Madison, WI, pp. 1085–1122.

Cao, L., Jiang, M., Zeng, Z., Du, A., Tan, H., Liu, Y., 2008. *Trichoderma atroviride* F6 improves phytoextraction efficiency of mustard *Brassica juncea* (L.) Coss var foliosa Bailey in Cd, Ni contaminated soils. Chemosphere 71 (9), 1769–1773.

Carman, E.P., Crossman, T.L., Gatliff, E.G., 1998. Phytoremediation of no. 2 fuel oil contaminated soil. Soil Sedim. Contam. 7 (4), 455–466.

Carucci, A., Cao, A., Fois, G., Muntoni, A., 2005. Phytoremediation of Zinc and Lead Contaminated Soils Using *Mirabilis jalapa*. Contaminated Soils, Sediments and Water, Part V. Springer, pp. 329–338.

Chigbo, C., Batty, L., 2013. Phytoremediation potential of *Brassica juncea* in Cu-pyrene co-contaminated soil: comparing freshly spiked soil with aged soil. J. Environ. Manage 129, 18–24.

Diab, E.A., 2008. Phytoremediation of oil contaminated desert soil using the rhizosphere effects of some plants. Res. J. Agric. Biol. Sci. 4 (6), 604–610.

Diab, A., Sandouka, M., 2010. Biodegradation of polycyclic aromatic hydrocarbons (PAHs) in the rhizosphere soil of *Cyperus conglomeratus*, an Egyptian wild desert plant. Nat. Sci. 8 (12), 144–153.

Garcia-Diaz, C., Ponce-Noyola, M.T., Esparza-Garcia, F., Rivera-Orduna, F., Barrera-Cortes, J., 2013. PAH removal of high molecular weight by characterized bacterial strains from different organic sources. Intr. Biodeter. Biodegrad. 85, 311–322.

Gautam, M., Agrawal, M., 2017. Phytoremediation of metals using vetiver (*Chrysopogon zizanioides* (L.) Roberty) grown under different levels of red mud in sludge amended soil. J. Geochem. Explor. 182, 218–227.

Gee, G.W., Bauder, J.W., Klute, A. (Eds.), 1986. Particle Size Analysis. Methods of Soil Analysis, Physical and Mineralogical Methods Part I, second ed. Soil Science Society of America Inc., American Society of Agronomy, Madison, pp. 383–411.

Hilyard, E.J., Jones-Meehan, J.M., Spargo, B.J., Hill, R.T., 2008. Enrichment, isolation and phylogenetic identification of polycyclic aromatic hydrocarbon-degrading bacteria from Elizabeth river sediments. Appl. Environ. Microbiol. 74, 1176–1182.

Jing, Y., He, Z., Yang, X., 2007. Role of soil rhizobacteria in phytoremediation of heavy metal contaminated soils. J. Zhejiang Univ. Sci. B 8 (3), 192–207.

Jung, S., Kim, M., Moon, H., Park, Y.K., Rinklebe, J., Park, C.J., Kwon, E.E., 2020. Valorization of phytoremediation byproduct via synthesis of biodiesel from cockspur grass (*Echinochloa crusgalli*) Seed. ACS Sustain. Chem. Eng. 8 (31), 11588–11595.

Kiyohara, H., Nagao, K., Yana, K., 1982. Rapid screen for bacteria degrading water insoluble, solid hydrocarbons on agar plates. Appl. Environ. Microbiol. 43, 454–457.

Koduah Owusu Ansah, 2012. Warm Season Turfgrass as Potential Candidates to Phytoremediate Arsenic Pollutants at Obuasi Gold Mine in Ghana. Master of Science, Thesis, Colorado State University, Colorado.

Lebkowska, M., Zborowska, E., Karwowska, E., Miaskiewicz-Peska, E., Muszynski, A., Tabernacka, A., Naumczyk, J., Jeczalik, M., 2011. Bioremediation of soil polluted with fuels by sequential multiple injection of native microorganisms: field scale processes in Poland. Ecol. Eng. 37, 1895–1900.

Liu, R., Jadeja, R.N., Zhou, Q., Liu, Z., 2012. Treatment and remediation of petroleum contaminated soils using selective ornamental plants. Environ. Eng. Sci. 29 (6), 494–501.
Liu, X., Shen, S., Zhang, X., Chen, X., Jin, R., Li, X., 2020. Effect of enhancers on the phytoremediation of soils polluted by pyrene and Ni using Sudan grass (Sorghum sudanense (Piper) Stapf.). Environ. Sci. Pollut. Res. 27 (33), 41639–41646.
Mahdavi, A., Khermandar, K., Asbchin, S.A., Tabaraki, R., 2014. Lead accumulation potential in *Acacia victoria*. Intr. J. Phytorem. 16, 582–592.
Merkl, N., Kraft, R.S., Infante, C., 2005. Phytoremediation in the tropics-influence of heavy crude oil on root morphological characteristics of graminoids. Environ. Pollut. 138, 86–91.
Muratova, A.Y., Dmitrieva, T.V., Panchenko, L.V., Turkovskaya, O.V., 2008. Phytoremediation of oil sludge contaminated soil. Intr. J. Phytorem. 10 (6), 486–502.
Njoku, K.L., Akinola, M.O., Oboh, B.O., 2009. Phytoremediation of crude oil contaminated soil: the effect of growth of *Glycine max* on the physic-chemistry and crude oil contents of soil. Nat. Sci. 7 (10), 79–87.
Olson, P.E., Castro, A., Joern, M., DuTeau, N.M., Pilon Smits, E.A.H., Reardon, K.F., 2007. Comparison of plant families in a greenhouse phytoremediation study on aged polycyclic aromatic hydrocarbon-contaminated soil. J. Environ. Qual. 36, 1461–1469.
Peer, W.A., Baxter, I.R., Richards, E.L., Freeman, J.L., Murphy, A.S., 2005. Phytoremediation and hyperaccumulator plants. In: Tamas, M., Martinoia, E. (Eds.), Molecular Biology of Metal Homeostasis and Detoxification. Topics in Current Genetics, vol. 14. Springer, Berlin, pp. 299–340.
Prathap, M.G., ZainUlIbad, C., Ram, S.H., Vivek, P., Rajasekaran, M., Sudarsan, J.S., Nithiyanantham, S., 2019. Effectiveness of phytoremediation to the removal of heavy metals using absorbents: wastewater treatment. Int. J. Energy Water Res. 3 (3), 263–267.
Rajkumar, M., Sandhya, S., Prasad, M.N.V., Freitas, H., 2012. Perspectives of plant-associated microbes in heavy metal phytoremediation. Biotech. Adv. 30, 1562–1574.
Ramachandran, V., D'Souza, T.J., 1998. Plant uptake of cadmium, zinc and manganese in soils amended with sewage sludge and city compost. Bull. Environ. Contam. Toxicol. 61, 347–354.
Razmjoo, K., Adavi, Z., 2012. Assessment of Bermuda grass cultivars for phytoremediation of petroleum contaminated soils. Intr. J. Phytorem. 14 (1), 14–23.
Revathi, K., Haribabu, T.E., Sudha, P.N., 2011. Phytoremediation of Chromium contaminated soil using Sorghum plant. Intr. J. Environ. Sci. 2 (2), 417–428.
Rezek, J., Wiesche, C.I.D., Mackova, M., Zadrazil, F., Macek, T., 2008. The effect of ryegrass (*Lolium perenne*) on decrease of PAH content in long term contaminated soil. Chemosphere 70, 1603–1608.
Sanusi, S.N.A., Abdullah, S.R.S., Idris, M., 2012. Preliminary test of phytoremediation of hydrocarbon contaminated soil using *Paspalum vaginatum* Sw. Aust. J. Basic Appl. Sci. 6 (1), 39–42.
Schwartz, G., Ben-Dor, E., Eshel, G., 2012. Quantitative analysis of total petroleum hydrocarbons in soils: comparison between reflectance spectroscopy and solvent extraction by 3 certified laboratories. Appl. Environ. Soil Sci. 2012. https://doi.org/10.1155/2012/751956. Article ID 751956, 11 pages.
Shi, G., Cai, Q., 2009. Cadmium tolerance and accumulation in eight potential energy crops. Biotechnol. Adv. 27 (5), 555–561.
Sinha, R.K., Dalsukh, V., Shanu, S., Shweta, S., Sunil, H., 2009. Bioremediation of contaminated sites: a low cost nature's biotechnology for environmental clean-up by versatile microbes, plants and earthworms. In: Timo, F., Johann, H. (Eds.), Solid Waste Management and Environmental Remediation: Environmental Remediation Technologies. Regulations and Safety, Nova Science Publishers, Chapter 1, pp. 1–72.
Sumia, K., Muhammad, A., Samina, I., Qaiser, M.K., 2013. Plant-bacteria partnerships for the remediation of hydrocarbon contaminated soils. Chemosphere 90 (4), 1317–1332.
Tang, J., Wang, R., Niu, X., Wang, M., Zhou, Q., 2010. Characterization on the rhizoremediation of petroleum contaminated soil as affected by different influencing factors. Biogeosci. Discuss. 7, 4665–4688.
Tangahu, B.V., Abdullah, S.R.S., Basri, H., Idris, M., Anuar, N., Mukhlisin, M., 2011. A review on heavy metals (As, Pb and Hg) uptake by plants through phytoremediation. Intr. J. Chem. Eng. 2011. https://doi.org/10.1155/2011/939161. Article ID 939161, 31 pages.
Techer, D., Chois, C.M., Gilly, P.L., Henry, S., Bennasroune, A., Marielle, D.I., Falla, J., 2012. Assessment of *Miscanthus×giganteus* for rhizoremediation of long term PAH contaminated soils. Appl. Soil Ecol. 62, 42–49.
Thomas, C., Butler, A., Larson, S., Medina, V., Begonia, M., 2014. Complexation of lead by bermuda grass root exudates in aqueous media. Intr. J. Phytorem. 16 (6), 634–640.
United States Environmental Protection Agency (USEPA), 1978. Method 418.1, Test Method for Evaluating Total Recoverable Petroleum Hydrocarbon, (Spectrophotometric, Infrared). Government Printing Office, Washington, DC, USA.
United States Environmental Protection Agency (USEPA), 2000. Method 3546: Microwave Extraction, Draft Version 0, SW-846. Washington, DC, USA.
Van Gestel, K., Mergaert, J., Swings, J., Coosemans, J., Ryckeboer, J., 2003. Bioremediation of diesel oil contaminated soil by composting with biowaste. Environ. Pollut. 125, 361–368.
Wang, J., Zhang, Z., Su, Y., He, W., He, F., Song, H., 2008. Phytoremediation of petroleum polluted soil. Petroleum Sci. 5, 167–171.
White, P.M., Wolf, D.C., Thoma, G.J., Reynolds, C.M., 2003. Influence of organic and inorganic soil amendments on plant growth in crude oil-contaminated soil. Intr. J. Phytorem. 5, 381–397.
Wiangkham, N., Prapagdee, B., 2018. Potential of Napier grass with cadmium-resistant bacterial inoculation on cadmium phytoremediation and its possibility to use as biomass fuel. Chemosphere 201, 511–518.
Wilde, E.W., Brigmon, R.L., Dunn, D.L., Heitkamp, M.A., Dagnan, D.C., 2005. Phytoextraction of lead from firing range soil by Vetiver grass. Chemosphere 61 (10), 1451–1457.
Yoon, J., Cao, X., Zhou, Q., Ma, L.Q., 2006. Accumulation of Pb, Cu and Zn in native plants growing on a contaminated Florida site. Sci. Total Environ. 368, 456–464.
Zand, A.D., Bidhendi, G.N., Mehrdadi, N., 2009. Phytoremediation of total petroleum hydrocarbons (TPHs) using plant species in Iran. Turkish J. Agric. Forest. 34, 429–438.

CHAPTER

13

Proteobacteria response to heavy metal pollution stress and their bioremediation potential

Veronica Fabian Nyoyoko
Department of Microbiology, University of Nigeria, Nsukka, Nigeria

OUTLINE

13.1 Introduction

The World Health Organization (WHO) has estimated that 4.9 million deaths (8.3 percent of total mortality worldwide) are attributable to environmental exposure and inappropriate serious management of toxic chemicals (Pruss-Ustun et al., 2011). Environmental pollution has been on the rise in the past few decades owing to increased human activities on energy reservoirs, unsafe agricultural practices, and rapid industrialization (Hadia-e-Fatima, 2018). Among the pollutants that are of environmental and public health concerns due to their toxicities are heavy metals, nuclear wastes, pesticides, greenhouse gases, and hydrocarbons. These toxic pollutants are discharged from specific locations worldwide and thus pollute specific regions. These pollutants have different toxicity and chemical behavior (Nibourg et al., 2013). Environmental contamination caused by heavy metals (HMs) has received increased attention worldwide (Gupta et al., 2016; Sharma, 2016; Ayangbenro and Babalola, 2017; Liu et al., 2018a,b). Heavy metal (HM) is any metallic element that has a relatively high density and is toxic or poisonous at low concentrations. Heavy metals are elements with an atomic number higher than 20, an atomic mass greater than 40 g, and a specific weight of more than 5 g/cm^3 (Canli and Atli, 2003; Hamsa et al., 2017). These elements often find their way into soil through environmental contaminants including the atmospheric pollutants in industrial regions (emissions from the rapidly expanding industrial areas), unlimited use of agricultural fertilizers, mine tailings, disposal of high metal wastes, leaded gasoline and paints, animal manures, sewage sludge, pesticides, wastewater irrigation, coal combustion res-

Cost Effective Technologies for Solid Waste and Wastewater Treatment
https://doi.org/10.1016/B978-0-12-822933-0.00010-3

idues, spillage of petrochemicals, atmospheric deposition, and municipal and industrial sewage systems in a nonreturnable fashion (Cobbina et al., 2015; Malik et al., 2017; Srivastava et al., 2017; Ramya and Joseph Thatheyus, 2018).

Activities such as the use of agrochemicals and long-term application of urban sewage sludge, industrial waste disposal, waste incineration, and vehicle exhausts are the main sources of HM in agricultural soils (Mishra et al., 2017). Heavy metals in the soil include mercury (Hg), lead (Pb), chromium (Cr), arsenic (As), zinc (Zn), cadmium (Cd), uranium (U), selenium (Se), silver (Ag), gold (Au), copper (Cu), and nickel (Ni) (Srivastava et al., 2017). The danger of heavy metals is intensified by their almost indefinite persistence in the environment due to their absolute nature which cannot be degraded (Gupta et al., 2016). Metals are nonbiodegradable but can be transformed through sorption, methylation, complexation, and changes in valence state (Anyanwu et al., 2011).

Toxic metals apply their toxicity in the displacement of essential metals from their normal binding sites of biological molecules, inhibition of enzymatic functioning and disruption of nucleic acid structure, oxidation stress, genotoxicity, and interfering with signaling pathways (Srivastava et al., 2017; Venkatachalam et al., 2017). Ecologically, the accumulation of heavy metals in soils is extremely hazardous because soil is a major link in the natural cycling of chemical elements; it is also a primary component of the trophic chain (Liu et al., 2012; Sagi and Yigit, 2012; Wyszkowska et al., 2013).

Pollution by heavy metal is a threat to the environment, and its remediation is a major challenge to environmental research. Heavy metal pollution is a serious global environmental problem as it adversely affects the biotic and abiotic components of the ecosystem and alters the composition and activity of soil microbial communities (Ayangbenro and Babalola, 2017). The non-biodegradability of heavy metals makes it hard to remove them from contaminated biological tissues and soil, and this is a major concern for global health because of their lethal nature.

Proteobacteria are named after Proteus, a Greek god of the sea, capable of assuming many different shapes, and it is therefore not named after the genus Proteus. Proteobacteria make up one of the largest phyla and most versatile in the bacteria domain. All proteobacteria are Gram-negative, with an outer membrane mainly composed of lipopolysaccharides. Many move about using flagella, but some are nonmotile or rely on bacterial gliding. The last include the myxobacteria, a unique group of bacteria that can aggregate to form multicellular fruiting bodies. There is also wide variety in the types of metabolism. Most members are facultatively or obligately anaerobic, chemoautotrophs, and heterotrophic, but there are numerous exceptions. A variety of genera, which are not closely related to each other, convert energy from light through photosynthesis. These are called purple bacteria, referring to their mostly reddish pigmentation.

The divisions of the proteobacteria were once regarded as subclasses (e.g., α-subclass of the Proteobacteria), but are now regarded as classes (e.g., the Alphaproteobacteria). These classes include Alphaproteobacteria, Betaproteobacteria, Gammaproteobacteria, Deltaproteobacteria, Epsilonproteobacteria, and Zetaproteobacteria.

Ammonium-oxidizing microorganisms are organism that carries out the first step in nitrification reaction (biochemical process of oxidation of ammonia (NH_4^+)). They belong to the phylum Proteobacteria. They include ammonia-oxidizing bacteria (AOB) *(Nitrosomonas, Nitrosococcus, Nitrosospira, Nitrosolobus, Nitrosovibrio)*, ammonia-oxidizing archaea (AOA), and heterotrophic bacteria (*Arthrobacter globiformis, Aerobacter aerogenes, Thiosphaera pantotropha, Streptomyces grisens*, and various *Pseudomonas* spp*) and* fungi (*Aspergillus flavus*) (Hamsa et al., 2017). Recent research on the metabolic pathways of heterotrophic ammonia oxidation has been conducted using *Paracoccus denitrificans* (Moir et al., 1996b), *Alcaligenes faecalis, Pseudomonas putida* (Daum et al., 1998), and a few other bacterial species.

Ammonium-oxidizing microbes (AOM) obtain their energy by oxidation of ammonia (NH_3) to nitrite (NO_2^-). These organisms utilize a few key enzymes such as ammonia monooxygenase (AMO) and hydroxylamine oxidoreductase (HAO) to bring about the conversion. The presence of ammonia monooxygenase subunit-A gene (*amoA*) encodes ammonia monooxygenase (AMO), a key enzyme that catalyzes the first step in ammonia oxidation. AOB was first reported in 1890 by Winogradsky, and several groups began isolating and cultivating AOB from a variety of environments such as marine waters, estuarine soils, and wastewater treatment systems (Reddy et al., 2015).

Nitrogen is an essential element for plants (Vimal et al., 2017). Nitrifying bacteria play important role in soil fertility, make available nitrate nitrogen to plants (common soil nutrient element required in large quantity by plants), aid in waste treatment plant, biogeochemical cycling of nitrogen compounds, and purification of the air. Nitrifying bacteria in polluted soil initiate a syntrophic pathway that provides intermediates for heterotrophic bacterial activity and thus are excellent candidates for remediation (John and Okpokwasili, 2012).

Microorganisms are very sensitive; they react quickly to any kind of changes (natural and anthropogenic) in the environment and quickly adapt themselves to new conditions. Microorganisms take heavy metals into the cell in significant amounts. This phenomenon leads to the intracellular accumulation of metal cations of the environment and is defined as bioaccumulation (Wolejko et al., 2016). Some bacterial plasmids contain specific genes for resistance to toxic heavy metal ions (Liu et al., 2018a,b; Pacwa-Płociniczak et al., 2018; Lukina et al., 2016; Sharma, 2016) and ability to solubilize phosphate (biofertilizers) (Ibiene and Okpokwasili, 2011; Gupta et al., 2014). Some microorganisms can adjust their metabolic activity or community structure to adapt to the harmful shock loadings. Microorganisms play important role in the stress environment and the derived ecosystem functions (Singh et al., 2016a,b; Vimal et al., 2017). Microorganisms can mobilize or immobilize metals by biosorption, sequestration, production of chelating agents, chemoorganotrophic and autotrophic leaching, methylation, and redox transformations. These mechanisms stem from prior exposure of microorganisms to metals which enable them to develop the resistance and tolerance useful for biological treatment (Viti and Giovannetti, 2003). Microbe-metal interaction in soil/waste disposal is of interest to environmentalists in order to use adapted microorganisms as a source of biomass for bioremediation of heavy metals (Sharma, 2016; Singh et al., 2016a,b). Metals detoxify through resistance and tolerance, and this resistance can be attributed to mechanisms of exclusion or tolerance (Klassen et al., 2000). In an endeavor to safeguard the susceptible cellular components, a cell is capable of building up resistance to metals. In an effective bioremediation process, microorganisms enzymatically attack the pollutants and convert them to harmless products through chemical, physical, and biological processes. Proteobacteria have been shown to tolerate heavy metal stress, including toxic salts of noble metals (Johnson et al., 2019).

13.2 Materials and methods

13.2.1 Sample collection

Surface soil samples at depth of 0–15 cm were collected at random from Akwa Ibom State University in Akwa Ibom State, and soil sample was collected from University of Nigeria, Nsukka, and from solid waste disposal site in Uyo, Akwa Ibom State. The soil was collected using sterile auger borer and into sterile polyethylene bag, merged to form a composite soil sample, and transferred to the laboratory for analysis.

13.2.2 Microbiological analysis

13.2.2.1 Preparation of samples for analyses

Precisely, 5 g of the sieved soil sample was suspended in 45 ml of sterile phosphate buffer containing 139 mg of K_2HPO_4 and 27 mg KH_2PO_4 per liter (pH 7.0) and shaked at 100 rpm for 2 h in order to liberate the organisms into the liquid medium (Deni and Penninck, 1999; John and Okpokwasili, 2012).

13.2.2.2 Preparation of media

Media preparation was carried out using Winogradsky broth medium for serial dilution of soil samples and Winogradsky solid medium for the inoculation of serially diluted soil suspension.

13.2.2.3 Preparation of Winogradsky broth

Winogradsky broth medium phase I was prepared with the following composition (g/L) in sterile distilled water: $(NH_4)_2SO_4$, 2.0; K_2HPO_4, 1; $MgSO_4$. $.7H_2O$, 0.5; NaCl, 2.0; $FeSO_4$ $.7H_2O$, 0.4; and $CaCO_3$, 0.01. Each of the 10 test tubes was filled with 9 ml of the Winogradsky broth media 1, autoclaved at 121°C at 15 psi for 15 min, and allowed to cool. The test tubes were used to carry out tenfold serial dilutions of the soil suspension (John and Okpokwasili, 2012).

13.2.2.4 Preparation of Winogradsky agar media

Winogradsky agar media for nitrification phases I was prepared by adding 15.0 g agar to 1000 mL of fresh broth and sterilized at 121°C at 15 psi for 15 min and allowed to cool to about 45°C before dispersing into sterile Petri dishes (John and Okpokwasili, 2012).

13.2.2.5 Isolation of AOB bacteria from soil sample

All the plates were aseptically inoculated with 0.1 mL of the appropriate dilution of the soil suspension using spread plate technique. All the inoculated Petri dishes were incubated aerobically at room temperature (28 + 2°C) for 1 week and examined for growth.

13.2.2.6 *Purification of isolates*

Discrete colonies that developed on Winogradsky agar media for nitrification phases 1 after 1 week of incubation were aseptically subcultured repeatedly on corresponding freshly prepared Winogradsky agar medium. All the inoculated Petri dishes were incubated aerobically at room temperature ($28 \pm 2°C$) for 3–5 days. The pure isolates were transferred to Winogradsky agar slants and stored in the refrigerator for further use.

13.2.2.7 *Identification of isolates*

The pure isolates from the corresponding agar slants were characterized and identified using morphological (cell and colonial morphology, shape, motility, and Gram reaction), biochemical, and physiology attributes (Holt et al., 1994; Cheesbrough, 2006). The molecular characterization was based on 16SrDNA sequencing (Saha et al., 2013).

13.2.2.8 *Physiological characterization of the isolate*

Nitrite determination by Griess method (Bhaskar and Charyulu, 2005).

Griess-Ilosvay reagent was prepared as follows:

Solution A: 0.6g of sulfanilic acid was dissolved in 70mL of hot distilled water and cooled; 20mL of concentrated HC1 was added; and the volume was made up to 100mL with distilled water.

Solution B: 0.6g of a-naphthylamine was dissolved in 10mL of distilled water containing 1mL of concentrated HC1, and the volume was made up to 100mL with distilled water. Solution C: 16.4g of sodium acetate was dissolved in 70mL of distilled water, and the volume was made up to 100mL with distilled water. The three solutions (A, B, and C) were stored in dark bottles and mixed in equal parts before use.

Ammonium-oxidizing **test for determination of nitrite.**

Five milliliters of Winogradsky mineral basal medium was prepared. The tubes were sterilized by autoclaving at 121°C at 15psi for 15min and allowed to cool. One loopful of each ammonium-oxidizing bacteria isolate was added into each tube and incubated aerobically for 5 days at room temperature. At the end of the incubation period, the presence of nitrite was tested using Griess-Ilosvay reagent. The reagent was added and observed for the development of purplish red/pink coloration within 5min (Joel and Amajuoyi, 2010; Hoang et al., 2016).

13.2.2.9 *Inoculum preparation and standardization*

Inocula were prepared by inoculating isolates onto prepared nutrient agar plates and incubating at 30°C for 24h. After incubation, colonies were suspended in test tubes containing sterile normal saline solution. The tubes were vortexed for 2min and then transferred into a sterile test tube. The cells' suspension was adjusted to a 0.5 McFarland standard (optical density of 0.5–1.0 at 600nm) using sterile normal saline to give a concentration of 108cfu/mL, to get the final inocula.

13.2.2.10 *Tolerance study*

Mineral salts medium of the following composition (g/L): $(NH_4)_2SO_4$, 1.0; KH_2PO_4, 1.0g; K_2HPO_4, 1.0g; $MgSO_4$, 0.2; $CaCl_2$, 0.02; $FeCl_3.6H_2O$; 0.004 for ammonium-oxidizing bacteria. Sterilized by autoclaving at 121°C at 15psi for 15min and allowed to cool.

Salts of copper (Cu), nickel (Ni), cadmium (Cd), and lead (Pb) were used as $CuSO_4.5H_2O$, $NiSO_4.6H_2O$, $CdCl_2.H_2O$, and $Pb(CH_2COO)_2)_2$. $(OH)_2$, respectively.

13.2.2.11 *Primary screening of heavy metal-resistant nitrifying proteobacteria*

An amount (0.1mL) of the standard inoculum was plated into mineral salt agar medium supplement with metal salt concentrations of 100mg/L. The plates were incubated at room temperature for 3–5 days (Jamaluddin et al., 2012; Pandit et al., 2013). The ammonia-oxidizing bacterial isolates were selected for tolerance study.

13.2.2.12 *Experimental setup*

Analytical grades of metal salts were used to prepare stock solutions. The mineral salt medium for ammonia-oxidizing bacteria was amended with the appropriate aliquot of the stock solution of the metal salt concentrations of 100mg/L, 200mg/L, 500mg/L, and 1000mg/L.

13.2.2.13 *Effect of heavy metal on the growth of nitrifying isolates*

Changes in the population of the nitrifying isolates were monitored following their exposure to heavy metals. About 1mL of the standard inoculum was introduced into each flask containing heavy metal salts amended in mineral salt medium (MSM). The cultures were incubated aerobically at room temperature (28 ± 2°C) for 7 days. The growth was measured by withdrawing samples from the medium every 24 h and absorbance of the turbidity measured at 600 nm using spectrophotometer (Bhaskar and Charyulu, 2005).

13.2.3 Biosorption of heavy metals

13.2.3.1 *Experimental setup for biosorption of heavy metals*

Biosorption of selected heavy metals was carried out in a flask. Analytical grades of metal salts were used to prepare stock solutions. The mineral salt medium for ammonia-oxidizing bacteria and nitrite-oxidizing bacteria was amended with the appropriate aliquot of the stock solution of the metal salt concentration of 100 mg/L.

13.2.3.2 *Bioremediation of copper, nickel, cadmium, and lead*

Bioremediation of the selected heavy metals by isolates was carried out in a 250-mL Erlenmeyer flask containing sterile minimal salt medium. In mitigation experiment with 100 mg/L concentrations of the metals. The experiment was conducted on a shaker incubator at 25°C and continuous shaking at 130 rpm. Mitigation was assessed by comparing the disappearance of the metals in the sample and controls over the period of microbial growth. The metal concentrations were monitored over a time to compare lag periods and bioaccumulation rates for different concentrations. The lag period was determined as the time during which the metal concentrations remained relatively constant. Microbial growth was observed in terms of CFU and O.D. (John and Okpokwasili, 2012). The samples (5 mL) were withdrawn hourly from 0 to 6 h and then every 24 h for 14 days. Samples were transferred to 10 ml vials and capped for AAS analysis. The physicochemical parameters such as pH and temperature were observed (Sharma, 2016).

Biosorption capacity (mg/g) of the biosorbent can be defined as the amount of biosorbate (metal ions) biosorbed per unit weight of the biosorbent and can be expressed by using the following mass balance equation:

$$q_e = \frac{(Ci - Ce)V}{M}$$

The% biosorption (R%) known as biosorption efficiency for the metal was evaluated from the following equation:

$$\%\text{Bioaccumulation} = \frac{a-b}{a} \times \frac{100}{1}$$

where a is the weight of heavy metal in before incubation control; b is the weight of heavy metal in the each case after incubation.

13.2.3.3 *DNA extraction using Zr fungal/bacterial DNA Miniprep (manufactured by Zymo research, cat number: D6005)*

Amount 50-100 mg (wet weight) fungal or bacterial cells that have been resuspended in up to 200ul of water or isotonic buffer (e.g., PBS) or up to 200 mg of tissue was added to a ZR Bashing™ Lysis Tube. Added 750 μL Lysis Solution to the tube. Secure in a bead fitted with 2 mL tube holder assembly and process at maximum speed for > 5 min. ZR BashingBead Lysis Tube was centrifuged in a microcentrifuge at > 10,000 × g for 1 min. Thereafter, 400 μL supernatant was transferred to a Zymo-Spin IV Spin Filter (orange top) in a Collection Tube and centrifuged at 7000 × g for 1 min. Note: Snap off the base of the Zymo-Spin Spin Filter prior to use. About 1200 μL of Fungal/Bacterial DNA Binding Buffer was added to the filtrate in the Collection Tube from Step 4. About 800 μL of the mixture from Step 5 was transferred to a Zymo-Spin IIC Column in the Collection Tube and centrifuged at 10,000 × g for 1 min. Note: The Zymo-Spin IIC Column has a maximum capacity of 800 μL. The flow was discarding from the Collection Tube and repeat Step 6. About 200 μL DNA Pre-Wash Buffer was added to the Zymo-Spin IIC Column in a new Collection Tube and centrifuged at 10,000 × g for 1 min. About 500 μL of Fungal/Bacterial DNA Wash Buffer was added to the Zymo-Spin IIC Column and centrifuged at 10,000 × g for 1 min. The Zymo-Spin IIC Column was transfer to a clean 1.5 mL microcentrifuge tube and add 100 μL (35 μL minimum) DNA Elution Buffer directly to the column matrix. Centrifuge at 10,000 × g for 30 s to elute the DNA

13.2.3.4 *PCR analysis*

PCR sequencing preparation cocktail consisted of 10 μL of 5 × GoTaq colorless reaction, 3 μL of 25 mM $MgCl_2$, 1 μL of 10 mM of dNTPs mix, 1 μL of 10 pmol each of the 16SrRNA gene forward primer (27F: AGAGTTTGATCMTGGCTCAG) and reverse primer (1525R: AAGGAGGTGWTCCARCCGCA) and 0.3 units of Taq DNA polymerase (Promega, USA) made up to 42 μL with sterile distilled water 8 μl DNA template. PCR was carried out in a GeneAmp 9700 PCR System Thermalcycler (Applied Biosystems Inc., USA) with a PCR profile consisting of an initial denaturation at 94°C for 5 min, followed by 30 cycles consisting of 94°C for 30 s, 30 s annealing of primer at 56°C and 72°C for 1 min 30 s, and a final termination at 72°C for 10 mins and cooling at 4°C.

13.2.3.5 *Gel integrity*

The integrity of the DNA and PCR amplification was checked on 1% and 1.5% agarose gel, respectively. The buffer (1XTBE buffer) was prepared and subsequently used to prepare agarose gel. The suspension was boiled in a microwave for 5 min. The molten agarose was allowed to cool to 60°C and stained with 3 μL of 0.5 g/mL ethidium bromide (which absorbs invisible UV light and transmits the energy as visible orange light). A comb was inserted into the slots of the casting tray, and the molten agarose was poured into the tray. The gel was allowed to solidify for 20 min to form the wells. The 1XTAE buffer was poured into the gel tank to barely submerge the gel. Two microliter (2 L) of 10 × blue gel loading dye (which gives color and density to the samples to make it easy to load into the wells and monitor the progress of the gel) was added to 4 μL of each PCR product and loaded into the wells after the DNA ladder was loaded into the well. The gel was electrophoresed at 120 V for 45 min visualized by ultraviolet transillumination and photographed. The sizes of the PCR products were estimated by comparison with the mobility of a hyper ladder1 that was ran alongside experimental samples in the gel.

13.2.3.6 *Purification of amplified product*

After gel integrity, the amplified fragments was ethanol purified in order to remove the PCR reagents.

Briefly, 7.6 μL of Na acetate 3 M and 240 μL of 95% ethanol were added to each about 40 μL PCR amplified product in a new sterile 1.5 μL tube eppendorf, mix thoroughly by vortexing and keep at – 20°C for at least 30 min. Centrifugation for 10 min at 13000 g and 4°C followed by removal of supernatant (invert tube on trash once) after which the pellet were washed by adding 150 μL of 70% ethanol and mix then centrifuge for 15 min at 7500 g and 4°C. Again, remove all the supernatant (invert tube on trash) and invert the tube on paper tissue and let it dry in the fume hood at room temperature for 10–15 min. Then, resuspend with 20 μL of sterile distilled water and keep in -20°C prior to sequencing. The purified fragment was checked on a 1.5% agarose gel ran on a voltage of 110 V for about 1 h as previous, to confirm the presence of the purified product, and quantified using a NanoDrop 2000 spectrophotometer (Thermo Scientific).

13.2.3.7 *Sequencing*

The amplified fragments were sequenced using a genetic Analyzer 3130 × l sequencer from applied Biosystems using manufacturer's manual while the sequencing kit used was that of big dye terminator v3.1 cycle sequencing kit. Bio-edit software and MEGA 6 were used for all genetic analysis.

13.2.3.8 *Phylogenetic analysis between ammonium-oxidizing and nitrite-oxidizing bacteria*

Phylogenetic relationship, similarity index, distance between all isolates and maximum likelihood pattern analysis outlined and phylogenetic tree were generated by Mega v5.05. The 16S rDNA sequences in this study were submitted to the GenBank database by using Sequin software (www.ncbi.nlm.nih.gov/projects/Sequin/) under nucleotide accession number (Saha et al., 2013).

13.2.3.9 *Statistical analysis*

Result was reported as mean ± standard deviation. All data were subjected to statistical analysis by analysis of variance (ANOVA). The means were separated with least significant difference. The result was considered significant at $P < 0.05$. Least significant difference test (LSD) was also being performed between each treatment and the control. The correlation (association) and regression (changes) analysis was done using Statistical Product and Service Solution (SPSS) for Windows version 20.

13.3 Results

13.3.1 Morphological and biochemical characterization of ammonia-oxidizing bacterial isolates

Table 13.1 shows the morphological and biological characteristics of ammonia-oxidizing isolates, and the five potential heavy metal-tolerant ammonia-oxidizing bacteria were characterized based on their cultural, morphological,

TABLE 13.1 Morphological and biochemical characterization of ammonia-oxidizing bacterial isolates.

Suspected organism	Colony	Cell shape	Gram stain	Spore stain	Cat	H2S	Ind	Cit	MR	Vp	Ur	Mot	Oxid	Nit red
A1	Colorless mucoid raise	Rod	–	–	+	–	–	+	–	–	+	+	+	+
A2	White mucoid raise	Rod	–	–	+	–	–	+	–	+	–	+	+	+
A3	White Mucoid raise	Rod	–	–	+	–	–	+	–	–	–	+	+	+
A4	Raise, brownish	Rod	–	–	+	+	–	+	–	+	–	–	–	+
A5	Colorless mucoid raise	Rod	–	–	+	+	+	–	+	–	+	+	+	+

Key: +, present (positive); –, absent (negative); Cat, catalase; Ind, indole; Cit, citrate; MR, methyl red; Vp, Voges-Proskauer; Ur, urease; Mot, motility; Oxid, oxidase; A1, *Achromobacter* spp.; A2, *Achromobacter xylosoxidans*; A3, *Aeromonas* spp.; A4, *Nitrosomonas* spp.; A5, *Achromobacter insolitus*.

and biochemical characteristics. Isolates were identified as Gram negative, catalase positive, indole negative, methylene red negative, Voges-Proskauer negative, urease negative, and oxidase negative. The isolates were compared with the standard description of Bergey's Manual of Determinative Bacteriology. A1, A2, A3, A4, and A5 represent different ammonia-oxidizing bacteria.

13.3.2 Tolerance of isolates to the copper

The results shown in Fig. 13.1 represented the rate of response of different ammonia-oxidizing bacteria to different concentrations of copper within 24, 72, 120, and 168h.

13.3.3 Tolerance of isolates to the nickel

The results shown in Fig. 13.2 represented the rate of response of different ammonia-oxidizing bacteria to different concentrations of nickel within 24, 72, 120, and 168h.

13.3.4 Tolerance of isolates to the lead

The results shown in Fig. 13.3 represented the rate of response of different ammonia-oxidizing bacteria to different concentrations of lead within 24, 72, 120, and 168h.

13.3.5 Tolerance of isolates to the cadmium

The results shown in Fig. 13.4 represented the rate of response of different ammonia-oxidizing bacteria to different concentrations of cadmium within 24, 72, 120, and 168h.

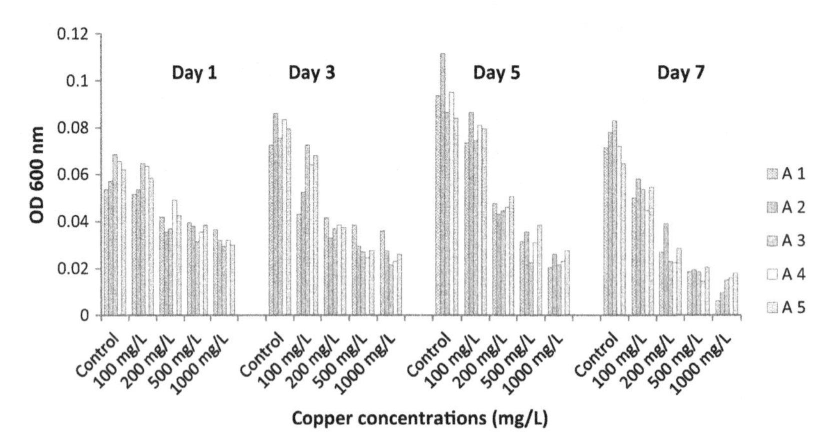

FIG. 13.1 Response of different ammonia-oxidizing bacteria to different concentrations of copper within 7 days.

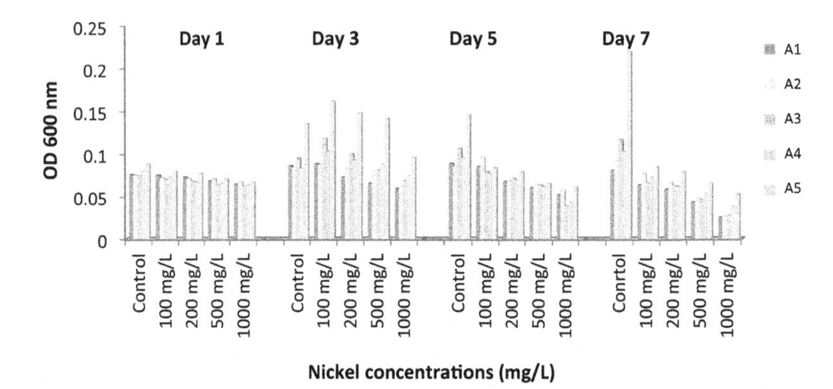

FIG. 13.2 Response of different ammonia-oxidizing bacteria to different concentrations of nickel within 7 days.

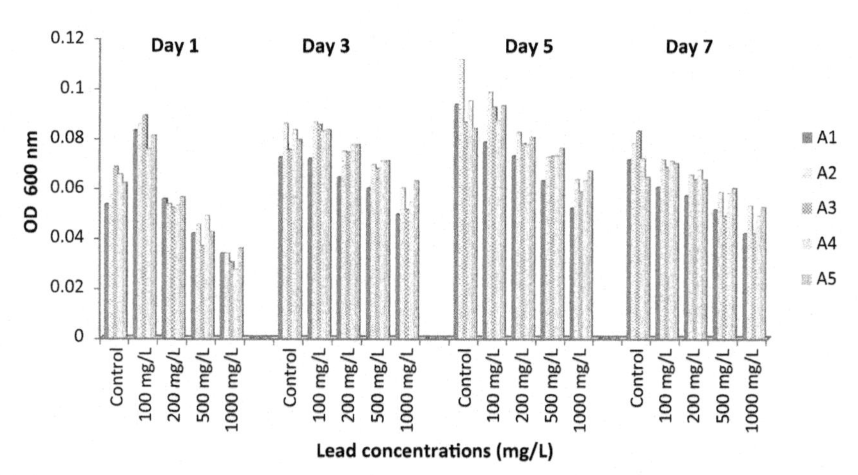

FIG. 13.3 Response of different ammonia-oxidizing bacteria to different concentrations of lead within 7 days.

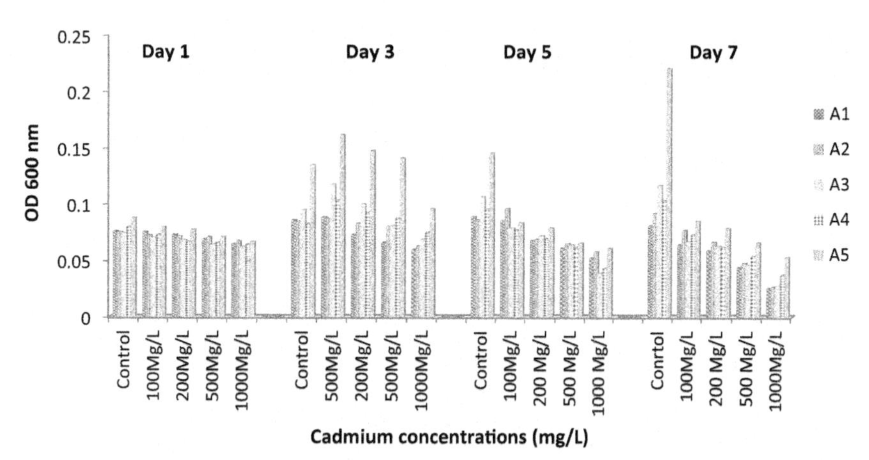

FIG. 13.4 Response of different ammonia-oxidizing bacteria to different concentrations of cadmium within 7 days.

13.3.6 Biosorption of heavy metal study

Achromobacter xylosoxidans (A2) mitigation of copper ranges from 10% to 55.5%; 67% to 81% and 90.1% at 1 day, 7 days, 14 days, 21 days, and 28 days, respectively. A2 biosorption of nickel was 12% on day 1, 48% on day 7, 77% on day 14, 84% on day 21, and 96.51% on day 28. A2 mitigation of lead ranges from 15% to 67%; 74% to 81.5% and 92% at 1 day, 7 days, 14 days, 21 days and 28 days, respectively. A2 biosorption of cadmium was 5% on day 1, 41.796% on day 7, 52% on day 14, 61% on day 21, and 84.82% on day 28.

Achromobacter insolitus (A5) mitigation of copper ranges from 9% to 52.67%; 68% to 86% and 90.04% at 1 day, 7 days, 14 days, 21 days, and 28 days, respectively. A5 biosorption of nickel was 7% on day 1, 49.64% on day 7, 67% on day 14, 80% on day 21, and 94.67% on day 28. A5 mitigation of lead range from 11% to 62.75%; 68% to 79% and 90.25% at 1 day, 7 days, 14 days, 21 days, and 28 days, respectively. A5 biosorption of cadmium was 6% on day 1, 48.56% on day 7, 54% on day 14, 73% on day 21, and 89.21% on day 28. Ammonia-oxidizing bacteria (A2 and A5) biosorption of copper, nickel, lead and cadmium (Fig. 13.5).

13.4 Discussion

All the bacterial isolates show the ability to grow in mineral salt agar medium incorporated with different heavy metal salts at different concentrations (100 ppm, 200 ppm, 500 ppm, and 1000 ppm). The cultures were incubated for 7 days and measured for optical density (at wavelength = 600 nm) in UV spectrophotometer. All bacteria showed high tendency to decrease optical density while increasing metal concentration in the medium when compared to control.

The growth of the ammonia-oxidizing organism increased successively throughout the period of five days and decrease at 7 days exposure time at 0 (Control), 100 mg/L, 200 mg/L, 500 mg/L, and 1000 mg/L concentration of

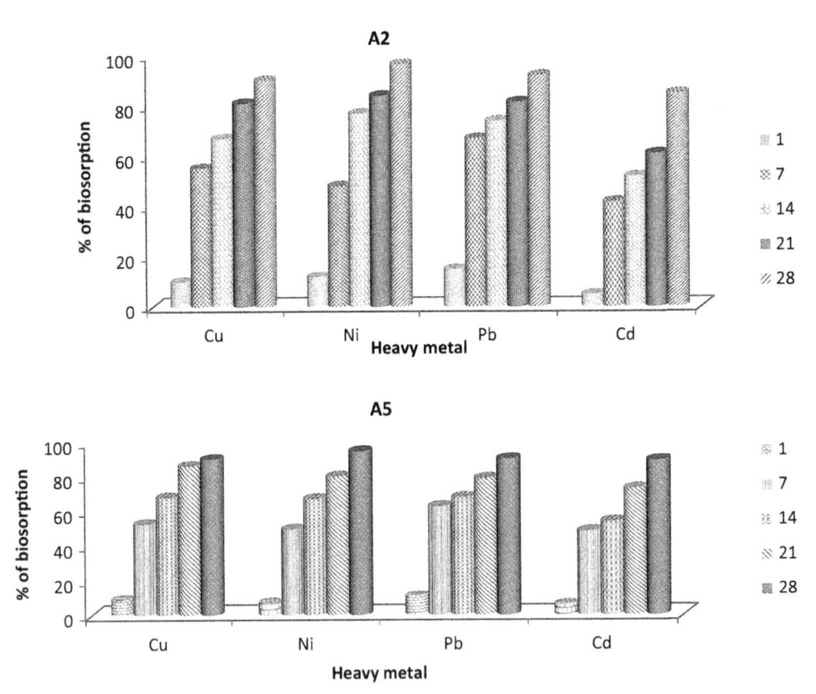

FIG. 13.5 Ammonia-oxidizing bacteria (A2 and A5) biosorption of copper, nickel, lead, and cadmium.

metal salt. The highest growth was observed, for all the isolates, in the medium with no metal ion amendment, which served as the control, followed by 100mg/L, 200mg/L, 500mg/L, and 1000mg/L, respectively. The increase in the growth of ammonia-oxidizing organism was observed throughout the period of five days and decrease mostly at 7 days exposure time at 0 (Control), 100mg/L, and 200mg/L. The sensitivity of the isolates to salts of copper (Cu), nickel (Ni), cadmium (Cd), and lead (Pb) was in the decreasing order of A1 > A3 > A4 > A2 > A5. All the isolates showed the high sensitivity to high concentration of 500 and 1000mg/L salts of copper (Cu), nickel (Ni), cadmium (Cd), and lead (Pb). A2 and A5 exhibited the greatest ability to tolerate the metal salts than A1, A3, and A4. Isolate A2 and A5 were identified as *Achromobacter xylosoxidans* and *Achromobacter insolitus*, respectively.

Tolerance for the metal salt was dependent on concentration, time, and the isolate tested. Khan et al. (2010); Subrahmanyam et al. (2014); and Wang et al. (2017) reported that the toxicological effects of metals on soil microbial community structure and activities depend largely on the type and concentration of metal and incubation time. This finding revealed that when tolerance to the four metals was compared with respect to each bacteria isolate, it was evident that the bacteria were more sensitive to Cu^{2+} and Cd^{2+} than the other two metals (Ni^{2+} and Pb^{2+}). The sensitivity of the metal was in the order of Cd > Cu > Ni > Pb.

Microorganisms are very sensitive; they react quickly to any kind of changes (natural and anthropogenic) in the environment and quickly adapt themselves to new conditions including high metal concentrations. Microorganisms take heavy metals into the cell in significant amounts. This phenomenon leads to the intracellular accumulation of metal cations of the environment and is defined as bioaccumulation (Wolejko et al., 2016). Some bacterial plasmids contain specific genes for resistance to toxic heavy metal ions (Liu et al., 2018a,b; Pacwa-Płociniczak et al., 2018; Lukina et al., 2016; Sharma, 2016), ability to produce siderophore, and ability to solubilize phosphate (biofertilizers) (Ibiene and Okpokwasili, 2011; Gupta et al., 2014). Some microorganisms can adjust their metabolic activity or community structure to adapt to the harmful shock loadings. Microorganisms play important role in stress environment and the derived ecosystem functions (Singh et al., 2016a,b; Vimal et al., 2017). Metals detoxify through resistance and tolerance, and this resistance can be attributed to the mechanisms of exclusion or tolerance (Klassen et al., 2000; Bruins et al., 2000). These mechanisms stem from prior exposure of microorganisms to metals which enable them to develop the resistance and tolerance useful for biological treatment (Viti and Giovannetti, 2003; Sharma, 2016; Singh et al., 2016a,b).

Nitrifying bacteria could be reduced by heavy metals and therefore considered as sensitive microbial process with regard to heavy metal stress (Smolders et al., 2001). Lee et al., 2011; Mertens et al., 2009; and Vasileiadis et al., 2012, have particularly emphasized the response of ammonia-oxidizing bacteria to heavy metals such as Zn, Cu, and Hg. Frey et al., 2008; Lee et al., 2011; and Vasileiadis et al., 2012, have reported a sensitive response of the AOB community to heavy metals. Heavy metals can affect diversity of certain microbial communities and related soil processes (Haferburg and Kothe, 2007; Li et al., 2009a,b). Li et al. (2009a,b) reported that an increase in available Cu (0–2400 $mg\,kg^{-1}$) was paralleled by a concomitant decrease in AOB amoA copy numbers. Heavy metals used in greater amounts result in metabolic disorders and suppress the growth of most plants and microorganisms (Ashraf and Ali, 2007; Zimmer et al., 2016).

Markedly different responses of AOB communities to metal pollution stress have been observed. Both metal-sensitive and metal-tolerant AOB populations have been observed in agricultural soils amended with metals (Stephen et al., 1999). Ferry reported that soil metal contamination did not decrease the abundance of AOB (Frey et al., 2008), while the research of Stefanowicz et al. (2008) and Qu et al. (2011) found that bacterial functional diversity was significantly decreased with increasing soil pollution. Bermudez et al. (2009) hypothesize that high organic matter contents of soil can bind metals and decrease their toxicity. Moreover, Kris et al. (2005) had demonstrated that the nitrifying community displays a tolerance to long-term Zn stress (Rotthauwe et al., 1997).

Bioremediation is an innovative and promising technology available for removal of heavy metals and recovery of the heavy metals in polluted water and lands. Since microorganisms have developed various strategies for their survival in heavy metal-polluted habitats, these organisms are known to develop and adopt different detoxifying mechanisms such as biosorption, bioaccumulation, biotransformation, and biomineralization, which can be exploited for bioremediation either ex situ or in situ. The main purpose of bioremediation is not to remove pollutants at all costs, despite the economic or technological reality, but to limit the risk of negative impact on the environment and thus on human health. A2 and A5 exhibited the greatest ability to tolerate the metal salts than the others which were used for biosorption study.

13.5 Conclusion

Most Proteobacteria were able to adapt and grow under various extreme conditions. *Achromobacter xylosoxidans* (A2) and *Achromobacter insolitus* (A5) showed a high level of tolerance for heavy metal tested and were able to mitigate heavy metal from the medium. Proteobacteria (*Achromobacter xylosoxidans* (A2) and *Achromobacter insolitus* (A5)) remain a good option for heavy metals' bioremediation since it is eco-friendly and efficient.

Acknowledgments

We thank the University of Nigeria Nsukka for giving us the opportunity to carry out the research, and we also appreciate Rev. Fr. Prof. Vincent Nyoyoko for funding the research.

Conflict of interest

The authors declare that they have no conflicts of interest.

References

Anyanwu, C.U., Nwankwo, S.C., Moneke, A.N., 2011. Soil bacterial response to introduced metal stress. Int. J. Basic Appl. Sci. 11 (01), 73–76.

Ashraf, R., Ali, T.A., 2007. Effect of heavy metals on soil microbial community and mung beans seed germination. Pak. J. Bot. 39 (2), 629.

Ayangbenro, A., Babalola, O., 2017. A new strategy for heavy metal polluted environments: a review of microbial biosorbents. Int. J. Environ. Res. Public Health 14, 94.

Bermudez, G.M.A., Moreno, M., Invernizzi, R., Pla, R., Pignata, M.L., 2009. Heavy metal pollution in topsoils near a cement plant: the role of organic matter and distance to the source to predict total and HCl extracted heavy metal concentrations. Chemosphere 78 (4), 375–381.

Bhaskar, K.V., Charyulu, P.B.B.N., 2005. Effect of environmental factors on nitrifying bacteria isolated from the rhizosphere of Setaria italica (L.) Beauv. Afr. J. Biotechnol. 4 (10), 1145–1146.

Bruins, M.R., Kapil, S., Oehme, F.W., 2000. Microbial resistance to metals in the environment. Ecotoxicol. Environ. Saf. 45 (3), 198–207.

Canli, M., Atli, G., 2003. The relationships between heavy metal (Cd, Cr, Cu, Fe, Pb, Zn) levels and the size of six Mediterranean fish species. Int. J. Environ. Pollut. 121 (1), 129–136.

Cheesbrough, M., 2006. District Laboratory Practice in Tropical Countries, Part 2. Cambridge University Press, United Kingdom, pp. 62–70.

Cobbina, S.J., Chen, Y., Zhou, Z., Wu, X., Zhao, T., Zhang, Z., 2015. Toxicity assessment due to sub-chronic exposure to individual and mixtures of four toxic heavy metals. J. Hazard. Mater., 294109–294120.

Daum, M., Zimmer, W., Papen, H., Kloos, K., Nawrath, K., Bothe, H., 1998. Physiological and molecular biological characteriza-tion of ammonia oxidation of the heterotrophic nitrifier Pseudomonas putida. Curr. Microbiol. 37, 281–288.

Deni, J., Penninck, M.J., 1999. Nitrification and autotrophic nitrifying bacteria in a hydrocarbon polluted soil. Appl. Environ. Microbiol. 65, 4008–4020.

Frey, B., Pesaro, M., Rudt, A., Widmer, F., 2008. Resilience of the rhizosphere pseudomonas and ammonia-oxidizing bacterial populations during phytoextraction of heavy metal polluted soil with poplar. Environ. Microbiol. 10, 1433–1449.

Gupta, D.K., Chatterjee, S., Datta, S., Veer, V., Walther, C., 2014. Role of phosphate fertilizers in heavy metal uptake and detoxification of toxic metals. Chemosphere 108, 134–144.

Gupta, A., Joia, J., Sood, A., Sood, R., Sidhu, C., Kaur, G., 2016. Microbes as potential tool for remediation of heavy metals: a review. J. Microb. Biochem. Technol. 8, 364–372.

Hadia-e-Fatima, A.A., 2018. Heavy metal pollution—a mini review. J. Bacteriol. Mycol. Open Access 6 (3), 179–181.

Haferburg, G., Kothe, E., 2007. Microbes and metals: interactions in the environment. J. Basic Microbiol. 47 (6), 453–467.

Hamsa, N., Yogesh, G.S., Koushik, U., Patil, L., 2017. Nitrogen transformation in soil: effect of heavy metals. sInt. J. Curr. Microbiol. App. Sci. 6, 816–832.

Hoang, P.H., Nguyen, Hong, T., Tran Trung, T., Tran, T.T., Do, L.P., Le, T.N.C., 2016. Isolation and selection of nitrifying bacteria with high biofilm formation for treatment of ammonium polluted aquaculture water. J. Vietnam. Environ. 8 (1), 33–40.

Holt, J., Krieg, G., Sneath, N.R., P. H. A., 1994. In: Staley, J.T., Williams, S.T. (Eds.), Bergey's Manual of Determinative Bacteriology, ninth ed. Williams & Wilkins, Baltimore, MD.

Ibiene, A.A., Okpokwasili, G.S.C., 2011. Comparative toxicities of three agro-insecticide formulations on nitrifying bacteria. Rep. Opin. 3 (12), 14–17.

Jamaluddin, H., Zaki, D.M., Ibrahim, Z., Tan, S., Zainudin, N., Yusoh, O., 2012. Isolation of metal tolerant bacteria from polluted wastewater. In: Paper presented at the Pertanika Journal of Tropical Agricultural Science.

Joel, O.F., Amajuoyi C. A (2010). Determination of the concentration of ammonia that could have lethal effect on fish pond. Asian Res. Publ. Netw. J. Eng. Appl. Sci.. Pg 1–5.

John, R.C., Okpokwasili, G.C., 2012. Crude oil-degradation and plasmid profile of nitrifying bacteria isolated from oil-impacted mangrove sediment in the Niger Delta of Nigeria. Bull. Environ. Contam. Toxicol. 88, 1020–1026.

Johnson, H., Cho, H., Choudhary, M., 2019. Bacterial heavy metal resistance genes and bioremediation potential. Comput. Mol. Biosci. 9, 1–12.

Khan, S., Hesham, E.L., Qiao, M., Rehman, S., He, J.Z., 2010. Effects of cd and Pb on soil microbial community structure and activities. Environ. Sci. Pollut. R 17 (2), 288–296.

Klassen, S.P., Malean, J.E., Grossl, P.R., Sims, R.C., 2000. Heavy metals in the environment. J. Environ. Qual. 29, 1826–1834.

Kris, B., Jelle, M., Erik, S., 2005. Toxicity of heavy metals in soil assessed with various soil microbial and plant growth assays: a comparative study. Environ. Toxicol. Chem. 24 (3), 634–640.

Lee, S., Cho, K., Lim, J., Kim, W., Hwang, S., 2011. Acclimation and activity of ammonia-oxidizing bacteria with respect to variations in zinc concentration, temperature, and microbial population. Bioresour. Technol. 102, 4196–4203.

Li, X., Zhu, Y.G., Cavagnaro, T.R., Chen, M., Sun, J., Chen, X., Qiao, M., 2009a. Do ammonia-oxidizing archaea respond to soil cu contamination similarly as ammonia-oxidizing bacteria? Plant Soil 324 (1–2), 209–217.

Li, F., Zheng, Y.M., He, J.Z., 2009b. Microbes influence the fractionation of arsenic in paddy soils with different fertilization regimes. Sci. Total Environ. 407 (8), 2631–2640.

Liu, Q., Liu, Y., Zhang, M., 2012. Mercury and cadmium contamination in traffic soil of Beijing, China. Bull. Environ. Contam. Toxicol. 88, 154–157.

Liu, J., Cao, W., Jiang, H., Jing Cui, J., Shi, C., Qiao, X., Zhao, J., Si, W., 2018a. Impact of heavy metal pollution on ammonia oxidizers in soils in the vicinity of a Tailings Dam, Baotou, China. Bull. Environ. Contam. Toxicol. 101 (1), 110–116.

Liu, S., Niu, G.Z., Liu, Y., et al., 2018b. Bioremediation mechanisms of combined pollution of PAHs and heavy metals by bacteria and fungi: a mini review. Bioresour. Technol. 224, 25–33.

Lukina, A.O., Boutin, C., Rowlan, O., et al., 2016. Evaluating trivalent chromium toxicity on wild terrestrial and wetland plants. Chemosphere 162, 355–364.

Malik, Z., Ahmad, M., Abassi, G.H., Dawood, M., Hussain, A., Jamil, M., 2017. Agrochemicals and soil microbes: interaction for soil health. In: Hashmi, M.Z. (Ed.), Xenobiotics in the Soil Environment: Monitoring, Toxicity and Management. Springer International Publishing, Cham, pp. 139–152.

Mertens, J., Broos, K., Wakelin, S.A., Kowalchuk, G.A., Springael, D., Smolders, E., 2009. Bacteria, not archaea, restore nitrification in a zinc-contaminated soil. ISME J. 3, 916–923.

Mishra, J., Singh, R., Arora, N.K., 2017. Alleviation of heavy metal stress in plants and remediation of soil by rhizosphere microorganisms. J. Front. Microbiol. 8, 1706.

Moir, J.W.B., Wehrfritz, J.M., Spiro, S., Richardson, D.J., 1996b. The biochemical characterization of novel noefn-hame iron hydroxylamine oxidase from *Paracoccus denitrificans GB17*. Biochem. J. 319, 823–827.

Nibourg, G.A.A., Hoekstra, R., Van der Hoever, T.V., Ackermans, M.T., Hakvoort, T.B.M., Van Gulik, T.M., Chamuleau, R.A.F.M., 2013. Effects of acute-liverfailure-plasma exposure, on hepatic functionality of Hepa RG-AMC-bioartificial liver. Liver Int. 33 (4), 516–524.

Pacwa-Płociniczak, M., Płociniczak, T., Yu, D., et al., 2018. Effect of Silene Vulgaris and Heavy Metal Pollution on Soil Microbial Diversity in Long-Term Contaminated Soil. Water Air Soil Pollut. 229 (1), 13.

Pandit, R., Patel, B., Kunjadia, P., Nagee, A., 2013. Isolation, characterization and molecular identification of heavy metal resistant bacteria from industrial effluents. Amala-khadiAnkleshwar, Gujarat. Int. J. Environ. Sci. 3 (5), 1689–1699.

Pruss-Ustun, A., Vickers, C., Haefliger, P., Bertollini, R., 2011. Knowns and unknowns on burden of disease due to chemicals: a systematic review. Environ. Health 10, 9.

Qu, J., Ren, G., Chen, B., Fan, J., Yong, E., 2011. Effects of lead and zinc mining contamination on bacterial community diversity and enzyme activities of vicinal cropland. J. Environ. Monitor. Assess. 182 (1–4), 597–606.

Ramya, D., Joseph Thatheyus, A., 2018. Microscopic investigations on the biosorption of heavy metals by bacterial cells: a review. Sci. Int. 6, 11–17.

Reddy, A.D., Subrahmanyam, G., NaveenKumar, S., Karunasagar, I., Karunasagar, I., 2015. Isolation of ammonia oxidizing bacteria (AOB) from fish processing effluents. Natl. Acad. Sci. Lett. 38 (5), 393–397.

Rotthauwe, J.H., Witzel, K.P., Liesack, W., 1997. The ammonia monooxygenase structural gene amoA as a functional marker: molecular fine-scale analysis of natural ammonia-oxidizing populations. Appl. Environ. Microbiol. 63, 4704–4712.

Sagi, Y., Yigit, S.A., 2012. Heavy metals in Yenicacga Lake and its potential sources: soil, water, sediment and plankton. Environ. Monit. Assess. 184, 1379–1389.

Saha, M., Sarkar, A., Bandhophadhyay, B., 2013. Development of molecular identification of nitrifying bacteria in water bodies of East Kolkata Wetland, West Bengal. J. Bioremed. Degrad. 5 (1), 1–5.

Sharma, J., 2016. Removal of heavy metals by indigenous microorganisms and identification of gene responsible for remediation. Int. J. Nano Stud. Technol. ISSN: 2167-8685.

Singh, J.S., Abhilash, P.C., Gupta, V.K., 2016a. Agriculturally important microbes in sustainable food production. Trends Biotechnol. 34, 773–775.

Singh, J.S., Kaushal, S., Kumar, A., Vimal, S.R., Gupta, V.K., 2016b. Book review: microbial inoculants in sustainable agricultural productivity-Vol. II: functional application. Front. Microbiol. 7, 2015.

Smolders, E., Brans, K., Coppens, F., Merckx, R., 2001. Potential nitrification rate as a tool for screening toxicity in metal contaminated soils. Environ. Toxicol. Chem. 20, 2469–2474.

Srivastava, V., Sarkar, A., Singh, S., Singh, P., de Araujo, A.S.F., Singh, R.P., 2017. Agroecological responses of heavy metal pollution with special emphasis on soil health and plant performances. Front. Environ. Sci. 5, 64.

Stefanowicz, A.M., Niklińska, M., Laskowski, R., 2008. Metals affect soil bacterial and fungal functional diversity differently. Environ. Toxicol. Chem. 27 (3), 591–598.

Stephen, J.R., Chang, Y.J., Macnaughton, S.J., Kowalchuk, G.A., Leung, K.T., Flemming, C.A., White, D.C., 1999. Effect of toxic metals on indigenous soil beta-subgroup proteobacterium ammonia oxidizer community structure and protection against toxicity by inoculated metal-resistant bacteria. Appl. Environ. Microbiol. 65 (1), 95–101.

Subrahmanyam, G., Hu, H.-W., Zheng, Y.-M., Gattupalli, A., He, J.-Z., Liu, Y.-R., 2014. Response of ammonia oxidizing microbes to the stresses of arsenic and copper in two acidic alfisols. Appl. Soil Ecol. 77 (59), 67.

Vasileiadis, S., Coppolecchia, D., Puglisi, E., Balloi, A., Mapelli, F., Hamon, R.E., Daniele, D., Trevisan, M., 2012. Response of ammonia oxidizing bacteria and archaea to acute zinc stress and different moisture regimes in soil. Microb. Ecol. 64, 1028–1037.

Venkatachalam, P., Jayalakshmi, N., Geetha, N., Sahi, S.V., Sharma, N.C., Rene, E.R., et al., 2017. Accumulation efficiency, genotoxicity and antioxidant defense mechanisms in medicinal plant *Acalypha indica* L. under lead stress. Chemosphere 171, 544–553.

Vimal, S.R., Singh, J.S., Arora, N.K., Singh, S., 2017. Soil-plant-microbe interactions in stressed agriculture management: a review. Pedosphere 27 (2), 177–192.

Viti, C., Giovannetti, L., 2003. The impact of chromium contamination on soil heterotrophic and photosynthetic microorganisms. Ann. Microbiol. 51, 201–213.

Wang, P., Di, H.J., Cameron, K.C., Tan, Q., Podolyan, A., Zhao, X., McLaren, R.G., Hu, C., 2017. The response of ammonia-oxidizing microorganisms to trace metals and urine in two grassland soils in New Zealand. Environ. Sci. Pollut. Res. 24, 2476–2483.

Wolejko, E., Wydro, U., Loboda, T., 2016. The ways to increase efficiency of soil bioremediation. Ecol. Chem. Eng. S 23 (1), 155–174.

Wyszkowska, J., Borowik, A., Kucharski, M., Kucharski, J., 2013. Effect of cadmium, copper and zinc on plants, soil microorganisms and soil enzymes. J. Element. School, 769–796.

Zimmer, S., Messmer, M., Haase, T., Piepho, P.H., Mindermann, A., Schulz, H., Habekuß, A., Ordon, F., Wilbois, K.P., Heß, J., 2016. Effects of soybean variety and Bradyrhizobium strains on yield, protein content and biological nitrogen fixation under cool growing conditions in Germany. Eur. J. Agron. 72, 38–46.

CHAPTER

14

Treatment of harvested rainwater and reuse: Practices, prospects, and challenges

Siril Singh[a,b], *Rajni Yadav*[b], *Srujana Kathi*[c], *and Anand Narain Singh*[b]

[a]Department of Environment Studies, Panjab University, Chandigarh, India [b]Soil Ecosystem and Restoration Ecology Lab, Department of Botany, Panjab University, Chandigarh, India [c]ICSSR Postdoctoral Research Fellow, Department of Applied Psychology, Pondicherry University, Puducherry, India

OUTLINE

14.1 Introduction

In the new global development, freshwater scarcity has become a central issue in sustainable development. This issue is becoming a threat and the most considerable global risk in its potential impact. The main driving forces for the rising global demand for freshwater are the growing world population, improving living standards, changing consumption patterns, and expanding irrigated agriculture (Ercin and Hoekstra, 2014). Also, the mismatch between freshwater demand and availability is the essence of global water scarcity. Therefore, several studies have been carried out to assess global water scarcity in physical, social, and economical aspects (Oki et al., 2001; Wolfe and Brooks, 2003). In order to reduce water scarcity consequences, the use of rainwater has been widely accepted as a reliable alternative today. Studies on the rainwater harvesting system (RWHS), particularly on the techniques and treatment system, have increased significantly in recent years (Vieira et al., 2014; Fonseca et al., 2017). Several studies describe RWH from a water saving and alternative water supply perspective for potable and nonpotable use (Farreny et al., 2011a; Fewkes, 2012). RWH can be used as an adaptation measure against periodic water scarcity and reduction of drinking water use, but Farreny et al. (2011a, b) label it as a sustainable strategy to be included in urban water cycle management. It may reduce the additional water demand, alleviate water stress on the area, reduce nonpoint source pollution, reduce

Cost Effective Technologies for Solid Waste and Wastewater Treatment.
https://doi.org/10.1016/B978-0-12-822933-0.00003-6

treatable urban runoff volume, prevent flooding, and alleviate climate change. RWHS have been used extensively by many countries worldwide, as it strengthens urban stormwater management, reduces municipal water supply stress simultaneously, and restores natural water cycles in the urban areas (Jones and Hunt, 2010; Vialle et al., 2015).

Rainwater harvesting (RWH) refers to collecting rain for beneficial uses before it drains away as runoff. The concept of RWH has a long history of use. The collection and storage of rainwater in earthen tanks for domestic and agricultural uses has been common practice in developing countries since historical times. However, after the implementation of dam and irrigation projects, the traditional knowledge and practice of RWH have largely been abandoned. Since the early 90s, there is a renewed interest in RWH projects.

Water harvesting refers to collecting and storing rainwater and other activities to harvest surface and groundwater, preventing losses through evaporation and seepage and all other hydrological studies and engineering inventions. It aims to conserve and efficiently utilize limited water physiographic unit such as a watershed. Rain is a primary source of water for all of us. There are two main techniques of RWH (Fig. 14.1):

- Storage of rainwater on surface for future use.
- Recharge to groundwater.

Directly collected rainwater can be stored for direct use or can be recharged into the groundwater. All the secondary sources of water like rivers, lakes, and groundwater are entirely dependent on rain as a primary source. The term "water harvesting" is understood to comprehend a wide range of concerns, including rainwater collection with both rooftop and surface runoff catchment, rainwater storage in small tanks and large-scale artificial reservoirs, groundwater recharge, and also protection of water sources against pollution. The objective of water harvesting differs between urban and rural areas. In urban areas, the emphasis is put on increasing groundwater recharge and managing stormwater. On the other hand, in rural areas securing water is more crucial.

14.2 History of rainwater harvesting

RWH has been used in developing countries such as India, the Middle East, and Africa throughout history and was the backbone of agriculture, especially in arid and semiarid areas worldwide. In India, RWH is considered as an ancient technique dating back some 4000–5000 years. In North America, the agriculture of many indigenous peoples in what are now the southern states was historically dependent on simple floodwater harvesting methods. In the early 20th century, conservation agencies' prime focus was soil erosion control to reduce soil losses; this progressed to soil and water conservation, based mainly on structural measures. The harvesting of runoff that went with some soil conservation measures was more or less a side effect whose potential was unacknowledged.

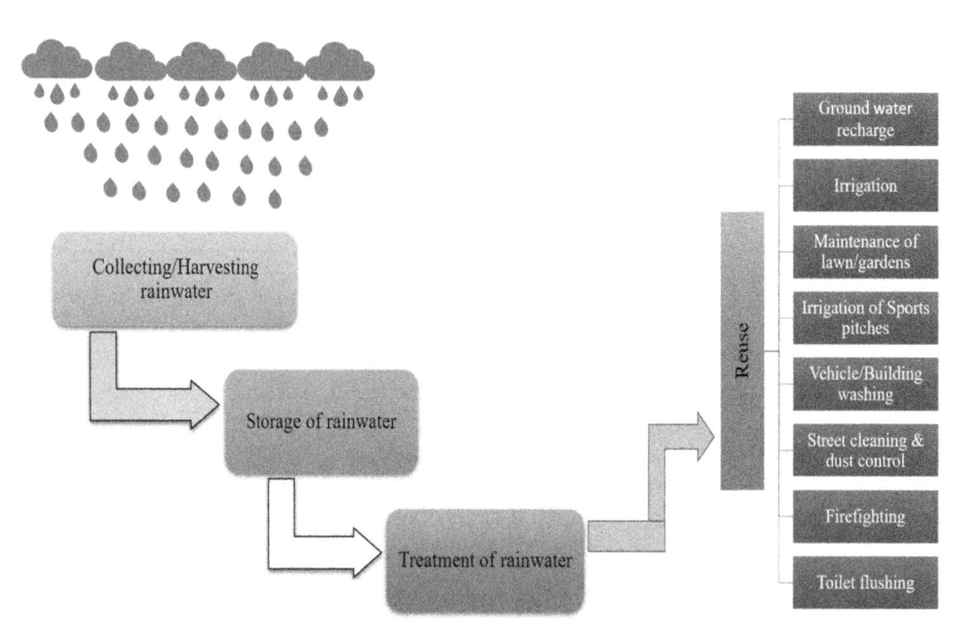

FIG. 14.1 Rainwater harvesting system and its applications.

Moreover, the green revolution's success, based on hybrid seeds, inorganic fertilizers, and pesticides, resulted in a rapid expansion of irrigated areas—and this was seen as the "modern" way forward to improving agricultural water management. However, this expansion soon reached its limits due to over-abstraction, declining water resources and increasing salinization in the water bodies, led to the further impoverishment and conflicts. Water scarcity and the widespread droughts in Africa led to a growing awareness of water harvesting potential for improved crop production in the 1970s. After a silent period in the late 1980s, water harvesting again became the subject of study and project implementation at the turn of the century, and indigenous practices regained credibility. In China today, water harvesting is seen as a significant component in reducing the rural migration and controlling severe soil erosion and is a subject of steadfast projects, intended at helping millions of people (Critchley and Siegert, 1991; Oweis et al., 2012; Falkenmark et al., 2001).

14.3 Aims/needs of rainwater harvesting

The aim of RWH is to collect rainwater from areas of surplus rain, store it, and make it available, where and when there is water shortage. This results in an increase in water availability by either:

- impeding and trapping surface runoff,
- maximizing water runoff storage, or
- trapping and subsurface harvesting water.

The need for RWH is due to the following reasons:

- To overcome the inadequacy of surface water to meet the daily water demands.
- To overcome the decline in groundwater levels.
- To enhance availability of ground and surface water at a specific area and time and utilize it for sustainable development.
- To increase ground water quality by dilution.
- To increase agriculture and forest production.
- To improve ecology of an area by increasing vegetation cover.

14.4 Principle and components

The basic principle of RWH is to capture the precipitation falling in one area and allocate it to another, thereby increasing the amount of water available in the latter. A RWHS's primary components are a catchment or collection area, the runoff conveyance or distribution system, a storage component, and an application area. In some cases, the components are adjacent to each other; a conveyance system connects them in other cases. The storage and application areas may also be the same, typically where water is concentrated in the soil for direct use by plants.

a) **Catchment or collection area**: this is the surface which directly receives the rainfall and provides water to the system, and here, the rainwater in the form of runoff is harvested. The catchment may be as small as a few square meters or as large as several square kilometers. It may be a rooftop, a paved road, compacted surfaces, rocky areas or open rangelands, cultivated or uncultivated land, and natural slopes.
b) **Conveyance system**: this is where harvested rainwater is conveyed through gutters, pipes or overland, rill, gully, or channel flow and either diverted onto cultivated fields (where water is stored in the soil) into specifically designed storage facilities. Both drain pipes and roof surfaces should be constructed of chemically inert materials such as plastic, aluminum, or fiberglass to avoid adverse effects on water quality.
c) **Storage component**: this is where harvested runoff water is stored until it is utilized for commercial or domestic purposes. Water may be stored in the soil profile as soil moisture, above ground, underground, or groundwater (near-surface aquifers) (Oweis et al., 2012). Storage tanks may be constructed as part of the building or built as a separate unit located some distance away from the building. Storage tanks may be of two types in rooftop RWHS: surface tanks and subsurface tanks. In areas, where concentrated runoff is directly diverted to the fields, the application area is identical to the storage area, as plants can directly use the accumulated soil water. A great variety of designed storage systems keep the water until it is used either adjacent to the storage facilities or further away.

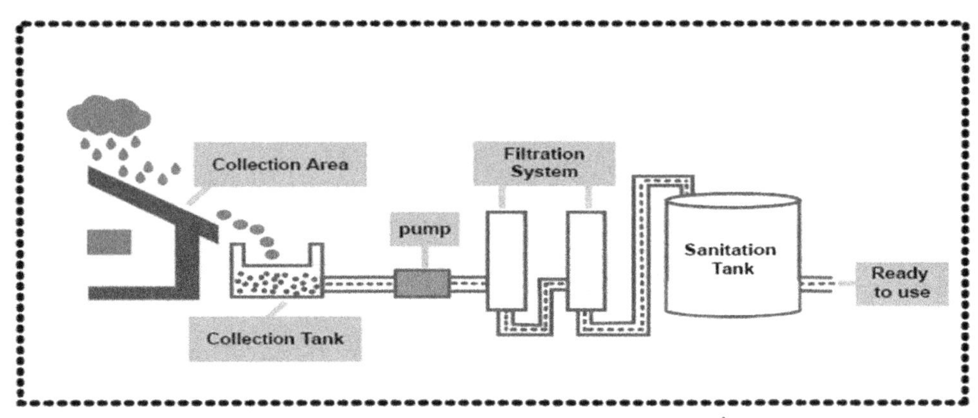

FIG. 14.2 Necessary components of the RWH system.

d) Application area or target area: this is where the harvested water is used either for domestic consumption (drinking and other household uses), livestock consumption, or agricultural use (including supplementary irrigation) or commercial use (Fig. 14.2).

14.5 Classification of RWH systems

The two most frequently used criteria to classify water harvesting systems are the catchment type and size and water storage method. Four groups are distinguished as follows: floodwater harvesting, macrocatchment systems, microcatchment systems, and domestic (rooftop/courtyard) water harvesting. This categorization considers the size of catchment and takes account of storage methods and end-use. It integrates the classifications used by Tuinhof et al. (2012).

i) Floodwater harvesting

Floodwater harvesting can be defined as collecting and storing ephemeral channel flow for irrigation of crops, fodder, and trees and groundwater recharge. The catchment area may be several kilometers long. In areas where evaporation exceeds rainfall, floodwater harvesting systems provide an option for optimal water use during flood events due to heavy rainfall.

ii) Macrocatchment water harvesting

Macrocatchment water harvesting is a method of harvesting runoff water from a natural catchment such as a mountain or hill slope. It may be runoff collection from shallow soils or sealed and compacted surfaces; direct diversion and spreading of overland surface water flow onto application area at the foot of hills or flat terrain (mainly cultivated areas); or impeding and collecting runoff through barriers and storage facilities.

iii) Microcatchment water harvesting

Microcatchment RWH collects surface runoff/sheet (and sometimes rill flow) from small short length catchments. Runoff water is concentrated in an adjacent application area and stored in the root zone for plant's direct use. The catchment and application areas alternate within the same field; thus, rainwater is concentrated within a confined area where plants are grown. Hence, the system is replicated many times in an identical pattern. Microcatchment RWH technologies are often combined with specific agronomic measures for annual crops or tree establishment, especially fertility management and pest management.

iv) Rooftop and courtyard water harvesting

Rooftop and courtyard RWH are getting more and more popular in both developed and emerging economy countries to secure/improve water supply for domestic use such as sanitation or garden irrigation and commercial use such as a cooling agent (Fig. 14.3).

a) Rooftop WH: Harvesting of rainwater can be from roofs of private, public, or commercial buildings (apartments, greenhouses, schools). The effective area of the roof and local annual rainfall determines the rainwater volume that can be captured. About 80–85 per cent of rainfall can be collected and stored (Oweis et al., 2012). Rainfall collected from roofs is used for drinking, especially in areas where tap water is unavailable or unreliable (Worm and van Hattum, 2006). These systems are used in most tropical and subtropical countries.

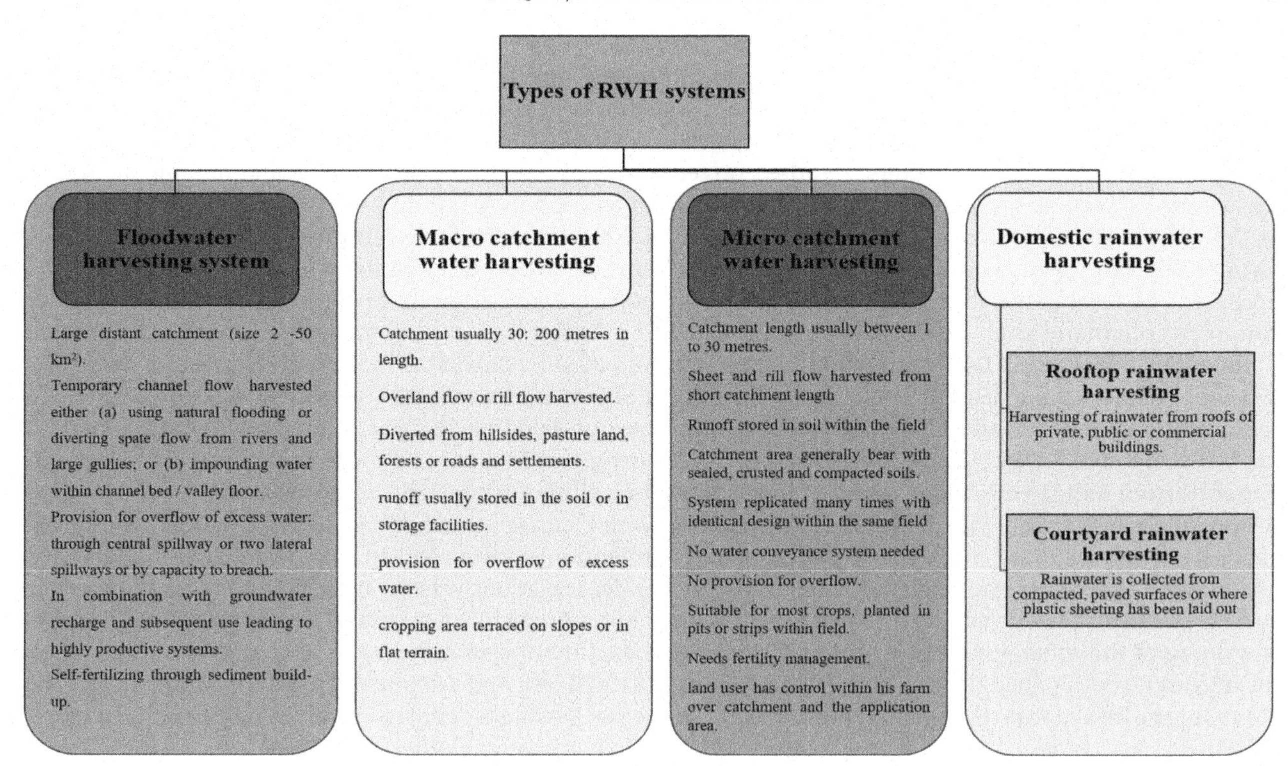

FIG. 14.3 Classification of rainwater harvesting categories.

b) Courtyard WH: Rainwater is collected from compacted, paved surfaces, or where plastic sheeting has been laid out. The slope and permeability affect the amount of rainwater that can be collected. The water may be stored above or below ground.

14.6 Quality assessment of harvested rainwater

Rainwater is one of the cleanest sources of natural water. However, rainwater can absorb gases such as hydrogen sulfides, nitrogen oxides, carbon dioxide, and sulfur dioxide from the atmosphere. It can also capture soot and other microscopic particulates as it falls through the sky. Although rainwater can be contaminated by absorbing air-borne chemicals, most of the chemicals present in harvested rainwater are introduced during collection, treatment, and distribution (TCEQ, 2007).

i) Nature and pathways of rainwater contamination

The quality of harvested rainwater can be altered by many features including the environmental activities from industries, the nature of the catchment system, automobiles and anthropogenic activities, the roof materials, the presence of debris, particles, and birds or rodents dropping on roofs, and the type of storage materials used for harvested rainwater (Ahmed et al., 2010a). Atmospheric pollutants including particles, microorganisms, heavy metals, and organic substances accumulate on the catchment areas as dry deposition and are washed out from the atmosphere during rainfall events. Rainwater in rural areas being situated far from atmospheric and industrial pollution is fairly clean except for some dissolved gases. On the other hand, urban areas are characterized by a high traffic and industry impact and are therefore heavily contaminated by dust particles, heavy metals, and organic air pollutants. Also, the catchment surfaces themselves may be a source of heavy metals and organic substances. Low polluted rainwater can be collected from roofs constructed with tiles, slates, and aluminum sheets. Roof tied with bamboo gutters are least suitable because of possible health hazards. Similarly, zinc and copper roofs or else roofs with metallic paint or other coatings are not recommended because of high heavy metal concentrations. If the catchment areas are roads, the rainwater may be polluted by heavy metals originating from brakes and tires and organic compounds like polycyclic aromatic hydrocarbons (PAH) and aliphatic hydrocarbons from incomplete combustion processes. To apply drinking water quality, the removal of these hazardous compounds is necessary. Bacteria, viruses, and protozoa may originate from fecal pollution by birds, mammals, and reptiles with access to catchments and rainwater storage tanks. Therefore, harvested rainwater is often unsuitable for drinking without any treatment. Disinfection should then be applied to improve microbiological quality. Roof runoff may contain a complex chemical and microbiological composition. Conventional chemical contaminants include nutrients, heavy metals, and organic compounds such as petroleum hydrocarbons (Heberer, 2002), while microbial contaminants include pathogens and disease vectors. Besides conventional contaminants, emerging contaminants including as pharmaceuticals such as hormonal or endocrine-disrupting compounds (Hoffman et al., 1982) and nanoparticles (Gao, 2008) have recently drawn many research attention. The nature and concentrations of the contaminants in roof water will largely depend on the source, anthropogenic activities, flow pathways, and interactions with the environment. Several studies have demonstrated roof runoff contamination by physical, chemical, and biological contaminants (Garnaud et al., 1999; Zhong et al., 2001) and microbiological contaminants (Corden and Millington, 2001; Jones and Harrison, 2004; Evans et al., 2006). Contamination mechanisms include air-borne and atmospheric deposition, direct leaching of metal constituents from metal sheets, colonization by various plants and consequently direct weathering of roof material and the leaching of the accumulated particulate organic matter and flora on roof surfaces (Chang et al., 2004; Adeniyi and Olabanji, 2005). King and Bedient (1982) attributed leaching of chemical compounds from roofs to the acidic nature of ambient rainfall, while Chang et al. (2004) reported that the rough surfaces and cracks associated with wood shingle roofs trap water, aerosols, and debris which promote plant growth, particularly fungi resulting in increased organic matter retention on the roof. In Nigeria, Adeniyi and Olabanji (2005) revealed that direct leaching of roof materials by chemical reactions was the main source of contamination on Al and Fe-Zn sheets. Also, concrete tiles and asbestos were more prone to colonization by various plants, leading to the softening of the roofing material. Although some studies did not establish clear correlations between contaminants and roof types (Chang et al., 2004; Mendez et al., 2010; Meera and Ahammed, 2006), Adeniyi and Olabanji (2005) managed to separate their roof types into two clusters. One cluster consisting of asbestos and concrete tiles was significantly correlated with Ca and bicarbonate ions. In contrast, the other cluster comprising thatch, Al and Fe-Zn sheets, was significantly correlated with K, Cl, sulfates, and nitrate contaminants. In summary, the risk for roof runoff contamination exists along the flow pathway of rainwater from the source to the point of consumption. Sources of contaminants include air-borne, atmospheric and animal/bird deposition on roof/catchment surfaces (Evans et al., 2006), storage and conveyance facilities, and humans and utensils point of consumptions. Various sources and pathways of contaminants demonstrate the complexity of maintaining and safeguarding public health when relying on roof water harvesting systems for domestic supply. The quality of rainwater is a key determinant of its suitability for various uses. Although rainwater can be used for various purposes including irrigation, livestock watering (Mwenge-Kahinda et al., 2007), laundry, bathing, and toilet use (Lynch and Dietsch, 2010; Ward et al., 2012), here we focus on its quality concerning drinking purposes. To safeguard public health, the water quality requirements for drinking purposes are more stringent than other uses. The quality of rainwater shows considerable spatial and temporal variability. Literature suggests that roof water quality's spatial and temporal variability depends on roof materials, catchment characteristics, precipitation properties, local weather, and pollutants' chemical properties (Vazquez et al., 2003; Abbasi and Abbasi, 2011). Several other studies have investigated the impacts of these factors and their interaction on roof water quality (Mendez et al., 2010; Meera and Ahammed, 2006). Understanding any trends will provide insightful information that can be extrapolated in the other regions where data are scarce.

ii) Quality assessment in various countries

The roof-harvested rainwater is frequently used for household activities and principally based on drinking water where town water is not available (Ahmed et al., 2008). However, minimum attention has been given to the quality of roof-harvested rainwater used for potable purposes. In most cases, roof-harvested rainwater is used without any form of water treatment. The reports on quality assessment of rainwater have been documented from South Africa (Kahinda et al., 2007), Greece (Gikas and Tsihrintzis, 2012; Sazaklia et al., 2007), Canada (Despins et al., 2009), Netherlands (Schets et al., 2010), Spain (Farreny et al., 2011a, b), Mediterranean (Angrill et al., 2012), the USA (Mendez et al., 2011), Palestine (Daoud et al., 2011), Australia (Ahmed et al., 2008, 2010a, 2010b, 2011), France (Vialle et al., 2011), South Korea (Lee et al., 2012), Tanzania (Mwamila et al., 2016), and Jordan (Radaideh et al., 2009). These studies collectively show how harvested rainwater quantity depends on geographical climate and storage tank size (Campisano et al., 2013). In contrast, harvested rainwater quality is dependent on the spatial and temporal inconsistency of rainfall (Evans et al., 2006), roofing constituents (Mendez et al., 2011), and first-flush diverters (Gikas and Tsihrintzis, 2012). Successful application of RWHSs, therefore, requires local data catalogue on the physical, chemical, and microbiological characteristics of rainwater.

To reduce the health risks from rainwater reuse, consideration of water treatment systems could be efficient. Continuous release of pollutants into the atmosphere has led to public health concern about the quality of rainwater in industrialized and oil-producing areas (Efe, 2006). The activities of petroleum industries and other related industries emit greenhouse gas, sulfur (IV) oxide, and metals such as lead, mercury, and copper into the atmosphere, thus depleting rainwater quality (Ezenwaji et al., 2013). Also, metals such as lead, mercury, and copper are released from some industries that consequently dissolve in rainwater. Their continuous release resulted in the increased concentration of these metals in the atmosphere. Though some heavy metal indicators (Zn, Na, K, and Ca) in this study were within the WHO guidelines, some locations revealed heavy metal (Cu, Fe, and Cd) concentrations slightly above the WHO guidelines. In the Brazilian semiarid region, the studies are focused on the impact of RWH on the population's health, mainly checking the presence of contaminants such as *Escherichia coli* (Table 14.1).

TABLE 14.1 Studies and findings on quality testing of harvested rainwater in various countries.

Sr. No.	References	Country	Findings
1	Abbasi and Abbasi, 2011	India	• The quality of harvested rainwater decreases as the storage period increases. • The inlet of the storage tank should be designed so that sediments on the tank's bottom are not disturbed when water flows into the tank. • It is best to extract water from the top of the tank via a floating inlet instead of the bottom of the tank, as this is where the water is dirtiest. • The overflow of the tank should be designed to exclude or discard the dirtiest water. Possible designing methods such that overflow can be taken from the bottom of the tank where the dirtiest water is, or design the inlet. Such that overflow bypasses the inlet and never enters the tank.
2	Abdulla and Al-Shareef, 2009	Jordan	• The quality of rainwater within an RWH system depends on how long it is stored and the patterns of use. • Performing regular maintenance, as well as efficient management and operation of the system, improves water quality. • Roof washing and first-flush diversion are encouraged to maintain good water quality. • The intensity of rainfall and the number of antecedent dry days significantly affect water quality. • 60 cisterns were tested, and samples met WHO guidelines for physical and chemical parameters, indicating harvested rainwater, including the first flush, usually provides safe drinking water.
3	Daoud et al., 2011	Palestine	• An increase in pH occurred due to the alkaline nature of the roofing and storage tank materials. • The overall quality of stored rainwater meets WHO drinking water guidelines, except for pH, turbidity, and lead.
4	Domènech and Saurí, 2011	Spain	• Acceptable chemical concentrations in all tanks sampled, except one system contained boron above the European maximum allowable concentration. • The cleanliness of the catchment surface, gutters, and storage tanks was a direct indicator of water quality within the system.
5	Vialle et al., 2012	France	• Ionic concentrations were low, indicating the rainwater had low level of mineralization. • Harvested rainwater had relatively good physicochemical quality, but the quality varied greatly and did not meet drinking water guidelines. • Concentrations of tap samples parameters were similar to those from tank samples where mains water was used to top up the system.
6	Despins et al., 2009	Canada	• Storing water in a cistern improves its quality primarily through sedimentation. • Storing water in a concrete tank tends to increase the pH, minimizing the potential for leaching metals. • There is some potential for leaching chemicals from the tank material into the stored water (e.g., zinc from metal tanks and organic compounds from plastic tanks). • Water quality parameters are susceptible to the design aspects of rainwater harvesting systems as well as environmental conditions. • pH is easily affected by the type of tank material. The pH of water stored in plastic tanks was slightly acidic, while water stored in concrete tanks was more basic. • Color, turbidity, and total nitrogen concentrations increased during dry periods. • More inferior water quality was observed during the summer and fall seasons, especially TOC. This may be due to cold climate conditions (i.e., low atmospheric deposition, decrease in plant/animal activity, or presence of snow that minimized transfer of pollutants to roof surface). • The selection of appropriate catchment and storage material and the proper application of pre- and post-storage treatment can ensure a consistently high quality of harvested rainwater.

Continued

TABLE 14.1 Studies and findings on quality testing of harvested rainwater in various countries.—cont'd

Sr. No.	References	Country	Findings
7	Morrow et al., 2010	Australia	• All parameters sampled from the cistern water were below Australia Drinking Water Guidelines for the entire sampling period. • Iron concentrations in 3 of the monitored systems were lower in the stored water than in the roof runoff, most likely due to flocculation, settling, or microbial processes occurring within the tank. • The elemental composition of harvested rainwater varies throughout the rainwater harvesting system and depends upon individual site characteristics.
8	Sazaklia et al., 2007	Greece	• The physical and chemical characteristics of stored water met the requirements for safe drinking water. • Catchment areas and storage tanks must be cleaned regularly to maintain good water quality, and first flush should be diverted.
9	Sung et al., 2010	Taiwan	• Storing water for long periods can decrease water quality. • Larger suspended sediment particles settled rapidly within the storage tank, thereby improving water quality. • TOC concentrations were very high, perhaps due to atmospheric deposition or long retention time.
10	Spinks et al., 2006	Australia	• Physicochemical results were compliant with Australia Drinking Water Guidelines for all parameters except pH. None of the systems sampled disinfected the water before consumption; only 9 employed first-flush diversion. • Elevated cadmium concentrations in some systems most likely due to the age and type of roofing material (galvanized iron). • High zinc concentrations may also be due to age and type of roof and tank material corrosion, pipes, and fittings. There was a significant correlation between elevated zinc levels and the use of a galvanized iron tank; however, concrete tanks were significantly correlated with zinc concentrations lower than the drinking water guidelines.

iii) Implications of using harvested rainwater

The enormous potential of water quality contamination throughout a RWHS necessitates the use of treatment options to produce water of suitable quality for potable and nonpotable uses. Potential treatment options for RWH systems include both prestorage (debris screens and filters and first-flush diversion) and poststorage measures (poststorage filtration, flocculation, and disinfection). The majority of studies on harvested rainwater quality acknowledge that first-flush diversion can significantly improve the quality of collected rainwater and recommend this as a staple in RWH system design (Kus et al., 2010b; Despins et al., 2009; Mendez et al., 2011). Diverting the first flush can retard the buildup of particulates and sediments within storage tanks, prevent odor and other aesthetic issues, and improve overall water quality (Lee et al., 2010; Abbasi and Abbasi, 2011). It is also highly recommended to decrease the concentrations of pesticides and other organic compounds that enter the storage tank (Zhu et al., 2004). While the recommendation for including first-flush diversion is universal, the diversion volume recommendation varies greatly. The exact volume that can be considered of first-flush quality at any given time is dependent upon several factors, including the number of preceding dry days, amount and type of debris present on the roof surface, season, and quality and type of roof surface. Debris screens and filters can be used between the roof surface and the storage tank to prevent particulate matter (and contaminants adsorbed in the particulate matter) from entering the tank (Abbasi and Abbasi, 2011). Abbasi and Abbasi (2011) recommend the following characteristics to maximize the effectiveness of debris screens when employed by an RWH system:

- Filter should be easy to clean or chiefly self-cleaning.
- Filter should not clog easily and should be easy to detect and repair.
- Filter should not provide an entrance for additional contamination.

Poststorage treatment can consist of in-line sediment filters on pumps, slow sand filtration, flocculation, and disinfection. Particle filtration (sediment filters, sand filtration, other types of filters) has been shown to remove particulates and heavy metals and improve turbidity (Despins et al., 2009). Adding a flocculant such as alum or calcium hydroxide to the storage tank promotes flocculation and suspended delicate particulate matter (Abbasi and Abbasi, 2011). Finally, disinfection methods include bleaching powder, potassium permanganate, iodine, heat (boiling water), chlorine, and ultraviolet light. Each of these options has pros and cons to its use; however, disinfection is predominantly used to improve rainwater's microbiological quality. Although first-flush diversion and prestorage filtration can substantially improve the quality of water stored in a RWHS, frequent maintenance of these systems is very important. Numerous studies have found that regular maintenance improves water quality (Abdulla and Al-Shareef, 2009; Magyar et al., 2007).

iv) Rainwater treatment technologies used in developing countries

Harvested rainwater treatment strategies may include chlorination (Lantagne et al., 2008; Keithley et al., 2012), metal/chemical additives (Nawaz et al., 2012), ozonation (Ha et al., 2013), filtration (Kim et al., 2005; Jordan et al., 2008), UV treatment (Jordan et al., 2008), SODIS (Amin and Han, 2009; Strauss et al., 2016), and solar pasteurization (Dobrowsky et al., 2015). Although varying degrees of treatment efficiency were obtained using these unique technologies (Hamilton et al., 2019), the successful implementation of these technologies will be dependent on cost, ease of management, and water access, among others (Hunter et al., 2010). Thus, it is crucial that these water supply determinants of good health be considered when developing efficient rainwater treatment strategies for developing countries. SODIS (solar disinfection) is considered an inexpensive and straightforward treatment system used daily by approximately five million people in Africa, Asia, and South America (McGuigan et al., 2012). In its simplest form, transparent bottles are filled with contaminated water and are exposed to natural sunlight for 6–48. The ultraviolet radiation inactivates microbial contaminants by reacting directly with the microbial cellular components or indirectly through the generation of reactive oxygen species within the water, which also damages cellular components (Nelson et al., 2018). The water temperature will also increase as water molecules absorb the UV radiation, with the increase in temperature (above 45°C) contributing to the disinfection process by leading to cell membrane damage (McGuigan et al., 2012). Recent research efforts have focused on increasing the treatment volume of traditional SODIS using solar mirrors/compound parabolic collectors, larger reactor tubes, or the addition of heterogeneous photocatalysts (titanium dioxide) to facilitate the disinfection process (Table 14.2).

v) Potential biological treatment strategies

TABLE 14.2 Treatment methods with merits and demerits of RWHS.

Sr. No.	Treatment method	Subcategory	Merits	Demerits
1	Filtration	Sand filter	• Can improve microbiological and physiochemical qualities effectively	• Cannot remove viruses • Performance varies with time, presence of oxygen, and temperature. Comparatively large area required
		Activated carbon filter	• Can remove Fe and H_2S • Can improve taste • Can remove odors and colors	• Does not remove bacteria and viruses
		Reverse osmosis filter	• Highly effective in removing microorganisms as well as leftover chemicals in the water	• Expensive • Require additional draining arrangements and power supply • Prefiltration required
		Membrane filter	• Can remove bacteria and sediment	• May not remove viruses • Expensive • Prefiltration required
2	UV disinfection		• Highly effective in removing bacteria, viruses, and protozoa • Requires less maintenance	• Prefiltration required for effective light penetration • Power supply required • Water must be used right after the treatment
3	Chlorination		• Highly effective for the removal of bacteria, viruses, and Giardia • Residual chlorine can protect the water quality for a more extended period	• Long-term consumption of chlorinated water may pose a health risk (Morris, 1995) • Hard to determine an appropriate dosage
4	Ozonation		• Minimal effect on the taste of water • Kills bacteria, parasites, and other harmful pathogens • Produce fewer chemical by-products	• Even if a smaller number of by-products are produced, they are susceptible to be a carcinogen • No residual effects • Treated water should be consumed right after the treatment

Biological treatment methods that allow for the targeted removal of persistent organisms may include predatory bacteria and bacteriophages. In contrast, biofilter systems display promise for the nonselective removal of microbial contaminants. Also, the use of microbially produced secondary metabolites (biosurfactants) displays promise for rainwater treatment. Contributing to the appeal of biological treatment is the ease with which these strategies may be combined with physical and chemical disinfection methods to treat rainwater. Biosurfactants are a diverse group of surface-active compounds synthesized by several microbial genera and may display antimicrobial activity against a wide range of pathogenic and opportunistic pathogenic microorganisms (Rahman et al., 2012a, b). Although biosurfactants' antimicrobial properties rely on different mechanisms to destroy microbial cells, they primarily destroy bacterial cells by disrupting the plasma membrane or cell wall (Basinger et al., 2010). In addition, biosurfactants not only inhibit biofilm formation on various surfaces (Mehrabadi et al., 2013). Some classes also have the potential to disrupt preformed biofilms (Hashim et al., 2013). As biofilm formation is one of the critical survival mechanisms used by opportunistic pathogens in water distribution and treatment systems, biosurfactants promise to use in rainwater treatment strategies. For example, polymeric surfaces may be coated with biosurfactants to prevent biofilm formation. In contrast, biosurfactants may also be embedded in filtration devices to act as an additional disinfectant during water treatment. In comparison, a nonselective biological water treatment strategy uses biofilters, which consist of a bed of media that allow for the removal of microorganisms as they are captured in the filter bed (biofilm formation). These biofilms can then capture and retain microorganisms, remove organic pollutants, and allow for predation of pathogens by other microorganisms within the biofilm matrix (Hudzori, 2017). Biofilters can thus effectively reduce microbial contaminants in harvested rainwater and may also easily be applied as a pretreatment to existing technologies (Belmeziti et al., 2014).

14.7 Challenges

Historically, challenges to the social acceptance of RWH have focused on water quality, risk perception, and health risk (Fewtrell and Kay, 2007; Ward et al., 2010; Rozin et al., 2015), as well as financial viability (Roebuck et al., 2011). Despite some households being resistant to using rainwater indoors (Mankad and Tapsuwan, 2011), it is now acknowledged that RWH is an acceptable source of nonpotable water compared to other types of water reuse for nonpotable purposes (Egyir et al., 2016; Dobrowksy et al., 2014). The focus on acceptability and financial returns to date has often detracted attention from broader challenges. These include evaluating social as well as financial benefits to engender more comprehensive institutional support and reflexive analysis of the international RWH niche to enable greater consideration of system efficacy and community participation, both of which will enhance the hydrosocial contract and diffusion of RWH into wider society (Stenekes et al., 2006; Getnet and MacAlister, 2012). Fig. 14.4 illustrates the key factors that are necessary for successful implementation of RWHS. Moving away from rhetoric, insights and costs enables the RWH sector to move toward a more optimistic and innovative space where challenges are redefined and answered by policy-makers, businesses, and communities.

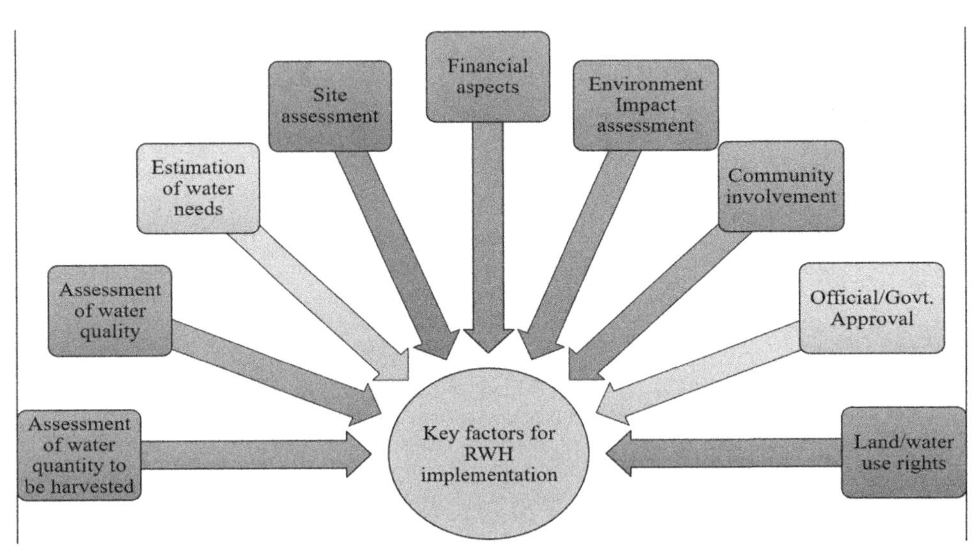

FIG. 14.4 Key factors to be considered for implementation of RWH system.

14.7.1 Cost

In developing countries, the cost is still a challenging issue when applying RWHS. Initial cost and maintenance are still debatable regarding how this system can be affordable for all societies, especially for people in the low-income category. This limitation is associated with national income and the low awareness of the community. In order to maximize the benefits, an optimum RWHS design is highly crucial. Also, material selection can reduce the initial cost. Designing RWHS using gravity can reduce the operation and maintenance cost compared to that system with pumping operation. Moreover, the government may provide subsidies to encourage the public to install the system. Also, training and awareness campaigns are highly beneficial for enhancing the interest of the community.

14.7.2 Treatment system

Most of the existing RWHS is for nonpotable uses, in which the water is used directly from the collection tank. Some parameters such as turbidity, lead, fecal coliforms, and total coliforms are present above the limit regulated by the World Health Organization (WHO). The facts reveal that a simple treatment still needs to be done before the rainwater can be widely used for potable uses. The selection of rainwater treatment method has implications for the installation and maintenance costs. For instance, nonpotable harvested rainwater uses, such as toilet flushing, landscape irrigation, and car washing, do not require treatment.

Conversely, harvested rainwater for potable uses such as drinking, cooking, shower, and cloth washing needs a cost-effective treatment method. Treatment is also necessary when the harvested rainwater is used for a cooling system. Therefore, it is crucial to provide a cheap treatment method to maximize its economic benefit. Also, a simple treatment system with less maintenance has additional benefits for installation in rural areas.

14.7.3 Environmental considerations

Environmental feasibility depends upon the amount and patterns of rainfall in an area and the duration of dry periods and availability of other water resources in that area. The rainfall patterns and quantity over a year play a crucial role in determining the success of RWHS. It was estimated that the percentage of reliability of RWHS for toilet flushing, laundry, and irrigation use increased from 40% to 71% for study locations having an average annual rainfall ranging from 743 to 1325 mm in Australia (Hajani and Rahman, 2014). As a general rule, rainfall should be 50 mm/month for at least six months in an area to make RWHS environmentally feasible and successful.

14.7.4 Policy

Setting up institutional and policy frameworks creates an enabling environment for the adoption of RWH, strengthening institutional capacities and collaboration and networking. Rules, regulations, and bylaws need to be established but must be relevant to be accepted and followed. Several improvements on policy implementation are necessary to gain wider acceptance of RWHS, which includes providing an appropriate incentive and regulating the excessive use of piped water. For the future, the RWHS policy should be extended to all buildings with large roof area such as commercial buildings, which are expected to have a more considerable economic benefit. A comprehensive study considering an optimum tank size according to the various roof sizes and climatic conditions in all the countries should be carried out for the foreseeable future as a scientific judgment before issuing a legal policy.

14.7.5 The material choice of RWHS

Rainwater is relatively clean but can be contaminated by the roof materials and deposition on the roof surfaces. In older systems, the commonly used roof materials were steel, copper, aluminum, zinc, or tin. Over the time, the roof materials become rusty and subjected to leaching by rainwater, which is usually quite acidic (Möller and Zierath, 1986). Thus, it became a source of a contaminant in the collected rainwater. Also, application of paint, tar, glue, sealant, and other protective materials to lengthen the roof life span may contribute additional forms of contaminant. Moreover, various types of tanks depend on the materials being used, such as polyethylene, concrete, galvanized steel, fiberglass, and stainless steel, which tend to rust over time and release certain chemicals. These shortcomings could be overcome by introducing more inert and environmentally friendly materials. For this purpose, natural resources such as rattan, bamboo, and oil palm in the form of fibers or particles can be used as composite materials. Natural materials have been proven to have physical and mechanical properties comparable to synthetic materials (Nikmatin et al., 2017). Therefore, a comprehensive study by applying natural materials is much needed. This knowledge is useful because better-harvested rainwater quality can be achieved using inert and environmentally friendly materials.

14.7.6 Public perception

Despite various government initiatives to promote RWHS, acceptance among people of many countries is still unsatisfactory. One of the main reasons for the flawed acceptance is because of low water tariff. In countries that receive abundant rainfall with rare occurrences of significant drought, the general public feel that there is no necessity to explore other alternative water resources. There is also a "yuck factor" associated with the use of harvested rainwater that needs to be rectified by creating awareness regarding treatment methods of harvested rainwater and superior quality end product among masses. The public is inadequately educated on the importance of rainwater utilization within the context of water demand management. Both strategies in terms of penalty and incentive are crucial for ensuring fuller implementation of RWH at residential, commercial, and industrial premises.

14.7.7 Lack of information and knowledge of RWHS

The lack of information and knowledge relating to RWH systems permeates through all of the categories identified above. For instance, Roaf (2006) suggests the lack of knowledge and information relating to water recycling and reuse systems among the various stakeholders as one of the main barriers to their more comprehensive application and use. This ranges from a lack of empirical data on which central government and regulators can formulate design guides and standards to a general lack of awareness and knowledge of reuse and recycling systems among architects, consultants, and developers. However, current research focuses on water savings, water quality, economics, and operational matters but rarely considers the broader environmental issues such as sustainable resource use. A proper awareness program is necessary to educate the public on how RWHS can be implemented to reduce domestic water supply dependency.

14.8 Benefits and applications of RWH

In general, the benefit of RWHS can be divided into two categories, namely environmental and economic. For environmental benefit, it can be used as an alternative water supply to supplement piped water. Large-scale RWHS can help reduce flash flood in urban area and minimize soil erosion. It also prevents entry of pollutants into the water bodies. It is known to reduce nonpotable water demand by storing rainwater from larger roof areas in Iran. Since the benefit of RWHS is highly dependent on water usage, system design, rainfall, and other uncertainty variables, its evaluation of long-term performance is needed better to understand the effects of each variable on its benefits. This is very useful as a basis for designing future RWHS (Fig. 14.5).

14.9 Recommendations to encourage RWH

A participatory approach contributes to creating an enabling environment for the adoption and sustainability of WH technologies. Different approaches are needed in different contexts, and it has to be acknowledged that apart from government intervention and donor investments, greater engagement of civil society and empowering stakeholders at grassroots is required. Approaches need to be developed, not selected, transferred, or copied: depending on the situation, the people involved, objectives, possible solutions, and resources available (Liniger et al., 2011).

14.9.1 Subsidies

It is noticed that RWH scheme is less attractive in many developing countries. This is because of the high installation and maintenance cost and low water tariff, which result in a long payback period. On the other hand, the success of RWHS in developed countries is contributed by the support from the government especially during the initial stages of implementation. Several countries have introduced subsidies for the premise owner who installed RWHS. For instance, in Spain, there are subsidies of up to $1200 for each house owner who has installed RWHS on their initiative (Domènech and Saurí, 2011). The Australian government has also launched the Home WaterWise Rebate Scheme, which provides subsidies to residents who have implemented RWHS for nonpotable domestics uses (Ahmed et al., 2011). In Germany, the government supports the installation of RWHS in new or existing households

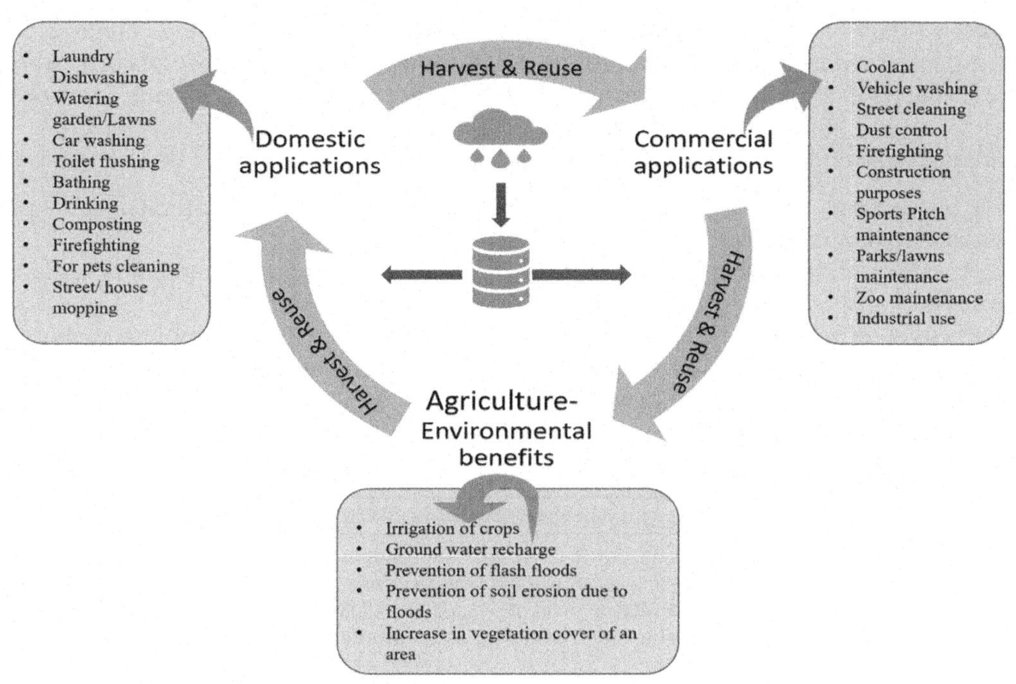

FIG. 14.5 Benefits and applications of harvested rainwater.

by subsidizing one-third of the total costs or up to $2000 (Schuetze, 2013). Other countries such as Japan, Uganda, USA, and Germany have also paid attention to encourage RWHS implementation by providing subsidy and low interest, subsidy for construction materials, rebates and tax exemptions, and exemptions from stormwater taxes, respectively (Table 14.3).

14.9.2 Implementation of strict laws for RWH

Specific regulations are imperative for guiding, restricting, and enforcing the targeted technology potential's adoption and attainment. These should be recognizable at the national level. Furthermore, national RWH technology guidelines and standards are vital. A policy framework for water harvesting systems arises mainly because the

TABLE 14.3 The incentives offered by countries for promotion and encouragement of RWHS.

Sr. No.	Country	Incentives	References
1	Japan	The government provides subsidy and a low-interest loan to premises for RWHS installation.	Furumai, 2008
2	Australia	The government offers up to $500 rebates to houses that install a RWHS.	Rahman et al., 2012a, b
3	Taiwan	A new guideline for RWHS as new water conservation alternative for domestic water use is issued for federal buildings.	Cheng et al., 2006
4	Uganda	The government provides subsidies to RWHS construction materials in rural areas.	Baguma and Loiskandl, 2010
5	Jordan	The government has incorporated RWHS in the water demand management policy.	Abdelkhaleq and Dziegielewski, 2006
6	Spain	The government has made it mandatory for new buildings with a particular garden area to install RWHS.	Domènech and Saurí, 2011
7	Brazil	The government promotes programs that aim to install one million cisterns in semiarid areas.	Domènech and Saurí, 2011
8	Belgium	The government has mandated for new buildings with a roof area greater than 100 m^2 to install RWHS.	Domènech and Saurí, 2011
9	USA	The government provides rebates and tax exemptions to foster rainwater use.	Domènech and Saurí, 2011
10	Germany	Premises with RWHS are exempted from stormwater taxes.	Herrmann and Schmida, 2000

general policy statements do not touch extensively upon the issue in many developing countries. There is a clear need to evolve a decentralized legal regime about RWH and its utilization, which empowers people and makes them real managers of resources. For promoting urban water harvesting, a policy should include a mix of incentives and penalties.

Measures that need to be undertaken include the following:

- RWH/recharge of groundwater system should be an essential town planning requirement and a prerequisite for permission to develop new colonies.
- Provision of rainwater and harvesting structures in all building plans should be mandatory for issuing of building permission.
- Appropriate rebates on property/fiscal incentives should be granted for effective implementation of RWHSs. Table 14.4 shows the policies and legislations implemented in various states and union territories for RWH in India (source: www.rainwaterharvesting.org).

TABLE 14.4 Status of laws and policies in Indian states for RWH.

Sr. No.	State/UT	Status of policies and laws for rain water harvesting (RWH)
1	Andhra Pradesh	Andhra Pradesh Water, Land and Tree Act, 2002′ stipulates mandatory provision to construct rainwater harvesting structures at new and existing constructions for all residential, commercial and other premises and open space having area of not less than 200 m^2 in the stipulated period, failing which the authority may get such rain water harvesting (RWH) structures constructed and recover the cost incurred along with the penalty as may be prescribed
2	Daman and Diu	Daman Municipal Building Model Bye-laws and Zoning Regulation, 2002 have provision for the construction of sump well for groundwater recharge. Instructions have been issued to local PWD for construction of RTRWH structures. Local bodies such as Municipality & District Panchayats have already initiated action in this regard
3	New Delhi	Modified Building Bye-laws, 1983 to incorporate the mandatory provision of roof-top RWH in a new building on plots of 100 m^2 or above. Through storage of rainwater run-off to recharge underground aquifer in NCR, Delhi exist. To encourage rainwater harvesting by Resident Welfare Associations/Group Housing Societies, the Govt. of NCR Delhi has launched a scheme for financial assistance in the Bhagidari concept, where 50% of the total cost of the project subject to a maximum of Rs. 50,000/- is being given to the RWAs as a grant if they adopt rainwater harvesting
4	Goa	PWD, Goa, has been asked to take up RWH structure for government buildings. The PWD, Goa, is studying various designs of rooftop RWH for taking up other existing/new coming up large government buildings
5	Gujarat	Metropolitan Areas have notified rules under which no new building plan is approved without corresponding rainwater harvesting structure. The D/o Roads & Buildings have been directed to ensure that all significant Govt. constructions, including educational institutions, had adequate rainwater harvesting facilities. The Urban Development and Urban Housing Department has issued necessary orders Gujarat Town Planning Act, 1976 to incorporate the rules for RWH
6	Haryana	Haryana Municipal Building Bye-laws 1982 has been amended to incorporate the provision of compulsory Roof Top RWH
7	Kerala	Rooftop RWH has become mandatory as per Kerala Municipality Building (Amendment) Rules, 2004 for all new buildings
8	Himachal Pradesh	Installation of RWH system has been made mandatory for all buildings to be constructed in urban areas of the State and no building plan without RWH system can be approved including schools, all Government buildings, and rest houses. Construction of RWH system has also been made mandatory for all schools, Govt. buildings and rest houses, upcoming industries, bus stands, etc.
9	Karnataka	The State has adopted an RWH policy to mandate this in all new construction. Bangalore City Corporation has already incorporated mandatory RWH in Building Bye-laws. Other ULB's are being encouraged to do so. Rural Development & Panchayati Raj department has issued orders to implement the rooftop RWH system in all government buildings and schools
10	Madhya Pradesh	The State Govt. vide Gazette notification dated 26.8.2006, has made rooftop RWH mandatory for all types of buildings having a plot size of more than 140 m^2. Govt. has also announced 6% rebate in property tax to individuals for the year in which the individual will install rooftop RWH structures
11	Maharashtra	Maharashtra Government is promoting RTRWH under the "Shivkalin Pani Sthawan Yojana." It provides that all houses should have provision for rainwater harvesting without which house construction plan should not be approved. Bombay Municipal Corporation and Pimpri-Chinchwad Municipal Corporation have made RWH mandatory by enacting building bye-laws

TABLE 14.4 Status of laws and policies in Indian states for RWH.—cont'd

Sr. No.	State/UT	Status of policies and laws for rain water harvesting (RWH)
12	Meghalaya	The State Government is considering to constitute State RWH Authority
13	Nagaland	The State Government has already made provision for rooftop rainwater compulsory for all new Government buildings
14	Puducherry	Approvals are issued to new constructions subject to the provision of RWH in building designs. PWD, Pondicherry has started constructing rooftop RWH structures in the Government buildings since 2002. The UT Administration has made rules for installation of the RWH system in all the new constructions
15	Rajasthan	Roof Top RWH has been made mandatory in State-owned buildings of plot size more than 300 Sq.mt. with effect from 03.01.2006. For violation of building by-laws, punitive measures, and viz. disconnection of water supply have also been made. The Govt. has made the compulsory installation of rainwater harvest system in all newly and existing construction building and Govt. offices vide order dated 31.05.2000 and 12.12.2005. The State Government is also considering to modify Municipal Corporation Act making provisions of RWH
16	Tamil Nadu	Vide Ordinance No. 4 of 2003 dated July, 2003 laws relating to Municipal Corporations and Municipalities in the State have been amended, making it mandatory for all the existing and new buildings to provide RWH facilities. The State has launched the RWH scheme's implementation on a massive scale in Government buildings, private houses/ Institutions and commercial buildings in urban & rural areas. The State Government has achieved cent percent coverage in rooftop RWH
17	Uttar Pradesh	Mandatory rules have been framed for compulsory installation of RWH system in all the new housing schemes/ plots/buildings of all uses, group housing schemes with a separate network of pipes for combined RWH/ Recharging system. Rooftop RWH has been made mandatory for plots of 100-200 m^2. In Govt. Buildings (both new and old), installation of RWH structures has been made mandatory
18	West Bengal	Vide Rule 171 of the West Bengal Municipal (Building) Rules, 2007, installation of RWH system has been made mandatory
19	Arunachal Pradesh	Building Bye-laws are being framed keeping provision for RWH as mandatory in Government Buildings
20	Punjab	Building Bye-laws amended to make RWH System mandatory in all buildings of above 200 sq. yards. The Punjab Urban Development Authority (PUDA) is amending the PUDA (Building) Rules 1996 for making this system mandatory. Municipal Corporation of Ludhiana and Jalandhar have framed Bye-laws to make RWH mandatory in new buildings
21	Jharkhand	The State Government has initiated action to construct RTRWH structures in Government/Public buildings in a phased manner. A promotional scheme has also been started to promote groundwater and artificial recharge by a grant of Rs. 25000/- for construction of artificial recharge structures. Ranchi Regional Development Authority (Jharkhand) has made Building Bye-laws for RWH
22	Uttarakhand	The Govt. of Uttarakhand (Awas evam Shahari Vikas) has made rules for the compulsory installation of RWH system and directed to adopt rules in building by-laws 15.11.2003. Accordingly, all the Development Authorities had made partial amendments in the prevalent House Building and Development Bye-laws/Regulations.
23	Tripura	As per Rule-110 of the Tripura Building Rules, 2004, water harvesting through starting of rainwater runoff is mandatory in all new buildings having plinth area more than 300 m^2 for all types of uses and in group housing of any size.
24	Bihar	The Bihar Ground Water (Regulation and Control of Development and Management) Act, 2006 has been enacted which provides mandatory provision of RTRWH structures in the building plan in an area of 1000 m^2 or more.

14.10 Conclusion & future prospects

With a growing population, climate change, higher food prices, and rising shortages of safe drinking water, collective emphasis must be put on integrated water management. In particular, water harvesting has a high potential for increasing crop production in dry areas and providing drinking, sanitation and household water, and water for livestock and commercial purposes. However, initiatives are still too dispersed, and experiences related to "best" RWH practices are rarely known. Policies, legal directives, and governmental budgets often lack water harvesting in integrated water resource management. To address the water scarcity and growing demands, water provision for drinking, domestic, and livestock use needs to be decentralized and water itself should be used more competently by harvesting local resources. Today, RWH is being increasingly promoted as a coping strategy. Both national and international organizations are beginning to invest more in RWH for domestic water supply, livestock consumption, and plant production. However, to support and stimulate this development more attention needs to be paid to:

- Capitalizing from local and traditional knowledge, as well as innovations by water users and research
- Facilitate and share knowledge and decision support for local execution and regional planning
- Upscaling the wealth of RWH knowledge and successful RWH practices based on informed decision-making
- Building up the useful and well-experienced extension and technical advice services
- Demonstrating the benefits of RWH, including cost and benefit assessments
- Mainstreaming RWH implementation into development projects, investment frameworks, national strategies, and action plans
- Encouraging coordination and collaboration among stakeholders
- Assuring an enabling framework from the policy level: especially securing land and resource use rights
- Supporting effective decentralization and good governance by offering capacity building and training.

References

Abbasi, T., Abbasi, S.A., 2011. Sources of pollution in roof-top rainwater harvesting systems and their control. Crit. Rev. Environ. Sci. Technol. 41 (23), 2097–2167.

Abdelkhaleq, R., Dziegielewski, B., 2006. A National Water Demand Management Policy in Jordan. Ministry of Water and Irrigation, Jordan.

Abdulla, F.A., Al-Shareef, A.W., 2009. Roof rainwater harvesting systems for household water supply in Jordan. Desalination 243 (1-3), 195–207.

Adeniyi, I.F., Olabanji, I.O., 2005. The physico-chemical and bacteriological quality of rainwater collected over different roofing materials in Ile-Ife, southwestern Nigeria. Chem. Ecol. 21, 149–166.

Ahmed, W., Gardner, T., Toze, S., 2011. Microbiological quality of roof-harvested rainwater and health risks: a review. J. Environ. Qual. 40 (1), 13–21.

Ahmed, W., Goonetilleke, A., Gardner, T., 2010b. Implications of faecal indicator bacteria for the microbiological assessment of roof-harvested rainwater quality in Southeast Queensland, Australia. Can. J. Microbiol. 56 (6), 471–479.

Ahmed, W., Huygens, F., Goonetilleke, A., Gardner, T., 2008. Real-time PCR detection of pathogenic microorganisms in roof-harvested rainwater in Southeast Queensland, Australia. Appl. Environ. Microbiol. 74, 5490–5496.

Ahmed, W., Vieritz, A., Goonetilleke, A., Gardner, T., 2010a. Health risk from the use of roof-harvested rainwater in Southeast Queensland, Australia, as potable or nonpotable water, determined using quantitative microbial risk assessment. Appl. Environ. Microbiol. 76 (22), 7382–7391.

Amin, M.T., Han, M.Y., 2009. Roof-harvested rainwater for potable purposes: application of solar disinfection (SODIS) and limitations. Water Sci. Technol. 2009 (60), 419–431.

Angrill, S., Farreny, R., Gasol, C.M., Gabarrell, X., Viñolas, B., Josa, A., Rieradevall, J., 2012. Environmental analysis of rainwater harvesting infrastructures in diffuse and compact urban models of Mediterranean climate. Int. J. Life Cycle Assess. 17 (1), 25–42.

Baguma, D., Loiskandl, W., 2010. Rainwater harvesting technologies and practises in rural Uganda: a case study. Mitig. Adapt. Strateg. Glob. Chang. 15 (4), 355–369.

Basinger, M., Montalto, F., Lall, U., 2010. A rainwater harvesting system reliability model based on nonparametric stochastic rainfall generator. J. Hydrol. 392 (3-4), 105–118.

Belmeziti, A., Coutard, O., de Gouvello, B., 2014. How much drinking water can be saved by using rainwater harvesting on a large urban area? Application to Paris agglomeration. Water Sci. Technol. 70 (11), 1782–1788.

Campisano, A., Gnecco, I., Modica, C., Palla, A., 2013. Designing domestic rainwater harvesting systems under different climatic regimes in Italy. Water Sci. Technol. 67, 2511–2518.

Chang, M., McBroom, M.W., Beasley, R.S., 2004. Roofing as a source of non-point water pollution. J. Environ. Manag. 73, 307–315.

Cheng, C.L., Liao, M.C., Lee, M.C., 2006. A quantitative evaluation method for rainwater use guideline. Build. Serv. Eng. Res. Technol. 27 (3), 209–218.

Corden, J.M., Millington, W.M., 2001. The long-term trends and seasonal variation of the aeroallergen Alternaria in Derby, UK. Aerobiologia 17 (2), 127–136.

Critchley, W., Siegert, K., 1991. Water Harvesting. A Manual for the Design and Construction of Water harvesting Schemes for Plant Production. FAO, Rome. http://www.fao.org/docrep/U3160E/U3160E00.htm#Contents.

Daoud, A.K., Swaileh, K.M., Hussein, R.M., Matani, M., 2011. Quality assessment of roof-harvested rainwater in the West Bank, Palestinian authority. J. Water Health 9, 525–534.

Despins, C., Farahbakhsh, K., Leidl, C., 2009. Assessment of rainwater quality from rainwater harvesting systems in Ontario, Canada. J. Water Supply Res Technol. 58, 117–135.

Dobrowksy, H., Mannel, D., De Kwaadsteniet, M., Prozesky, H., Khan, W., Cloete, T.E., 2014. Quality assessment and primary uses of harvested rainwater in Kleinmond, South Africa. Water SA 40 (3), 401–406.

Dobrowsky, P.H., Lombard, M., Cloete, W.J., Saayman, M., Cloete, T.E., Carstens, M., Khan, S., Khan, W., 2015. Efficiency of microfiltration systems for the removal of bacterial and viral contaminants from surface and rainwater. Water Air Soil Pollut. 226, 33.

Domènech, L., Saurí, D., 2011. A comparative appraisal of the use of rainwater harvesting in single and multi-family buildings of the Metropolitan Area of Barcelona (Spain): social experience, drinking water savings and economic costs. J. Clean. Prod. 19 (6-7), 598–608.

Efe, S., 2006. Quality of rainwater harvesting for rural communities of Delta State, Nigeria. Environmentalist 26, 175–181.

Egyir, S.N., Brown, C., Arthur, S., 2016. Rainwater as a domestic water supplement in Scotland: attitudes and perceptions. Br. J. Environ. Clim. Change 6 (3), 160–169.

Ercin, A.E., Hoekstra, A.Y., 2014. Water footprint scenarios for 2050: a global analysis. Environ. Int. 64, 71–82.

Evans, C.A., Coombes, P.J., Dunstan, R.H., 2006. Wind, rain and bacteria: the effect of weather on the microbial composition of roof-harvested rainwater. Water Res. 40, 37–44.

Ezenwaji, E.E., Okoye, A.C., Otti, V.I., 2013. Effects of gas flaring on rainwater quality in Bayelsa State, Eastern Niger-Delta region, Nigeria. J. Toxicol. Environ. Health Sci. 5 (6), 97–105.

Falkenmark, M., Fox, P., Persson, G., Rockström, J., 2001. Water Harvesting for Upgrading of Rainfed Agriculture: Problem Analysis and Research Needs. Stockholm International Water Institute (SIWI) Report 11, Stockholm.

Farreny, R., Morales-Pinzon, T., Guisasola, A., Taya, C., Rieradevall, J., Gabarrell, X., 2011a. Roof selection for rainwater harvesting: quantity and quality assessments in Spain. Water Res. 45, 3245–3254.
Farreny, R., Gabarrell, X., Rieradevall, J., 2011b. Cost-efficiency of rainwater harvesting strategies in dense Mediterranean neighbourhoods. Resour. Conserv. Recycl. 55, 686–694.
Fewkes, A., 2012. A review of rainwater harvesting in the UK. Struct. Surv. 30, 174–194.
Fewtrell, L., Kay, D., 2007. Microbial quality of rainwater supplies in developed countries: a review. Urban Water J. 4 (4), 253–260.
Fonseca, C.R., Hidalgo, V., Díaz-Delgado, C., Vilchis-Francés, A.Y., Gallego, I., 2017. Design of optimal tank size for rainwater harvesting systems through use of a web application and geo-referenced rainfall patterns. J. Clean. Prod. 145, 323–335.
Furumai, H., 2008. Rainwater and reclaimed wastewater for sustainable urban water use. Phys. Chem. Earth A/B/C 33 (5), 340–346.
Gao, L., 2008. Potential Environmental Implications of Manufactured Nanomaterials: Toxicity, Mobility and Nanowastes in Aquatic and Soil Systems. PhD Dissertation, Florida USA, University of Florida.
Garnaud, S., Mouchel, J.-M., Chebbo, G., Thevenot, D.R., 1999. Heavy metal concentrations in dry and wet atmospheric deposits in Paris district: comparison with urban run-off. Sci. Total Environ. 235, 235–245.
Getnet, K., MacAlister, C., 2012. Integrated innovations and recommendation domains: paradigm for developing, scaling-out, and targeting rainwater management innovations. Ecol. Econ. 76, 34–41.
Gikas, G.D., Tsihrintzis, V.A., 2012. Assessment of water quality of first-flush roof run-off and harvested rainwater. J. Hydrol. 466–467, 115–126.
Ha, S.A., Kim, I.S., Son, K.S., Wang, J.P., 2013. Development of rainwater purification and reclaimed water treatment systems using a high-efficiency air-cooled ozone generator. Appl. Mech. Mater. 423–426, 1383–1387.
Hajani, E., Rahman, A., 2014. Reliability and cost analysis of a rainwater harvesting system in peri-urban regions of Greater Sydney, Australia. Water 6 (4), 945–960.
Hamilton, K.A., Reyneke, B., Waso, M., Clements, T.L., Ndlovu, T., Khan, W., Digiovanni, K., Rakestraw, E., Montalto, F., Haas, C.N., 2019. A global review of the microbiological quality and potential health risks associated with roof-harvested rainwater tanks. Npj Clean Water 2, 7.
Hashim, H., Hudzori, A., Yusop, Z., Ho, W.S., 2013. Simulation based programming for optimization of large-scale rainwater harvesting system: Malaysia case study. Resour. Conserv. Recycl. 80, 1–9.
Heberer, T., 2002. Occurrence, fate, and removal of pharmaceutical residues in the aquatic environment: a review of recent research data. Toxicol. Lett. 131, 5–17.
Herrmann, T., Schmida, U., 2000. Rainwater utilization in Germany: efficiency, dimensioning, hydraulic and environmental aspects. Urban Water 1 (4), 307–316.
Hoffman, E.J., Latimer, J., Mills, G.L., Quinn, J.G., 1982. Petroleum hydrocarbons in urban run-off from a commercial land use area. J. Water Pollut. Cont. Fed. 5 (11), 1517–1525.
Hudzori, A.B., 2017. Optimization model of large-scale rainwater harvesting system. In: Chemical Engineering. Universiti Teknologi Malaysia, Skudai, Malaysia.
Hunter, P.R., MacDonald, A.M., Carter, R.C., 2010. Water supply and health. PLoS Med. https://doi.org/10.1371/journal.pmed.1000361. e1000361.
Jones, A.M., Harrison, R.M., 2004. The effects of meteorological factors on atmospheric bioaerosol concentrations—a review. Sci. Total Environ. 326, 151–180.
Jones, M.P., Hunt, W.F., 2010. Performance of rainwater harvesting systems in the southeastern United States. Resour. Conserv. Recycl. 54, 623–629.
Jordan, F.L., Seaman, R., Riley, J.J., Yoklic, M.R., 2008. Effective removal of microbial contamination from harvested rainwater using a simple point of use filtration and UV-disinfection device. Urban Water J. 5, 209–218.
Kahinda, J.M.M., Rockström, J., Taigbenu, A.E., Dimes, J., 2007. Rainwater harvesting to enhance water productivity of rainfed agriculture in the semi-arid Zimbabwe. Phys. Chem. Earth A/B/C 32 (15-18), 1068–1073.
Keithley, S.E., Kiristis, M.J., Kinney, K.A., 2012. The Effect of Treatment on the Quality of Harvested Rainwater. MSc Thesis, University of Texas at Austin.
Kim, R., Lee, S., Kim, J., 2005. Application of a metal membrane for rainwater utilization: filtration characteristics and membrane fouling. Desalination 2005 (177), 121–132.
King, T.L., Bedient, P.B., 1982. Effect of acid rain upon cistern water quality. In: Proceedings of an International Conference on Rainwater Cistern Systems. University of Hawaii at Manoa, pp. 244–248.
Kus, B., Kandasamy, J., Vigneswaran, S., Shon, H.K., 2010b. Analysis of first flush to improve the water quality in rainwater tanks. Water Sci. Technol. 61 (2), 421–428.
Lantagne, D.S., Blount, B.C., Cardinali, F., Quick, R., 2008. Disinfection by-product formation and mitigation strategies in point-of-use chlorination of turbid and non-turbid waters in western Kenya. J. Water Health 6, 67–82.
Lee, J.Y., Bak, G., Han, M., 2012. Quality of roof-harvested rainwater and comparison of different roofing materials. Environ. Pollut. 162, 422–429.
Lee, J.Y., Yang, J.S., Han, M., Choi, J., 2010. Comparison of the microbiological and chemical characterization of harvested rainwater and reservoir water as alternative water resources. Sci. Total Environ. 408 (4), 896–905.
Liniger, H., Mekdaschi, R., Hauert, C., Gurtner, M., 2011. Sustainable Land Management in Practice: Guidelines and Best Practices for Sub-Saharan Africa. FAO, Rome, Italy.
Lynch, D.F., Dietsch, D.K., 2010. Water efficiency measures at Emory University. J. Green Build. 5 (2), 41–54.
Magyar, M.I., Mitchell, V.G., Ladson, A.R., Diaper, C., 2007. An investigation of rainwater tanks quality and sediment dynamics. Water Sci. Technol. 56 (9), 21–28.
Mankad, A., Tapsuwan, S., 2011. Review of socio-economic drivers of community acceptance and adoption of decentralized water systems. J. Environ. Manag. 92 (3), 380–391.
McGuigan, K.G., Conroy, R.M., Mosler, H., Du Preez, M., UbombaJaswa, E., Fernandez-Ibañez, P., 2012. Solar water disinfection (SODIS): a review from bench-top to roof-top. J. Hazard. Mater. 235–236, 29–46.
Meera, V., Ahammed, M.M., 2006. Water quality of roof-top rainwater harvesting systems: a review. J. Water Supply Res. Technol. AQUA 55 (4), 257–268.
Mehrabadi, M.H.R., Saghafian, B., Fashi, F.H., 2013. Assessment of residential rainwater harvesting efficiency for meeting non-potable water demands in three climate conditions. Resour. Conserv. Recycl. 73, 86–93.
Mendez, C.B.J., Klenzendorfa, B., Afshar, B.R., Simmons, M.T., Barrett, M.E., Kinney, K.A., Kirisits, M.J., 2011. The effect of roofing material on the quality of harvested rainwater. Water Res. 45, 2049–2059.

Mendez, C.B., Afshar, B.R., Kinney, K.A., Barrett, M.E., Kirisits, M.J., 2010. Effect of Roof Material on Water Quality for Rainwater Harvesting Systems. Texas Water Development Board Report, Texas Water Development Board, Austin, Texas, p. 46.
Möller, D., Zierath, R., 1986. On the composition of precipitation water and its acidity. Tellus B 38 (1), 44–50.
Morris, R.D., 1995. Drinking water and cancer. Environ. Health Perspect. 103 (Suppl 8), 225–231.
Morrow, A.C., Dunstan, R.H., Coombes, P.J., 2010. Elemental composition at different points of the rainwater harvesting system. Sci. Total Environ. 408 (20), 4542–4548.
Mwamila, T.B., Katambara, Z., Han, M.Y., 2016. Strategies for household water supply improvement with rainwater harvesting. J. Geosci. Environ. Protect. 4, 146–158.
Mwenge-Kahinda, J., Taigbenu, A.E., Boroto, R.J., 2007. Domestic rainwater harvesting to improve water supply in rural South Africa. Phys. Chem. Earth 32, 1050–1057.
Nawaz, M., Han, M.Y., Kim, T., Mazoor, U., Amin, M.T., 2012. Silver disinfection of *Pseudomonas aeruginosa* and *E. coli* in roof-top harvested rainwater for potable purposes. Sci. Total Environ. 431, 20–25.
Nelson, K.L., Boehm, A.B., Davies-Colley, R.J., Dodd, M.C., Kohn, T., Linden, K.G., et al., 2018. Sunlight-mediated inactivation of health-relevant microorganisms in water: a review of mechanisms and modeling approaches. Environ. Sci. Process. Impacts 20 (8), 1089–1122.
Nikmatin, S., Syafiuddin, A., Kueh, A.B.H., Maddu, A., 2017. Physical, thermal, and mechanical properties of polypropylene composites filled with rattan nanoparticles. J. Appl. Res. Technol. 15 (4), 386–395.
Oki, T., Agata, Y., Kanae, S., Saruhashi, T., Yang, D., Musiake, K., 2001. Global assessment of current water resources using total run-off integrating pathways. Hydrol. Sci. J. 46, 983–995.
Oweis, T.Y., Prinz, D., Hachum, A.Y., 2012. Water Harvesting for Agriculture in the DA. ICARDA, CRC Press, Balkema, Leiden, the Netherlands.
Radaideh, J., Al-Zboon, K., Al-Harahsheh, A., Al-Adamat, R., 2009. Quality assessment of harvested rainwater for domestic uses. Jordan J. Earth Environ. Sci. 2, 26–31.
Rahman, A., Keane, J., Imteaz, M.A., 2012a. Rainwater harvesting in Greater Sydney: water savings, reliability and economic benefits. Resour. Conserv. Recycl. 61, 16–21.
Rahman, A., Dbais, J., Islam, S.M., Eroksuz, E., Haddad, K., 2012b. Rainwater harvesting in large residential buildings in Australia. In: Polyzos, S. (Ed.), Urban Development. 2012. InTech, Rijeka, Croatia, pp. 159–178.
Roaf, S., 2006. Drivers and barriers for water conservation and reuse in the UK. Water Demand Manag., 215–235.
Roebuck, R.M., Oltean-Dumbrava, C., Tait, S., 2011. Whole life cost performance of domestic rainwater harvesting systems in the United Kingdom. Water Environ. J. 25 (3), 355–365.
Rozin, P., Haddad, B., Nemeroff, C., Slovic, P., 2015. Psychological aspects of the rejection of recycled water: contamination, purification and disgust. Judgem. Decis. Mak. 10 (1), 50–63.
Sazaklia, E., Alexopoulos, A., Leotsinidis, M., 2007. Rainwater harvesting, quality assessment and utilization in Kefalonia Island, Greece. Water Res. 41 (9), 2039–2047.
Schets, F.M., Italiaander, R., van den Berg, H.H.J.L., de Roda Husman, A.M., 2010. Rainwater harvesting: quality assessment and utilization in The Netherlands. J. Water Health 8, 224–236.
Schuetze, T., 2013. Rainwater harvesting and management-policy and regulations in Germany. Water Sci. Technol. Water Supply 13 (2), 376–385.
Spinks, A.T., Dunstan, R.H., Harrison, T., Coombes, P., Kuczera, G., 2006. Thermal inactivation of water-borne pathogenic and indicator bacteria at sub-boiling temperatures. Water Res 40 (6), 1326–1332.
Stenekes, N., Colebatch, H.K., Waite, T.D., Ashbolt, N.J., 2006. Risk and governance in water recycling: Public acceptance revisited. Sci. Technol. Hum. Values 31 (2), 107–134.
Strauss, A., Dobrowsky, P.H., Ndlovu, T., Reyneke, B., Khan, W., 2016. Comparative analysis of solar pasteurization versus solar disinfection for the treatment of harvested rainwater. BMC Microbiol. 16, 289.
Sung, M., Kan, C., Wan, M., Yang, C., Wang, J., Yu, K., Lee, S., 2010. Rainwater harvesting in schools in Taiwan: system characteristics and water quality. Water Sci. Technol. 61 (7), 1767–1778.
TCEQ, 2007. Harvesting, Storing, and Treating Rainwater for Domestic Indoor Use. Texas Commission on Environmental Quality (TCEQ). GI366.
Tuinhof, A., van Steenbergen, F., Vos, P., Tolk, L., 2012. Profit From Storage: The Costs and Benefits of Water Buffering. 3R Water Secretariat, Wageningen, The Netherlands.
Vazquez, A., Cotstoya, M., Pena, R.M., Garca, S., Herrero, C., 2003. A rainwater quality monitoring study of the composition of rainwater in Galicia. Chemosphere 51 (5), 375–385.
Vialle, C., Busset, G., Tanfin, L., Montrejaud-Vignoles, M., Huau, M.C., Sablayrolles, C., 2015. Environmental analysis of a domestic rainwater harvesting system: a case study in France. Resour. Conserv. Recycl. 102, 178–184.
Vialle, C., Sablayrolles, C., Lovera, M., Huau, M.C., Jacob, S., Montréjaud-Vignoles, M., 2012. Water quality monitoring and hydraulic evaluation of a household roof runoff harvesting system in France. Water Resour. Manag. 26 (8), 2233–2241.
Vialle, C., Sablayrolles, C., Lovera, M., Jacob, S., Huau, M.C., Montrejaud-Vignoles, M., 2011. Monitoring of water quality from roof run-off: interpretation using multivariate analysis. Water Res. 45 (12), 3765–3775.
Vieira, A.S., Beal, C.D., Ghisi, E., Stewart, R.A., 2014. Energy intensity of rainwater harvesting systems: a review. Renew. Sust. Energ. Rev. 34, 225–242.
Ward, S., Memon, F.A., Butler, D., 2010. Harvested rainwater quality: the importance of appropriate design. Water Sci. Technol. 61 (7), 1707–1714.
Ward, S., Memon, F.A., Butler, D., 2012. Performance of a large building rainwater harvesting system. Water Res. 46 (16), 5127–5134.
Wolfe, S., Brooks, D.B., 2003. Water scarcity: an alternative view and its implications for policy and capacity building. Nat. Resour. Forum 27, 99–107.
Worm, J.T., van Hattum, 2006. Rainwater Harvesting for Domestic Use. AGROMISA and CTA/RAIN (Rainwater Harvesting Implementation Network), Wageningen, The Netherlands.
Zhong, Z.C., Victor, T., Balasubramanian, R., 2001. Measurement of major organic acids in rainwater in Southeast Asia during burning and non-burning periods. Water Air Soil Pollut. 130 (1-4), 457–462.
Zhu, K., Zhang, L., Hart, W., Liu, M., Chen, H., 2004. Quality issues in harvested rainwater in arid and semi-arid Loess Plateau of northern China. J. Arid Environ. 57 (4), 487–505.

CHAPTER

15

Phytoremediation: A wonderful cost-effective tool

Rajni Yadav[a], *Siril Singh*[b], *Abhishek Kumar*[a], *and Anand Narain Singh*[a]

[a]Soil Ecosystem and Restoration Ecology Lab, Department of Botany, Panjab University, Chandigarh, India [b]Department of Environment Studies, Panjab University, Chandigarh, India

OUTLINE

15.1 Introduction

Contamination can be defined as the presence of some undesirable elements that spoil a healthy environment and its functioning or present an unacceptable risk to humans. Humans are at risk from contaminated soils via dermal contact, inhalation of dust or vapors, and ingestion of food grown in contaminated areas (Nathanail and Earl, 2001). Soils may fail to support vegetation due to the phytotoxic effects of contaminated or disrupted biological cycling of nutrients (Belyaeva et al., 2005; Siddiqui et al., 2001). These soils may also

Cost Effective Technologies for Solid Waste and Wastewater Treatment
https://doi.org/10.1016/B978-0-12-822933-0.00008-5

affect the hydrosphere and compromise the quality of drinking water resources and aquatic ecosystems. Bilek (2004) reported that approximately 5000 km^2 in Germany affected by lignite mining threatens the surface water and aquifers with acidification. Contaminants can enter and accumulate further in the soils from various anthropogenic sources. Frequent, human-assisted routes for entry of organic contaminants and heavy metals in particular, into the agricultural and nonagricultural land, are via application of sewage sludge, disposal of industrial effluents, military operations, mining, landfill operations, use of agricultural chemicals like pesticides, herbicides, and inorganic fertilizers, etc.

The dispersal of wastes like sewage sludge or other biosolids to land may be a problem, particularly where these wastes have been applied frequently over several years. Some wastes (pulverized fuel ash, dredging, mine spoils) form new "soil-forming" materials (Bright and Healey, 2003). Nalbandian (2012) reported that the natural fossil fuel (coal) consists of an array of heavy metals containing Hg, Pb, Cr, Cd, Cu, Co, Ni, and Zn in the concentration range of about 0.1 to 18 mg/kg, and these heavy metals are released into the environment in the form of ash, particulate matter, and vapor form. Soils may become contaminated through atmospheric deposition from traffic, incinerator, and metal smelting emissions over a long time. The other reported causes of soil contamination include spillage of chemicals through flooding, petroleum oil, industrial solvents, and irrigation with contaminated wastewater (Collins et al., 2002). Historically, the contamination of soil has been of restricted concern. As a result, there are various examples of severe and widespread contamination. For example, during the nineteenth century, Cu and Ni metal mining waste disposal and smelting emissions widespread up to 30 km from the source of contamination and more than 50,000 ha of land become inadequate of supporting vegetation or very limited vegetation cover in the Sudbury region of Canada. In this region, remediation of soil and water occurs in more than 40 years (Winterhalder, 1996). Mulligan et al. (2001) have reported that 1,372,000 Mt of Zn and 22,000 Mt of Cd were accumulated globally on soils during the 1980, and Fengxiang et al. (2003) reported a cumulative increase in the concentration in the surface soil via anthropogenic activity which is equal to 2.18 mg/kg and an accelerating trend shown up to the year 2000. Nowadays, to remove contaminants from agricultural soil, phytoremediation is a widely accepted method of the in situ remediation technique. Restoring polluted land is a time-consuming process due to the difficulty inherent in the process. Phytoremediation is to grow plants, namely *Helianthus annuus*, *Zea mays*, *Brassica campestris*, *Pisum sativum*, and *Pteris vittata*, in contaminated soil to remove heavy metals and persistent organic pollutants (POPs) or stabilize them into a harmless state. It is a plant-based, operationally simple technology, widely accepted and economically viable, as compared to physical and chemical treatments, which irreversibly change soil properties. Phytoremediation commonly improves the physical, chemical, and biological condition of contaminated soil. Since the 1970s, several plants have been tested and used in treating various types of contamination in soils, water, and wetlands. In the 1980s, phytoremediation technique was adopted by the government and commercial users to clean up contaminated sites (Van Nevel et al., 2007; Pinto et al., 2015; Sarwar et al., 2017). Mainly phytoremediation is classified into two broad techniques: phytoextraction through which contaminants are absorbed by plants from soil and accumulated in roots, shoots, and leaves; and phytostabilization, in which contaminants are immobilized in the soil by plant roots. Recently, plant growth regulators have been considered as an appropriate method for improving the efficiency of phytoremediation (Rostami and Azhdarpoor, 2019). Phytoremediation includes the use of green plants for the restoration of contaminated soil. It is an advanced engineering technique of reclamation that is cost-effective and helps in the conservation of soil properties. Moreover, it has the least adverse impacts on the environment and soil properties (Rostami et al., 2017; Parseh et al., 2018).

In this chapter, we have explained in detail about general introduction on phytoremediation, types of contamination, conventional methods of remediation techniques, types of phytoremediation, mechanisms of phytoremediation, quantification of phytoremediation efficiency, the perspective and limitations of phytoremediation, and future trends in phytoremediation.

Contamination by inorganic and organic compounds in soil has become a serious concern worldwide. This problem has been increasing due to anthropogenic activities that affect the physical, chemical, and biological properties of the soil at exceptional rates. Two major issues of consideration are as follows:

1. Contaminants are persistent and remain in the environment for several years. Therefore, the current evaluation of risk is questionable (uncertain).
2. It should be acknowledged that the current environmental conditions are in a state of constant change worldwide and increase in temperature, future climate change, flooding, or drought may be more common or acute, and vegetation and other ecological communities may change that could impact the form in which the contaminants occur and all these factors alter environmental conditions.

15.2 Soil contamination and remediation technologies

Several organic and inorganic compounds are naturally present in the soil, but whenever their concentration exceeds beyond a certain limit, it impairs the proper functioning of soil along with posing risks to the human health and environment; it is considered soil contamination. Human-derived activities like mining, industrialization, urbanization, and agriculture have led to soil contamination over the past century. Several waste products disposed of industries and the excessive use of herbicides and pesticides in agriculture are the major sources of soil contamination and degradation in developing countries like India. Soil contaminants like heavy metals and persistent organic pollutants (pesticides and polycyclic aromatic hydrocarbons) are often discharged as wastewater from industries that are used for irrigation by poor farmers. This wastewater not only contaminates the agricultural soil but also affects the quality of the surface or even the groundwater. Similarly, sewage effluents are considered a rich source of organic matter and other nutrients, but they also elevate the levels of heavy metals, such as Fe, Mn, Cu, Zn, Pb, Cr, Ni, Cd, and Co and persistent organic pollutants in the receiving soils.

In agroecosystems, the uncontrolled use of herbicides and pesticides is a major source of soil contamination. Although the use of these compounds is intended to increase the yield of crops, they are also increasing the concentration of toxic compounds in the soil leading to soil contamination and degradation. Thus, humans are contributing to the adverse environmental and health consequences as soil contamination affects the whole ecosystem via food chain transfer and biomagnification (Bhatnagar, 2001). The fate of pesticides applied to the soil largely depends on their persistence and solubility. Once applied to the croplands, pesticides undergo processes such as adsorption, transfer, breakdown, and degradation into harmless compounds. Adsorption is the binding of pesticides to soil particles. The amount a pesticide is adsorbed to the soil varies with the type of pesticide, soil, moisture, soil pH, and soil texture. Pesticides are strongly adsorbed to soils that are high in clay or organic matter. Chlorinated pesticides are nonbiodegradable and remain in a living system for thousands of years and cause many health hazards and alter the whole ecosystem. Most potential contaminants are essential for agriculture production, but they become hazardous when an excessive amount is present in the soil.

15.2.1 Contamination by heavy metals

Heavy metals have relatively high density and often are toxic to biological forms at lower concentrations. Although the criteria for heavy metals may vary with authors and context, generally it has been accepted that metals with a relative density of more than $5.0\,g/cm^3$ are considered as heavy metals Zn (7.1), Cr (7.2), Ni (8.7), Cd (8.6), Co (8.9), Pb (11.4), and Hg (13.5). These metals or their compounds form the major proportion of inorganic contaminants. Interestingly, some of these metals, such as Zn, Ni, Co, and Mo, are essential for normal plant growth when present in small quantities but become toxic at higher concentrations. Other nonnutrient heavy metals, such as Cd, Cr, and Pb, on the other hand, are phytotoxic even at low concentrations and, therefore, pose a higher risk of soil contamination rather than organic contaminants.

The concentration of these toxic elements in soil may increase from various sources, including anthropogenic pollution, industrial discharge, excessive use of pesticides and herbicides, weathering of natural high background rocks, and metal deposits. In the last five decades, the concentration of heavy metals reached up to 22,000 tons for Cd, 783,000 tons for Pb, 939,000 tons for Cu, and 1,350,000 tons for Zn worldwide. Sources of heavy metal contaminants in soils, sewage sludge treatment, waste disposal sites, agriculture fertilizers, and military training include metalliferous mining and smelting, metallurgical industries, and electronic industries (Alloway, 1995). Ground transportation also causes metal contamination. Highway traffic, maintenance, and deicing operations generate continuous groundwater and surface water-contaminant sources. Heavy metal contaminants in roadside soil originate from engine and brake pad wear (Cd, Cu, and Ni) (Viklander, 1998), lubricants (Cd, Cu, and Zn) (Birch and Scollen, 2003; Turer et al., 2001), exhaust emission (Pb) (Sutherland et al., 2003), and tire abrasion (Zn) (Smolders and Degryse, 2002). The problems of heavy metal toxicity are further aggravated by the persistence of metals in the environment. For example, Pb can persist in the environment for 150–5000 years (Friedland, 1989). There is a crucial need to deal with the issue of excess metals present in the soil and to prevent future contamination.

Many heavy metals are found in contaminated soils; however, some of them are potentially hazardous to plants, animals, and human beings too. All heavy metals impart toxicity at higher concentrations primarily due to the formation of free radicals, which leads to oxidative stress. Further, many heavy metals disrupt physiological functions by displacing the essential metals in enzymes or pigments. Thus, heavy metals adversely affect plant growth and the associated biodiversity of an ecosystem. Not only plants but also humans are affected by heavy metal contamination in soils. The mutagenic ability of some heavy metals causes DNA damage, which often leads to cancer in humans (Baudouin et al., 2002).

15.2.2 Contamination by persistent organic pollutants

Apart from inorganic contaminants, several organic compounds that are constituents of herbicides and pesticides are increasingly recognized as contaminants of the soils. Organic compounds like organochlorines, organophosphorus, and polycyclic aromatic hydrocarbons remain in the soil for a long period due to their slow degradability, and such compounds are known as persistent organic pollutants. However, the continuous use of these compounds in the form of herbicides and pesticides has led to the contamination of agricultural soils. On the other hand, urban development and industrialization are increasing their concentration in the environment, which is responsible for soil contamination. These compounds can easily enter living cells due to their lipophilic nature and therefore pose high toxicity to biological forms. Modern agricultural practices reveal an increase in the use of pesticides and fertilizers to meet the food demand of the increasing population. Not only there were various incidents of acute toxicity in the short term, but also that residues of some persistent pesticides remain in the environment for decades after their excessive use in agriculture. Pesticides enter the soil through direct treatment or being washed off from plant surfaces during rainfall. It is predicted that an average of 35%–50% of the plant protection material is accumulated on soil immediately after spraying and it depends on the phenotype and density of the plant. The obvious reason lies in the research data that say only 0.1% of pesticide application targets the pest and rest 99.9% remain and seep in the environment (Bhardwaj and Sharma, 2013). The behavior of pesticides in soil includes persistence, movement, and metabolism. The formation of residues in soil primarily depends on the binding capacity of both inorganic and organic elements of soil and water solubility. Among the different types of pesticides, organochlorines are of concern because of their recalcitrance and persistence in the environment. Recently, the World Health Organization's (WHO) estimate indicated 25 million cases of acute occupational pesticide poisoning in developing countries and 20,000 deaths worldwide each year.

However, residues can also be detected in every environmental compartment around the world. Traces of the number of pesticides (mostly Organochlorines) have even been detected in regions, like arctic regions, where these chemicals were never used. In addition to this diffuse contamination, severe risks of further environmental pollution exist in the form of huge stockpiles of obsolete pesticides in many countries. Their bioaccumulation and bioconcentration in the food chain have since posed long-term risk to nontarget species, including humans.

However, extensive use of chemical pesticides over the years has adversely affected human health, nontarget organisms, and the environment and has also enhanced the development of pesticide resistance among pest species. Besides these alarming numbers, long-term exposure at even low concentrations causes serious health problems such as immune suppression, hormone disruption, diminished intelligence, reproductive abnormalities, and cancer (Abhilash and Singh, 2009; Yadav et al., 2015).

15.2.3 Conventional remediation technologies

The selection of a suitable remediation technology to mitigate the problems caused by soil contamination is based on various criteria, like characteristics of contaminated land, form and concentration of the contaminants, including availability and effectiveness of the remediation technologies. However, the cost and performance of the excellent technology rely on whether the technology is employed in situ or ex situ. In most cases, contaminated land that is remediated by using an appropriate technology also plays a major role in the selection of suitable technology, as most conventional methods restore infertile soil and land unsuitable for agricultural purposes.

Conventional technologies convenient for soil and water remediation can be generally classified based on whether they are employed in situ or ex situ. The most frequently used technologies for in situ and ex situ remediation are pneumatic fracturing, soil flushing, electrokinetic extraction, chemical oxidation/reduction, excavation, retrieval, and offsite disposal, surface capping, encapsulation, stabilization, phytoremediation, bioremediation, landfilling, soil washing, solidification, and vitrification (Liu et al., 2018).

15.2.3.1 In situ remediation techniques for contaminated soil

In situ remediation does not need mining, transport, and removal of the contaminated soil to offsite treatment facilities and minimizing soil disturbance. However, specific field conditions need to have careful consideration, such as contamination depth, soil permeability, weather, and potential deep leaching of chemicals (Olexsey and Parker, 2006). Soil flushing, pneumatic fracturing, electrokinetic extraction, chemical oxidation/reduction, excavation, retrieval, and offsite disposal, surface capping, encapsulation, stabilization, phytoremediation, and bioremediation are the most frequently used technologies for in situ remediation.

- *Soil flushing*: This process involves physical separation by horizontal or vertical leaching using a fluid (water or an aqueous solution containing chelators), which is followed by the collection and treatment of the leachates in basins or trench infiltration system. The average cleanup time for 20,000 tons of contaminated soil sites using this approach is more than 3 years. Reddy et al. (2011) reported that 0.2 M EDTA was a more effective extraction fluid as compared to water and in the industry-contaminated loam sand, cyclodextrin, and surfactants, flushing out 25%–75% of the Cu, Pb, and Zn. Jiang et al. (2011) reported that the biodegradable chelating agent chitosan (pH 3.3) performed better than EDTA (pH 3.1) at $2.0\,g\,l^{-1}$ to extract Cu and Ni from a pH 5.0 clay loam. The removal efficiency was the lowest for Cu (25%) and the highest for Pb (75%).
- *Pneumatic fracturing*: This process involves injecting pressurized air into the soil to develop cracks in low permeability areas, thereby enhancing the extraction efficiencies of other in situ technologies.
- *Electrokinetic extraction*: The contaminants are mobilized as charged species move toward polarized electrodes placed in the soil. The migrated contaminants can be treated in situ. When low-density direct current (DC) electricity is given by electrodes inserted in the contaminated ground, cations in the solution phase of the contaminated soil migrate to the cathode and anions migrate to the anode at the attractive force of the established electrical field. Metal contaminants accumulated at the polarized electrodes are later removed by electroplating, solution pumping, precipitation, and ion exchange resin complexation (Alshawabkeh, 2009; FRTR, 2012).
- *Chemical reduction/oxidation*: In this remediation process, the contaminants are converted into less hazardous, less mobile, more stable, and inert forms. Chemical immobilization does not extract contaminants from soil. Instead, the concentration of heavy metals and their mobility in soil pore water are terribly reduced, minimizing their potential transport to microorganisms, plants, and water (Tajudin et al., 2016).
- *Excavation, retrieval, and offsite disposal*: In this process, contaminated soil is removed and transported to an offsite treatment where a disposal facility is available.
- *Surface capping*: In this remediation process, the contaminated site is simply covered with a layer of waterproof material to form a stable protection surface. This technique is not a true soil "remediation" method because no efforts are made to extract heavy metals from contaminated sites. By this process, we can reduce their reactivity in the soil. This method has been widely practiced to remediate small soil-contaminated areas by organic pollutants and heavy metals. The surface capping project cost is reflected by the materials, labor, essential engineering design, and follow-up operations (maintenance, inspection). The recent cost ranges from $\$20\,m^{-2}$ to $\$90\,m^{-2}$ in the USA (NJDEP, 2014).
- *Encapsulation*: This remediation process is similar to surface capping and encapsulation is also limited to a small area that has shallow and serious contaminated sites. Encapsulation is termed as "barrier wall," "cutoff wall," or "liner" method. The technique is to contain contaminated soil according to designed physical barrier system consisting of low permeability caps, enclosing underground barriers, and in rare situations, barrier floors. Contaminated sites are isolated and enclosed, eliminating offsite dispersion of the contaminants and onsite bioexposure to the contaminants. The technology is frequently selected to manage sites contaminated by polycyclic aromatic hydrocarbons (PAHs), radionuclides, asbestos, heavy petroleum hydrocarbons, and mixed wastes when other cost-effective excavation and treatment remediation technologies are not available (Meuser, 2013).
- *Phytoremediation*: Phytoremediation is plant-based technology in which plants are grown in contaminated soils to remove heavy metals (phytoextraction and phytovolatilization) or stabilize them into nonhazardous forms (phytoimmobilization and phytostabilization) (Mahmood et al., 2015). This plant-based technology is operationally simple, eco-friendly, esthetically preferable, economically viable, and widely accepted. The physical and chemical treatments alter soil properties, but phytoremediation improves the physical, chemical, and biological quality of contaminated soils.
- *Bioremediation:* In bioremediation, instead of plants, microorganisms are used to decontaminate soil. The technique is more usually applied to detoxify organic contaminants in groundwater and soil (FRTR, 2012). Microorganisms can also detoxify metals through biosorption (to the cell surface), valence transformation (Cr (VI) to Cr (III), SeO_4^{2-} to Se), volatilization (dimethyl selenide, trimethylarsine, and Hg vapor), and extracellular chemical precipitation (by S^{2-} from sulfur-reducing bacteria) (Garbisu and Alkorta, 2003).

15.2.3.2 Ex situ remediation techniques for contaminated soil

Ex situ soil remediation involves the mining of soil from the contaminated site, transport of the contaminated soil to an offsite treatment facility, and disposal of the treated soil at permitted locations. Ex situ remediation needs additional costs for digging, transport, disposal, and site refilling as compared to in situ remediation, but the treatment can be accelerated and controlled, so that better results can be achieved in a shorter time.

- *Landfilling*: Landfilling, or "dig and haul," is the simplest soil remediation technique. In this technique, contaminated soil is removed from its original site and transported to a secure landfill for disposal, which is an engineered structure with impermeable liners, leachate drains, and dike enclosures. Landfilling is a well-known technique for the purification of hazardous waste sites. It was the most frequent waste disposal method used in the USA before 1984. Due to a large distance between the contaminated site and the secure landfill, the overall cost of soil landfilling in the USA ranges from \$300 ton^{-1} to \$500 ton^{-1} (FRTR, 2012).
- *Soil washing*: This process refers to the separation of contaminants absorbed to find soil particles using an aqueous solution through size separation, gravity separation, or attrition scrubbing. Soil washing depends on washing solutions to mobilize heavy metals by altering the solution's ionic strength, soil acidity, redox potential, and complexation. An ideal washing solution should effectively improve the solubility and mobility of heavy metal contaminants but interact weakly with soil constituents and should be biodegradable and nontoxic. An array of chemicals, such as acetic acid, citric acid, hexafluorosilicic acid, formic acid, hydrochloric acid, sulfuric acid, nitric acid, phosphoric acid, oxalic acid, polyglutamic acid, tartaric acid, EDTA (ethylenediaminetetraacetic acid), DTPA (diethylenetriaminepentaacetic acid), NTA (nitrilotriacetic acid), EDDS (ethylenediamine disuccinic acid), carbonate/bicarbonate, sodium hydroxide, calcium chloride, ferric chloride, ammonium chloride, dithionite, isopropyl alcohol, ammonium acetate, subcritical water, to formulate effective washing solutions (Zhu et al., 2015; Alghanmi et al., 2015; Bilgin and Tulun, 2016; Yang et al., 2018), have been tested.
- *Vitrification*: This technology utilizes thermal energy to melt the soil to enable physical or chemical stabilization. Vitrification does not apply to soils with high organic matter content (7%) and high moisture content (10%). It is not applicable to soils heavily contaminated by volatile or flammable organics. A sufficient amount of monovalent alkaline cations (Na^+ and K^+) should be present in the soil (Meuser, 2013).
- *Solidification/stabilization*: In this process, the contaminant is physically enclosed in a stabilized mass or through chemical interactions induced between the stabilizing agent and the contaminant. This technology is also termed as "microencapsulation." In this technique, stabilizing agents are used to immobilize contaminants through chemical reactions. Through this process, an average of 20,000 tons of contaminated soil remediates in about 1 year (FRTR, 2012).

In general, all of these conventional technologies, colloquially termed as "pump and treat" and "dig and dump" techniques, are limited in their applicability to small areas and have their inherent limitations. At sites where the contaminants are slightly higher than the industrial criterion (governmental regulations), the use of conventional technologies is not economically viable (Table 15.1 and Figs. 15.1 and 15.2). So far, irrespective of the technologies being selected, the cost estimates for utilizing conventional remediation techniques have remained high. The overall remediation budget includes design, construction, operation, and maintenance costs of the process associated with each technology in addition to mobilization, demobilization, and pre- and posttreatment costs, which are determined on a site-to-site basis. Also, in the case of most ex situ treatment technologies, excavations and transport costs need to be factored in to arrive at the final cost for remediating a contaminant site. A survey indicated that the world remediation market ranged between the USD 15 and 18 billion in 1997 and the environmental remediation market is expected to reach USD 123.13 billion by 2022, at a CAGR of 7.62% between 2016 and 2022. The base year used for this study is 2015, and the forecast period is considered for the period between 2016 and 2022 (Environmental Remediation Market Report, 2016).

15.2.4 What is phytoremediation?

Phytoremediation is an advanced remediation technology, which uses plants and their associated rhizospheric microorganisms to extract various contaminants from contaminated soils, groundwater, surface water, and sediments. Due to urbanization and industrialization, large amounts of hazardous compounds are discharged into the biosphere, among which heavy metals are common such as mercury, cadmium, lead, zinc, arsenic, and nickel. To remediate the contaminated soil, several physical, chemical, and biological processes are already reported in the literature. Remediation is mandatory to reclaim the contaminated areas and to ensure the reduced entry of hazardous elements into the food chains. Certain plants may be used to treat various types of contaminants,

TABLE 15.1 Evaluation and comparative budgeting of remediation technologies.

Technology	Cost estimates in US $	Duration
Phytoremediation	2000–5000/acre (land based)	Long
Solidification/stabilization	50–330/m cube	Medium
Soil flushing	Inadequate information	Medium
Bioremediation	30–100/m cube	Long
Electro kinetics	Inadequate information	Medium
Chemical reduction/oxidation	190–660/m cube	Short
Soil washing	120–200/ton	Short
Low-temperature thermal desorption	45–200/ton	Medium
Incineration	200–600/ton	Short
Vitrification	700/ton	Short
Pneumatic fracturing	8–12/ton	Short
Excavation/retrieval disposal	270–460/ton	Short
Disposal alone	35–60/ton	Short
Landfill disposal	150–200/ton	Short

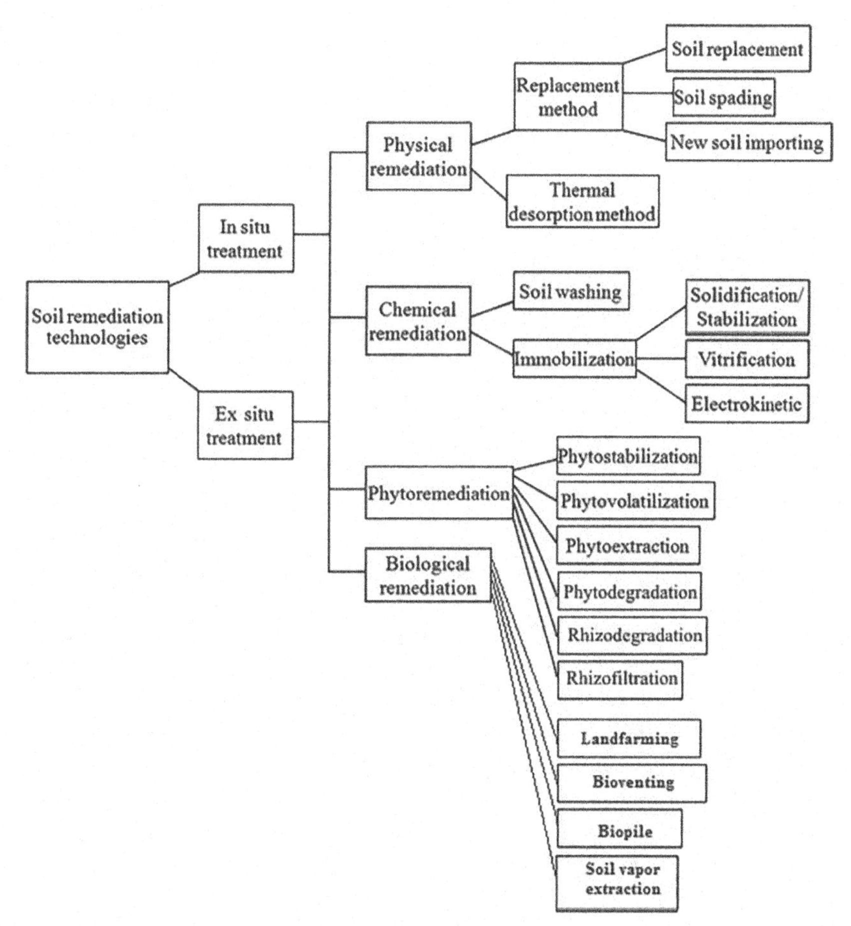

FIG. 15.1 Flowchart depicting soil remediation technologies. *From Nejad, Z.D., Jung, M.C., Kim, K.H., 2018. Remediation of soils contaminated with heavy metals with an emphasis on immobilization technology. Environ. Geochem. Health 40 (3), 927–953; Rostami, S., Azhdarpoor, A., 2019. The application of plant growth regulators to improve phytoremediation of contaminated soils: a review. Chemosphere 220, 818–827.*

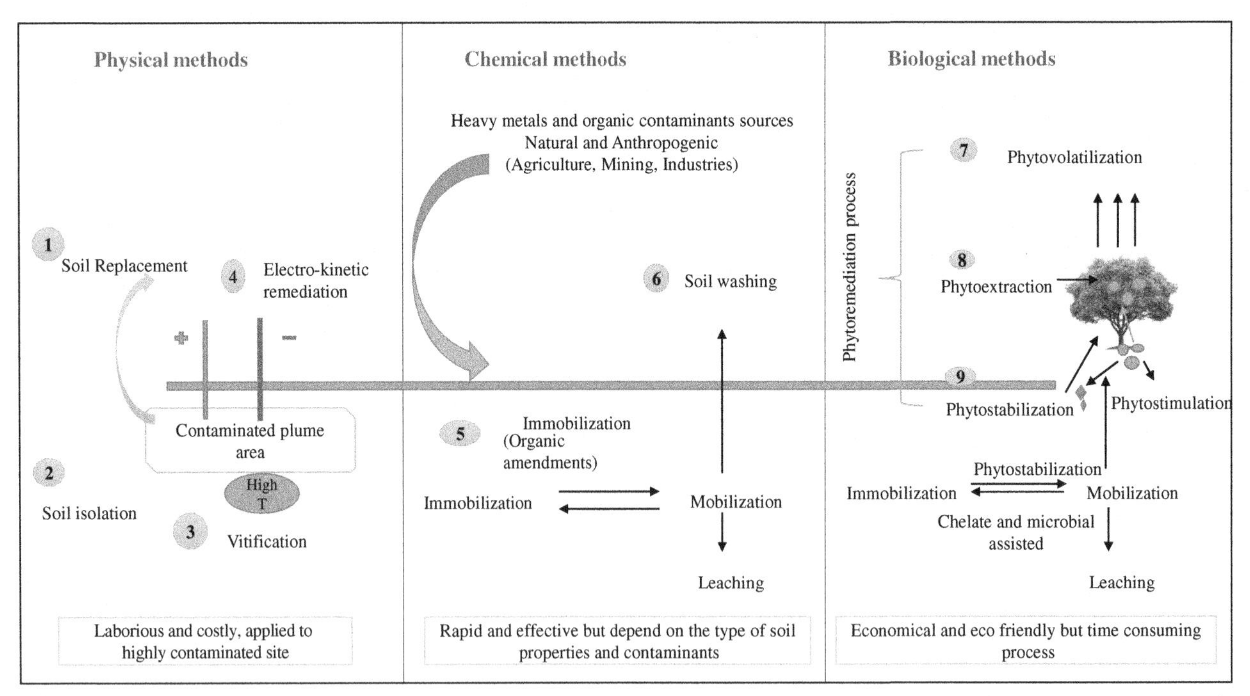

FIG. 15.2 Comparison of different soil cleanup methods. Soil remediation methods can be broadly divided into three categories: physical, chemical, and biological. Physical remediation methods include (1) soil replacement, (2) soil isolation, (3) vitrification, and (4) electrokinetic; chemical methods contain (5) immobilization and (6) soil washing; and biological methods generally include (7) photoevaporation, (8) phytoextraction, and (9) phytostabilization. However, biological and chemical methods can be applied jointly depending on the type of metal, soil, plant, and chemical reagent. Moreover, the effectiveness of different phytoremediation techniques can be enhanced by microbial, chelate, and genetic-assisted remediation.

including heavy metals, pesticides, polycyclic aromatic hydrocarbons, explosives, a chlorinated solvent, petroleum hydrocarbon, radionuclides, and excess nutrients. Phytoremediation is an eco-friendly and cost-effective technique that utilizes the plant's natural abilities to extract, degrade, and accumulate contaminants from the environment. The principle of phytoremediation is that the plant roots either break down the contaminants in the soil or absorb the contaminant storing it in the stems and leaves of the plant. It is a novel technology and is perceived as a cost-effective, efficient, eco-friendly, and solar-driven technology. It is also a thrust area to explore. As many plant species, namely *Helianthus annuus, Zea mays, Brassica campestris, Pisum sativum,* and *Pteris vittata,* have been identified as hyperaccumulators of various heavy metals, i.e., Cd, Cu, Co, Ni, Cr, Pb, and as present in the soil (Tariq and Ashraf, 2016).

15.2.4.1 *Phytoremediation of heavy metals*

The plants used in phytoremediation are commonly selected based on their biomass, growth rate, the depth of their root zone, their potential to tolerate and bioaccumulate particular contaminants, and their ability of groundwater evapotranspiration (Lasat, 2002). Due to the growth habitat and adaptability to a wide range of edaphic and climatic conditions, grasses are more convenient for remediation of mine spoils where heavy metals contamination is present. Many plants have the inherent capacity to break down and metabolize certain heavy metals and stabilize several metal ions by acting as traps or filters. The uptake and accumulation of heavy metals differ from plant to plant and also from species to species within a genus. Some plant species act as powerful remediators of the primary chemical contaminants and some remediate only the chemical species of the primary contaminants (Kramer and Chardonnens, 2001). Some plants which grow on metalliferous soils have developed the ability to accumulate massive amounts of indigenous metals in their tissues without symptoms of toxicity. The idea of using plants to extract metals from contaminated soils was reintroduced and developed by Utsunamyia (1980). The first field trial on Zn and Cd phytoextraction was conducted by Baker et al. (1991).

Some plant families seem to play a promising role in the absorption of one or more heavy metals, but the problem is that the accumulation capacity of each plant species does not depend solely on its genetic properties but also on the paleoclimatic and microbiological conditions of the site where they thrive. Another important aspect concerns the location in plants where the accumulation of heavy metal takes place. Some species have a specific absorption capacity for different heavy metals, such as some metals are stored in roots, stems, and leaves or fruits. The knowledge of the storage characteristics of each species allows knowing if the metals are stored in no edible parts of the plants. The success of remediation of heavy metal-polluted soil employing plants depends upon soil properties, metal contents, climatic factors, and the plant used.

Heavy metal bonding in soils is closely related to pH with weak or negligible binding at acidic pH and strong bonding at neutral to alkaline pH (Sauve et al., 2000; Sukreeyapongse et al., 2002). Other soil properties such as organic matter content and mineral composition also affect heavy metal retention (Holm et al., 2003). The suitability of a certain plant for heavy metal remediation is determined by various plant properties such as heavy metal tolerance, size, growth rate and rooting depth, heavy metal accumulation in aboveground plant parts, climatic adaptation, and pest resistance (Kirkham, 2006; Pulford and Watson, 2003).

15.2.4.2 Phytoremediation for pesticides contamination

The uptake of pesticides by plants is dependent on the physicochemical properties of the compound, mode of application, soil type, climatic factors, and plant species. The compounds absorbed through plant roots may be translocated to other parts via xylem (Bromilow and Chamberlain, 1995). The permeation from plant roots to xylem is optimal for those compounds that are only slightly hydrophobic. More hydrophobic compounds tend to bind with lipid membranes in the roots. The hydrophilic compounds, on the other hand, have a limited absorption through the leaf cuticle waxes. The microbial activities in the rhizosphere also play a crucial role in transforming the contaminants, which may aid in root uptake and further degradation in plants. The microbial transformation in the rhizosphere, therefore, must be considered as an integral component of phytoremediation. Several studies have attempted to determine the absorption and translocation of persistent pesticides, and it has been reported that rape seedlings are grown from seeds treated with lindane or maize plants grown in lindane-treated soils absorbed and translocated the pesticides (Heinrich and Schulz, 1996).

Subsequently, phytoremediation became well-established technology to remediate soil and groundwater contaminated by organic compounds, which include pesticides, dioxins, chlorinated solvents, pesticides, explosives, polyaromatic hydrocarbons (PAHs), and polychlorinated biphenyls (PCBs) (Aken et al., 2010).

15.3 Types of phytoremediation

Phytoremediation is the use of living green plants to fix or adsorb contaminants, and cleaning the contaminants, or making their risk reduction or disappearance. The rhizofiltration, phytostabilization, phytovolatilization, phytoextraction, and phytodegradation are the main mechanisms of phytoremediation. Phytoremediation is specifically suitable for treating large, diffusely, and superficially polluted areas with fine-textured, high organic matter content soils (Chaney and Baklanov, 2017). Currently, the technique is still at its beginning stage. Further research and development are warranted to understand the soil-metal-plant interactions in the rhizosphere and the mechanisms of plants absorbing, translocating, and accumulating heavy metals.

15.3.1 Rhizofiltration

In the rhizofiltration, both aquatic and terrestrial plants are used to absorb, concentrate, and precipitate contaminants from contaminated sources. Rhizofiltration can partially or completely treat the agricultural runoff, industrial discharge, acid mine drainage, and many other kinds of liquid wastes released as a result of anthropogenic activities. This process is mainly used for those metals and organic contaminants that are primarily retained in roots, such as lead, copper, cadmium, zinc, and nickel. The advantages of rhizofiltration include its ability to be used as in situ or ex situ applications and species other than hyperaccumulators can also be used. Plants such as Indian mustard, sunflower, tobacco, spinach, corn, and rye have been studied for their ability to extract lead from effluent and sunflower having the greatest ability to remove lead from contaminated land (Table 15.2).

15.3.2 Phytostabilization

It is commonly used for the remediation of soil, sludges, and sediment and depends on the root's ability to limit contaminant mobility and bioavailability in the soil. Phytostabilization can occur via sorption, complexation, precipitation, and metal valence reduction. The plant's primary purpose is to reduce the amount of water

TABLE 15.2 Rhizofiltration efficiency of different plant species against heavy metal.

Plant species	Metals	Metal accumulation (mg/kg)	Metal accumulation part of the plant	Duration of treatment	Concentration of metal	Medium	References
Pteris vittata	As	5947	Frond and Root	198 days	357 mg/kg	Soil	Kalve et al. (2011)
Arabis paniculata	Cd Zn	8400 12,400	Root	4 months	267 μM 1223 μM	Water	Tang et al. (2009)
Thlaspi caerulescens	Cd	263	Shoot	391 days	19 mg/kg	Soil	Lombi et al. (2001)
Lolium italicum	Pb	218 7232	Shoot Root	–	–	Soil	Rizzi et al. (2004)
Brassica juncea	Ni	3916	Shoot	60 days	254 mg/kg	Soil	Saraswat and Rai (2009)
Alyssum heldreichii	Ni	1180	Leaves	–	1070 mg/kg 3280 mg/kg	Soil	Bani et al. (2010)

percolating via the soil matrix, which may prevent soil erosion and dispersion of the toxic metal to other areas. A dense root system stabilizes the soil and prevents erosion. It is most effective when rapid immobilization is required to preserve ground and surface water and the disposal of biomass is not needed. However, the major disadvantage is that the contaminant remains in soil as it is and therefore requires regular monitoring. As phytostabilization is similar to a meadow, soil amendments analogous to those used in agriculture may also be applied and assume a role of great importance by helping to inactivate metal contaminants, preventing plant uptake, and reducing biological activity. Generally, soil amendments may be easy to apply and to handle and it is also easy to produce, safe to workers handling the amendment, nontoxic to the plants, and inexpensive. Soil amendments are cost-effective as compared to other amendments (Marques et al., 2008). Attention should also be given to the capacity of the amendments to reduce the leaching of metals, as this may be an important advantage in an in situ stabilization process and play an important role in groundwater protection and reduction of metal dispersion (Ruttens et al., 2006a,b).

Marques et al. (2008) have shown that the application of organic matter amendments, like manure or compost, to metal-contaminated soil led to a significant reduction in the amount of Zn, leached through the soil in combination with plants, the reduction in metal exposure up to 80%. A range of inorganic and organic compounds, such as lime, phosphate, and other low economic value organic materials such as biosolids, compost, litter, and manure, may be used. Liming has been examined as a significant management tool in decreasing the toxicity of metals in soils (Adriano et al., 2004; Gray et al., 2006; Madejon et al., 2006). Bolan et al. (2003) reported that the application of lime or phosphate is effective in reducing Cd in contaminated soils. The use of organic amendments, such as manure (Chiu et al., 2006; Clemente et al., 2006; Walker et al., 2004), compost (Cao and Ma, 2004; Clemente et al., 2006; Marques et al., 2008), and other biowastes (Karaca, 2004; Madejon et al., 2006), is a standing practice used for the restoration of contaminated sites. Walker et al. (2004) reported lower Zn tissue concentration in *Chenopodium album* L. when plants grown in compost or manure-amended soils, and Marques et al. (2008) showed that the addition of manure or compost to the soils induced reductions in the Zn accumulation in *Solanum nigrum* up to 40%–80% (Table 15.3).

Due to their low commercial cost, organic matter amendments are the most promising additives for soil remediation purposes. Organic matter amendment application improves soil physical properties, water holding capacity, and water infiltration and provides essential nutrients for plant growth. Immobilization of metals via such amendments is achieved by adsorption, complexation, and redox reactions between organic matter and heavy metals to make strong complexes. The presence of phosphates, aluminum, and other inorganic minerals in several organic amendments is also believed to be responsible for the retention of metals (Adriano et al., 2004). Additionally, organic matter amendment degradation may alter the soil pH and thereby indirectly affect the bioavailability of metals because metal solubility is determined by the pH (Yoo and James, 2002; Karaca, 2004). The research in soil amelioration using metal-immobilizing amendments is now also focusing on the application of another type of compounds, like cyclonic ashes (Ruttens et al., 2006a,b), calcium carbonate (Lee et al., 2004), red mud (bauxite residue) (Gray et al., 2006), leonardite (Madejon et al., 2006), and steel shots (Ruttens et al., 2006a,b), with positive effects on the reduction of soluble concentrations of heavy metals in soils (Table 15.4).

TABLE 15.3 Plants used for phytostabilization and the metal contaminants removed by them.

Plant species	Contaminants	References
Zea mays	Cd	Redjala et al. (2011)
Jatropha curcas	Zn, Fe	Abioye et al. (2010)
Tamarindus indica	Cd	Udoka et al. (2014)
Amaranthus spinosus L.	Cu, Pb, and Cd	Chinmayee et al. (2012)
Myriophyllum aquaticum, Ludwigia palustris, Mentha aquatica	Fe, Zn, Cu and Hg	Kamal et al. (2004)
Pteris vittata	Cr	Su et al. (2005)
Ricinus communis L.	Ni	Adhikari and Kumar (2012)
Populus cathayana, P. przewalskii, P. yunnanensis	Cd	Chen et al. (2011)
Solanum nigrum L. and *Spinacia oleracea* L.	Pb, Cu, Cd, and Cr	Dinesh et al. (2014)

TABLE 15.4 Phytoremediation includes the following processes and mechanisms of contaminant removal.

Process	Mechanism	Contaminant
Rhizofiltration	Rhizospheric accumulation of heavy metals and microbial assisted degradation	Organics/inorganics
Phytostabilization	Reduction of the mobility of contaminants in soil	Inorganics
Phytoextraction	Hyperaccumulation	Inorganics
Phytovolatilization	Conversion of contaminants into volatile form and then transpiring them in the atmosphere	Organics/inorganics
Phytodegradation	Degradation in plant tissues	Organics

15.3.3 Phytoextraction

It is the best approach to extract the contamination primarily from the soil and segregate it without destroying the soil structure and fertility. It is also referred as phytoaccumulation, as the plant absorbs, concentrates and precipitates the toxic metals and radionuclides from contaminated soils into the biomass. It is best suited for the remediation of diffusely polluted areas, where pollutants occur only at relatively low concentration and superficially. Various approaches have been used, but the two basic strategies of phytoextraction, which have finally developed, are as follows:

(i) Chelate-assisted phytoextraction or chelate-induced phytoextraction, in which artificial chelates are added to increase the mobility and uptake of metal contaminants.
(ii) Continuous phytoextraction, in which the removal of metal depends on the natural ability of the plant to remediate; only the number of plant growth repetitions is controlled.

A major factor driving up the availability of metallic ions, solubility, depends on several soil physicochemical factors, like the degree of complexation with soluble ligands, the type and density of the charge on soil colloids, the reactive surface area, and also the soil pH (Petrangeli Papini et al., 2001). Soil colloidal particles provide a large interface and specific surface areas, which play an important role in regulating the concentrations of many trace elements and heavy metals in natural soils. In the soil, metal availability to plant roots decreases as the soil pH increases, as shown by Wang et al. (2006) for *Thlaspi caerulescens* growing in a Cd- and Zn-contaminated soil. McBride and Martinez (2000) have reported that the solubility of Mo, As, Cd, Pb, and Cu was decreased by the addition of an amendment consisting of hydroxides with high reactive surface area, whereas the solubility of Zn and Ni remained unchanged. These physiochemical factors are dependent upon soil properties, including metal concentration and form, particle size distribution, quantity and reactivity of hydrous oxides, mineralogy, and degree of aeration and microbial activity (Magnuson et al., 2002). The limited bioavailability of various metallic ions, due to their low solubility in water and strong binding to soil particles, restricts their uptake/accumulation by plants. The plant itself can enhance metal bioavailability. For example, plants can extrude H^+ via ATPases, which replace cations at soil cation exchange capacity (CEC) sites, making metal cations more bioavailable (Taiz and Zeiger, 2002). Plant species vary significantly in the ability to accumulate metals from contaminated soils, as a balance between the uptake of essential metal ions to maintain growth and development and the ability to protect sensitive cellular activity and structures from excessive levels of essential and nonessential metals is required (Garbisu and Alkorta, 2001). The ideal plant to be used in phytoextraction should have the following characteristics: tolerant to high levels of the metal, profuse root system, rapid growth rate, potential to produce high biomass in the field and accumulate high levels of the metal in the harvestable parts, as generally the harvestable portion of most plants is limited to the aboveground parts (although the roots of some crops may also be harvestable).

15.3.4 Phytovolatilization

Phytovolatilization involves the utilization of plants to take up contaminants from the soil, transforming them into simpler volatile form and transpiring them into the atmosphere. Phytovolatilization occurs as growing trees and other plants take up water, organic, and inorganic contaminants. Some of these contaminants can pass through the plants to the leaves and volatilize into the atmosphere at comparatively low concentrations. Phytovolatilization has been primarily used for the removal of mercury; the mercuric ion is transformed into less toxic elemental mercury (Ghosh and Singh, 2005). The disadvantage is that mercury released into the atmosphere is probably going to be recycled by precipitation and then redeposited back into the ecosystem. USDA-ARS has found that some plants that grow in high selenium media produce volatile selenium in the form of dimethyl selenide and dimethyl diselenide (Banuelos, 2000). Phytovolatilization has been successful in tritium (3H), a radioactive isotope of hydrogen; it is decayed to stable helium with a half-life of about 12 years reported by Dushenkov (2003) (Table 15.5).

15.3.5 Phytodegradation

In the phytoremediation of organics, plant metabolism contributes to the contaminant reduction by transformation, stabilization, or volatilization and breaks down the contaminant compounds from soil and groundwater. Phytodegradation is the breakdown of mainly organic contaminants, which are further taken up by the plants and transformed into simpler molecules that are incorporated into the plant tissues. Plants contain enzymes that can break down and convert contaminants into simpler forms that are nontoxic to living cells. The enzymes are usually oxygenases, dehalogenases, and reductases. Rhizodegradation is the breakdown of organics in the soil through the microbial activity of the root zone (rhizosphere) and is a much slower process than phytodegradation. Bacteria, fungi, and other microorganisms consume and digest organic substances like fuels and solvents. All phytoremediation technologies are not exclusive and may be used simultaneously, but the metal extraction depends on its bioavailable fraction in soil (Table 15.6).

TABLE 15.5 Plants reported for phytovolatilization and the metal contaminants removed by them.

Plant species	Contaminants	References
Lolium perenne	Cu, Pb, and Zn	Alvarenga et al. (2008)
Sorghum bicolor L.	Cd and Zn	Soudek et al. (2012)
Alternanthera philoxeroides, *Artemisia princeps*, *Bidens frondosa*, *Bidens pilosa*, *Cynodon dactylon*, *Digitaria sanguinalis*, *Erigeron canadensis*, and *Setaria plicata*	Cd, Mn, Pb, and Zn	Yang et al. (2014)
Festuca rubra	Cu	Radziemska et al. (2017)
Agrostis castellana	Cu, Pb, and Zn	Pastor et al. (2015)
Athyrium wardii	Pb	Zhao et al. (2016)
Quercus ilex	Cd	Dominguez et al. (2009)

TABLE 15.6 Plants reported for phytodegradation and the metals contaminants removed by them.

Plants	Contaminants	References
Pteris vittata	As	Sakakibara et al. (2010)
Liriodendron tulipifera	Methyl Hg	Greipsson (2011)
Arundo donax L.	As	Mirza et al. (2011)
Azolla caroliniana	Hg	Bennicelli et al. (2004)
Canna glauca L., *Colocasia esculenta* L. Schott, *Cyperus papyrus* L., and *Typha angustifolia* L.	As	Jomjun et al. (2010)

15.4 Mechanism of phytoremediation

The processes that are assumed to be influencing contaminant accumulation rates in plants are mobilization and uptake from the soil, compartmentalization and sequestration within the root, efficiency of xylem loading and transport, distribution between metal sinks in the aerial parts, sequestration, and storage in leaf cells (Fig. 15.3). At every level, concentration and affinities of chelating molecules, as well as the presence and selectivity of transport activities, influence metal accumulation rate. Important insights have been obtained from the study of model metallophytes, often in comparison with closely related nonmetallophyte plants. The major model metallophytes being studied are several Zn, Ni, or Cd hyperaccumulating ecotypes of *Thlaspi caerulescens*, some Ni hyperaccumulating taxa of the genus *Alyssum*, and Zn hyperaccumulating populations of *Arabidopsis halleri*.

15.4.1 Mobilization

The elements essential for life are also among the most abundant on Earth (Silva and Williams, 2001). However, the actual bioavailability of some metals is limited due to low solubility in oxygenated water and strong binding to soil particles. Iron, the most extreme example, is mainly present as Fe (III), which forms insoluble hydroxides. The availability of Zn, which is required in similar quantities as Fe, is less restricted, whereas the bioavailability of some of the target metals in phytoremediation, particularly Pb, is limited. With the notable exception of Fe, which is solubilized by either reduction to Fe (II) or extrusion of Fe (III)-chelating phytosiderophores, little is known about the active mobilization of trace elements by plant roots (Marschner, 1995). Possibly, mechanisms assisting in the acquisition of phosphorus contribute to increasing the bioavailability of certain micronutrients. Both the acidification of the rhizosphere and the exudation of carboxylates are considered potential targets for enhancing metal accumulation.

15.4.2 Uptake and sequestration

Following mobilization, metal has to be captured by root cells. Metals are first bound by the cell wall, ion exchange of comparatively low affinity and low selectivity. Transport systems and intracellular high-affinity binding sites then mediate and drive uptake across the plasma membrane. Uptake of metal ions is likely to take place through secondary transporters like channel proteins and H^+-coupled carrier proteins. The membrane potential, which is negative on the inside of the plasma membrane and might exceed 200 mV in root epidermal cells, provides a strong driving force for the uptake of cations through secondary transporters (Hirsch et al., 1998).

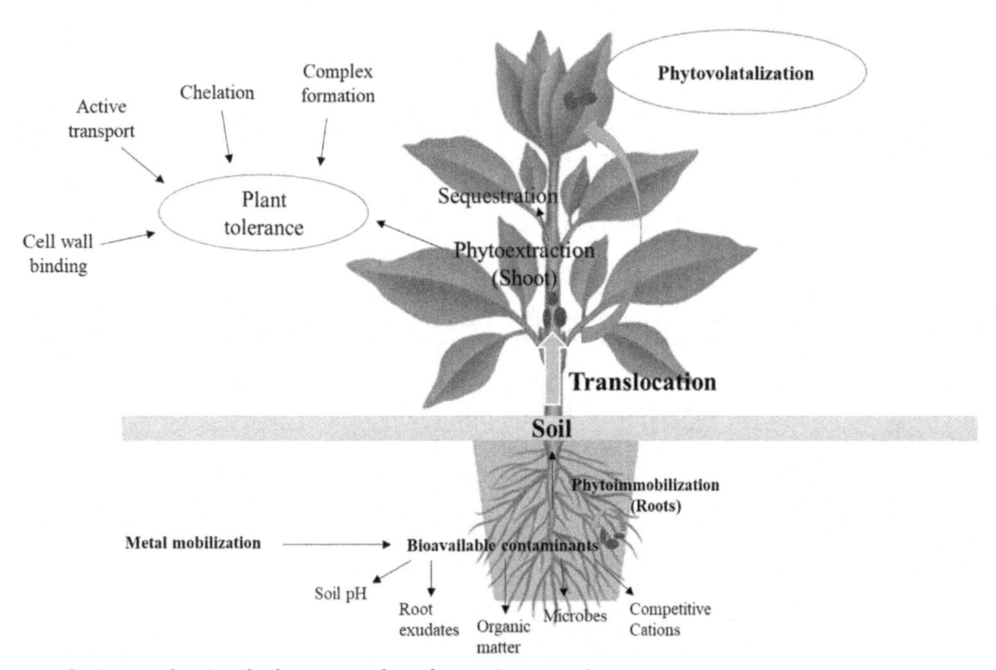

FIG. 15.3 Phytoremediation mechanism for heavy metals and organic contaminants.

Also, little is known about the actual physiological substrates for uptake. For example, the metals might be taken up as hydrated ions and in metal-chelate complexes. At least three different engineering approaches to enhance metal uptake can be envisioned. The number of uptake sites could be increased, the specificity of uptake systems could be altered, so that competition by unwanted cations is reduced, and the sequestration capacity could be enhanced by increasing the number of intracellular high-affinity binding sites or the rates of transport into organelles (Lasat et al., 2000).

15.4.3 Xylem transport

The apoplast continuum of the root epidermis and cortex is readily permeable for solutes. The cell walls of the endodermal cell layer act as a barrier for apoplastic diffusion into the vascular system. In general, solutes have to be taken up into the root symplasm before they can enter the xylem (Tester and Leigh, 2001). After metal uptake into the root symplasm, three processes govern the movement of metals from the root into the xylem, sequestration of metals inside root cells, symplastic transport into the stele, and release into the xylem. The transport of ions into the xylem is generally a tightly controlled process mediated by membrane transport proteins. The metal transporters involved in xylem loading have yet to be identified (Gaymard et al., 1998).

Inside the root, the presence of elevated amounts of exogenously supplied histidine, which can chelate Ni^{2+}, results in a 50-fold increase in the rate of transport of Ni into the xylem of *Alyssum montanum* (Kramer et al., 1996). Various Ni hyperaccumulators respond to Ni^{2+} exposure by a large dose-dependent increase in histidine concentrations in the xylem. Thus, chelation with certain ligands, for example, histidine, nicotianamine, and citrate, appear to route metals primarily to the xylem (Senden et al., 1995).

Because an elevated histidine content, whether of exogenous or endogenous origin, is known to increase Ni tolerance, this approach targets at least two of the determinants of plant metal accumulation. The identification of novel proteins involved in the root-to-shoot transport of metals or its regulation should provide researchers with novel targets for engineering metal accumulation (Kramer et al., 1996).

15.4.4 Unloading, trafficking, and storage

Transition metals reach the apoplast of leaves in the xylem sap, from where they have to be scavenged by leaf cells (Marschner, 1995). Transporters mediate uptake into the symplast, and distribution within the leaf occurs via the apoplast or the symplast (Karley et al., 2000). Trafficking of metals occurs inside every plant cell, maintaining the concentrations within the specific physiological ranges in each organelle and ensuring delivery of metals to metal-requiring proteins. Metallochaperones, such as CCH1, and pumps, such as P-type ATPases, are probably involved in these processes (Himelblau et al., 1998). Excess essential metals, as well as nonessential metals, are sequestered in leaf cell vacuoles. Intriguingly, different leaf cell types show pronounced differential accumulation. The distribution pattern varies with plant species and elements. Zn accumulation in *Thlaspi caerulescens* leaves is 5- to 6.5-fold higher in epidermis cells than in mesophyll cells, whereas in metal-treated *Arabidopsis halleri*, the mesophyll cells are thought to contain more Zn and Cd than the epidermal cells (Kupper et al., 2000). Furthermore, trichomes play a major role in the storage and detoxification of metals. For the phytoremediation candidate species *Brassica juncea*, Cd accumulation was reported to be >40-fold higher in trichomes compared with the leaf total (Salt et al., 1998). A significant proportion of Ni and Zn are found in trichomes of the hyperaccumulators *Alyssum lesbiacum* and *Arabidopsis halleri*, respectively (Kupper et al., 1999; Kramer et al., 1996).

15.5 Factors affecting uptake mechanisms of contaminants

Several factors affect the uptake mechanism of contaminants, as shown in Fig. 15.4. Through the knowledge about all these important factors, the uptake mechanisms by the plant may be improved. Factors that improve the uptake mechanism are the plant species, properties of the medium, environmental conditions, root zone, the bioavailability of the metal, chemical properties of contaminants, and chelating agent.

15.5.1 The plant species

Plant species are screened; those species have superior remediation properties selected and used to remediate the contaminated area (Prasad and de Oliveira Freitas, 2003). The selection of a plant species is a critical decision for phytoremediation. Grasses are thought to be an excellent candidate due to the fibrous rooting system and provide

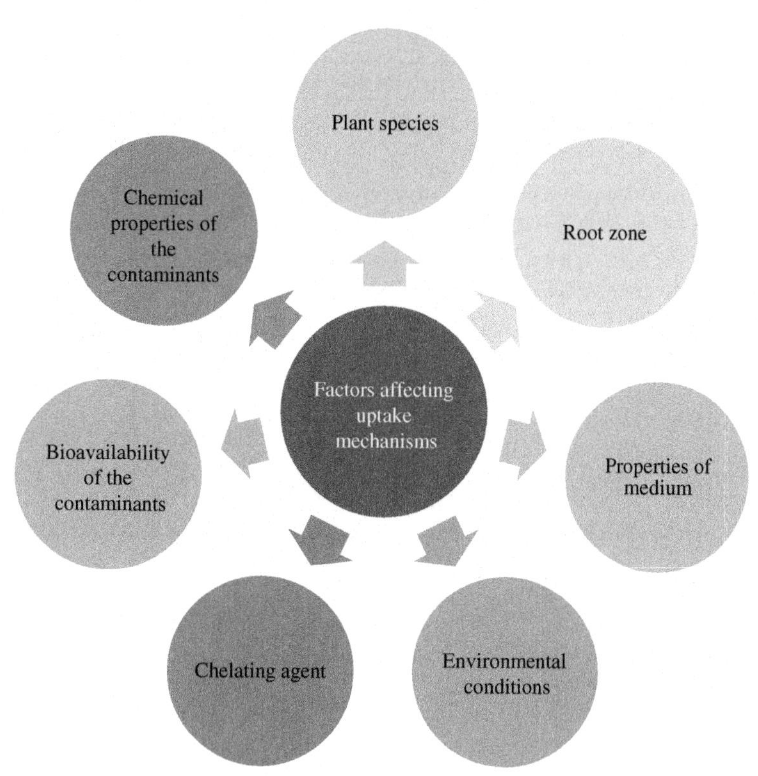

FIG. 15.4 The uptake mechanisms of contaminants depend on several factors.

a large surface area for root and soil contact (Kulakow et al., 2000). The uptake of a compound is affected by plant species characteristics, and phytoextraction technique success is mainly dependent on the identification of suitable plant species that hyperaccumulate contaminants and produce large amounts of biomass using established crop production and management practices (Rodriguez et al., 2005).

15.5.2 Properties of the medium

Agronomical practices such as pH adjustment, the addition of chelators, and fertilizers are used to enhance remediation. For example, the amount of contaminant uptake by the plants is affected by the pH, organic matter, and the phosphorus content of the soil. To reduce lead uptake by plants, the pH of the soil is adjusted with lime to a level of 6.5–7.0 (Prasad and de Oliveira Freitas, 2003; Traunfeld and Clement, 2001).

15.5.3 The root zone

In phytoremediation, the root zone is specially focused because it absorbs contaminants and metabolizes inside the plant tissue. Plant enzymes degrade the contaminants in the soil via roots, and exudation is also another phytoremediation mechanism. An increase in root diameter and reduced root elongation is a morphological adaptation of drought stress plant to less permeability of dried soil (Merkl et al., 2005).

15.5.4 Vegetative uptake and the bioavailability of the contaminants

Vegetative uptake is mainly affected by environmental conditions. The temperature affects growth substances and, consequently, root length (Merkl et al., 2005). The contaminant uptake by plants mainly depends on *the bioavailability* of the contaminants present in the soil in different phases, which in turn depends on the retention time of the contaminants, as well as the interaction with other elements. For example, when metals have been bound to the soil, the pH, redox potential, and organic matter content will all affect the tendency of the metal to exist in ionic and plant-available form. Plants will affect the soil via their ability to lower the pH and oxygenate the sediment, which affects the availability of the metals, increasing the bioavailability of metals by the addition of biodegradable physicochemical factors, such as chelating agents and micronutrients (Fritioff and Greger, 2003).

TABLE 15.7 Some case studies of phytoremediation of pesticides.

Site	Contaminants	Plants	Phytomechanism	Results	Source
Kazakhstan	HCH, DDT, DDE, DDD	Asteraceae and Chenopodiaceae	Phytoextraction	17 of 123 plant species were pesticide tolerant	Nurzhanova et al. (2013)
Czech Republic	PCBs	Austrian pines, Black locust, Willow trees	Rhizodegradation	Black locust and Austrian pine enhance PCB degradation	Leigh et al. (2006)
Canada	DDT and metabolites	Zucchini, Tall fescue ryegrass, Alfalfa, Pumpkin	Phytoextraction	Soil—3700 g/kg Zucchini roots—2043 g/kg Pumpkin shoot—57,536 g/kg	Lunney et al. (2004)
Canada	PCBs	Pumpkin, tall fescue	Phytoextraction	Pumpkin: Shoot—16.8 g/kg Root—730 g/kg Tall fescue: Shoot—6.2 g/kg Root—440 g/kg	Aslund et al. (2007)
–	PCBs, DDE, Chlordane	*Cucurbita pepo*	Phytoextraction	24% of contaminant extracted	White et al. (2003) and Mattina et al. (2004) White and Mattina (2004)
Riyadh, Saudi Arabica	Dimethoate, malathion	*Amaranthus caudate, Lactuca sativa, Nasturtium officinale, Phaseolus vulgaris*	Phytoextraction	90% was degraded after 42 days	Al-Qurainy and Abdel-Megeed (2009)
New Delhi, India	Endosulfan	*Brassica campestris, Zea mays*	Phytoextraction	34.5% and 47.2% extracted by plants, respectively	Mukherjee and Kumar (2012)
Ghaziabad, Uttar Pradesh, India	Endosulfan	*Digitaria longiflora, Sphenoclea zeylanica, Vetiveria zizanioides*	Phytoextraction	Plant—14 to 343 ng/g Soil—13 to 938 ng/g	Singh and Singh (2014)

15.5.5 Addition of chelating agent

The increase of the uptake of contaminants such as metals and pesticides by the crops may be influenced by increasing the bioavailability of contaminants via the addition of biodegradable physicochemical factors like chelating agents and micronutrients and also by stimulating the contaminant uptake capacity of the microbial community in and around the plant. This faster uptake of metals will result in shorter and, therefore, less expensive remediation. The use of chelating agents in contaminated soils could promote leaching of the contaminants into the soil (Fritioff and Greger, 2003; Van Ginneken et al., 2007). Such chelating agent EDTA is used for a longer period (2 weeks) in the contaminated soil could improve contaminant translocation in the plant tissue as well as the overall phytoextraction performance. The application of a synthetic chelating agent (EDTA) at 5 mmol/kg yielded positive results (Roy et al., 2005) (Table 15.7).

15.6 Quantification of phytoremediation efficiency

It was known as early as 1855 that the "calamine violet," *Viola calaminaria*, found on zinc-rich soils near Aachen, Germany, contained an unusually high zinc concentration (Forchhammer, 1855). As per the literature, the data on *Viola calaminaria* and its other species including "*Thlaspi alpestre* var. *calaminare*," showed the concentration of trace elements about 1.2% in dry leaves and 0.13% in roots (Sachs, 1865) depicts an extraordinary response by a plant not showing apparent ill effects from such abnormal uptake of trace element (Zn). The Zn levels reported in the dry matter of *Viola calaminaria* (typically about 800–1000 mg/kg or 0.08%–0.10%) are not now regarded as particularly remarkable, although higher values have since been reported. Observations of unusual accumulation of other metals emerged only during the twentieth century. There are records of strikingly high plant levels of lead dating from the 1920s, selenium from the 1930s, nickel from the 1940s, cobalt and copper from the 1960s, and cadmium, arsenic, and manganese from the 1970s. The majority of studies regarding the phytoremediation of heavy metals included in this chapter focus on lead, nickel, zinc, cadmium, copper, and cobalt. Various plants belonging to different families have been studied for phytoremediation of various heavy metals (lead, nickel, zinc, copper, cadmium, and cobalt). Metal concentrations in plant tissues also differed among different plant species grown on the same soil, indicating their different capacities for metal uptake (Table 15.4). Out of 25 case studies of phytoremediation of Pb, maximum number of plants (20%) were from Asteraceae family, followed by Brassicaceae (12%); Poaceae, Buddlejaceae, Polygonaceae, and Fabaceae (8%); and Chenopodiaceae, Liliaceae, Asclepiadaceae, Euphorbiaceae, Urticaceae, Caryophyllaceae, Solanaceae, Tamaricaceae, and Pteridaceae (4%). It has also been reviewed that plants belonging to various life forms are capable of phytoremediation lead-contaminated soil. Out of all the reported cases, 57.8% of plants were herbs, 36.8% were shrubs, and 5.26% were ferns. Maximum cases were reported from Thailand (30.43%), followed by China (26.1%); Greece, and USA (8.69%); and Austria, Iran, California, New Jersey, India, and Japan (4.35%).

Values of translocation factor (TF), bioaccumulation coefficient (BAC), and bioconcentration factor (BCF) have also been evaluated for various plant species (Table 15.4).

TF = metal concentration in plant shoot/metal concentration in plant roots.
BAC = metal concentration in plant shoot/metal concentration in soil.
BCF = metal concentration in plant roots/metal concentration in soil.

It has been reviewed that plants that have a value of BAC > 1 are suitable for phytoremediation and the plants that have BCF > 1 and low TF value have the potential for phytostabilization. Out of 30 case studies of phytoremediation of lead-contaminated soil, 20% of the plant species have TF > 1, 40% have TF $> 0.5 < 1$, and the rest 40% of the species have TF < 0.5. *Fagopyrum esculentum* has the maximum value of the translocation factor (3.0). Hence, it is concluded that *Fagopyrum esculentum* is a very good excluder. Similarly, BAC values were also calculated for the same plant species. 61.5% of the plant species had BAC values less than 0.1, 26.9% had BAC less than 0.5, and only 11.5% of the species had a BAC greater than 0.5. *Debregeasia orientalis* was reported to have the maximum value of BAC (0.79). Another phytoremediation parameter, BCF, was also calculated. 41.67% of the plant species had BCF less than 0.1, 41.6% of plant species had BCF less than 0.5, and 16.67% were found to have BCF greater than 0.5 but less than 1.

Similarly, 13 cases of phytoremediation of zinc-contaminated soils were reviewed. Out of 13 cases, 23.1% of the plant species used for phytoremediation belonged to the Asteraceae family, followed by Brassicaceae and Polygonaceae (15.4%), Caryophyllaceae, Violaceae, Apocynaceae, Pinaceae, Euphorbiaceae, and Poaceae. Plants belonging to different life forms have different potential for phytoremediation of metal-contaminated soils. Of the studied cases, the maximum number of plants is herbs (63.6%) followed by shrubs (27.3%) and trees (9.1%). Maximum cases were reported from Iran (25%), followed by UK, Spain, and Mexico (16.7%) and Germany, Belgium, and Thailand (8.3%). The three phytoremediation parameters were also calculated for the phytoremediation of zinc-contaminated soil. Out of all the species, 31.5% have TF greater than 1, 21.0% have TF greater than 0.5 and less than 1, and 47.3% have TF less than 0.5. *Chondrilla juncea* has a maximum TF value (1.36). Hence, it can act as a good excluder. 1 species has BAC greater than 1 (5.3%), 52.6% of the studied species have BAC less than 0.1, 36.8% have BAC greater than 0.1 but less than 0.5, and 5.3% have BAC greater than 0.5 but less than 1. *Rumex crispus* has a maximum value of BAC (1.33) and hence can act as a good accumulator. 15.8% of the species have BCF less than 0.1, 52.6% have BAC greater than 0.1 but less than 0.5, and 31.6% of them have BAC greater than 0.5 but less than 1. None of the species have BCF greater than 1. The maximum value (0.96) is reported in the *Rumex crispus*. Hence, it has the potential for acting as a good photostabilizer in zinc-contaminated soil.

Eight case studies for phytoremediation of nickel-contaminated soil are reviewed, out of which a maximum number of studied plants belonged to the family Brassicaceae (25%) followed by Acanthaceae, Clusiaceae, Fabaceae, Equisetaceae, Salicaceae, and Asteraceae (12.5% each). The maximum number of cases is reported from Switzerland (37.5%), followed by Brazil, Europe, the USA, Cuba, and Zimbabwe.

Similarly, for copper-contaminated soil, 14 phytoremediation cases are reviewed, out of which a maximum number of plants belonging to Fabaceae, Asteraceae, Poaceae (14.4% each) followed by Pteridaceae, Tiliaceae, Apiaceae, Melastomaceae, Oleaceae, Phytolaccaceae, Zingiberaceae, and Chenopodiaceae (7.1% each). 50% of the plants reported were herbs, and the rest 50% were shrubs and trees (25% each). Maximum cases are reported from China (64.3%), followed by Congo (21.4%) and Iran (14.3%). In the case of phytoremediation parameters, 22.2% of the plant species have TF greater than 1, 33.3% have TF greater than 0.5 but less than 1, and 44.4% have TF less than 0.5. The maximum value of TF is reported in *Osmanthus fordii* (1.05). 22.2% of the species have BAC less than 0.1 and 77.8% have BAC greater than 0.1 but less than 0.5. Similarly, the values of BCF were also calculated. 11.1% of the reported plant species have BCF less than 0.1 and 88.9% have BCF greater than 0.1 but less than 0.5. Hence, none of the species is reported with high BAC and BCF values.

In the case of cadmium-contaminated soil, it has been reviewed that a maximum number of phytoremediation plants belong to family Brassicaceae followed by Polygonaceae, Poaceae, Caryophyllaceae, Melastomaceae, Zingiberaceae, and Crassulaceae (12.5% each). And the maximum number of studied plants is herbs (70%), followed by trees (20%) and shrubs (10%). The maximum number of cases is reported from China. 37.5% of the species have TF greater than 1, 25.6 have TF greater than 0.5 but less than 1, and 37.5% of species have TF less than 0.5. The maximum value of TF is evaluated in *Viola baoshanensis* (1.56). 28.6% of the studied plant species have a BAC greater than 1, 28.6 have BAC greater than 0.5 but less than 1, and 42.8% have BAC less than 0.5. The maximum value of BAC is recorded in *Rumex crispus* (1.13); in the case of BCF, 66.7% of the species have BCF greater than 1 and 33.3% have BCF less than 0.5. The maximum value of BCF is again recorded in *Rumex crispus* (1.35).

The studies regarding the phytoremediation of pesticides mentioned in this chapter mainly focus on DDT, PCBs, and atrazine (Table 15.5). The primary mechanism used in these studies was phytoextraction. Rhizodegradation was also used frequently. The rest of the studies discuss a mix of mechanisms, including phytodegradation, and phytovolatilization, which was the least used. Plants for phytoremediation of heavy metal- and pesticide-contaminated soil are selected based on various parameters, which are described below.

15.6.1.1 Translocation factor

It is defined as the ratio of metal concentration in plant shoots to root. Hence, the plants that have a greater value of TF can accumulate more amount of heavy metal or pesticide contaminants in shoots than root, acting as good accumulators rather than phytostabilizers. In the case of zinc, six species have the value of TF > 1, and out of all of them, *Chondrilla juncea* has the highest value (1.36). Hence, it has the potential for acting as a good accumulator of zinc. Similarly, in the case of lead, *Fagopyrum esculentum* has the highest value of TF (3.03). In the case of copper, *Osmanthus fordii* (0.119) and for cadmium it is *Viola baoshanensis* (1.56) has highest value of TF.

15.6.1.2 Bioaccumulation coefficient

It is defined as the ratio of metal concentration in plant shoots to the soil. All the plants that have a value of BAC greater than one are suitable for phytoremediation. In the case of zinc, out of all the reviewed cases, *Rumex crispus* has a value of BAC > 1 (1.28) and in the case of lead, none of the species have a value of BAC > 1; hence, none of them have a very good potential for phytoremediation. Similarly, in the case of copper also, none of the studied species have a value of BAC > 1. In the case of cadmium, *Rumex crispus* has a value of BAC > 1 (1.13); hence, it can act as a good phytoremediator.

15.6.1.3 Bioconcentration factor

It is defined as the ratio of metal concentration in plant roots to the soil. The plants which have a greater value of BCF have a greater potential for phytostabilization. In the case of phytoremediation of zinc-contaminated soil, none of the species have a value of BCF > 1, but *Rumex crispus* has the maximum value, closer to one (0.96); hence, it can act as good phytostabilizer. Similarly, in the case of lead and zinc, none of the species have a value of BCF > 1. In the case of phytoremediation of cadmium, out of 10 studied species, 3 species have a value of BCF > 1. *Vetiveria zizanioides* have a maximum value (1.39); hence, it has the potential for acting as a good stabilizer in cadmium-contaminated soil.

However, many research articles are available in the literature but mostly belonging to the other continents except for Indian perspectives. In Indian perspectives, very limited information is available even how many species are hyperaccumulators, excluders, and extractors, and accumulators are not explored evidently. Moreover, there is no database of metallophytes available in India, while more than 400 plant species have been identified to have the potential for soil and water remediation as hyperaccumulators from the rest of the world. Among them, only the mustard plant is belonging to Indian origin. Although there are many limitations of introducing exotic plants for this purpose, it will open another problematic gate, such as invasion and ecological problems in a functioning ecosystem.

15.6.1.4 Phytoremediation efficiency

Phytoremediation efficiency is the ability of the plant to remediate the contaminated soil.

It has been reviewed and recalculated for different plant species growing in different types of contaminated soil (Figs. 15.5 and 15.6).

$$\text{Phytoremediation efficiency} = \text{TF} + \text{BAC} + \text{BCF}$$

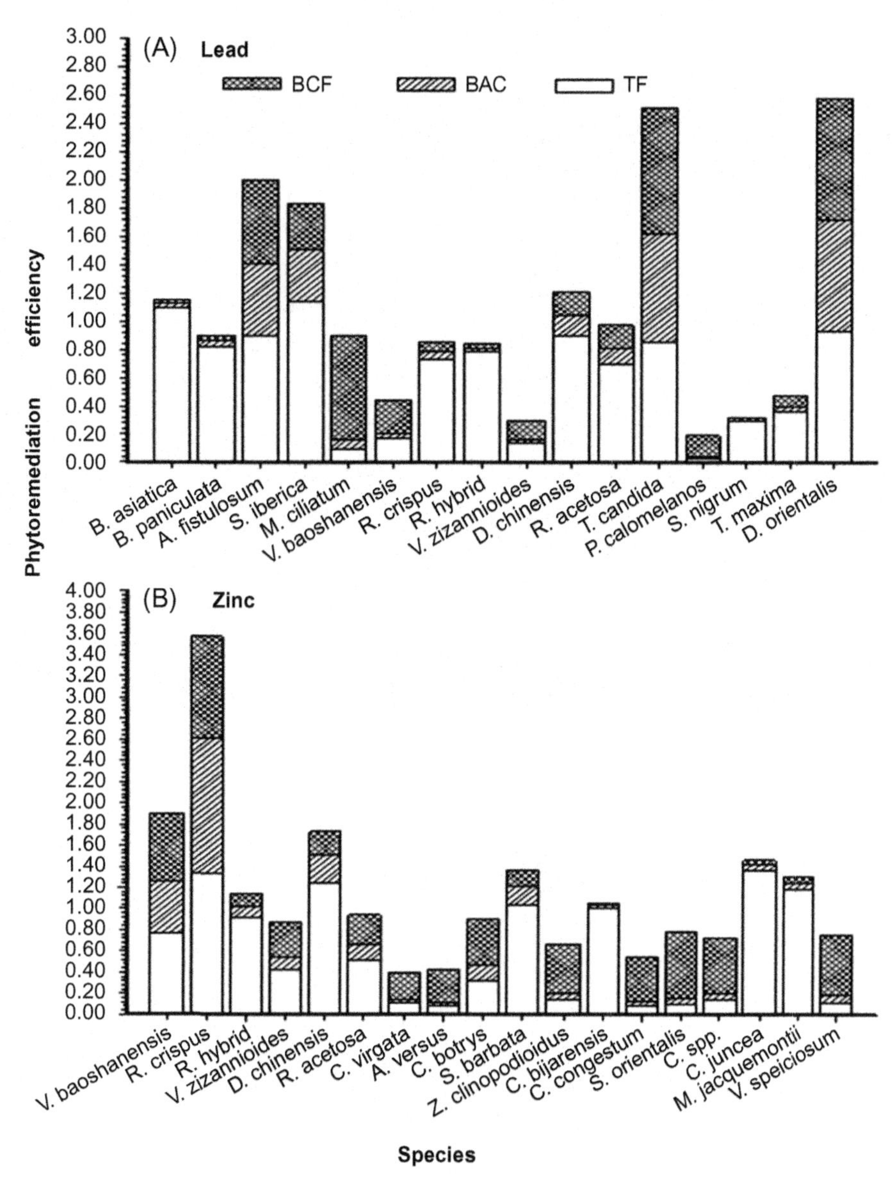

FIG. 15.5 Phytoremediation efficiency of certain plant species for lead (A) and zinc (B), heavy metals reported from various contaminated soils.

In the case of phytoremediation of lead-contaminated soil (Fig. 15.5A), *Debregeasia orientalis*, a shrub belonging to the family Urticaceae, has the maximum potential of phytoremediation among all the studied species. It means it has maximum phytoremediation efficiency and can act as good remediators of lead-contaminated soil, whereas *Pityrogramma calomelanos*, a fern belonging to the family Pteridaceae, has been reported with the least phytoremediation efficiency. Similarly, in the case of phytoremediation of zinc-contaminated soil (Fig. 15.5B), *Rumex crispus*, a herb belonging to the family Polygonaceae, has the maximum phytoremediation efficiency among all the reviewed species and *Centaurea virgata*, a herb belonging to the family Asteraceae, has the least potential for phytoremediation of zinc (Table 15.8).

In the case of phytoremediation of copper (Fig. 15.6A), *Melastoma dodecandrum*, a semi-herb belonging to the family Melastomaceae, has maximum phytoremediation efficiency among all the nine reviewed case studies, whereas *Parthenocissus heterophylla*, a shrub belonging to the family Apiaceae, has the least phytoremediation efficiency. Hence, it can be concluded that *Parthenocissus heterophylla* cannot act as a very good remediator for zinc-contaminated soil.

In the case of phytoremediation for cadmium-contaminated soil, a similar calculation was performed (Fig. 15.6B) and it was found that *Rumex crispus*, an herb belonging to the family Polygonaceae, has the highest phytoremediation efficiency, whereas *Elaeocarpus decipiens*, a tree, has the least potential for phytoremediation (Table 15.9).

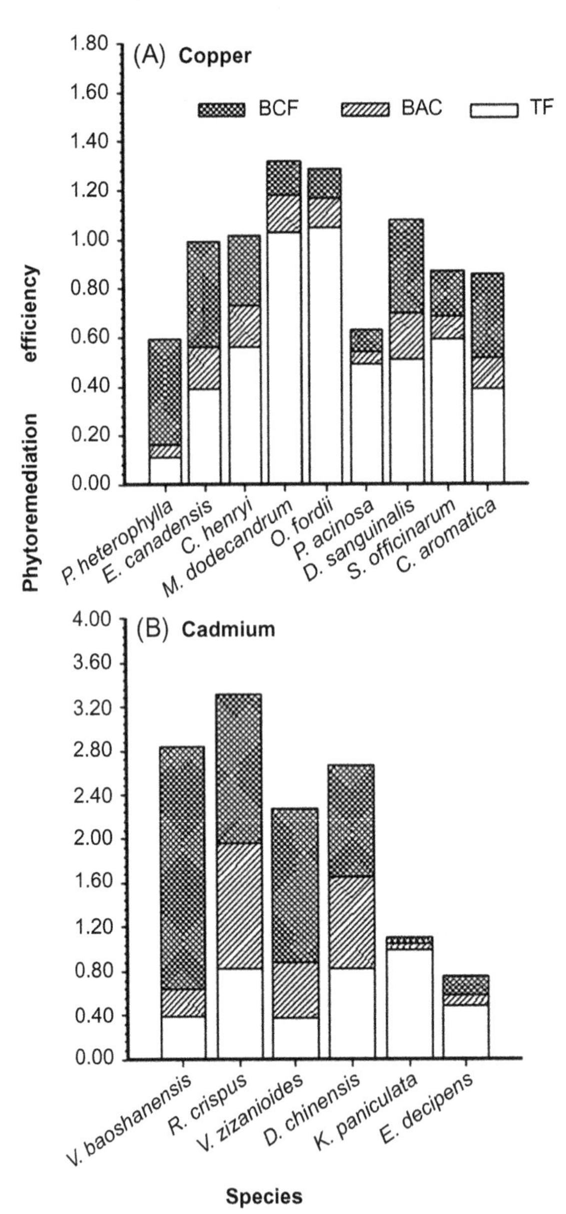

FIG. 15.6 Phytoremediation efficiency of certain plant species for copper (A) and cadmium (B), heavy metals reported from various contaminated soils.

TABLE 15.8 Evaluation of TF, BAC, and BCF for various plant species growing in contaminated soil.

Metal	Plant used	Stem	Leaves	Shoot	Root	Soil	TF	BAC	BCF
Zinc	*Viola baoshanensis*	–	–	514	669	1050	0.77	0.49	0.64
	Rumex crispus	–	–	1340	1007	1050	1.33	1.28	0.96
	Rumex k-1	–	–	114	125	1050	0.91	0.11	0.12
	Vetiveria zizanioides	–	–	144	339	1050	0.42	0.13	0.32
	Dianthus chinensis	–	–	282	228	1050	1.24	0.27	0.22
	Rumex acetosa DSL	–	–	151	289	1050	0.52	0.14	0.28
	Rumex acetosa JQW	–	–	708	866	1050	0.82	0.67	0.82
	Centaurea virgate	–	–	56.7	483.3	1955.5	0.12	0.03	0.25
	Astragalus versus	–	–	51.3	627.3	1955.5	0.08	0.03	0.32
	Chenopodium botrys	–	–	276	841.7	1955.5	0.33	0.14	0.43

TABLE 15.8 Evaluation of TF, BAC, and BCF for various plant species growing in contaminated soil—cont'd

Metal	Plant used	Stem	Leaves	Shoot	Root	Soil	TF	BAC	BCF
	Stipa barbata	–	–	329.3	315.7	1955.5	1.04	0.17	0.16
	Ziziphora clinopodioides	–	–	126	916.3	1955.5	0.14	0.06	0.47
	Cousinia bijarensis	–	–	46.7	46.3	1955.5	1.01	0.02	0.02
	Cirsium congestum	–	–	74.7	805.3	1955.5	0.09	0.04	0.41
	Scariola orientalis	–	–	117.3	1208.3	1955.5	0.10	0.06	0.62
	Cousinia species	–	–	145	1017.7	1955.5	0.14	0.07	0.52
	Chondrilla juncea	–	–	116.3	85.3	1955.5	1.36	0.06	0.04
	Melica jacquemontii	–	–	133.3	113.2	1955.5	1.18	0.07	0.06
	Verbascum speciosum	–	–	129.5	1086.7	1955.5	0.12	0.07	0.56
Lead	*Helianthus annuus*	843	3611	4454	2668	–	1.67	–	–
	Buddleja asiatica	–	–	2552.3	2325.5	94,584	1.10	0.03	0.02
	Buddleja paniculata	–	–	3485.5	4275.4	94,584	0.82	0.04	0.04
	Allium fistulosum	7.2	2.7	9.9	11.1	18.9	0.89	0.52	0.59
	Salsola iberica	–	–	30.4	26.7	83.16	1.14	0.37	0.32
	Microstegium ciliatum	–	–	12,200	128,830	175,170	0.09	0.07	0.74
	Dactyloctenium aegyptium	–	–	8100	5930	–	1.36	–	–
	Pennisetum polystachion	–	–	6205	24,705	–	0.251	–	–
	Viola baoshanensis	–	–	37	223	960	0.165	0.0385	0.232
	Rumex crispus	–	–	52	71	960	0.732	0.0541	0.0739
	Rumex k-1	–	–	19	24	960	0.791	0.0197	0.025
	Vetiveria zizanioides	–	–	19	136	960	0.139	0.0197	0.1416
	Dianthus chinensis	–	–	146	163	960	0.895	0.152	0.169
	Rumex acetosa DSL	–	–	107	152	960	0.703	0.111	0.158
	Rumex acetosa JQW	–	–	41	86	960	0.476	0.042	0.089
	Sedum alfredii	–	–	104	–	960	0.108	–	–
	Tephrosia candida	–	–	1689	1980	2207	0.853	0.765	0.897
	Tephrosia candida	2000	8000	10,000	3300	–		–	–
	Pityrogramma calomelanos	–	402.7	402.7	14,161.1	92,900	0.0284	0.00433	0.152
	Solanum nigrum	–	–	694	2320	120,080	0.299	0.0057	0.0193
	Thysanolaena maxima	–	–	3520	9870	118,967	0.3566	0.0295	0.0829
	Debregeasia orientalis	–	–	1763	1903	2217	0.926	0.795	0.858
Copper	*Parthenocissus heterophylla*	–	–	2.9	27.3	62.8	0.106	0.046	0.434
	Erigeron Canadensis	25.7	–	10.6	26.7	62.8	0.397	0.168	0.425
	Castanea henryi	–	–	10.4	18.5	62.8	0.562	0.165	0.294
	Melastoma dodecandrum	22.3	–	9.2	8.9	62.8	1.033	0.146	0.141
	Osmanthus fordii	7.8	–	7.9	7.5	62.8	1.053	0.125	0.119
	Phytolacca acinosa	10.3	–	2.9	5.9	62.8	0.491	0.046	0.0939
	Digitaria sanguinalis	13.7	–	12.3	24	62.8	0.512	0.195	0.382
	Saccharum officinarum	–	–	6.5	11.1	62.8	0.5855	0.103	0.176
	Curcuma aromatic	26.7	–	8.4	21.5	62.8	0.390	0.133	0.342

Continued

TABLE 15.8 Evaluation of TF, BAC, and BCF for various plant species growing in contaminated soil—cont'd

Metal	Plant used	Stem	Leaves	Shoot	Root	Soil	TF	BAC	BCF
Cadmium	*Rorippa globosa*	107.1	150.1	257.1	–	–	–	–	–
	Viola baoshanensis	–	–	28	18	72	1.55	0.388	0.25
	Rumex crispus	–	–	8.1	9.7	72	0.835	0.112	0.134
	Vetiveria zizanioides	–	–	3.7	10	72	0.37	0.051	0.1388
	Dianthus chinensis	–	–	6	7.2	72	0.833	0.0833	0.1
	Melastoma dodecandrum	–	2.03	1.26	1.15	–	1.09	–	–
	Curcuma aromatic	–	0.24	0.26	2.26	–	0.115	–	–
	Koelreuteria paniculata	0.19	0.51	0.7	0.7	13.15	1	0.05	0.05
	Elaeocarpus decipiens	0.24	0.92	1.16	2.37	13.15	0.489	0.088	0.180
	Populus europea	–	–	0.07	–	1.44	–	0.0486	–

TABLE 15.9 List of models used for quantification of phytoremediation.

Name of the model	Equation	Used for	Reference
Stirred-flow model	$S=S_e+S_k+S_s$ Where S_e (mg/kg) amount retained on equilibrium sites S_k (mg/kg) amount retained on kinetic type sites S_s (mg/kg) amount retained irreversibly S (mg/kg) total sorbed amount	Soil	Sun and Selim (2019)
Prediction model	$ME=1- [\Sigma (Y_{model}-Y_{actual})^2 / \Sigma (Y_{model}-C)^2]$ ME – Regression model Y_{model} is a predicted response Y_{actual} is a measured response C is a mean metal concentration in plant tissue $MANE=\Sigma (\mid Y_{model}-Y_{actual} \mid / Y_{actual}) / n$ MANE – mean average normalizing error n – total number of observations per replicate	*Pistia stratiotes* (Lettuce)	Kumar et al. (2019)
Sigmoid model	M_{HM} Plant $=M_{max} / 1+e^{-Km (t-t_{0.5m})} X\ C_{HM}, BAF_{HM}$ M_{HM} Plant is the total mass of heavy metal accumulation in a certain plant C_{HM} is the heavy metal concentration in a certain plant Km is a constant M_{max} is a maximum plant biomass t is growth time $t_{0.5m}$ is the infection time BAF_{HM} is the bioaccumulation factor of target heavy metal	*Helianthus annuus*	Zhao et al. (2019)
Kinetics model	$\log K_{di}-\log K_{dj}=½ (\log K_{mj}-\log K_{mi})$ K_{mj} and K_{mi} are metal-binding constants for sites i and j, respectively K_{di} among different specific binding sites is well constrained	Soil	Liu et al. (2019)
Regression model	$ME=1-\{\Sigma (C_{model}-C_{measured})^2 / \Sigma (C_{measured}-C_{mean})^2\}$ ME – Regression model $MNAE=\{\Sigma (\mid C_{model}-C_{measured} \mid / C_{measured})\} / n$ MNAE – mean normalized average error	*Faba sativa* Bernh.	Eid et al. (2019)

15.7 Advantages, limitation, and future perspective of phytoremediation

15.7.1 Advantages of phytoremediation

Phytoremediation has various advantages and wonderful market opportunities on a commercial scale. In upcoming times, the remediation cost of waste disposal and contaminated soil may rise many times over;

phytoremediation offers an eco-friendly and cost-effective solution for in situ remediation of contaminated sites compared to conventional techniques such as soil flushing, pneumatic fracturing, electrokinetic extraction, chemical oxidation/reduction, excavation, retrieval, and offsite disposal, surface capping, encapsulation, and stabilization. It achieves the desired results without disturbing the native soil microflora and fauna and involves profit from the natural ability of the environment to restore itself. This technique has the potential to treat a wide range of contaminants, such as heavy metals, pesticides, industrial effluents, petroleum hydrocarbons. Plants also support the activities of microbial communities in the rhizosphere via root exudates, which enable the microorganisms to degrade a wide range of organic compounds. Phytoremediation also encompasses several demerits as it mainly depends upon the depth of the roots and the tolerance of the plant to the contaminants. In Europe, there is no significant commercial use of phytoremediation, but this may develop soon because interest and funding for phytoremediation research are increasing rapidly, and many polluted sites in the new European Union countries (Eastern Europe) await remediation. Phytoremediation may also become a technology of choice for remediation projects in developing countries because it is cost-efficient and easy to implement.

15.7.2 Limitations of phytoremediation

Phytoremediation has gained much attention and acceptance during recent years. Its primary advantages are cost-effectiveness, feasibility, and eco-friendly approach. Although phytoremediation has many advantages in comparison with other conventional remediation technologies, it has some limitations too (Chandra and Kumar, 2017; Ekta and Modi, 2018). Phytoremediation has some inherent technical limitations, such as the contaminant must be within or nearby rhizospheric zone of the plants that are actively growing. Metal contaminants that are tightly bound with soil are tough to remediate by plants due to the low bioavailability of metals. Various synthetic chelating agents (citric acid, EDTA, EGTA, CDTA, DTPA, etc.) are used to overcome the problem of metal mobilization and bioavailability. However, nondegradable and toxic nature and persistence of these agents in the soil can create problems in the future (Oh et al., 2014). It must not present imminent danger to human health or further environmental harm. The phytoremediation is a slow process that depends on the biological cycles and needs a few months to several years and multiple crop cycles to remediate soil (Khalid et al., 2017). There may also be a considerable time differential between phytoremediation techniques and "dig and dump" techniques. Research in this area is spurred by current engineering technologies that tend to be clumsy, costly, and disruptive. The research community, both basic and applied, working in multidisciplinary teams, has a unique opportunity to produce a needed technology that is low cost, low impact, visually benign, and environmentally sound. Plant biologists could play a pivotal role in providing knowledge of basic plant processes on which genetic modification and breeding efforts depend.

15.7.3 Future perspective of phytoremediation

The foremost requirement for the success of phytoremediation technology, at this point, is the need to establish appropriate governmental regulations to monitor and foster the growth of the phytoremediation industry. Although the application of phytoremediation as a decontamination strategy poses many challenges, the benefits are far more than the limitations.

Future research in this field should focus on identifying naturally occurring metal hyperaccumulating plants with added characteristics such as:

1. Multimetal tolerance
2. Ability to hyperaccumulate several metals
3. Ability to perform in mixed contaminant sites, along with
4. A fast growth habit and an efficient root system to enhance metal tolerance, uptake, translocation, and sequestration.

Through genetic engineering technologies, identifying new species of organisms or by plant genetic manipulation of existing microorganisms that can coinhabit with plants would facilitate metal uptake at sites of metal contamination (Table 15.10).

Hence, utility and application of phytoremediation can be enhanced with the help of biotechnology.

TABLE 15.10 Selected inorganic phytoremediation case studies involving transgenic plants, with a focus on phytochelatins and metallothionein transgenic lines.

Target plant	Gene(s)	Source	Degradation	References
Arabidopsis thaliana	MerC	Bacteria	Hg	Uraguchi et al. (2019)
A. thaliana	AdPCs1–2-3	*Arundo donax*	Cd	Li et al. (2019)
Brassica napus	CKX2	*Arabidopsis thaliana*	Cd, Zn	Nehnevajova et al. (2019)
–	OsPCs5/OsPCs15	*Oryza sativa*	Cd	Park et al. (2019)
A. thaliana	PCs1	*Brassica napus*	Cd	Bai et al. (2019)
Sedum plumbizincicola	SpHMA1	*Sedum alfredii*	Cd	Zhao et al. (2019)
Nicotiana tabacum	LmSAP	*Lobularia maritima*	Cd, Cu, Mn, Zn	Saad et al. (2018)
A. thaliana	PCs1	*Vicia sativa*	Cd	Zhang et al. (2018)
N. tabacum	PtoEXPA12	*Populus tomentosa*	Cd	Zhang et al. (2018)
B. napus	OsMyb4	*Oryza sativa*	Cu, Zn	Raldugina et al. (2018)
A. thaliana	VsCCoAOMT	*Vicia sativa*	Cd	Xia et al. (2018)
N. tabacum	PtoHMA5	*Populus tomentosa*	Cd	Wang et al. (2018)
A. thaliana N. tabacum	PCs	*Morus alba*	Cd, Zn	Fan et al. (2018)
A. thaliana Populus tomentosa	PtABCC1	*Populus trichocarpa*	Hg	Sun et al. (2018b)
A. thaliana	SaCAD	*Sedum alfredii*		Qiu et al. (2018)
Wild-type rice	OsARM1	*Oryza sativa*	As	Wang et al. (2017)
A. thaliana	SsMT2	*Suaeda salsa*	Cd, Na	Jin et al. (2017)
N. tabacum	AtACR2	*Arabidopsis thaliana*	As	Nahar et al. (2017)
Urtica dioica	CUP	*Bacillus shackletonii Streptomyces badius*	Cd, Pb, Zn	Viktorova et al. (2017)
A. thaliana	SaNramp6	*Sedum alfredii*	Cd	Chen et al. (2017)
N. tabacum	PjHMT	*Prosopis juliflora*	Cd	Keeran et al. (2017)
O. sativa	ricMT	*Oryza sativa*	Cd, Cu	Zhang et al. (2017)
N. tabacum	OsMTP1	*Oryza sativa*	As, Cd	Das et al. (2016)
A. thaliana	IlMT2b	*Iris lactea*	Cu	Gu et al. (2015)
A. thaliana	OsMT2c	*Oryza sativa*	Cu	Liu et al. (2015b)
A. thaliana	MAN3	*Arabidopsis thaliana*	Cd	Chen et al. (2015a)
A. thaliana	CePCs1	*Caenorhabditis elegans*	Cd	Kühnlenz et al. (2015)
Beta vulgaris	StGCS-GS	*Streptococcus thermophilus*	Cd, Cu, Zn	Liu et al. (2015a)
N. tabacum	PtPCs	*Populus tomentosa*	Cd	Chen et al. (2015b)
P. tremula X P. alba	γ-glutamylcysteine synthetase	*Escherichia coli*	Cd	He et al. (2015)
Brassica juncea	AtACBP1 AtACBP4	*Arabidopsis thaliana*	Pb	Du et al. (2015)

15.8 Conclusion

The global problem concerning the contamination of various ecosystems as a consequence of human activities is increasing. Environmental pollution affects the quality of air, water, and soil in the biosphere. Great efforts have been made in the last two decades to reduce pollution sources and the polluted soil and water resources by several physical, chemical, and biological procedures, of which phytoremediation (use of suitable hardy nature plants that could be tolerable, survivable, and sequestering power) emerges as cost-effective with fewer side effects than physical and chemical approaches. Phytoremediation is an emerging onsite, in situ remediation technology to treat contaminated soils, sediments, plumes, and waters that can be achieved by several processes, i.e., rhizofiltration, stabilization, immobilization, extraction, and biodegradation or volatilization. Phytostabilization and immobilization are technologies based on mechanical stabilization of soil, reduced leaching rates, absorption and precipitation of pollutants in the rhizosphere. Phytoextraction, phytodegradation, and phytovolatilization are cleanup technologies that lead to partial removal of pollutants from polluted media; phytoextraction applies to metals, metalloids, halides, radionuclides, and organic pollutants, while phytodegradation is exclusively working with organics.

In the present chapter, we have discussed several aspects of phytoremediation, their mechanism, pattern, process, and the opportunities in a changing environment. Therefore, an intensive and ecological study is highly recommended for the identification and characterization of suitable species for phytoremediation purposes based on individual needs and mechanisms in a changing environment. It is also expected that ecological evaluation of the desirable and most applicable, especially indigenous, plants will play a promising role in the development of a database for future research aspects in remediation and decontamination of limited air, water, and soil resources of the biosphere.

References

Abhilash, P.C., Singh, N., 2009. Pesticide use and application: an Indian scenario. J. Hazard. Mater. 165 (1–3), 1–12.

Abioye, O.P., Agamuthu, P., Abdul Aziz, A., 2010. Phytoaccumulation of zinc and iron by *Jatropha curcas* grown in used lubricating oil contaminated soil. Malays. J. Med. Sci. 29, 207–213.

Adhikari, T., Kumar, A., 2012. Phytoaccumulation and tolerance of *Riccinus communis* L. to nickel. Int. J. Phytoremediation 14, 481–492.

Adriano, D.C., Wenzel, W.W., Vangronsveld, J., Bolan, N.S., 2004. Role of assisted natural remediation in environmental clean-up. Geoderma 122 (2–4), 121–142.

Aken, B.V., Correa, P.A., Schnoor, J.L., 2010. Phytoremediation of polychlorinated biphenyls: new trends and promises. Environ. Sci. Technol. 44 (8), 2767–2776.

Alghanmi, S.I., Al Sulami, A.F., El-Zayat, T.A., Alhogbi, B.G., Salam, M.A., 2015. Acid leaching of heavy metals from contaminated soil collected from Jeddah, Saudi Arabia: kinetic and thermodynamics studies. Int. Soil Water Conserv. Res. 3 (3), 196–208.

Alloway, B.J., 1995. Soil processes and the behaviour of metals. In: Alloway, B.J. (Ed.), Heavy Metals in Soils. Blackie, London, pp. 38–57.

Al-Qurainy, F., Abdel-Megeed, A., 2009. Phytoremediation and detoxification of two organophosphorous pesticides residues in Riyadh area. World Appl. Sci. J. 6 (7), 987–998.

Alshawabkeh, A.N., 2009. Electrokinetic soil remediation: challenges and opportunities. Sep. Sci. Technol. 44 (10), 2171–2187.

Alvarenga, P., Gonçalves, A.P., Fernandes, R.M., 2008. Evaluation of composts and liming materials in the phytostabilization of a mine soil using perennial ryegrass. Sci. Total Environ. 406, 43–56.

Aslund, M.L.W., Zeeb, B.A., Rutter, A., Reimer, K.J., 2007. In situ phytoextraction of polychlorinated biphenyl-(PCB) contaminated soil. Sci. Total Environ. 374 (1), 1–12.

Bai, J., Wang, X., Wang, R., Wang, J., Le, S., Zhao, Y., 2019. Overexpression of three duplicated BnPCS genes enhanced Cd accumulation and translocation in *Arabidopsis thaliana* mutant cad1–3. Bull. Environ. Contam. Toxicol. 102, 146–152.

Baker, A.J.M., Reeves, R.D., McGrath, S.P., 1991. In Situ Decontamination of Heavy Metal Polluted Soils Using Crops of Metal-Accumulating Plants-A Feasibility Study. In Situ Bioreclamation. Butterworth-Heinemann, pp. 600–605.

Bani, A., Pavlova, D., Echevarria, G., 2010. Nickel hyperaccumulation by the species of *Alyssum* and *Thlaspi* (Brassicaceae) from the ultramafic soils of the Balkans. Botanica Serbica 34, 3–14.

Banuelos, G.S., 2000. Phytoextraction of selenium from soils irrigated with selenium-laden effluent. Plant Soil 224 (2), 251–258.

Baudouin, C., Charveron, M., Tarrouse, R., Gall, Y., 2002. Environmental pollutants and skin cancer. Cell Biol. Toxicol. 18, 341–348.

Belyaeva, O.N., Haynes, R.J., Birukova, O.A., 2005. Barley yield and soil microbial and enzyme activities as affected by contamination of two soils with lead, zinc or copper. Biol. Fertil. Soils 41 (2), 85–94.

Bennicelli, R., Stępniewska, Z., Banach, A., 2004. The ability of *Azolla caroliniana* to remove heavy metals (hg (II), Cr (III), Cr (VI)) from municipal waste water. Chemosphere 55, 141–146.

Bhardwaj, T., Sharma, J.P., 2013. Impact of pesticides application in agricultural industry: an Indian scenario. Int. J. Agric. Food Sci. Technol. 4 (8), 817–822.

Bhatnagar, V.K., 2001. Pesticides pollution: trends and perspectives. ICMR Bull. 31, 87–88.

Bilek, F., 2004. Prediction of ground water quality affected by acid mine drainage to accompany in situ remediation. Appl. Earth Sci. 113 (1), 31–42.

Bilgin, M., Tulun, S., 2016. Removal of heavy metals (Cu, Cd and Zn) from contaminated soils using EDTA and $FeCl_3$. Global Nest J. 18, 98–107.

Birch, G.E., Scollen, A., 2003. Heavy metals in road dust, gully pots and parkland soils in a highly urbanised subcatchment of Port Jackson, Australia. Aust. J. Soil Res. 41, 1329–1342.

Bolan, N.S., Adriano, D.C., Naidu, R., 2003. Role of phosphorus in mobilization and bioavailability of heavy metals in the soil-plant system. Rev. Environ. Contam. Toxicol. 177, 1–44.

Bright, D.A., Healey, N., 2003. Contaminant risks from biosolids land application: contemporary organic contaminant levels in digested sewage sludge from five treatment plants in Greater Vancouver, British Columbia. Environ. Pollut. 126 (1), 39–49.

Bromilow, R.H., Chamberlain, K., 1995. Principles Governing Uptake and Transport of Chemicals Plant Contamination, Modeling and Simulation of Organic Chemical Processes. 37 Lewis Publishers, Boca Raton, FL, p. 68.

Cao, X., Ma, L.Q., 2004. Effects of compost and phosphate on plant arsenic accumulation from soils near pressure-treated wood. Environ. Pollut. 132 (3), 435–442.

Chandra, R., Kumar, V., 2017. Phytoextraction of heavy metals by potential native plants and their microscopic observation of root growing on stabilised distillery sludge as a prospective tool for in situ phytoremediation of industrial waste. Environ. Sci. Pollut. Res. 24 (3), 2605–2619.
Chaney, R.L., Baklanov, I.A., 2017. Phytoremediation and phytomining: status and promise. Adv. Bot. Res. 83, 189–221. Academic Press.
Chen, S., Han, X., Fang, J., Lu, Z., Qiu, W., Liu, M., Sang, J., Jiang, J., Zhuo, R., 2017. Sedum alfredii SaNramp6 metal transporter contributes to cadmium accumulation in transgenic *Arabidopsis thaliana*. Sci. Rep. 7, 13318.
Chen, L., Han, Y., Jiang, H., Korpelainen, H., Li, C., 2011. Nitrogen nutrient status induces sexual differences in responses to cadmium in *Populus yunnanensis*. J. Exp. Bot. 62 (14), 5037–5050.
Chen, Y., Liu, Y., Ding, Y., Wang, X., Xu, J., 2015b. Overexpression of PtPCS enhances cadmium tolerance and cadmium accumulation in tobacco. Plant Cell Tissue Organ Culture (PCTOC) 121 (2), 389–396.
Chen, J., Yang, L., Gu, J., Bai, X., Ren, Y., Fan, T., Cao, S., 2015a. MAN 3 gene regulate cadmium tolerance through the glutathione-dependent pathway in Arabidopsis thaliana. New Phytol. 205 (2), 570–582.
Chinmayee, M.D., Mahesh, B., Pradesh, S., 2012. The assessment of phytoremediation potential of invasive weed *Amaranthus spinosus* L. Appl. Biochem. Biotechnol. 167, 1550–1559.
Chiu, K.K., Ye, Z.H., Wong, M.H., 2006. Growth of *Vetiveria zizanioides* and *Phragmities australis* on Pb/Zn and cu mine tailings amended with manure compost and sewage sludge: a greenhouse study. Bioresour. Technol. 97 (1), 158–170.
Clemente, R., Almela, C., Bernal, M.P., 2006. A remediation strategy based on active phytoremediation followed by natural attenuation in a soil contaminated by pyrite waste. Environ. Pollut. 143 (3), 397–406.
Collins, C., Laturnus, F., Nepovim, A., 2002. Remediation of BTEX and trichloroethene-current knowledge with special emphasis on phytoremediation. Environ. Sci. Pollut. Res. 9, 86–94.
Das, N., Bhattacharya, S., Maiti, M.K., 2016. Enhanced cadmium accumulation and tolerance in transgenic tobacco overexpressing rice metal tolerance protein gene OsMTP1 is promising for phytoremediation. Plant Physiol. Biochem. 105, 297–309.
Dinesh, M., Kumar, M.V., Neeraj, P., Shiv, B., 2014. Phytoaccumulation of heavy metals in contaminated soil using Makoy (*Solenum nigrum* L.) and spinach (*Spinacia oleracea* L.) plant. Int. J. Life Sci. 2, 350–354.
Dominguez, M.T., Madrid, F., Maranon, T., Murillo, J.M., 2009. Cadmium availability in soil and retention in oak roots: potential for phytostabilization. Chemosphere 76, 480–486.
Du, Z.Y., Chen, M.X., Chen, Q.F., Gu, J.D., Chye, M.L., 2015. Expression of *Arabidopsis* acyl-CoA-binding proteins AtACBP 1 and AtACBP 4 confers P b (II) accumulation in *Brassica juncea* roots. Plant Cell Environ. 38 (1), 101–117.
Dushenkov, D., 2003. Trends in phytoremediation of radionuclides. Plant Soil 249, 167–175.
Eid, E.M., Alrumman, S.A., Galal, T.M., El-Bebany, A.F., 2019. Regression models for monitoring trace metal accumulations by *Faba sativa* Bernh. plants grown in soils amended with different rates of sewage sludge. Sci. Rep. 9 (1), 5443.
Ekta, P., Modi, N.R., 2018. A review of phytoremediation. J. Pharmacogn. Phytochem. 7 (4), 1485–1489.
Environmental Remediation Market Report, 2016. Environmental remediation market by environmental medium (soil and ground water), Technology (Bioremediation, pump and treat, soil vapor extraction, thermal treatment, soil washing) application and geography-Global forecast to 2022.
Fan, W., Guo, Q., Ying, C., Xueqin, L., Zhang, L.M., Long, D., Xiang, Z., Zhao, A., 2018. Two mulberry phytochelatin synthase genes confer zinc/cadmium tolerance and accumulation in transgenic *Arabidopsis* and tobacco. Gene 645, 95–104.
Fengxiang, X.H., Ti, S., Monts, D.L., Plodinec, M.J., Banin, A., Triplett, G.E., 2003. Assessment of global industrial-age anthropogenic arsenic contamination. Naturwissenschaften 90, 395–401.
Forchhammer, J.G., 1855. About influence of table salt on formation of minerals. Ann. Phys. 171 (5), 60–96.
Friedland, A.J., 1989. The movement of metals through soils and ecosystems. In: Heavy Metal Tolerance in Plants: Evolutionary Aspects, pp. 7–19.
Fritioff, A., Greger, M., 2003. Aquatic and terrestrial plant species with potential to remove heavy metals from stormwater. Int.J. Phytoremediation 5 (3), 211–224.
FRTR, 2012. Remediation Technologies Screening Matrix and Reference Guide, Version4.0. Federal Remediation Technologies Roundtable, Washington, DC.
Garbisu, C., Alkorta, I., 2001. Phytoextraction: a cost-effective plant-based technology for the removal of metals from the environment. Bioresour. Technol. 77 (3), 229–236.
Garbisu, C., Alkorta, I., 2003. Basic concepts on heavy metal soil bioremediation. Eur. J. Miner. Process. Environ. Prot. 3 (1), 58–66.
Gaymard, F., Pilot, G., Lacombe, B., Bouchez, D., Bruneau, D., Boucherez, J., Sentenac, H., 1998. Identification and disruption of a plant shaker-like outward channel involved in K^+ release into the xylem sap. Cell 94 (5), 647–655.
Ghosh, M., Singh, S.P., 2005. A review on phytoremediation of heavy metals and utilization of it's by products. Asian J. Energy Environ. 6 (4), 214–231.
Gray, C.W., Dunham, S.J., Dennis, P.G., Zhao, F.J., McGrath, S.P., 2006. Field evaluation of in situ remediation of a heavy metal contaminated soil using lime and red-mud. Environ. Pollut. 142, 530–539.
Greipsson, S., 2011. Phytoremediation. Nat. Educat. Knowl. 3, 7.
Gu, C.S., Liu, L.Q., Deng, Y.M., Zhu, X.D., Huang, S.Z., Lu, X.Q., 2015. The heterologous expression of the Iris lactea var. chinensis type 2 metallothionein IlMT2b gene enhances copper tolerance in *Arabidopsis thaliana*. Bull. Environ. Contam. Toxicol. 94 (2), 247–253.
He, J., Li, H., Ma, C., Zhang, Y., Polle, A., Rennenberg, H., Luo, Z.B., 2015. Overexpression of bacterial γ-glutamylcysteine synthetase mediates changes in cadmium influx, allocation and detoxification in poplar. New Phytol. 205 (1), 240–254.
Heinrich, K., Schulz, E., 1996. Uptake of selected organochlorine pesticides from a sandy soil (deep loam grey soil) by maize in a pot experiment. Mitteilgn. Dtsch. Bodenkundl. Gesellsch. 79, 283–286.
Himelblau, E., Mira, H., Lin, S.J., Culotta, V.C., Penarrubia, L., Amasino, R.M., 1998. Identification of a functional homolog of the yeast copper homeostasis gene ATX1 from Arabidopsis. Plant Physiol. 117 (4), 1227–1234.
Hirsch, R.E., Lewis, B.D., Spalding, E.P., Sussman, M.R., 1998. A role for the AKT1 potassium channel in plant nutrition. Science 280 (5365), 918–921.
Holm, P.E., Rootzen, H., Borggaard, O.K., Moberg, J.P., Christensen, T.H., 2003. Correlation of cadmium distribution coefficients to soil characteristics. J. Environ. Qual. 32 (1), 138–145.

Jiang, W., Tao, T., Liao, Z.M., 2011. Removal of heavy metal from contaminated soil with chelating agents. Open J. Soil Sci. 1 (02), 70.

Jin, S., Xu, C., Li, G., Sun, D., Li, Y., Wang, X., Liu, S., 2017. Functional characterization of a type 2 metallothionein gene, SsMT2, from alkaline-tolerant *Suaeda salsa*. Sci. Rep. 7, 17914.

Jomjun, N., Siripen, T., Maliwan, S., 2010. Phytoremediation of arsenic in submerged soil by wetland plants. Int.J. Phytoremediation 13, 35–46.

Kalve, S., Sarangi, B.K., Pandey, R.A., Chakrabarti, T., 2011. Arsenic and chromium hyperaccumulation by an ecotype of *Pteris vittata* prospective for phytoextraction from contaminated water and soil. Curr. Sci. 100, 888–894.

Kamal, M., Ghaly, A.E., Mahmoud, N., Cotecote, R., 2004. Phytoaccumulation of heavy metals by aquatic plants. Environ. Int. 29, 1029–1039.

Karaca, A., 2004. Effect of organic wastes on the extractability of cadmium, copper, nickel, and zinc in soil. Geoderma 122 (2–4), 297–303.

Karley, A.J., Leigh, R.A., Sanders, D., 2000. Where do all the ions go? The cellular basis of differential ion accumulation in leaf cells. Trends Plant Sci. 5 (11), 465–470.

Keeran, N.S., Ganesan, G., Parida, A.K., 2017. A novel heavy metal ATPase peptide from *Prosopis juliflora* is involved in metal uptake in yeast and tobacco. Trans. Res. 26 (2), 247–261.

Khalid, S., Shahid, M., Niazi, N.K., Murtaza, B., Bibi, I., Dumat, C., 2017. A comparison of technologies for remediation of heavy metal contaminated soils. J. Geochem. Explor. 182, 247–268.

Kirkham, M.B., 2006. Cadmium in plants on polluted soils: effects of soil factors, hyperaccumulation, and amendments. Geoderma 137 (1–2), 19–32.

Kramer, U., Chardonnens, A., 2001. The use of transgenic plants in the bioremediation of soils contaminated with trace elements. Appl. Microbiol. Biotechnol. 55 (6), 661–672.

Kramer, U., Cotter-Howells, J.D., Charnock, J.M., Baker, A.J., Smith, J.A.C., 1996. Free histidine as a metal chelator in plants that accumulate nickel. Nature 379 (6566), 635.

Kühnlenz, T., Westphal, L., Schmidt, H., Scheel, D., Clemens, S., 2015. Expression of *Caenorhabditis elegans* PCS in the AtPCS 1-deficient *Arabidopsis thaliana* cad1-3 mutant separates the metal tolerance and non-host resistance functions of phytochelatin synthases. Plant Cell Environ. 38 (11), 2239–2247.

Kulakow, P.A., Schwab, A.P., Banks, M.K., 2000. Screening plant species for growth on weathered, petroleum hydrocarbon-contaminated sediments. Int.J. Phytoremediation 2 (4), 297–317.

Kumar, V., Singh, J., Kumar, P., 2019. Heavy metal uptake by water lettuce (*Pistia stratiotes* L.) from paper mill effluent (PME): experimental and prediction modeling studies. Environ. Sci. Pollut. Res. 26 (14), 14400–14413.

Kupper, H., Zhao, F.J., McGrath, S.P., 1999. Cellular compartmentation of zinc in leaves of the hyperaccumulator *Thlaspi caerulescens*. Plant Physiol. 119 (1), 305–312.

Kupper, H., Lombi, E., Zhao, F.J., McGrath, S.P., 2000. Cellular compartmentation of cadmium and zinc in relation to other elements in the hyperaccumulator *Arabidopsis halleri*. Planta 212 (1), 75–84.

Lasat, M.M., 2002. Phytoextraction of toxic metals. J. Environ. Qual. 31 (1), 109–120.

Lasat, M.M., Pence, N.S., Garvin, D.F., Ebbs, S.D., Kochian, L.V., 2000. Molecular physiology of zinc transport in the Zn hyperaccumulator *Thlaspi caerulescens*. J. Exp. Bot. 51 (342), 71–79.

Lee, T.M., Lai, H.Y., Chen, Z.S., 2004. Effect of chemical amendments on the concentration of cadmium and lead in long-term contaminated soils. Chemosphere 57 (10), 1459–1471.

Leigh, M.B., Prouzova, P., Mackova, M., Macek, T., Nagle, D.P., Fletcher, J.S., 2006. Polychlorinated biphenyl (PCB)-degrading bacteria associated with trees in a PCB-contaminated site. Appl. Environ. Microbiol. 72 (4), 2331–2342.

Li, M., Stragliati, L., Bellini, E., Ricci, A., Saba, A., di Toppi, L.S., Varotto, C., 2019. Evolution and functional differentiation of recently diverged phytochelatin synthase genes from *Arundo donax* L. J. Exp. Bot. 70 (19), 5391–5405.

Liu, D., An, Z., Mao, Z., Ma, L., Lu, Z., 2015a. Enhanced heavy metal tolerance and accumulation by transgenic sugar beets expressing Streptococcus thermophilus StGCS-GS in the presence of Cd, Zn and Cu alone or in combination. PLoS One 10 (6), e0128824.

Liu, L., Li, W., Song, W., Guo, M., 2018. Remediation techniques for heavy metal-contaminated soils: principles and applicability. Sci. Total Environ. 633, 206–219.

Liu, J., Shi, X., Qian, M., Zheng, L., Lian, C., Xia, Y., Shen, Z., 2015b. Copper-induced hydrogen peroxide upregulation of a metallothionein gene, OsMT2c, from Oryza sativa L. confers copper tolerance in Arabidopsis thaliana. J. Hazard. Mater. 294, 99–108.

Liu, P., Wang, P., Lu, Y., Ding, Y., Lu, G., Dang, Z., Shi, Z., 2019. Modeling kinetics of heavy metal release from field-contaminated soils: roles of soil adsorbents and binding sites. Chem. Geol. 506, 187–196.

Lombi, E., Zhao, F.J., Dunham, S.J., McGrath, S.P., 2001. Phytoremediation of heavy metal-contaminated soils. J. Environ. Qual. 30, 1919.

Lunney, A.I., Zeeb, B.A., Reimer, K.J., 2004. Uptake of weathered DDT in vascular plants: potential for phytoremediation. Environ. Sci. Technol. 38 (22), 6147–6154.

Madejon, E., Perez-de-Mora, A.P., Felipe, E., Burgos, P., Cabrera, F., 2006. Soil amendments reduce trace element solubility in a contaminated soil and allow re-growth of natural vegetation. Environ. Pollut. 139, 40–52.

Magnuson, M.L., Kelty, K.C., Kelty, C.A., 2002. Trace metal leaching behavior studied through the use of parametric modeling of water borne soil particles fractionated with a split-flow thin cell. Environ. Sci. Technol. 36 (20), 4288–4294.

Mahmood, Q., Mirza, N., Shaheen, S., 2015. Phytoremediation using algae and macrophytes: I. In: Phytoremediation. Springer, Cham, pp. 265–289.

Marques, A.P., Oliveira, R.S., Rangel, A.O., Castro, P.M., 2008. Application of manure and compost to contaminated soils and its effect on zinc accumulation by Solanum nigrum inoculated with arbuscular mycorrhizal fungi. Environ. Pollut. 151 (3), 608–620.

Marschner, H., 1995. Mineral Nutrition of Higher Plants. Academic Press, London, p. 889.

Mattina, M.I., Eitzer, B.D., Iannucci-Berger, W., Lee, W.Y., White, J.C., 2004. Plant uptake and translocation of highly weathered, soil-bound technical chlordane residues: data from field and rhizotron studies. Environ. Toxicol. Chem. 23 (11), 2756–2762.

McBride, M.B., Martinez, C.E., 2000. Copper phytotoxicity in a contaminated soil: remediation tests with adsorptive materials. Environ. Sci. Technol. 34 (20), 4386–4391.

Merkl, N., Schultze-Kraft, R., Infante, C., 2005. Phytoremediation in the tropics–influence of heavy crude oil on root morphological characteristics of graminoids. Environ. Pollut. 138 (1), 86–91.

Meuser, H., 2013. Soil Remediation and Rehabilitation: Treatment of Contaminated and Disturbed Land. Springer, Dordrecht, The Netherlands.

Mirza, N., Pervez, A., Mahmood, Q., 2011. Ecological restoration of arsenic contaminated soil by *Arundo donax* L. J. Ecol. Eng. 37, 1949–1956.

Mukherjee, I., Kumar, A., 2012. Phytoextraction of endosulfan a remediation technique. Bull. Environ. Contam. Toxicol. 88 (2), 250–254.

Mulligan, C.N., Yong, R.N., Gibbs, B.F., 2001. Remediation technologies for metal-contaminated soils and groundwater: an evaluation. Eng. Geol. 60 (1–4), 193–207.

Nahar, N., Rahman, A., Nawani, N.N., Ghosh, S., Mandal, A., 2017. Phytoremediation of arsenic from the contaminated soil using transgenic tobacco plants expressing ACR2 gene of *Arabidopsis thaliana*. J. Plant Physiol. 218, 121–126.

Nalbandian, H., 2012. Trace Element Emissions From Coal. 601 IEA Clean Coal Centre, p. 89.

Nathanail, C.P., Earl, N., 2001. Human health risk assessment: guideline values and magic numbers. In: Assessment and Reclamation of Contaminated Land. Thomas Telford Publishing, pp. 85–102.

Nehnevajova, E., Ramireddy, E., Stolz, A., Knorck, M.G., Novak, O., Strnad, M., Schmulling, M., 2019. Root enhancement in cytokinin-deficient oilseed rape causes leaf mineral enrichment, increases the chlorophyll concentration under nutrient limitation and enhances the phytoremediation capacity. BMC Plant Biol. 19, 83.

NJDEP, 2014. Technical Guidance on the Capping of Sites Undergoing Remediation. New Jersey Department of Environmental Protection, Trenton, NJ.

Nurzhanova, A., Kalugin, S., Zhambakin, K., 2013. Obsolete pesticides and application of colonizing plant species for remediation of contaminated soil in Kazakhstan. Environ. Sci. Pollut. Res. 20 (4), 2054–2063.

Oh, K., Cao, T., Li, T., Cheng, H., 2014. Study on application of phytoremediation technology in management and remediation of contaminated soils. J. Clean Energy Technol. 2 (3), 216–220.

Olexsey, R.A., Parker, R.A., 2006. Current and future in situ treatment techniques for the remediation of hazardous substances in soil, sediments, and groundwater. In: Soil and Water Pollution Monitoring, Protection and Remediation. Springer, Dordrecht, pp. 211–219.

Park, H.C., Hwang, J.E., Jiang, Y., Kim, Y.J., Kim, S.H., Nguyen, X.C., Kim, C.Y., Chung, W.S., 2019. Functional characterisation of two phytochelatin synthases in rice (*Oryza sativa* cv. Milyang 117) that respond to cadmium stress. Plant Biol. 21 (5), 854–861.

Parseh, I., Teiri, H., Hajizadeh, Y., Ebrahimpour, K., 2018. Phytoremediation of benzene vapors from indoor air by Schefflera arboricola and Spathiphyllum wallisii plants. Atmos. Pollut. Res. 9 (6), 1083–1087.

Pastor, J., Gutierrez-gines, M.J., Hernandez, A.J., 2015. Heavy-metal phytostabilizing potential of *Agrostis castellana* Boiss. and reuter. Int.J. Phytoremediation 17, 988–998.

Petrangeli Papini, M., Majone, M., Rolle, E., 2001. Kaolinite sorption of Cd, Ni and Cu from landfill leachates: influence of leachate composition. Water Sci. Technol. 44 (2–3), 343–350.

Pinto, A.P., De Varennes, A., Fonseca, R., Teixeira, D.M., 2015. Phytoremediation of soils contaminated with heavy metals: Techniques and strategies. In: Phytoremediation. Springer, Cham, pp. 133–155.

Prasad, M.N.V., de Oliveira Freitas, H.M., 2003. Metal hyperaccumulation in plants: biodiversity prospecting for phytoremediation technology. Electron. J. Biotechnol. 6 (3), 285–321.

Pulford, I.D., Watson, C., 2003. Phytoremediation of heavy metal-contaminated land by trees-a review. Environ. Int. 29 (4), 529–540.

Qiu, W., Song, X., Han, X., Liu, M., Qiao, G., Zhuo, R., 2018. Overexpression of *Sedum alfredii* cinnamyl alcohol dehydrogenase increases the tolerance and accumulation of cadmium in *Arabidopsis*. Environ. Exp. Bot. 155, 566–577.

Radziemska, M., Vaverkova, M.D., Baryła, A., 2017. Phytostabilization management strategy for stabilizing trace elements in contaminated soils. Int. J. Environ. Res. Public Health 14, 958.

Raldugina, G.N., Maree, M., Mattana, M., Shumkova, G., Mapelli, S., Kholodova, V.P., Karpichev, I.V., Kuznetzov, V.V., 2018. Expression of rice OsMyb4 transcription factor improves tolerance to copper or zinc in canola plants. Biol. Plant. 62, 511–520.

Reddy, K.R., Al-Hamdan, A.Z., Ala, P., 2011. Enhanced soil flushing for simultaneous removal of PAHs and heavy metals from industrial contaminated soil. J. Hazard. Toxic Radioact. Waste 15 (3), 166–174.

Redjala, T., Zelko, I., Sterckeman, T., 2011. Relationship between root structure and root cadmium uptake in maize. Environ. Exp. Bot. 71, 241–248.

Rizzi, L., Petruzzelli, G., Poggio, G., Guidi, G.V., 2004. Soil physical changes and plant availability of Zn and Pb in a treatability test of phytostabilization. Chemosphere 57, 1039–1046.

Rodriguez, L., Lopez-Bellido, F.J., Carnicer, A., Recreo, F., Tallos, A., Monteagudo, J.M., 2005. Mercury recovery from soils by phytoremediation. In: Environmental Chemistry. Springer, Berlin, Heidelberg, pp. 197–204.

Rostami, S., Azhdarpoor, A., 2019. The application of plant growth regulators to improve phytoremediation of contaminated soils: a review. Chemosphere 220, 818–827.

Rostami, S., Azhdarpoor, A., Samaei, M.R., 2017. Removal of pyrene from soil using phytobioremediation (Sorghum Bicolor-Pseudomonas). Health Scope 6 (4), 62153.

Roy, S., Labelle, S., Mehta, P., Mihoc, A., Fortin, N., Masson, C., Olsen, C., 2005. Phytoremediation of heavy metal and PAH-contaminated brownfield sites. Plant Soil 272 (1–2), 277–290.

Ruttens, A., Colpaert, J.V., Mench, M., Boisson, J., Carleer, R., Vangronsveld, J., 2006a. Phytostabilization of a metal contaminated sandy soil, II: influence of compost and/or inorganic metal immobilizing soil amendments on metal leaching. Environ. Pollut. 144, 533–539.

Ruttens, A., Mench, M., Colpaert, J.V., Boisson, J., Carleer, R., Vangronsveld, J., 2006b. Phytostabilization of a metal contaminated sandy soil, I: influence of compost and/or inorganic metal immobilizing soil amendments on phytotoxicity and plant availability of metals. Environ. Pollut. 144, 524–532.

Saad, R.B., Hsouna, A.B., Saibi, W., Hamed, K.B., Brini, F., Ghneim-Herrerad, T., 2018. A stress-associated protein, LmSAP, from the halophyte *Lobularia maritima* provides tolerance to heavy metals in tobacco through increased ROS scavenging and metal detoxification processes. J. Plant Physiol. 231, 234–243.

Sachs, J., 1865. Handbuch der Experimental-Physiologie der Pflanzen. In: Hofmeister, W. (Ed.), Handbuch der Physiologischen Botanik. Vol. IV. Engelmann, Leipzig, pp. 153–154.

Sakakibara, M., Watanabe, A., Inoue, M., 2010. Phytoextraction and phytovolatilization of arsenic from As-contaminated soils by *Pteris vittata*. In: Proceedings of the Annual International Conference on Soils, Sediments, Water and Energy, p. 26.

Salt, D.E., Smith, R.D., Raskin, I., 1998. Phytoremediation. Annu. Rev. Plant Biol. 49 (1), 643–668.

Saraswat, S., Rai, J.P.N., 2009. Chemistry and ecology phytoextraction potential of six plant species grown in multimetal contaminated soil. J. Chem. Ecol. 25, 1–11.
Sarwar, N., Imran, M., Shaheen, M.R., Ishaque, W., Kamran, M.A., Matloob, A., Hussain, S., 2017. Phytoremediation strategies for soils contaminated with heavy metals: modifications and future perspectives. Chemosphere 171, 710–721.
Sauve, S., Hendershot, W., Allen, H.E., 2000. Solid-solution partitioning of metals in contaminated soils: dependence on pH, total metal burden, and organic matter. Environ. Sci. Technol. 34 (7), 1125–1131.
Senden, M.H.M.N., Van Der Meer, A.J.G.M., Verburg, T.G., Wolterbeek, H.T., 1995. Citric acid in tomato plant roots and its effect on cadmium uptake and distribution. Plant Soil 171 (2), 333–339.
Siddiqui, S., Adams, W.A., Schollion, J., 2001. The phytotoxicity and degradation of diesel hydrocarbons in soil. J. Plant Nutr. Soil Sci. 164 (6), 631–635.
Silva, J.F., Williams, R.J.P., 2001. The Biological Chemistry of the Elements: The Inorganic Chemistry of Life. Oxford University Press.
Singh, V., Singh, N., 2014. Uptake and accumulation of endosulfan isomers and its metabolite endosulfan sulfate in naturally growing plants of contaminated area. Ecotoxicol. Environ. Saf. 104, 189–193.
Smolders, E., Degryse, F., 2002. Fate and effect of zinc from tire debris in soil. Environ. Sci. Technol. 36, 3706–3710.
Soudek, P., Petrova, S., Vanek, T., 2012. Phytostabilization or accumulation of heavy metals by using of energy crop *Sorghum* sp. In: 3rd International Conference on Biology, Environment and Chemistry IPCBEE. IACSIT Press, Singapore.
Su, Y., Han, F.X., Sridhar, B.B.M., Monts, D.L., 2005. Phytotoxicity and phytoaccumulation of trivalent and hexavalent chromium in brake fern. Environ. Toxicol. Chem. 24, 2019–2026.
Sukreeyapongse, O., Holm, P.E., Strobel, B.W., Panichsakpatana, S., Magid, J., Hansen, H.C.B., 2002. pH-dependent release of cadmium, copper, and lead from natural and sludge-amended soils. J. Environ. Qual. 31 (6), 1901–1909.
Sun, L., Ma, Y., Wang, H., Huang, W., Wang, X., Han, L., Sun, W., Han, E., Wang, B., 2018. Overexpression of PtABCC1 contributes to mercury tolerance and accumulation in Arabidopsis and poplar. Biochem. Biophys. Res. Commun. 497 (4), 997–1002.
Sun, W., Selim, H.M., 2019. A general stirred-flow model for time-dependent adsorption and desorption of heavy metal in soils. Geoderma 347, 25–31.
Sutherland, R.A., Day, J.P., Bussen, J.O., 2003. Lead concentrations, isotope ratios, and source apportionment in road deposited sediments, Honolulu, Oahu, Hawaii. Water Air Soil Pollut. 142 (1–4), 165–186.
Taiz, L., Zeiger, E., 2002. Plant Physiology, third ed. Sinauer Associates Inc., Sunderland, MA.
Tajudin, S.A.A., Azmi, M.A.M., Nabila, A.T.A., 2016. Stabilization/solidification remediation method for contaminated soil: a review. Conf. Ser.: Mater. Sci. Eng. 136 (1), 012043. IOP Publishing.
Tang, Y.T., Qiu, R.L., Zeng, X.W., et al., 2009. Lead, zinc, cadmium hyperaccumulation and growth stimulation in Arabis paniculata Franch. Environ. Exp. Bot. 66, 126–134.
Tariq, S.R., Ashraf, A., 2016. Comparative evaluation of phytoremediation of metal contaminated soil of firing range by four different plant species. Arab. J. Chem. 9 (6), 806–814.
Tester, M., Leigh, R.A., 2001. Partitioning of nutrient transport processes in roots. J. Exp. Bot. 52 (1), 445–457.
Traunfeld, J.H., Clement, D.L., 2001. Lead in Garden Soils. Home and Garden. Maryland Cooperative Extension, University of Maryland. http://www.hgic.umd.edu/media/documents/hg18.pdf.
Turer, D., Maynard, J.B., Sansalone, J.J., 2001. Heavy metal contamination in soils of urban highways: comparison between runoff and soil concentrations at Cincinnati, Ohio. Water Air Soil Pollut. 132, 293–314.
Udoka, O.C., Ekanem, E.O., Harami, M.D., Tafawa, A., 2014. Phytoaccumulation potentials of *Tamarindus indica*. Int. J. Innov. Sci. Res. 11, 72–78.
Uraguchi, S., Sone, Y., Kamezawa, M., Tanabe, M., Hirakawa, M., Nakamura, R., Takanezawa, Y., Kiyono, M., 2019. Ectopic expression of a bacterial mercury transporter MerC in root epidermis for efficient mercury accumulation in shoots of Arabidopsis plants. Sci. Rep. 9, 4347.
Utsunamyia, T., 1980. Japanese Patent Application No. 55-72959.
Van Ginneken, L., Meers, E., Guisson, R., Ruttens, A., Elst, K., Tack, F.M., Dejonghe, W., 2007. Phytoremediation for heavy metal-contaminated soils combined with bioenergy production. J. Environ. Eng. Landsc. Manag. 15 (4), 227–236.
Van Nevel, L., Mertens, J., Oorts, K., Verheyen, K., 2007. Phytoextraction of metals from soils: how far from practice? Environ. Pollut. 150 (1), 34–40.
Viklander, M., 1998. Particle size distribution and metal content in street sediments. J. Environ. Eng. 124, 761–766.
Viktorova, J., Jandova, Z., Madlenakova, M., Prouzova, P., Bartunek, V., Vrchotova, B., Lovecka, P., Musilova, L., Macek, T., 2017. Native phytoremediation potential of urtica dioica for removal of PCBs and heavy metals can be improved by genetic manipulations using constitutive CaMV 35S promoter. PLoS One 12 (10), e0187053.
Walker, D.J., Clemente, R., Bernal, M.P., 2004. Contrasting effects of manure and compost on soil pH, heavy metal availability and growth of Chenopodium album L. in a soil contaminated by pyritic mine waste. Chemosphere 57 (3), 215–224.
Wang, A.S., Angle, J.S., Chaney, R.L., Delorme, T.A., Reeves, R.D., 2006. Soil pH effects on uptake of Cd and Zn by *Thlaspi caerulescens*. Plant Soil 281 (1–2), 325–337.
Wang, F.Z., Chen, M.X., Yu, L.J., Xie, Li-JuanL.J., Yuan, L.B., Qi, H., Xiao, M., Guo, W., Chen, Z., Yi, K., Zhang, J., Qiu, R., Shu, W., Xiao, S., Chen, Q.F., 2017. OsARM1, an R2R3 MYB transcription factor, is involved in regulation of the response to arsenic stress in rice. Front. Plant. Sci. 8, 1868.
Wang, X., Zhi, J., Liu, X., Zhang, H., Liu, H., Xu, J., 2018. Transgenic tobacco plants expressing a P1B-ATPase gene from *Populus tomentosa* Carr. (PtoHMA5) demonstrate improved cadmium transport. Int. J. Biol. Macromol. 1 (113), 655–661.
White, J., Mattina, M., 2004. Phytoextraction of recalcitrant organic pollutants from soil by agricultural species. In: Proceedings of the Fourth International Conference on Remediation of Chlorinated and Recalcitrant Compounds, Monterey, CA, p. 4E-13.
White, J.C., Mattina, M.I., Lee, W.Y., Eitzer, B.D., Iannucci-Berger, W., 2003. Role of organic acids in enhancing the desorption and uptake of weathered p, p′-DDE by Cucurbita pepo. Environ. Pollut. 124 (1), 71–80.
Winterhalder, K., 1996. Environmental degradation and rehabilitation of the landscape around Sudbury, a major mining and smelting area. Environ. Rev. 4 (3), 185–224.
Xia, Y., Liu, J., Wang, Y., Zhang, X., Shen, Z., Hu, Z., 2018. Ectopic expression of *Vicia sativa* Caffeoyl-CoA O-methyltransferase (VsCCoAOMT) increases the uptake and tolerance of cadmium in *Arabidopsis*. Environ. Exp. 145, 47–53.

Yadav, I.C., Devi, N.L., Syed, J.H., Cheng, Z., Li, J., Zhang, G., Jones, K.C., 2015. Current status of persistent organic pesticides residues in air, water, and soil, and their possible effect on neighboring countries: a comprehensive review of India. Sci. Total Environ. 511, 123–137.
Yang, S., Liang, S., Yi, L., 2014. Heavy metal accumulation and phytostabilization potential of dominant plant species growing on manganese mine tailings. Front. Environ. Sci. Eng. 8, 394–404.
Yang, Z.H., Dong, C.D., Chen, C.W., Sheu, Y.T., Kao, C.M., 2018. Using poly-glutamic acid as soil-washing agent to remediate heavy metal-contaminated soils. Environ. Sci. Pollut. Res. 25 (6), 5231–5242.
Yoo, M.S., James, B.R., 2002. Zinc extractability as a function of pH in organic waste-amended soils. Soil Sci. 167 (4), 246–259.
Zhao, L., Li, T., Zhang, X., 2016. Pb uptake and phytostabilization potential of the mining ecotype of *Athyrium wardii* (Hook.) grown in Pb-contaminated soil. Clean-Soil Air Water 44, 1184–1190.
Zhang, H., Lv, S., Xu, H., Hou, D., Li, Y., Wang, F., 2017. H_2O_2 is involved in the metallothionein-mediated rice tolerance to copper and cadmium toxicity. Int. J. Mol. Sci. 18 (10), 2083.
Zhang, X., Rui, H., Zhang, F., Hu, Z., Xia, Y., Shen, Z., 2018. Overexpression of a functional *Vicia sativa* PCS1 homolog increases cadmium tolerance and phytochelatins synthesis in *Arabidopsis*. Front. Plant. Sci. 9, 107.
Zhao, X., Joo, J.C., Lee, J.K., Kim, J.Y., 2019. Mathematical estimation of heavy metal accumulations in *Helianthus annuus* L. with a sigmoid heavy metal uptake model. Chemosphere 220, 965–973.
Zhu, G., Guo, Q., Yang, J., Zhang, H., Wei, R., Wang, C., Yang, J., 2015. Washing out heavy metals from contaminated soils from an iron and steel smelting site. Front. Environ. Sci. Eng. 9 (4), 634–641.

CHAPTER

16

Vermicomposting: An efficient technology for the stabilization and bioremediation of pulp and paper mill sludge

Subpiramaniyam Sivakumar[a], *Kaliannan Thamaraiselvi*[b], *Duraisamy Prabha*[c], *Thyagarajan Lakshmi Priya*[d], *Hong Sung-Chul*[a], *Yi Pyoung-In*[a], *Jang Seong-Ho*[a], *and Suh Jeong-Min*[a]

[a]Department of Bioenvironmental Energy, College of Natural Resources and Life Science, Pusan National University, Miryang, Gyeongsangnam-do, Republic of Korea [b]Professor of Environmental Biotechnology at Bharathidasan University, Tiruchirappalli, India [c]Department of Environmental Sciences, Bharathiar University, Coimbatore, Tamil Nadu, India [d]Department of Civil Engineering, Environmental Engineering Division, Government College of Technology, Coimbatore, India

OUTLINE

16.1 Introduction

The worldwide growth of pulp and paper mill production is increasing every year. Pulp and paper mills generate several types of solid wastes, such as ashes, dregs, grits, lime mud, and pulp mill sludge (PMS) (Simão et al., 2018; Monte et al., 2009; Wirojanagud et al., 2004). One tonne of paper production generates 40–50 kg of sludge, where 70% is primary sludge, and 30% is secondary sludge (Bajpai and Bajpai, 2015). The primary sludge is generated from the production of virgin wood fiber, and the secondary activated sludge comes from the secondary wastewater treatment system (Geng et al., 2007). The primary sludge consists of wood fibers (cellulose, hemicellulose, and lignin), papermaking filters (kaolin and calcium carbonate), pitch, lignin by-products, and ash (Ochoa de Alda, 2008). The secondary sludge mostly consists of microbial biomass along with cell-decay products and nonbiodegradable lignin

Cost Effective Technologies for Solid Waste and Wastewater Treatment
https://doi.org/10.1016/B978-0-12-822933-0.00001-2

precipitates (Puhakka et al., 1992). The amount and the composition of sludge depend on the paper grade produced, the raw materials used, the process techniques applied, and the paper properties to be achieved (Priya et al., 2012). The dry mass of sludge might also vary from 20% (newsprint mill) to 40% (tissue mill) (Bajpai and Bajpai, 2015).

In wastewater treatment, sludge constitutes the largest residual waste stream generated by the pulp and paper mills in terms of volume (Krigstin and Sain, 2006). However, improper disposal and the production of large quantities of this waste can lead to major environmental and disposal problems (Bajpai and Bajpai, 2015). Landfills and incinerators are the common disposal methods for PMS. In landfill disposal methods, the hazardous substances in the leachate of solid waste might contaminate groundwater. Similarly, the combusting of the solid waste in an incinerator is also problematic because of the generation of hazardous chemicals containing by-products (char and ash), which are dumped into the environment. The contaminants in sludge, such as metals (Cd, Cu, Ni, Zn, and Fe), might be biomagnified during the application to an agricultural land (Świerk et al., 2007). Therefore, safe disposal methods are required for maintaining a healthy environment (Hackett et al., 1999). In addition, solid waste from the pulp and paper industry is an important source of organic materials and nutrients (Kunzler, 2001). The paper mill dregs are useful soil amendments, and lime mud can be used in agriculture as well as in wastewater treatment technologies (Simão et al., 2018). Grits are used for construction purposes, and PMS is used in agriculture, construction, and energy processes. Although the PMS contains high levels of organic matter and nutrients, such as nitrogen, potassium, phosphorus, and calcium, these components are generally not in forms that are directly or immediately available to crops (Sharma et al., 2002). Therefore, various methods have been applied in efforts to produce value-added products from paper mill sludge, including pyrolysis (Cho et al., 2017), anerobic digestion (Meyer and Edwards, 2014), ethanol production by fermentation (Boshoff et al., 2016), composting (Hackett et al., 1999; Hazarika et al., 2017), and vermicomposting (Bustamante et al., 2012; Negi and Suthar, 2013; Sahariah et al., 2014).

16.2 Advantages of vermicomposting technology

In anerobic digestion, the high viscosity of PMS might affect the pumping, mixing, and sludge dewatering process (Baudez et al., 2011; Örmeci, 2007; Lindorfer and Demmig, 2016). Energy consumption is higher as in the case of pyrolysis and ethanol production by fermentation methods (Reckamp et al., 2014; Chen et al., 2014; Devi and Saroha, 2013; Khalili et al., 2002; Jiang and Ma, 2011). However, between these technologies, vermicomposting is an efficient and economically feasible process, where the minerals are released when the sludge is broken down by the earthworms and microorganisms (Sivakumar et al., 2009). During the composting process, earthworms act as a process driver, conditioning the substrate and altering the biological activity; however, microorganisms are responsible for the biodegradation of organic waste (Sim and Wu, 2010; Sivakumar et al., 2012). The effectiveness of vermicomposting technology has been proved by many studies using agroindustrial sludge (Suthar, 2010), beverage industry sludge (Singh et al., 2010), distillery sludge (Suthar, 2007; Suthar and Singh, 2008), food industry sludge (Yadav and Garg, 2009), sago sludge (Banu et al., 2008), sugar mill sludge (Sangwan et al., 2008), tannery sludge (Cardoso-Vigueros and Ramirez-Camperos, 2006; Dheepa et al., 2006), and textile industry sludge (Garg et al., 2006). Therefore, many studies are currently focusing on the vermicomposting of PMS. Moreover, it has been suggested that the vermistabilization of PMS using earthworms is an excellent method in the management of industrial sludge (Elvira et al., 1998).

16.3 Physicochemical characteristics of vermicompost at different materials mixed with pulp and paper mill sludge

The physicochemical characteristics of the substrates mixed with PMS are essential for accessing the nutrient quality of the end product and suitability of the substrate mix for vermiconversion. Therefore, different physicochemical characteristics of the organic substrate before and after treatment are discussed here.

16.3.1 Vermicompost-induced changes in pH

The comparison of the pH of the substrates (PMS mixed with various substrates) before and after vermitreatments is given in Table 16.1. Significantly or nonsignificantly, the initial alkaline pH of the mixed substrate was reduced to neutral in the final compost, where PMS is mixed with various ratios of cow dung (Suthar et al., 2014; Yuvaraj et al., 2018), cow dung and green manure plants (Karmegam et al., 2019), cattle dung (Elvira et al., 1998), and citronella

TABLE 16.1 Changes in pH and EC of various waste mixtures before and after vermitreatment.

Earthworm species	Substrate used	Duration of experiment	Sludge pH (minimum to maximum)[a]		EC (minimum to maximum)[a]		Reference
			Before treatment	After treatment	Before treatment (ds/m)	After treatment	
Eisenia fetida	PMS, cow dung, green manure	60 days	8.26–8.63	7.28–7.60	0.98–1.34	2.19–3.37	(Karmegam et al., 2019)
Perionyx excavatus	PMS, cow dung, straw	60 days	6.39–6.89	8.05–8.23	2.65–2.95	1.45–1.75	(Ganguly and Chakraborty, 2019)
Eisenia fetida	PMS, citronella bagasse	45 days	8.51	6.45	2.41	3.0	(Boruah et al., 2019)
Eisenia fetida	PMS, cow dung, brown-rot fungi	28 days	PMS 7.59, cow dung 8.12	7.97–8.40	PMS 3.11, cow dung 3.36 (μS)	3.19–4.36 (μS)	(Negi and Suthar, 2018)
Eisenia fetida	PMS, cow dung	60 days	7.99–8.87	7.40–8.53	0.11–0.7 (μS)	0.95–2.35 (μS)	(Suthar et al., 2014)
Eisenia fetida	PMS, cow dung	56 days	7.58–8.08	8.44–8.66	2.49–2.78 (μS)	3.37–4.54 (μS)	(Negi and Suthar, 2013)
Eisenia fetida	PMS, cattle dung	150 days	6.56–7.21	7.26–8.30	1.15–3.13	1.73–2.2	(Kaur et al., 2010)
Eisenia andrei	PMS, cattle dung, dairy sludge	70 days	8.3 and 8.6	8.1 and 7.9	2.0 and 2.1	0.42 and 0.45	(Elvira et al., 1998)
Eisenia andrei	PMS, primary sewage sludge	40 days	8.1	8.8	–	–	(Elvira et al., 1996)
Eudrilus eugeniae	PMS, cow dung, saw dust, effective microorganisms	30 days	6.50–6.53	7.0–7.07	1.8–2.10 (μmhos/cm)	0.89–0.92 (μmhos/cm)	(Priya et al., 2012)
Eisenia fetida	PMS, tomato plant debris	6 months	7.8 and 8.6	8.9 and 9.3	4 and 5.3	2.2 and 2.6	(Fernández-Gómez et al., 2015)
Perionyx excavatus	PMS, cow dung	60 days	8.15–8.45	7.35–7.55	0.53–0.96	0.74–1.80	(Yuvaraj et al., 2018)

[a] *The range of values (minimum and maximum) indicates the PMS with the addition of a different proportion of different substrates.*

bagasse (Boruah et al., 2019). The release of organic acids is responsible for this reduction during the bioconversion process (Sharma and Garg, 2018; Yuvaraj et al., 2019). However, the initial acidic or neutral pH of the mixed substrate was increased in the final compost, where PMS mixed with the primary sewage sludge (Elvira et al., 1996), cow dung (Negi and Suthar, 2013), cow dung and straw (Ganguly and Chakraborty, 2019), cattle dung (Kaur et al., 2010), cow dung and effective microorganisms (Priya et al., 2012), and tomato plant debris (Fernández-Gómez et al., 2015) might be due to the breakdown of short-chain fatty acids and precipitation of mineral salts along with cation-complexed humic acids (Tognetti et al., 2007). The alkaline pH of the initial substrate was increased, and the alkaline pH was maintained when the PMS was supplemented with cow dung and brown-rot fungi (Negi and Suthar, 2018), possibly due to the formation and accumulation of NH_3 during the decomposition of the mixtures, which forms alkaline salt (NH_3^+) after dissolving with water (Wong and Fang, 2000).

16.3.2 Vermitreatment-induced changes in electrical conductivity

The electrical conductivity (EC) values of various vermitreatments are given in Table 16.1. The amount of total soluble salts (salinity) is measured by EC. Generally, EC is an indicator of soil health but it does not provide the measurement of specific ions. After vermitreatments, PMS supplemented with or not supplemented with cow dung could improve the EC; however, the improvement was less in PMS without cow dung when compared to PMS with cow dung (Negi and Suthar, 2013; Suthar et al., 2014). With either the addition of cow dung or not, PMS supplemented with green manure plants promotes the EC in the vermicomposting process (Yuvaraj et al., 2018; Karmegam et al., 2019; Boruah et al., 2019), possibly due to loss of organic matter and release of ammonium, potassium, and phosphate salts during the process (Negi and Suthar, 2018; Fang et al., 1999; Yadav and Garg, 2011). EC was significantly higher in PMS supplemented with both cow dung and brown-rot fungi than in PMS supplemented with cow dung or brown-rot fungi alone (Negi and Suthar, 2018). At the same time, the high value of EC recorded in the initial feed significantly decreased after vermitreatment, indicating the stability of vermicompost (Priya et al., 2012; Singh et al., 2010; Elvira et al., 1998; Ganguly and Chakraborty, 2019; Kaur et al., 2010; Fernández-Gómez et al., 2015). In all cases, it must be noted that the level of EC (> 4 Sm/cm) could be maintained within the safe limits of phytotoxicity by vermitreatment process (Li et al., 2012).

16.3.3 Vermitreatment-induced changes in total organic carbon

The total organic carbon (TOC) values of vermitreatments are given in Table 16.1. The measurement of TOC is a widely used method to determine the volume of humus and organic matter in composted materials. Low nitrogen and high TOC in organic residues are phytotoxic to seedlings due to the presence of short-chain organic acids in the residues (Kaur et al., 2010). In addition, they immobilize soil nitrogen, phosphorus, and other nutrients, and make them nonavailable to plants (Kaur et al., 2010). Therefore, the reduction of TOC in organic residues is an important aspect to increase the yield of agricultural plants, which is achieved by vermitreatment methods. The addition of cow dung, cattle manure, effective microorganisms, and green manure in PMS was positively supported by the reduction of TOC (Elvira et al., 1998; Yuvaraj et al., 2018; Karmegam et al., 2019; Elvira et al., 1996; Fernández-Gómez et al., 2015). In PMS, there is a huge difference between the values of TOC in the primary and secondary sludge. After mixing with cow dung (TOC, 260 gm/kg), the initial value of TOC in the PMS was 735 mg/kg for the primary sludge mixture and 390 mg/kg for the secondary sludge mixture (Ganguly and Chakraborty, 2019). Nearly, 50%–80% of the TOC was reduced after vermitreatment (Boruah et al., 2019; Ganguly and Chakraborty, 2019). The degradation rate was high when the PMS was supplemented with cow dung and brown-rot fungi, possibly because of a greater loss of carbon than without the addition of fungi (Negi and Suthar, 2018). The reason for reduction of TOC during vermitreatments is the utilization of organic carbon by microorganisms and earthworms and the release of carbon in the form of CO_2 by microbial respiration (Yadav and Garg, 2009; Sharma and Garg, 2018; Negi and Suthar, 2018). Furthermore, the addition of other organic substrates to the PMS might accelerate both microbial and earthworm activity, which might lead to loss of CO_2, thereby resulting in a greater decrease of TOC (Negi and Suthar, 2013; Suthar et al., 2014; Boruah et al., 2019; Kaur et al., 2010; Negi and Suthar, 2018).

16.3.4 Vermitreatment-induced changes in total Kjeldahl nitrogen

The total Kjeldahl nitrogen (TKN) values of vermitreatments are given in Table 16.1. Nitrogen is organically bound with waste mixtures, and during vermicomposting, the nitrogen content is transformed into a soluble form. Various studies have shown that TKN is significantly increased when PMS is mixed with various organic wastes, such as cow dung (Negi and Suthar, 2013; Suthar et al., 2014; Yuvaraj et al., 2018; Ganguly and Chakraborty, 2019); cow dung and green manure (Karmegam et al., 2019); citronella bagasse (Boruah et al., 2019); cow dung and brown-rot fungi (Negi and Suthar, 2018); cattle dung (Kaur et al., 2010); cow dung, cattle manure, and dairy sludge (Elvira et al., 1998); primary sewage sludge (Elvira et al., 1996); and cow dung, sawdust, and effective microorganisms (Priya et al., 2012). This significant increase was higher than that seen with the normal composting method (Karmegam et al., 2019; Deka et al., 2011). Also, the degree of increase varied between various studies, indicating that the TKN results is associated with the type of worms used and the nature of the substrate mixed with the PMS. The increase of TKN in vermicompost is due to the mineralization of organic nitrogen through ammonification and ammonium volatilization (Karnchanawong et al., 2017); immobilization of nitrogen in composted materials (Huang et al., 2004); nitrogen in raw materials and dead worms tissues (Suthar, 2009; Bhat et al., 2015; Zuberi et al., 2015); excretion of nitrogenous substances, hormones, and enzymes by alive worms (Hobson et al., 2005; Suthar, 2006); and the reduction of the dry mass of composted material due to organic matter mineralization and CO_2 emission (Negi and Suthar, 2018).

16.3.5 Vermitreatment-induced changes in the C/N ratio

The values of the C/N ratio in various studies are given in Table 16.2. The production of a better quality of vermicompost is indicated by the C/N ratio. A value < 20 is considered as the availability of plant growth promoters in composted materials, and a value < 15 indicates suitability for agronomic use (Edwards and Bohlen, 1996). The final value of the C/N ratio in composted materials is based on the increase in TKN and concurrent decrease in TOC during the composting process by microbial and earthworm activities (Kaur et al., 2010). Generally, the reduction of the C/N ratio is greater in vermicompost than in common compost methods (Boruah et al., 2019). Also, the reduction

TABLE 16.2 Changes in TOC, TKN, and the C/N ratio of various waste mixtures before and after vermitreatment.

			Total organic carbon (TOC) (mg/kg) (minimum to maximum)[a]		TKN (g/kg) (minimum to maximum)[a]		C/N ratio (minimum to maximum)[a]		
Earthworm species	Substrate used	Duration of experiment	Before treatment	After treatment	Before treatment	After treatment	Before treatment	After treatment	Reference
Eisenia fetida	PMS, cow dung, green manure	60 days	498.0–630.4	389.3–448.4	9.98–14.15	18.39–31.26	36.42–56.33	13.45–22.33	(Karmegam et al., 2019)
Perionyx excavatus[1]	PMS, cow dung, straw	60 days	390.23 and 735	280.6 and 160.36	4.26 and 5.26	23.28 and 23.36	132.9 and 91.60	12.06 and 6.58	(Ganguly and Chakraborty, 2019)
Eisenia fetida	PMS, citronella bagasse	45 days	140	96.9	9.67	15.2	> 50	< 10	(Boruah et al., 2019)
Eisenia fetida	PMS, cow dung, brown-rot fungi	28 days	PMS 615.17, cow dung 677.0	515.7–550.6	PMS 615, cow dung 1.79	5.88–10.1	PMS 257.8, cow dung 379.2	54.51–98.4	(Negi and Suthar, 2018)
Eisenia fetida	PMS, cow dung	60 days	550–698	467–595	5.59–6.53	13.38–14.59	85.3–132.2	31.0–44.6	(Suthar et al., 2014)
Eisenia fetida	PMS, cow dung	56 days	598.3–666.8	453.40–506.43	2.39–12.24	7.06–34.33	51.89–257.31	13.34–71.73	(Negi and Suthar, 2013)
Eisenia fetida	PMS, cattle dung	150 days	271–348	150–268	4.1–13.03	6.11–21.8	20.84–84.95	7.5–43.97	(Kaur et al., 2010)
Eisenia andrei	PMS, cattle dung, dairy sludge	70 days	229 and 266 (g/kg)	151 and 185 (g/kg)	7 and 11	12 and 14	23 and 41	13 and 16	(Elvira et al., 1998)
Eisenia andrei	PMS, primary sewage sludge	40 days	424 (g/kg)	247 (g/kg)	11	38	40	6.4	(Elvira et al., 1996)
Eudrilus eugeniae	PMS, cow dung, saw dust, effective microorganisms	30 days	22.31–32.54 (%)	10.12–13.94 (%)	0.62–0.68 (%)	1.38–1.86 (%)	33.29–51.08	5.44–9.29	(Priya et al., 2012)
Eisenia fetida, Eudrilus eugeniae, Perionyx excavatus	PMS, saw dust	45 days	–	–	–	–	> 35	< 20	(Sonowal et al., 2014)
Eisenia fetida	PMS, tomato plant debris	6 months	232 and 283	163 and 179	11.9 and 16.1	10.6 and 12.6	19.4 and 17.6	14.2 and 15.3	(Fernández-Gómez et al., 2015)
Perionyx excavatus	PMS, cow dung	60 days	27.0–29.0	13.9–18.1	0.22–0.42 (%)	0.36–0.58 (%)	64.75–129.1	23.25–48.25	(Yuvaraj et al., 2018)

[a] *The range of values (minimum and maximum) indicates the PMS with the addition of a different proportion of different substrates.*

of the C/N ratio depends on the nature of waste (Negi and Suthar, 2018) and is strongly influenced by the addition of other organic substrates with PMS (Boruah et al., 2019; Negi and Suthar, 2018; Sonowal et al., 2014). The addition of cow dung (Negi and Suthar, 2013); cow dung along with green manure (Karmegam et al., 2019); cow dung and straw (Ganguly and Chakraborty, 2019); citronella bagasse (Boruah et al., 2019); cattle dung (Kaur et al., 2010); cattle dung and dairy sludge (Elvira et al., 1998); primary sewage sludge (Elvira et al., 1996); cow dung, sawdust, and effective microorganisms (Priya et al., 2012); and tomato plant debris (Fernández-Gómez et al., 2015) to the PMS can reduce the C/N ratio to < 15 within the maximum experimental period of 60 days. However, Negi and Suthar (Negi and Suthar, 2018) reported that the reduction was slow, and the C/N ratio reached > 50 at the end of 28 days experimental period, possibly because of the high content of cellulose, hemicellulose, and lignin in the PMS, thereby delaying the microbial-mediated carbon mineralization. Similar findings have been reported by Suthar et al. and Yuvaraj et al. (Suthar et al., 2014; Yuvaraj et al., 2018). Here, allowing a substrate mix greater than 28 days might improve the C/N ratio.

16.3.6 Vermistabilization of metals in PMS

PMS contains various toxic metals, such as Cr, Cd, Ni, and Pb (Chandra et al., 2011). Vermistabilization has been proven to be an efficient method to decrease the level of metals in PMS (Negi and Suthar, 2013). The order of metal removal efficiency based on the maximum rate of removal is Pb > Cr > Cu > Cd (Yuvaraj et al., 2018), Pb > Cu > Cr > Cd (Suthar et al., 2014) and Pb > Cd > Ni > Cu (Priya et al., 2012) (Table 16.3). In vermitreatments, metal mobility is reduced in PMS due to the bioaccumulation of metals in the chloragogen tissues of earthworms (Suthar et al., 2014; Yuvaraj et al., 2018). The highest removal efficiency of Cd, Cu, Pb, and Cr was found in the mixture of PMS with cow dung in the ratio of 1:1 (Suthar et al., 2014; Yuvaraj et al., 2018), suggesting that the functional groups in the cow dung (e.g., carboxylic acid and hydroxyl groups) are binding the metals. Furthermore, in various studies, the addition of various organic substrates, such as cow dung and plant debris, to PMS showed that the pH of the substrates is alkaline because of the release of anions by the organic substrates (Table 16.1). The negative charge of these anions absorbs the metals and might provide an opportunity for these to accumulate in the earthworm tissues (Wang et al., 2013). Secondly, the removal is due to the formation of oganometal complexes by the organic acids that are produced from the combined action of earthworms and their microbial enzymes in the gut on organic materials in the gut (Swati and Hait, 2017; Sivakumar, 2015; Evangelou et al., 2004). The studies concluded that, after vermitreatments, the metal-contaminated PMS is safe for land application in agroecosystems.

TABLE 16.3 Changes in metal contents of initial feed mixtures and vermiculated end products.

Earthworm species	Substrate used	Duration of experiment	Metals tested	Treatment (mg/kg) (minimum to maximum)[a]		Removal efficiency (%) (minimum to maximum)[a]	References
				Before	After		
Perionyx excavatus	PMS, cow dung	60 days	Cd	9.20–10.60	6.35–8.60	13.6–27.8	(Yuvaraj et al., 2018)
			Cu	107.25–128.65	61.85–97.55	24.2–42.3	
			Pb	61.70–76.90	30.0–53.65	27.1–56.3	
			Cr	101.95–122.3	57.85–95.05	19.4–46.2	
Eisenia fetida	PMS, cow dung	60 days	Cd	8.64–8.77	7.89–7.91	8.7–10	(Suthar et al., 2014)
			Cu	114.5–144.4	16.79–36.06	68.8–88.4	
			Pb	42.3–62.5	1.41–2.0	95.3–97.5	
			Cr	144.2–187.3	23.2–43.6	23.2–43.6	
Eudrilus eugeniae	PMS, cow dung, effective microorganisms	30 days	Cd	0.41–0.62	0.35–0.50	14.6–29.3	(Priya et al., 2012)
			Cu	32.2–38.7	29.1–33.4	9.6–19.2	
			Pb	13.2–16.8	9.1–15.8	4.2–31.1	
			Ni	21.5–26.8	17.5–22.4	16.4–20.2	

[a] *The range of values (minimum and maximum) indicates the PMS with the addition of a different proportion of different substrates.*

16.4 Conclusion

The PMS from pulp and paper mills should not be discarded as waste. Instead, they can be recycled into value-added end products. Also, the suitability of the agronomic use of PMS has been tested by various studies after vermicomposting of PMS as part of an organic waste mixture. From the above review, it is clear that PMS has recycling value and can be part of a sustainable waste management system.

References

Bajpai, P., Bajpai, P., 2015. Generation of waste in pulp and paper mills. In: Manag. Pulp Pap. Mill Waste, pp. 9–17, https://doi.org/10.1007/978-3-319-11788-1_2.

Banu, R.J., Yeom, I.T., Esakkiraj, Kumar, N., Lee, Y.W., Vallinayagam, S., 2008. Biomanagement of sago-sludge using an earthworm, *Lampito mauritii*. J. Environ. Biol. 29, 753–757.

Baudez, J.C., Markis, F., Eshtiaghi, N., Slatter, P., 2011. The rheological behaviour of anaerobic digested sludge. Water Res. 45, 5675–5680. https://doi.org/10.1016/j.watres.2011.08.035.

Bhat, S.A., Singh, J., Vig, A.P., 2015. Potential utilization of bagasse as feed material for earthworm *Eisenia fetida* and production of vermicompost. Springerplus 4. https://doi.org/10.1186/s40064-014-0780-y.

Boruah, T., Barman, A., Kalita, P., Lahkar, J., Deka, H., 2019. Vermicomposting of citronella bagasse and paper mill sludge mixture employing *Eisenia fetida*. Bioresour. Technol. 294. https://doi.org/10.1016/j.biortech.2019.122147.

Boshoff, S., Gottumukkala, L.D., van Rensburg, E., Görgens, J., 2016. Paper sludge (PS) to bioethanol: evaluation of virgin and recycle mill sludge for low enzyme, high-solids fermentation. Bioresour. Technol. 203, 103–111. https://doi.org/10.1016/j.biortech.2015.12.028.

Bustamante, M.A., Alburquerque, J.A., Restrepo, A.P., de la Fuente, C., Paredes, C., Moral, R., Bernal, M.P., 2012. Co-composting of the solid fraction of anaerobic digestates, to obtain added-value materials for use in agriculture. Biomass Bioenergy 43, 26–35. https://doi.org/10.1016/j.biombioe.2012.04.010.

Cardoso-Vigueros, L., Ramirez-Camperos, E., 2006. Tannery wastes and sewage sludge biodegradation by composting and vermicomposting process. Ing. Hidraul. En Mex. 21, 93–103.

Chandra, R., Abhishek, A., Sankhwar, M., 2011. Bacterial decolorization and detoxification of black liquor from rayon grade pulp manufacturing paper industry and detection of their metabolic products. Bioresour. Technol. 102, 6429–6436. https://doi.org/10.1016/j.biortech.2011.03.048.

Chen, H., Venditti, R., Gonzalez, R., Phillips, R., Jameel, H., Park, S., 2014. Economic evaluation of the conversion of industrial paper sludge to ethanol. Energy Econ. 44, 281–290. https://doi.org/10.1016/j.eneco.2014.04.018.

Cho, D.W., Kwon, E.E., Kwon, G., Zhang, S., Lee, S.R., Song, H., 2017. Co-pyrolysis of paper mill sludge and spend coffee ground using CO_2 as reaction medium. J. CO_2 Util. https://doi.org/10.1016/j.jcou.2017.09.003.

Deka, H., Deka, S., Baruah, C.K., Das, J., Hoque, S., Sarma, H., Sarma, N.S., 2011. Vermicomposting potentiality of *Perionyx excavatus* for recycling of waste biomass of java citronella—an aromatic oil yielding plant. Bioresour. Technol. 102, 11212–11217. https://doi.org/10.1016/j.biortech.2011.09.102.

Devi, P., Saroha, A.K., 2013. Effect of temperature on biochar properties during paper mill sludge pyrolysis. Int. J. ChemTech Res. 5, 682–687.

Dheepa, D., Subash, A., Parvatham, R., 2006. Selected physicochemical and microbiological analysis of vermicomposted lime and chrome sludge wastes from tannery industry. J. Ind. Pollut. Control. 22, 341–344.

Edwards, C.A., Bohlen, P.J., 1996. Biology and Ecology of Earthworms, third ed.

Elvira, C., Goicoechea, M., Sampedro, L., Mato, S., Nogales, R., 1996. Bioconversion of solid paper-pulp mill sludge by earthworms. Bioresour. Technol. 57, 173–177. https://doi.org/10.1016/0960-8524(96)00065-X.

Elvira, C., Sampedro, L., Benítez, E., Nogales, R., 1998. Vermicomposting of sludges from paper mill and dairy industries with *Eisena andrei*: a pilot-scale study. Bioresour. Technol. 63, 205–211. https://doi.org/10.1016/S0960-8524(97)00145-4.

Evangelou, M.W.H., Daghan, H., Schaeffer, A., 2004. The influence of humic acids on the phytoextraction of cadmium from soil. Chemosphere. https://doi.org/10.1016/j.chemosphere.2004.06.017.

Fang, M., Wong, J.W.C., Ma, K.K., Wong, M.H., 1999. Co-composting of sewage sludge and coal fly ash: nutrient transformations. Bioresour. Technol. 67, 19–24. https://doi.org/10.1016/S0960-8524(99)00095-4.

Fernández-Gómez, M.J., Nogales, R., Plante, A., Plaza, C., Fernández, J.M., 2015. Application of a set of complementary techniques to understand how varying the proportion of two wastes affects humic acids produced by vermicomposting. Waste Manag. https://doi.org/10.1016/j.wasman.2014.09.022.

Ganguly, R.K., Chakraborty, S.K., 2019. Assessment of qualitative enrichment of organic paper mill wastes through vermicomposting: humification factor and time of maturity. Heliyon 5. https://doi.org/10.1016/j.heliyon.2019.e01638.

Garg, V.K., Kaushik, P., Dilbaghi, N., 2006. Vermiconversion of wastewater sludge from textile mill mixed with anaerobically digested biogas plant slurry employing *Eisenia foetida*. Ecotoxicol. Environ. Saf. 65, 412–419. https://doi.org/10.1016/j.ecoenv.2005.03.002.

Geng, X., Zhang, S.Y., Deng, J., 2007. Characteristics of paper mill sludge and its utilization for the manufacture of medium density fiberboard. Wood Fiber Sci. 39, 345–351.

Hackett, G.A.R., Easton, C.A., Duff, S.J.B., 1999. Composting of pulp and paper mill fly ash with wastewater treatment sludge. Bioresour. Technol. 70, 217–224. https://doi.org/10.1016/S0960-8524(99)00048-6.

Hazarika, J., Ghosh, U., Kalamdhad, A.S., Khwairakpam, M., Singh, J., 2017. Transformation of elemental toxic metals into immobile fractions in paper mill sludge through rotary drum composting. Ecol. Eng. 101, 185–192. https://doi.org/10.1016/j.ecoleng.2017.02.005.

Hobson, A.M., Frederickson, J., Dise, N.B., 2005. CH_4 and N_2O from mechanically turned windrow and vermicomposting systems following in-vessel pre-treatment. Waste Manag., 345–352. https://doi.org/10.1016/j.wasman.2005.02.015.

Huang, G.F., Wong, J.W.C., Wu, Q.T., Nagar, B.B., 2004. Effect of C/N on composting of pig manure with sawdust. Waste Manag. 24, 805–813. https://doi.org/10.1016/j.wasman.2004.03.011.

Jiang, J., Ma, X., 2011. Experimental research of microwave pyrolysis about paper mill sludge. Appl. Therm. Eng., 3897–3903. https://doi.org/10.1016/j.applthermaleng.2011.07.037.

Karmegam, N., Vijayan, P., Prakash, M., John Paul, J.A., 2019. Vermicomposting of paper industry sludge with cowdung and green manure plants using *Eisenia fetida*: a viable option for cleaner and enriched vermicompost production. J. Clean. Prod. 228, 718–728. https://doi.org/10.1016/j.jclepro.2019.04.313.

Karnchanawong, S., Mongkontep, T., Praphunsri, K., 2017. Effect of green waste pretreatment by sodium hydroxide and biomass fly ash on composting process. J. Clean. Prod. 146, 14–19. https://doi.org/10.1016/j.jclepro.2016.07.126.

Kaur, A., Singh, J., Vig, A.P., Dhaliwal, S.S., Rup, P.J., 2010. Cocomposting with and without *Eisenia fetida* for conversion of toxic paper mill sludge to a soil conditioner. Bioresour. Technol. 101, 8192–8198. https://doi.org/10.1016/j.biortech.2010.05.041.

Khalili, N.R., Vyas, J.D., Weangkaew, W., Westfall, S.J., Parulekar, S.J., Sherwood, R., 2002. Synthesis and characterization of activated carbon and bioactive adsorbent produced from paper mill sludge. Sep. Purif. Technol. 26, 295–304. https://doi.org/10.1016/S1383-5866(01)00184-8.

Krigstin, S., Sain, M., 2006. Characterization and potential utilization of recycled paper mill sludge. Pulp Pap. Canada 107, 29–32.

Kunzler, C., 2001. Pulp and paper industry's diverse organics stream. Biocycle 42, 30–33.

Li, R., Wang, J.J., Zhang, Z., Shen, F., Zhang, G., Qin, R., Li, X., Xiao, R., 2012. Nutrient transformations during composting of pig manure with bentonite. Bioresour. Technol. 121, 362–368. https://doi.org/10.1016/j.biortech.2012.06.065.

Lindorfer, H., Demmig, C., 2016. Foam formation in biogas plants—a survey on causes and control strategies. Chem. Eng. Technol. 39, 620–626. https://doi.org/10.1002/ceat.201500297.

Meyer, T., Edwards, E.A., 2014. Anaerobic digestion of pulp and paper mill wastewater and sludge. Water Res. 65, 321–349. https://doi.org/10.1016/j.watres.2014.07.022.

Monte, M.C., Fuente, E., Blanco, A., Negro, C., 2009. Waste management from pulp and paper production in the European Union. Waste Manag. 29, 293–308. https://doi.org/10.1016/j.wasman.2008.02.002.

Negi, R., Suthar, S., 2013. Vermistabilization of paper mill wastewater sludge using *Eisenia fetida*. Bioresour. Technol. 128, 193–198. https://doi.org/10.1016/j.biortech.2012.10.022.

Negi, R., Suthar, S., 2018. Degradation of paper mill wastewater sludge and cow dung by brown-rot fungi *Oligoporus placenta* and earthworm (*Eisenia fetida*) during vermicomposting. J. Clean. Prod. 201, 842–852. https://doi.org/10.1016/j.jclepro.2018.08.068.

Ochoa de Alda, J.A.G., 2008. Feasibility of recycling pulp and paper mill sludge in the paper and board industries. Resour. Conserv. Recycl. 52, 965–972. https://doi.org/10.1016/j.resconrec.2008.02.005.

Örmeci, B., 2007. Optimization of a full-scale dewatering operation based on the rheological characteristics of wastewater sludge. Water Res. 41, 1243–1252. https://doi.org/10.1016/j.watres.2006.12.043.

Priya, T.L., Uma, R.N., Sivakumar, S., Meenambalm, T., Son, H.-K., 2012. Bioremediation of pulp and paper industry secondary sludge spiked with cow dung and effective microorganisms using epigeic earthworm *Eudrilus eugeniae* (Kinberg). In: Karmegam, N. (Ed.), Glob. Sci. Books. Vermitechn, Global Science Books, pp. 78–82.

Puhakka, J.A., Alavakeri, M., Shieh, W.K., 1992. Anaerobic treatment of kraft pulp-mill waste activated-sludge: gas production and solids reduction. Bioresour. Technol. 39, 61–68. https://doi.org/10.1016/0960-8524(92)90057-5.

Reckamp, J.M., Garrido, R.A., Satrio, J.A., 2014. Selective pyrolysis of paper mill sludge by using pretreatment processes to enhance the quality of bio-oil and biochar products. Biomass Bioenergy 71, 235–244. https://doi.org/10.1016/j.biombioe.2014.10.003.

Sahariah, B., Sinha, I., Sharma, P., Goswami, L., Bhattacharyya, P., Gogoi, N., Bhattacharya, S.S., 2014. Efficacy of bioconversion of paper mill bamboo sludge and lime waste by composting and vermiconversion technologies. Chemosphere 109, 77–83. https://doi.org/10.1016/j.chemosphere.2014.02.063.

Sangwan, P., Kaushik, C.P., Garg, V.K., 2008. Vermiconversion of industrial sludge for recycling the nutrients. Bioresour. Technol. 99, 8699–8704. https://doi.org/10.1016/j.biortech.2008.04.022.

Sharma, K., Garg, V.K., 2018. Vermicomposting: A Green Technology for Organic Waste Management. pp. 199–235, https://doi.org/10.1007/978-981-10-7431-8_10.

Sharma, R., Sharma, D., Rao, K.S., Jain, R.C., 2002. Experimental studies on waste paper pulp biodegradation. Indian J. Environ. Health 44, 181–188.

Sim, E.Y.S., Wu, T.Y., 2010. The potential reuse of biodegradable municipal solid wastes (MSW) as feedstocks in vermicomposting. J. Sci. Food Agric. 90, 2153–2162. https://doi.org/10.1002/jsfa.4127.

Simão, L., Hotza, D., Raupp-Pereira, F., Labrincha, J.A., Montedo, O.R.K., 2018. Wastes from pulp and paper mills—a review of generation and recycling alternatives. Ceramica 64, 443–453. https://doi.org/10.1590/0366-69132018643712414.

Singh, J., Kaur, A., Vig, A.P., Rup, P.J., 2010. Role of *Eisenia fetida* in rapid recycling of nutrients from bio sludge of beverage industry. Ecotoxicol. Environ. Saf. 73, 430–435. https://doi.org/10.1016/j.ecoenv.2009.08.019.

Sivakumar, S., 2015. Effects of metals on earthworm life cycles: a review. Environ. Monit. Assess. 187. https://doi.org/10.1007/s10661-015-4742-9.

Sivakumar, S., Kasthuri, H., Prabha, D., Senthilkumar, P., Subbhuraam, C.V., Song, Y.C., 2009. Efficiency of composting parthenium plant and neem leaves in the presence and absence of an oligochaete, *Eisenia fetida*. Iran. J. Environ. Heal. Sci. Eng. 6, 201–208.

Sivakumar, S., Nityanandi, D., Barathi, S., Prabha, D., Rajeshwari, S., Son, H.K., Subbhuraam, C.V., 2012. Selected enzyme activities of urban heavy metal-polluted soils in the presence and absence of an oligochaete, *Lampito mauritii* (Kinberg). J. Hazard. Mater. 227–228, 179–184. https://doi.org/10.1016/j.jhazmat.2012.05.030.

Sonowal, P., Khwairakpam, M., Kalamdhad, A.S., 2014. Stability analysis of dewatered sludge of pulp and paper mill during vermicomposting. Waste Biomass Valorizat. 5, 19–26. https://doi.org/10.1007/s12649-013-9225-z.

Suthar, S., 2006. Potential utilization of guar gum industrial waste in vermicompost production. Bioresour. Technol. 97, 2474–2477. https://doi.org/10.1016/j.biortech.2005.10.018.

Suthar, S., 2007. Vermicomposting potential of *Perionyx sansibaricus* (Perrier) in different waste materials. Bioresour. Technol. 98, 1231–1237. https://doi.org/10.1016/j.biortech.2006.05.008.

Suthar, S., 2009. Vermicomposting of vegetable-market solid waste using *Eisenia fetida*: impact of bulking material on earthworm growth and decomposition rate. Ecol. Eng. 35, 914–920. https://doi.org/10.1016/j.ecoleng.2008.12.019.

Suthar, S., 2010. Recycling of agro-industrial sludge through vermitechnology. Ecol. Eng. 36, 1028–1036. https://doi.org/10.1016/j.ecoleng.2010.04.015.

Suthar, S., Singh, S., 2008. Feasibility of vermicomposting in biostabilization of sludge from a distillery industry. Sci. Total Environ. 394, 237–243. https://doi.org/10.1016/j.scitotenv.2008.02.005.

Suthar, S., Sajwan, P., Kumar, K., 2014. Vermiremediation of heavy metals in wastewater sludge from paper and pulp industry using earthworm *Eisenia fetida*. Ecotoxicol. Environ. Saf. 109, 177–184. https://doi.org/10.1016/j.ecoenv.2014.07.030.

Swati, A., Hait, S., 2017. Fate and bioavailability of heavy metals during vermicomposting of various organic wastes—a review. Process Saf. Environ. Prot. 109, 30–45. https://doi.org/10.1016/j.psep.2017.03.031.

Świerk, K., Bielicka, A., Bojanowska, I., Maćkiewicz, Z., 2007. Investigation of heavy metals leaching from industrial wastewater sludge. Polish J. Environ. Stud. 16, 447–451.

Tognetti, C., Mazzarino, M.J., Laos, F., 2007. Cocomposting biosolids and municipal organic waste: effects of process management on stabilization and quality. Biol. Fertil. Soils. 43, 387–397. https://doi.org/10.1007/s00374-006-0164-8.

Wang, L., Zhang, Y., Lian, J., Chao, J., Gao, Y., Yang, F., Zhang, L., 2013. Impact of fly ash and phosphatic rock on metal stabilization and bioavailability during sewage sludge vermicomposting. Bioresour. Technol. 136, 281–287. https://doi.org/10.1016/j.biortech.2013.03.039.

Wirojanagud, W., Tantemsapya, N., Tantriratna, P., 2004. Precipitation of heavy metals by lime mud waste of pulp and paper mill. Songklanakarin J. Sci. Technol. 26, 45–53. http://www.doaj.org/doaj?func=openurl&issn=01253395&date=2004&volume=26&issue=Suppl.&spage=45&genre=article.

Wong, J.W.C., Fang, M., 2000. Effects of lime addition on sewage sludge composting process. Water Res. 34, 3691–3698. https://doi.org/10.1016/S0043-1354(00)00116-0.

Yadav, A., Garg, V.K., 2009. Feasibility of nutrient recovery from industrial sludge by vermicomposting technology. J. Hazard. Mater. https://doi.org/10.1016/j.jhazmat.2009.02.035.

Yadav, A., Garg, V.K., 2011. Vermicomposting—an effective tool for the management of invasive weed *Parthenium hysterophorus*. Bioresour. Technol. 102, 5891–5895. https://doi.org/10.1016/j.biortech.2011.02.062.

Yuvaraj, A., Karmegam, N., Thangaraj, R., 2018. Vermistabilization of paper mill sludge by an epigeic earthworm *Perionyx excavatus*: mitigation strategies for sustainable environmental management. Ecol. Eng. 120, 187–197. https://doi.org/10.1016/j.ecoleng.2018.06.008.

Yuvaraj, A., Thangaraj, R., Maheswaran, R., 2019. Decomposition of poultry litter through vermicomposting using earthworm *Drawida sulcata* and its effect on plant growth. Int. J. Environ. Sci. Technol. 16, 7241–7254. https://doi.org/10.1007/s13762-018-2083-2.

Zuberi, M.J.S., Torkmahalleh, M.A., Ali, S.M.H., 2015. A comparative study of biomass resources utilization for power generation and transportation in Pakistan. Int. J. Hydrogen Energy 40, 11154–11160. https://doi.org/10.1016/j.ijhydene.2015.05.166.

CHAPTER

17

Potential of solid waste prevention and minimization strategies

Anbarashan Padmavathy[a] *and Munisamy Anbarashan*[b]

[a]Department of Ecology and Environmental Sciences, Pondicherry University, Puducherry, India [b]Department of Ecology, French Institute of Pondicherry, Puducherry, India

OUTLINE

17.1 Introduction

Solid waste management is a local problem with universal consequences and it is a growing concern globally, especially in developing countries like India, Africa etc., where only 30% of generated waste is poised and disposed properly (Ziraba et al., 2016). The constant increase in world's population tends to increase the rate of waste generation. In 2015, worldwide solid waste generation was 2 billion metric tons and is expected to increase up to 3.4 billion metric tons by 2050, whereas in developing and under developed countries, the amount of waste generation is expected to be tripled by 2050 (Kaza et al., 2018). It is a necessity to have an effective solid waste management system in every place. Inadequate solid waste management plants might lead to severe damage and serious risks to environment, human health and their livelihoods.

Hasty population growth amplified the degree of urbanization and industrialization ensuing in more waste generation. About 30% of the world's population lived in urban areas in 1950 and it is assumed to increase to 66% by 2050. In 2002, worldwide 2.9 billion urban inhabitants produced about 0.64 kg of waste per person per day; in 2012,

Cost Effective Technologies for Solid Waste and Wastewater Treatment
https://doi.org/10.1016/B978-0-12-822933-0.00019-X

the rate increased to 1.2 kg of waste per day, and by 2025, it is estimated to be 1.42 kg of waste per day per person for about 4.3 billion of world's population who are living in urban areas (World Bank Group, 2021). Waste management or waste disposal includes all the activities and actions involved from waste inception to its final disposal, thus it includes the collection, transport, treatment, and disposal of waste organized with intensive care and directive of the waste management process. Earlier, waste management intended to minimize or avoid adverse effects of waste on human health, the environment, or aesthetics (Chartsbin, 2013). These management practices differ and vary from countries to countries (developed and developing nations), regions to regions (urban and rural areas), and from sector to sector (industry, agriculture). Proper and effective waste management is important for reliable society but it remains a hurdle for most of the developing countries as it is quite expensive by 20%–50% of municipal budgets.

Solid waste generation and urbanization are closely related and therefore, it is significant to have a better management plan. However, local governments face many challenges regarding the solid waste management and issue related to environmental and public health. However, while several causal linkages between exposure to waste and health outcomes for particular types of waste are well established, others remain unclear or not prioritized as public health issues. It became necessary for us to adopt a new strategy called waste minimization or prevention of waste.

Waste minimization is a set of practices and methods proposed to decrease the amount of waste produced. Either by eliminating or reducing the generation of destructive and tenacious wastes, waste minimization ensures a more sustainable society (Song et al., 2015). This process includes restructuring of processes and products and altering general forms of consumption and production. The most economically efficient and environmentally resourceful way to manage the waste is to address the problem in first phase itself, and by doing so, we can develop a very standard significant management strategy and this process requires substantial amount of resources and time. Generally, the waste management system involves recycling, re-use, and conversion of waste to energy but the concept of waste minimization primarily focuses on preventing the waste creation during production itself. In order to achieve this, the management and the manger must have adequate cradle-to-grave knowledge in production and composition of the waste (Hoornweg and Bhada-Tata, 2012).

Currently, every country in the world is in search for innovative tactics that could address ever-increasing forms of anthropogenic wastes. Solid waste management (SWM) is an essential service and a mandatory function of multiple authorities throughout the world, yet it is still being poorly managed in most countries giving rise to environmental degradation and serious health problems especially for women and children (Giusti, 2009; Krystosik et al., 2020). This clearly underlines the need for preparing a more strategic and detailed SWM plan that can be adopted by everyone. This chapter discusses various potentials of waste minimization and their management strategies. The quantum of waste generated varies mainly due to different lifestyles, which is directly proportional to socio economic status of the urban population.

17.2 Solid waste and solid waste management (SWM)

Solid wastes are useless or unwanted solid materials generated from residential, commercial, or industrial areas due to human activities. These wastes are classified by their source, properties, and risks to the environment and human health, thus it requires proper transport and management practices. As per the World Bank, wastes are classified into eight major types as mentioned in Table 17.1.

All this waste generation from different sectors indicates that the society receives, takes raw material and energy from the environment as inputs, and gives solid waste as an output, this ultimately causes an input-output imbalance in the environment which leads to degradation; thus, it became mandatory to adopt a certain waste management strategy to secure the environmental health (CPHEEO, 2016).

The major concern associated with solid waste management was to reduce or eliminate the hostile impact on the environment and human health. A number of processes are involved in traditional waste management like monitoring, collection, transport, processing, recycling, and disposal methods. These traditional practices are not apt to the ever-growing population, as waste generation is also growing with such practices (Nwachukwu et al., 2017). Hence, it is necessary to change from the previous management practices to new methods and adopt strategies for waste reduction or prevention.

17.2.1 Impacts of improper SWM on environmental and social parameters

Quality of soil, water, and air

1. Contamination of soil and water surface/groundwater.
2. High levels of toxic fumes and particulate matter from burning waste.

TABLE 17.1 Types of solid wastes.

Source	Waste generators	Types of generated wastes
Residential	Single or multifamily residences	Food wastes, cardboard, paper, textiles, yard wastes, metals, wood, leather, plastics, glass, ashes, special wastes, i.e., consumer electronics, batteries, oil, tires) and hazardous household wastes
Industrial	Light and heavy manufacturing, fabrication, construction sites, power and chemical plants	Housekeeping wastes, packaging, food wastes, construction and, demolition materials, hazardous wastes, ashes, special wastes
Commercial	Stores, hotels, restaurants, markets, office buildings, etc.	Paper, cardboard, plastics, wood, food wastes, glass, metals, special wastes, hazardous wastes
Institutional	Schools, hospitals, prisons, government centers	Same as commercial
Municipal services	Street cleaning, landscaping, parks, other recreational areas, water and wastewater treatment plants	Street sweepings; landscape and tree trimmings; general wastes from parks, beaches, and other recreational areas; sludge
Process	Heavy and light manufacturing, refineries, chemical plants, power plants, mineral extraction and processing	Industrial process wastes, scrap materials, off-specification products, slag, tailings
Construction and demolition	New construction sites, road repair, renovation sites, demolition of buildings	Wood, steel, concrete, dirt, etc.
Agriculture	Crops, orchards, vineyards, dairies, feedlots, farms	Spoiled food wastes, agricultural wastes, hazardous wastes (e.g., pesticides)

Health Safety

1. Risks of water-borne and communicable diseases
2. Potential breeding site for vectors and pests might cause dengue, malaria etc.
3. Chances to contact with bio-hazardous/eco-toxic wastes from clinics and industries
4. Cause asphyxiation or explosions from methane gas build-up
5. Collapsing of deposited wastes due to their instability (high and steep slopes)

Socio-Economic and Aesthetic values

1. Unpleasant scenery, visual blight, and bad odors from uncovered waste affect aesthetic values
2. Improper waste disposal reduces water/flood storage capacity of wetlands
3. Improper wase disposal and deposition causes blocking in the drainage ways

Ecology

1. Loss of terrestrial/wetland habitats along with their flora and fauna
2. Ecological balance in the area might change.

17.2.2 Objective of solid waste management

The major objective of solid waste management was to recover the materials source and energy from solid waste, and by doing this, the quantity of solid waste disposed off on land is reduced. This in fact requires less raw material and energy as inputs in technological processes (Bartl, 2014). Such management strategy techniques have to be adopted in each and every solid waste generating activity to achieve prevention or minimization of solid waste from the society.

17.3 Hierarchy in waste management

The most long-recognized waste management hierarchy (Fig. 17.1) consists of the order of preference of prevention, minimization, recycling and reuse, biological treatment, incineration, and landfill disposal.

17.3.1 Principles of solid waste management

Solid waste management involves the application and principle of Integrated Solid Waste Management (ISWM). The application of suitable strategy, techniques, and technologies to achieve waste reduction and to have a maximum resource efficiency and resource conservation is followed in ISWM (Hasmori et al., 2020).

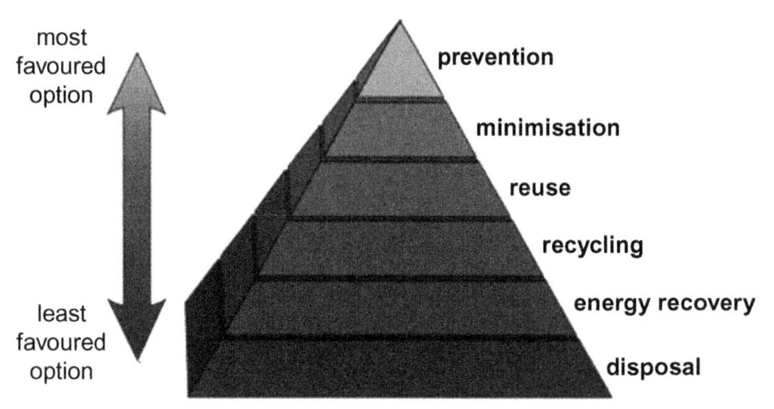

FIG. 17.1 Hierarchy in waste management. *Source: Wikimedia commons.*

17.4 Integrated solid waste management (ISWM)

The operations of this waste management occur according to their economic, energy, and environmental impacts in each step as, i.e., Tier 1: Source reduction or waste prevention, which includes reuse, is considered the best approach; Tier 2: recycling; Tier 3: recovery of material and composting of organic waste; Tier 4: Energy recovery-components of waste that cannot be prevented or recycled are processed, and finally Tier 5: disposal in sanitary landfill, least preferred option (Agarwal et al., 2015; CPHEEO, 2016). Based on local conditions, available technology, and appropriate system, ISWM hierarchy will be practised (Fig. 17.2).

17.4.1 Waste prevention

The ideal waste management alternative is to prevent waste generation in the first place. Hence, waste prevention at various stages like including in the design, production, packaging, and use is the basic and most preferred option for waste management in the ISWM. Waste prevention helps in reducing both environmental impacts (leachate, air emissions, and GHG emission) and cost expenses (reduces the handling, treatment, and disposal costs). Prevention or minimization of waste generation at source and products reuse are the most preferred potential waste management strategy (Frost, 2012).

Numerous technologies can be employed throughout the processes starting from manufacturing to post- product life cycles to prevent or reduce pollution and health problems (Antwi et al., 2015). Some strategies even include manufacturing methods that incorporate less hazardous or harmful materials, new chemical neutralization techniques to

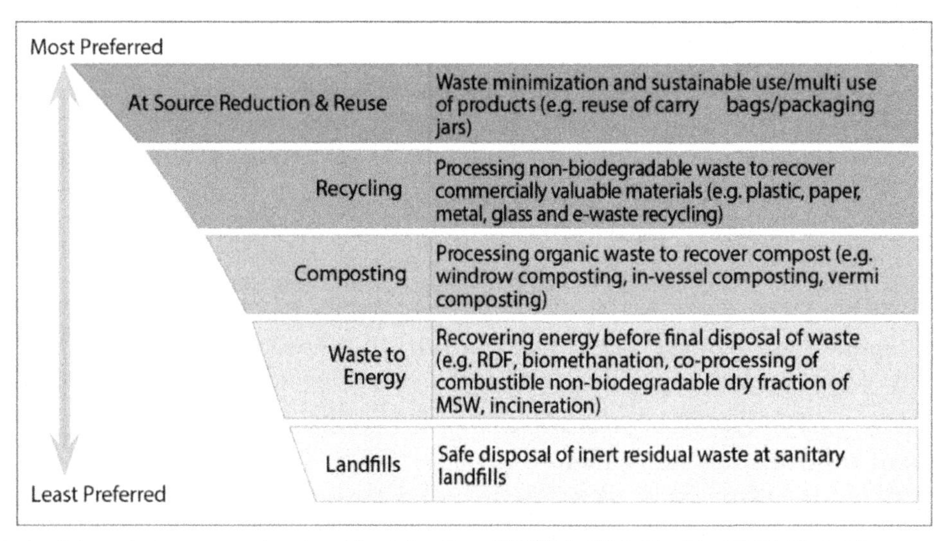

FIG. 17.2 Integrated solid waste management system hierarchy. *From CPHEEO, 2016. Municipal Solid Waste Management Manual Part II: The Manual Swachh Bharat Mission, Government of India, Agement Plan, http://cpheeo.gov.in/upload/uploadfiles/files/Part2.pdf, Accessed February 11, 2021.*

reduce reactivity, or water conserving technologies that reduce fresh water inputs and innovative leakage detection systems for material storage.

The ISWM hierarchy ranks strategies of waste management according to their environmental benefits.

17.5 Waste minimization and its strategy

In many cases, complete wastes elimination is not possible, so numerous waste reduction or minimization strategies can be implemented. Waste minimization refers to the collective of design and fabrication of products that minimize or reduce the generated waste and its toxicity (Zacho and Mosgaard, 2016). Often, these efforts come by identifying specific products or trend that may cause problems and the subsequent steps will be taken to halt these problems. In case of industrial sector, the waste generation can be cut-off by reducing or reusing the materials, using less hazardous substitute materials, or modifying design and processing. Source reduction leads to many benefits, including reduced use of natural resources and wastes toxicity (Kaza et al., 2018).

17.6 Resource and energy recovery

17.6.1 Waste recycling or reuse

The next preferred option in the ISWM hierarchy is recycling of waste to recover material resources through segregation, collection, and re-processing to create new products. In the waste management, composting of organic material is considered as a recovery process and composting of inorganic waste is considered as a recycling process (Robinson, 1986).

Recycling means recovery of useful materials like paper, plastics, glass, wood, and metals, which are recovered from the waste stream and they are incorporated into the fabrication of new products. By doing this, the required use of raw materials for new identical applications will be reduced. Recycling reduces exploitation of natural resources for raw materials, but it also allows waste materials to be recovered and utilized as valuable resource materials (Gibbs and Soell, 2010). Recycling of wastes directly conserves natural resources, reduces energy consumption and emissions generated by extraction of virgin materials and their subsequent manufacture into finished products, reduces overall energy consumption and greenhouse gas emissions that contribute to the global climate change, and reduces the incineration or landfilling of the materials that have been recycled (Nwachukwu et al., 2017). Moreover, recycling creates several economic benefits, including the potential to create job in markets and drive growth.

Common recycled materials include paper, plastics, glass, aluminum, steel, and wood. Additionally, many construction materials can be reused, including concrete, asphalt materials, masonry, and reinforcing steel. "Green" plant-based wastes are often recovered and immediately reused for mulch or fertilizer applications (CCAC and ISWA, 2016). Many industries also recover various by-products and/or refine and "re-generate" solvents for reuse. Examples include copper and nickel recovery from metal finishing processes; the recovery of oils, fats, and plasticizers by solvent extraction from filter media such as activated carbon and clays; and acid recovery by spray roasting, ion exchange, or crystallization. Furthermore, a range of used food-based oils are being recovered and utilized in "biodiesel" applications (Malinauskaite et al., 2017).

17.6.2 Waste to energy (WTE)

When recycling or material recovery is not possible, energy recovery from waste through production of heat, electricity, or fuel is preferred. Production of refuse-derived fuel (RDF), bio methanation, waste incineration, co-processing of combustible nonbiodegradable pyrolysis, or gasification are some waste-to-energy technologies that can be adopted.

Turning waste into energy is an action aiming at sustainable consumption and production patterns under the commitments of Agenda for Sustainable Development, 2030. By adopting this strategy, we can increase the value of products, materials, and resources and minimize resource use and waste generated. This method also enhances and increases energy efficiency, renewable energy, reduces the dependence on virgin resources, and by doing so, it provides economic opportunities by long-term competitiveness, thus it supports the principles of circular economy (Yagasa and Gamaralalage, 2019).

The waste-to-energy sector is currently developing dynamically worldwide. In developed countries it is well practised and in other countries it is at emerging state. Most of the WTE plants are built in medium and large cities, since the majority of the population lives in cities (Tong et al., 2021). Additionally, waste generation is lower in rural areas; the residents buy fewer goods and have higher levels of reuse and recycling. Moreover, the location of waste incineration plants in the big cities gives better capabilities for energy utilization, because the heat can be distributed more easily in the city (Fig. 17.3).

17.6.3 Biological treatment

Biodegradation using aerobic composting, anaerobic digestion, or mechanical biological treatment (MBT) methods is a resource recovery method for organic wastes. When organic fraction separated from inorganic material, aerobic

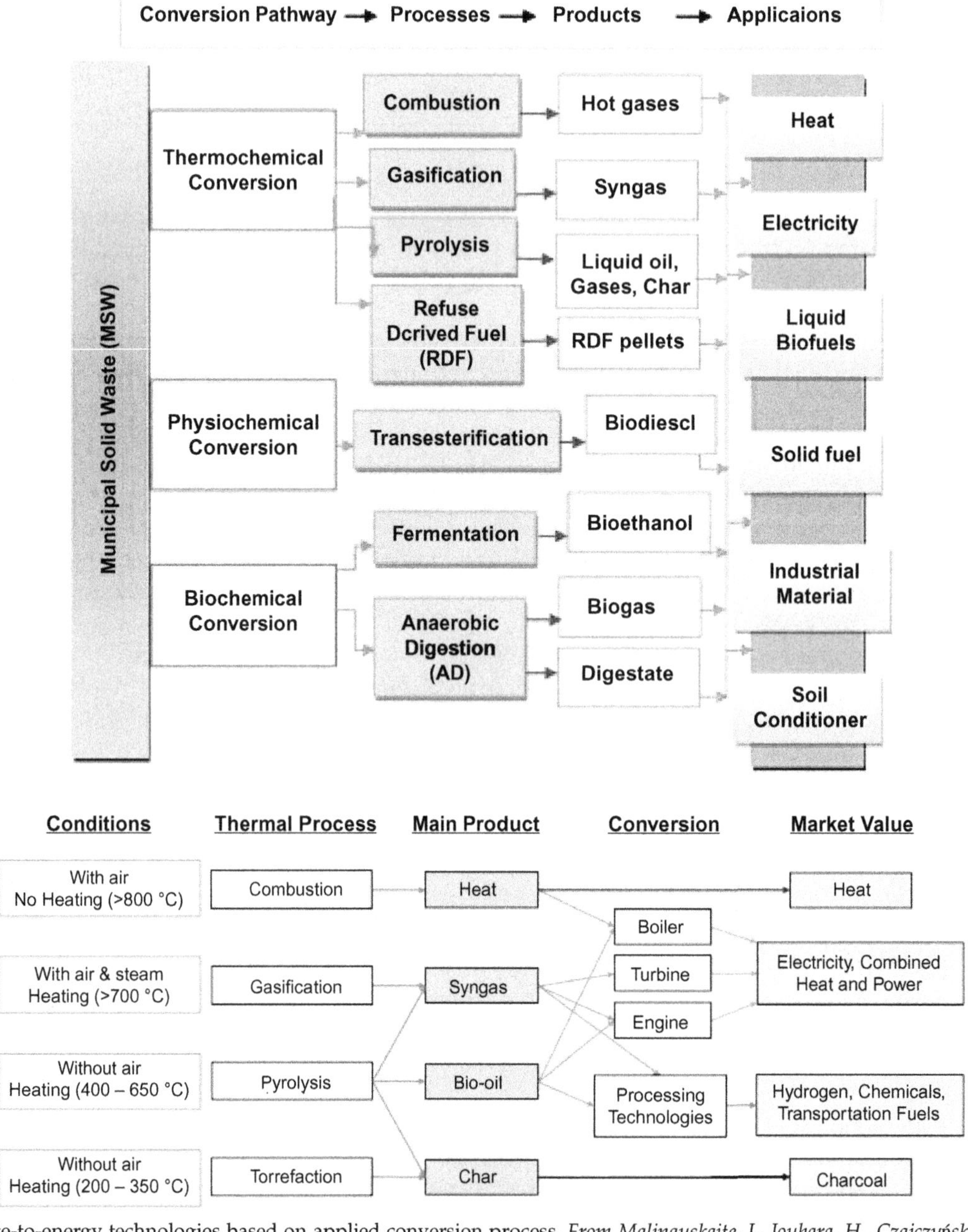

FIG. 17.3 Waste-to-energy technologies based on applied conversion process. *From Malinauskaite, J., Jouhara, H., Czajczyńska, D., Stanchev, P., Katsou, E., Rostkowski, P. et al., 2017. Municipal solid waste management and waste-to-energy in the context of a circular economy and energy recycling in Europe. Energy 141:2013–2044.*

composting or anaerobic digestion can be used to degrade the waste and convert it into usable compost. For example, organic wastes such as food waste, yard waste, and animal manure that consist of naturally degrading bacteria can be converted under controlled conditions into compost, which can then be utilized as a natural fertilizer. Aerobic composting is accomplished by placing selected proportions of organic waste into piles, rows, or vessels, either in open conditions or within closed buildings fitted with gas collection and treatment systems (CCAC and ISWA, 2016). During the process, bulking agents such as wood chips are added to the waste material to enhance the aerobic degradation of organic materials. Finally, the material is allowed to stabilize and mature during a curing process, where pathogens are concurrently destroyed. The end-products of the composting process include carbon dioxide, water, and the usable compost material (Yagasa and Gamaralalage, 2019.).

Compost material may be used in a variety of applications. In addition to its use as a soil amendment for plant cultivation, compost can be used to remediate soils, groundwater, and stormwater. Composting can be labor-intensive, and the quality of the compost is heavily dependent on proper control of the composting process (ADB, 2010). Inadequate control of the operating conditions can result in compost that is unsuitable for beneficial applications.

In some cases, aerobic processes are not feasible. As an alternative, anaerobic processes may be utilized. Anaerobic digestion consists of degrading mixed or sorted organic wastes in vessels under anaerobic conditions. The anaerobic degradation process produces a combination of methane and carbon dioxide (biogas) and residuals (biosolids). Biogas can be used for heating and electricity production, while residuals can be used as fertilizers and soil amendments (GMI, 2020). Anaerobic digestion is a preferred degradation for wet wastes as compared to the preference of composting for dry wastes. The advantage of anaerobic digestion is biogas collection; this collection and subsequent beneficial utilization make it a preferred alternative to landfill disposal of wastes. Also, waste is degraded faster through anaerobic digestion as compared to landfill disposal (EPA, 2020).

Another waste treatment alternative is mechanical biological treatment (MBT). However, this alternative is widely used in Europe. During implementation of this method, waste material is subjected to a combination of mechanical and biological operations that reduce volume through the degradation of organic fractions in the waste. Mechanical operations such as sorting, shredding, and crushing prepare the waste for subsequent biological treatment, consisting of either aerobic composting or anaerobic digestion (Aparcana, 2017). Following the biological processes, the reduced waste mass may be subjected to incineration.

17.6.4 Waste disposal

Residual inert wastes at the end of the hierarchy are to be incinerated or disposed in sanitary lined landfills. All over the world, landfills which integrate the capture and use of methane are preferred over landfills which do not capture the landfill gas. As per the hierarchy, the least preferred option is the disposal of waste in open dumpsites (Coursera, 2019).

17.6.5 Incineration

Waste degradation not only produces useful solid end-products (such as compost), but degradation of by-products can also be used as a valuable energy source. As discussed above, anaerobic digestion of waste can generate biogas, which can be captured and incorporated into electricity generation. Alternatively, waste can be directly incinerated to produce energy. Incineration consists of waste combustion at very high temperatures to produce electrical energy. The by-product of incineration is ash, which requires proper characterization prior to disposal, or in some cases, beneficial re-use. It is widely used in developed countries due to landfill space limitations. It is estimated that about 130 million tons of wastes are annually combusted in more than 600 plants in 35 countries (World Bank Group, 2021). Furthermore, incineration is often used to effectively mitigate hazardous wastes such as chlorinated hydrocarbons, oils, solvents, medical wastes, and pesticides.

17.6.6 Landfill disposal

Despite advances in reuse and recycling, landfill disposal remains the primary waste disposal worldwide. As previously mentioned, the rate of waste generation continues to increase, but overall landfill capacity is decreasing. New regulations regarding suitable waste disposal and the use of innovative liner systems to reduce the potential of groundwater contamination from leachate infiltration and migration have resulted in a substantial increase in the costs of landfill disposal.

ISWM is closely linked to the 3R approach (reduce, reuse, and recycle), which also preliminarily emphasizes the importance of waste reduction, reuse, and recycling over other forms of waste processing or management. The adoption of these principles helps in minimizing the amount of waste to be disposed, thus also diminishing the public health and environmental risks associated with it (Tong et al., 2021). Expansion of resource recovery at all stages of solid waste management is advocated by both approaches and general solid waste management must have cradle to grave approach.

17.7 Solid waste reduction benefits

By achieving a resource-efficient, socially inclusive, and low-carbon economy by tapping waste as a resource, the life-cycle of valuable materials is extended and the use of secondary materials is increased (CPHEEO, 2016; Kaza et al., 2018).

17.7.1 Better environment and health

- Preventing environmental impacts on air, water, soil, wildlife, and the marine environment
- Protecting human health in communities and at waste management facilities • Minimizing risks associated with hazardous waste
- Improving occupational health
- Reducing greenhouse gas emissions
- Reducing litter and odor
- Avoiding flood risks

Saves natural resources: With increased resource efficiency, the demand for primary raw materials is reduced along with the threat of their reduction. Waste is not just produced when people throw items away. Throughout the life cycle of a product or package—from extraction of raw materials, to transportation, to processing and manufacturing facilities, to manufacture and use—waste is generated. Reusing items or making them with fewer materials decreases waste dramatically. Ultimately, less materials will need to be recycled or collected and sent to disposal sites or waste combustion facilities.

Reduces toxicity of waste: Selecting non-hazardous or less hazardous items is another important component of source reduction. Using less hazardous alternatives for certain items (e.g., cleaning products and pesticides), sharing products that comprise hazardous chemicals instead of throwing out leftovers, reading label directions carefully, and using the least amount necessary are ways to decrease waste toxicity.

Reduces costs: The benefits of preventing waste go beyond reducing reliance on other forms of waste disposal. Preventing waste can also mean economic savings for businesses, schools, and individual consumers.

17.7.2 Economic opportunities and benefits

When efficient practices are introduced into production and consumption, it will

- Increase business opportunities
- Contribute to GDP
- Provide savings to businesses, especially in resource extraction and use, by waste prevention actions, recovery, and/or recycling activities
- Achieve economic savings by improvements in human health and the environment, leading to higher productivity, lower medical costs, better environmental quality, and the maintenance of ecosystem services
- Create employment, including low, medium, and high-skilled jobs

17.7.3 Social opportunities and benefits

- Integrating and professionalizing employment in the informal sector (the route to addressing equity and poverty issues)
- Delivering more attractive and pleasant human settlements and better social amenity
- Encouraging changes in community attitudes and behaviors

When communities are lifted out of poverty, health problems are solved or lessened.

17.8 Guidelines for waste management

The life-cycle—waste management—is governed by the concept of environmentally sound management, which signifies that waste is to be managed so as to protect human health and the environment against adverse effects.

17.8.1 Guidelines and policy for waste management (US, EPA, 2020)

Delivering a waste management system that is effective, equitable, robust, and sustainable under the prevailing conditions.

- Provision of public services (e.g., waste collection, transport, treatment, and disposal) suited to the needs of and afforded by local users
- Protection of public and occupational health and the environment
- Contributing to sustainable use of natural resources, e.g., though materials recovery and recycling, soil improvement, energy generation
- Contributing to economic development, including through fostering resource efficient production and developing waste recovery and recycling operations
- Providing employment and enterprise development opportunities
- Deploying technologies appropriate to prevailing conditions
- Building the capacities of those forming part of the waste management system
- Encouraging and inviting research and development into technologies and governance approaches for sustainable resource and waste management.
- A country should have an overall goal for waste management. This is the final destination of the strategy over the selected timescale, such as achieving a particular level of reduction in waste generation by a particular date.
- Targets, which are steps toward the overall goal, should be chosen for particular waste streams or particular waste management challenges. Targets can be set over different timescales (e.g., short-, medium-, and long-term).
- There may be complementary objectives set in other areas, or in other strategies, such as with respect to employment in the sector, since progress in waste management also contributes to broader health, environmental, economic, and social objectives.

17.9 Conclusion

Due to poor waste prevention strategies implementation, existing policies, and interferences, observation is almost non-existent. The present research is also limited, particularly in assessing exposure risk and health outcomes. Recognizing the extent of the challenge, and acknowledging the limited resources, there is a need to engage strategically at various levels to generate evidence that will help highlight the problem and also feed into advocacy plans for the sensitization of the public, public health officials, employers, and all those at heightened risk of ill-health from solid waste. Based on the foregoing discussion, the following recommendations are proposed: Allotting money to integrated waste management, engage several stakeholders in the management schemes or process, public awareness and education toward a cleaner environment on waste prevention and segregation at home level, and finally moving toward comprehensive management plan and implementation than policy-making. Promote use of technology in activities such as energy generation from waste.

References

ADB, 2010. Sustainable Urban Development in the People's Republic of China: Municipal Solid Waste Treatment: Case Study of Public-Private Partnerships (PPPs) in Wenzhou. Asian Development Bank. https://www.adb.org/sites/default/files/publication/27864/urbandev-prc-nov2010-waste.pdf. Accessed January 30, 2021.

Agarwal, R., Chaudhary, M., Singh, J., 2015. Waste management initiatives in India for human well-being. Eur. Sci. J.

Antwi, S.O., Eckert, E.C., Sabaque, C.V., Leof, E.R., Hawthorne, K.M., Bamlet, W.R., Chaffee, K.G., Oberg, A.L., Petersen, G.M., 2015. Exposure to environmental chemicals and heavy metals, and risk of pancreatic cancer. Cancer Causes Control 26 (11), 1583–1591.

Aparcana, S., 2017. Approaches to formalization of the informal waste sector into municipal solid waste management systems in low-and middle-income countries: review of barriers and success factors. Waste Manag. 61, 593–607.

Bartl, A., 2014. Moving from recycling to waste prevention: a review of barriers and enables. Waste Manag. Res. 32, 3–18.

CCAC, ISWA, 2016. Strategy for Organic Waste Diversion – Collection, Treatment, Recycling and Their Challenges and Opportunities for the City of Sao Paulo, Brazil. Climate and Clean Air Coalition and International Solid Waste Association. https://www.ccacoalition.org/en/resources/strategy-organic-waste-diversion-collection-treatmentrecycling-and-their-challenges-and. Accessed February 7, 2021.

Chartsbin, 2013. Global Illegal Waste Dumping by Country. Available http://chartsbin.com/view/576.

Coursera, 2019. Municipal Solid Waste Management in Developing Countries. Online course. https://www.coursera.org/learn/solid-waste-management. Accessed January 28, 2021.

CPHEEO, 2016. Municipal Solid Waste Management Manual Part II: The Manual Swachh Bharat Mission. Government of India, Agement Plan. http://cpheeo.gov.in/upload/uploadfiles/files/Part2.pdf. Accessed February 11, 2021.

EPA, 2020. Best Practices for Solid Waste Management: A Guide for Decision-Makers in Developing Countries.

Frost, S., 2012. The Global industrial waste recycling & services markets. In: TEKES Growth Workshop in Helsinki on 2 October. Available https://tapahtumat.tekes.fi/uploads/c8ffe124/TekesGG_Workshop_021012_global_industrial_waste_presentation-9175.pdf.

Gibbs, Soell, 2010. Sense and Sustainability Among Consumers and Fortune 1000 Executives, Municipal Solid Waste (Management and Handling) Rules 2000. Gibbs and Soell, Inc.

Giusti, L.A., 2009. Review of waste management practices and their impact on human health. Waste Manag. 29 (8), 2227–2239.

GMI, 2020. Biogas Sector Tools and Resources. Global Methane Initiative. https://www.globalmethane.org/toolsresources/resources_filtered.aspx?s=biogas. Accessed January 31.

Hasmori, M.F., et al., 2020. IOP Conf. Ser.: Mater. Sci. Eng. 713, 012038.

Hoornweg, D., Bhada-Tata, P., 2012. What a waste: a global review of solid waste management. In: Urban Development Series, Knowledge Papers. World Bank, Washington.

Kaza, S., Yao, L., Bhada-Tata, P., Van Woerden, F., 2018. What a waste 2.0: a global snapshot of solid waste management to 2050. World Bank Publications. https://openknowledge.worldbank.org/handle/10986/30317. License: CC BY 3.0 IGO.

Krystosik, A., Njoroge, G., Odhiambo, L., Forsyth, J.E., Mutuku, F., LaBeaud, A.D., 2020. Solid wastes provide breeding sites, burrows, and food for biological disease vectors, and urban zoonotic reservoirs: a call to action for solutions-based research. Front. Public Health 7, 405.

Malinauskaite, J., Jouhara, H., Czajczyńska, D., Stanchev, P., Katsou, E., Rostkowski, P., Thorne, R.J., Colon, J., Ponsá, S., Al-Mansour, F., Anguilano, L., 2017. Municipal solid waste management and waste-to-energy in the context of a circular economy and energy recycling in Europe. Energy 141, 2013–2044.

Nwachukwu, M.A., Ronald, M., Feng, H., 2017. Global capacity, potentials and trends of solid waste research and management. Waste Manag. Res. 35, 923–934.

Robinson, W.D., 1986. The Solid Waste Handbook: A Practical Guide. John Wiley and Sons, Chichester.

Song, Q., Li, J., Zeng, X., 2015. Minimizing the increasing solid waste through zero waste strategy. J. Clean. Prod. 104, 199–210.

Tong, Y.D., Huynh, T.D., Khong, T.D., 2021. Understanding the role of informal sector for sustainable development of municipal solid waste management system: a case study in Vietnam. Waste Manag. 124, 118–127.

World Bank Group, 2021. What a Waste 2.0: A global snapshot of solid waste management towards 2050. https://datatopics.worldbank.org/what-a-waste/. (Accessed February 2021).

Yagasa, R., Gamaralalage, P., 2019. Ecology Note – Towards a Clean and Beautiful Capital City. Institute for Global Environmental Strategies. https://www.iges.or.jp/en/pub/ecology-note-towards-clean-green-and-beautiful/en. Accessed April 27, 2020.

Zacho, K.O., Mosgaard, M.A., 2016. Understanding the role of waste prevention in local waste management: a literature review. Waste Manag. Res. 34 (10), 980–994.

Ziraba, A.K., Haregu, T.N., Mberu, B.A., 2016. Review and framework for understanding the potential impact of poor solid waste management on health in developing countries. Arch. Public Health 74, 55.

CHAPTER

18

Nanoremediation of pollutants: A conspectus of heavy metals degradation by nanomaterials

NT Nandhini and Mythili Sathiavelu

School of Bio Sciences and Technology, Vellore Institute of Technology, Vellore, India

OUTLINE

18.1 Introduction

Healthy beings mean a clean environment. Improper disposal and lack of waste treatment open the gate to a world of pollutants. An enormous amount of pollutants has been generated by various human activities. Lack of awareness about the dark sides of the pollutants and long-term impacts had led to a series of environmental threats. Anthropogenic emergence increases the difficulty in eradicating hazardous pollutants (Grieger et al., 2015). Heavy metals are omnipresent pollutants which have been generated by tanning, usage of fertilizers and pesticides in agricultural land, chemicals manufacturing industries, metal mining, development of industries, and urban development (Sun et al., 2019). The aggregation of high concentrations of essential and nonessential heavy metals in the soil near rural areas disturbs soil microorganisms and affects plant growth (Gil-díaz et al., 2016). Existence of heavy metals in the environment has led to serious health complications including cancer, malfunctioning of endocrine glands, skin and nervous system related problems and also affects people who are not directly involved in handling heavy metals (Joneidi et al., 2019; Sharma et al., 2018). Though the exposure of heavy metals had been gradually decreasing, the concentration of heavy metals in the environment is high and still objectionable (Lindqvist, 2000).

The requirement of newer and efficient technologies to mitigate heavy metals has been increasing (Xue et al., 2018b). Since physical, chemical and biological remediation methods have been gearing up to bring down the number of heavy metal pollutants, instability of the mixture of heavy metals due to various external factors makes them complex to degrade. Physical and chemical methods are expensive and instigate secondary pollutants. Biological methods like phytoremediation and microbial remediation take longer remediation time (Vitkova et al., 2015; Zhu et al., 2016). The major drawback of microbial remediation is found to be that heavy metals could be lethal to microorganisms when remediating organisms exposed to high metal concentrations (Mary et al., 2018).

Cost Effective Technologies for Solid Waste and Wastewater Treatment
https://doi.org/10.1016/B978-0-12-822933-0.00002-4

Being a promising field, nanotechnology is an efficient alternative for existing methods to eradicate heavy metals. Nanomaterials are highly reactive which have a higher surface to volume ratio, defined shape, and nanoscale size (Sharma, 2019). Nanomaterials target specific contaminants by inducing microbial activity to deteriorate specific contaminants (Mary et al., 2018). The existing nanomaterials such as dendrimers, zeolites, carbonaceous nanomaterials, metal and metal oxide nanoparticles have been utilized in water purification (Mittal, 2013). Nanomaterials could reduce the costs, contaminant concentration and cleanup time for highly contaminated sites (Karn et al., 2009). This chapter elucidates the causes, impacts, and removal of heavy metals by various nanomaterials and the mechanism behind the removal.

18.2 Toxicity and environmental impacts of heavy metals

Once a pollutant-free continent has now become a place with the traces of heavy metals. Antarctica has been reported to have found higher levels of heavy metals such as Cu (copper), Pb (lead), and Hg (mercury). The emergence of heavy metals in the continent might be due to oil slicks, incineration of wastes, volcanic eruption, effluent disposal, and combustion of fuel (Chu et al., 2019). Continuous exposures to the polluted environment lead to deteriorating male reproductive function. Ying yang et al performed an experiment in *Passer montanus* (tree sparrow, male) by exposing them to heavy metals contaminated sites. At the end of the experiment, Cd (cadmium) accumulation in the testes of sparrow was observed. Simultaneously, a drop in the level of superoxide dismutase and malondialdehyde (MDA) has been noted (Yang et al., 2020). Sunyani municipal waste dumpsite had been tested for the existence/presence of heavy metals by a group of scientists. Soil samples were collected and tested for the presence of heavy metals. Therefore, the results and analysis showed the existence of heavy metals such as Fe (iron), As (arsenic), Cd (cadmium), Pb (lead), and Zn (zinc). The experiment put forward a statement that due to the presence of heavy metals, the soil around dumpsites is not suitable for the production of crops, where it might be absorbed by the crops. This study clearly states that polluted soil will not be suitable for agricultural purposes (Amerh et al., 2020).

Comparably, a Cu-Mo (copper-molybdenum) mining site in Kajaran, Armenia had been chosen for the study. With the help of geostatistical analysis and geochemical mapping, the presence of heavy metals such as Ti (titanium), Mn (manganese), Co (cobalt), Fe, Cu, Zn, Mo, and Pb have been found. The study also revealed that Ti, Mn, Fe, and Co were naturally predominant contents and Cu, Zn, Mo, and Pb were due to anthropogenic activities. Human health risk assessment revealed that heavy metals such as Fe, Mn, Co, Cu, Pb, and Mo have been posing serious health risks in children. The noncarcinogenic health risk factor was found to be higher than 1, which states that the area around the site can cause serious health implications to children (Tepanosyan et al., 2018). The residues from mining sites are a major source of heavy metals. During precipitation, water percolates through the soil and leaches out the heavy metals to the agricultural lands making them less fertile and also affects the quality of ground/surface water. Leachate could cause accumulation of heavy metals in the agricultural land which may affect crop productivity. The residues of mine sites consist of greater than 5% of Zn and Cd and greater than 1% of Cr (chromium) and Cu. The presence of these heavy metals could cause various environmental impacts (Wang et al., 2019a). The sources and impacts of heavy metal tailings to living beings are given in Fig. 18.1.

18.3 Heavy metals degradation by nanomaterials

Nanomaterials play a crucial role in remediating hazardous heavy metals. Owing to higher adsorption efficiency, reactivity, and stability, nanomaterials are being effectively utilized in mitigating recalcitrant heavy metal contaminants (Cai et al., 2019; Wu et al., 2019b). Likewise, varieties of nanomaterials were used to remove heavy metals that have been given in Table 18.1.

18.3.1 Nanoparticles

Ferrous sulfide (FeS) nanoparticles were exploited to eliminate Cr(VI), which is one of the highly toxic and movable pollutants. The mechanisms of remediation comprised of co-precipitation, adsorption, and reduction with the removal efficiency of 1046.1 mg/g (Wang et al., 2019b). Likewise, FeS nanoparticles were used in the reduction of Cd. FeS nanoparticles have been stabilized with sodium carboxymethyl cellulose (CMC). At pH 7.0, the reduction of 1 mg/g Cd was observed to be 93% within 4 h when FeS nanoparticles with the concentration of 100 mg/L were used (Tian et al., 2020). Pb^{2+} and Cu^{2+} were successfully reduced using super paramagnetic maghemite (γ-Fe_2O_3)

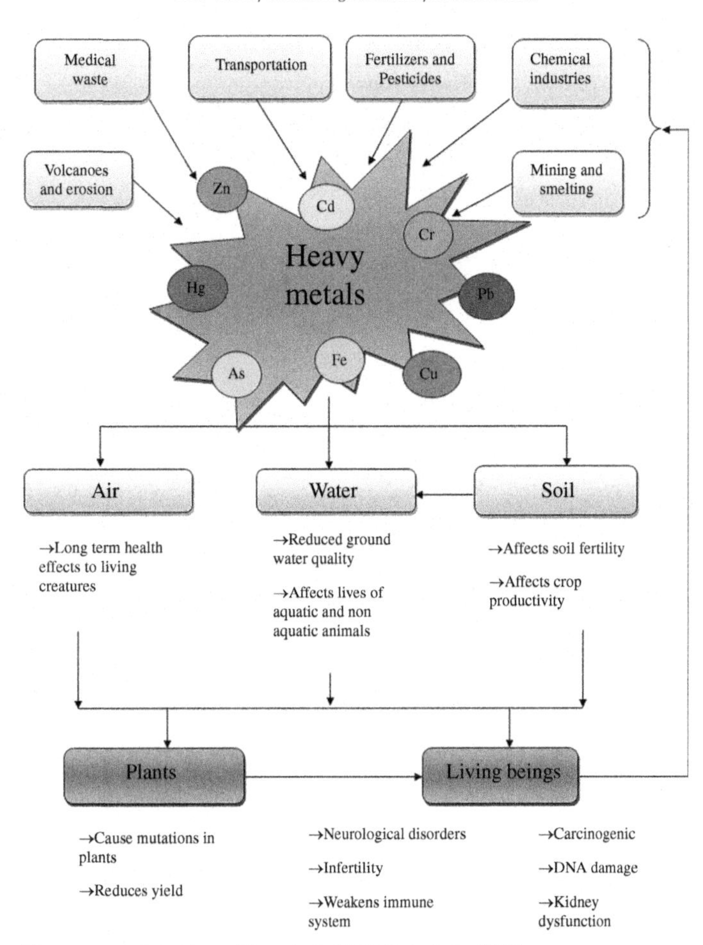

FIG. 18.1 Sources and impacts of heavy metal residues (Das et al., 2018; Rai et al., 2019).

TABLE 18.1 Heavy metals degradation by nanomaterials.

S. no.	Nanomaterial	Characterization technique	Heavy metal/ contaminant	Degradation rate	Reference
1	Kaolin-Fe/Ni/Cu nanoparticles	SEM, EDS, XRD, XPS	Copper (II) and Nitrate (co-contaminants)	42.5% of nitrate (presence of copper), 26.9% of nitrate (absence of cooper)	(Cai et al., 2014)
2	Goethite nanospheres (nGoethite) and zero valent iron nanoparticles (nZVI)	EDX, TEM, DLS	As	89.5% decrease (dose-2%) and 82.5% decrease (dose-0.2%) in the concentration of As	(Alonso et al., 2020)
3	CMC (carboxymethyl cellulose) stabilized Binary Fe-Mn oxide nanoparticles	XRD, FTIR	As(III)	78% of the reduction was observed	(An and Zhao, 2012)
4	Ce-Ti nanoparticles	ESEM, XRD	As(V)	Average sorption capacity is found to be 9.4 mg/g	(Deng et al., 2010)
5	nZVI (nano zerovalent iron)	–	As, Cd, Cr, Pb, Zn	At 10% nZVI—As, Cr and Pb (more than 82%), Zn (31%–75%), Cd (13%–42%) reduction were observed	(Pinilla et al., 2017)
6	nZVI/Cu bimetallic nanoparticles	SEM, XRD, XPS	Cr(VI)	›99% removal efficiency in 10 min	(Zhu et al., 2016)

Continued

TABLE 18.1 Heavy metals degradation by nanomaterials.—cont'd

S. no.	Nanomaterial	Characterization technique	Heavy metal/ contaminant	Degradation rate	Reference
7	TiO_2/SiO_2 nanoparticles	SEM, TEM, XRD, UV-Vis	Cr(III), Co(II), Pb(II)	The removal efficiencies of Cr, Co, and Pb were found to be 90.0%, 91.9%, and 98.6%, respectively	(Harraz et al., 2013)
8	Hybrid magnetic nanoparticles (lignin based)	XRD, FTIR, XPS, SEM, TEM	Pb^{2+}, Cu^{2+}	Adsorption capacities have been observed to be 150.33 mg/g (pb^{2+}) and 70.69 mg/g (Cu^{2+})	(Zhang et al., 2019)
9	Ni/Fe nanoparticles (calcium alginate encapsulated)	SEM, EDS, XRD	Cu (II)	Encapsulation increased Cu (II) removal from 83.9% to 86.7%	(Kuang et al., 2015a)
10	FeS-CFFO NPs/PVDF nanocomposite	SEM, TEM, AFM, FTIR, XRD, BET, XPS	Cr (VI), Cd^{2+}, Pb^{2+}, As	88% for Cr(VI), 95% for As, 99% for Cd^{2+}, and 99% for Pb^{2+} removal rates were observed	(Mishra et al., 2020)
11	$BiOBr/Ti_3C_2$ nanocomposite	XRD, XPS, SEM, UV-Vis DRS, PL spectra	Cr (VI)	Complete removal was achieved in 80 min	(Huang et al., 2019b)
12	ZIF-67@Fe_3O_4@ESM nanocomposite	XRD, FTIR, SEM/ EDS/Mapping, VSM, BET	Cu^{2+}	Maximum adsorption efficiency was found to be 344.82 mg/g	(Mohammad et al., 2019)
13	Silver-yttrium oxide nanocomposite (SYON)	XDR, TEM, SEM, EDX, TGA, FTIR	Cu(II), Cr(VI)	Maximum adsorption of Cu(II) and Cr(VI) was found to be 773 ± 4.94 and 720 ± 8.48 mg/g	(Kumar et al., 2017)
14	PAMAM/CNT nanocomposite	FTIR, TEM, SEM	Ni^{2+}, Zn^{2+}, As^{3+}, Co^{2+}	90% of pollutants removal was achieved in 15 min	(Hayati et al., 2016)
15	CF/G (Cobalt ferrite nanoparticles/alginate) and T/G (titanate nanotubes/ alginate)	FESEM, XRD, HRTEM, FTIR, VSM	Cu^{2+}, Fe^{3+}, As^{3+}	For Cu: CF/G and T/G showed 95% and 98% removal, respectively For Fe^{3+}: CF/G and T/G showed 60% and 82% removal respectively For As^{3+}: Only T/G showed 98% removal	(Esmat et al., 2017)
16	Nanocellulose iron oxide nanobiocomposites (NIONs) [Rice Husk (RH)-NIONS and Sugar Bagasse (SB)-NIONs]	TEM, XRD, FTIR, XPS	As, Pb, Cu, Zn, Mn	RH-NIONs and SB-NIONs showed minimal removal efficiency in As (5%–9%), RH-NIONs exhibited removal efficiency of 68% (Mn), 74% (Cu), 92% (Pb), and 95% (Zn), SB-NIONs exhibited removal efficiency of 65% (Mn), 71% (Cu), 89% (Zn), and 97% (Pb)	(Baruah et al., 2020)
17	Alendronate hydroxyapatite (AL-HAP) nanomaterial	XRD, FTIR, BET (Brunauer-Emmett-Teller) method, BJH (Barrett-Joyner-Halenda) model, SEM, HRTEM, DM (Digital Micrograph), XPS	Pb^{2+}, Cd^{2+}, Cu^{2+}	10% doping of AL in AL-HAP showed 226.6 mg/g (Cu^{2+}), 469 mg/g (Cd^{2+}), and 1431.8 mg/g (Pb^{2+}) maximum adsorption capacity	(Ma et al., 2020)
18	NMn_3O_4-OA nano adsorbent	FTIR, TGA, XRD, HRTEM	Zn(II), Co(II)	Maximum adsorptive capacities were 100 mg/g and 202 mg/g for Co^{II} and Zn^{II}, respectively under optimal conditions pH 7 and reaction time (5–20 min)	(Mahmoud et al., 2020)

TABLE 18.1 Heavy metals degradation by nanomaterials.—cont'd

S. no.	Nanomaterial	Characterization technique	Heavy metal/ contaminant	Degradation rate	Reference
19	PL/IO-NC (palm leaves/iron oxide-nanocomposite)	SEM, EDS, FTIR, XRD	Mn(II), Co(II), Pb(II)	PL showed adsorption capacity of 23.5, 23.7, and 58.1 mg/g for Co(II), Pb(II), and Mn(II) respectively. PL/IO-NC showed increased adsorption capacity of 153, 197, and 159 for Co(II), Pb(II), and Mn(II), respectively	(Alothman et al., 2020)
20	Magnetite graphene oxide	SEM, TEM, XRD, XPS, FTIR, VSM (vibrating sample magnetometer), AAS (atomic absorption spectrometry), ZP (Zeta potential)	Cd(II), As(V)	Maximum adsorption efficiencies of MGO were found to be 14mg/g for As(V) and 234 mg/g for Cd(II)	(Huang et al., 2019c)
21	L-nZVI@Ab (Organosolv lignin-nZVI-aluminum bentonite)	FTIR, VSM, EDS,TEM, BET technique, XRD, XPS	Cr(VI)	L-nZVI@Ab showed a removal efficiency of 46.2 mg/g of Cr(V)	(Chi et al., 2020)
22	3D-PSN (three-dimensional pollution-free starch based nanomaterial)	XRD, SEM, TGA, ZP, FTIR, XPS	Pb(II), Cu(II), Hg(II), Cd(II)	3D-PSN has the maximum adsorption capacity of 238.39, 354.15, 381.47, and 532.28 mg/g for Pb(II), Cu(II), Hg(II), and Cd(II), respectively	(Fang et al., 2020)
23	Fe_3O_4-NH_2/PEI-EDTA (EDTA functionalized, amine terminated magnetic iron oxide nanoparticles)	XRD, FTIR, TEM, VSM	Pb^{2+}, Cd^{2+}, Cu^{2+}	Pb^{2+} was removed much higher than Cd^{2+} and Cu^{2+}. The removal efficiency of Pb^{2+} was found to be 98.8% (concentration brought down from 10 to 0.12 mg/L at pH 5.05)	(Zhang et al., 2011)
24	Graphene oxide (GO), silica (SiO_2), tin oxide (SnO_2)	SEM, EDS, FTIR, ATR	Zn^{2+}	The reduction rates of GO, SiO_2 and SnO_2 were 93.1% ± 02.1%, 99.1% ± 0.3% and 83.2% ± 3.5% respectively	(Ahmed et al., 2019)
25	nZVI/Cu	SEM, XRD, XPS	Cr(VI)	0.06 g of nZVI/Cu showed 99% of Cr(VI) reduction (at pH 5, temperature 303 K)	(Zhu et al., 2016)
26	Z-nZVI (Zeolite supported)	SEM-EDX, FTIR, XRD, XPS	As(III), Cd(II), Pb(II)	At pH 6, the adsorption rates of As(III), Cd(II), and Pb(II) were found to be 11.52, 48.63, and 85.37 mg/g	(Li et al., 2018)
27	GONRs (graphene oxide nanoribbons)	TEM, EDX, SEM, FTIR	As(V), Hg(II)	The adsorption efficiencies were 33.02 and 155.61 mg/g for Hg(II) and As(V), respectively	(Sadeghi et al., 2020)
28	BNP (bimetallic nanoparticles). Aluminum coated iron (Al/Fe) BNPs and iron coated aluminum (Fe/Al) BNPs	ICP-AES (inductively coupled plasma atomic emission spectrometer), ESEM (environmental scanning electron microscope)/EDS, XRD, DLS (dynamic light scattering)	Cr^{6+}	Both (Al/Fe) BNPs and (Fe/Al) BNPs reduce Cr^{6+} to Cr^{3+} with the removal rates of 1.47 g/g BNP and 0.07 g/g BNP	(Ou et al., 2020)
29	Z-Fe/Pd (nano zerovalent iron/palladium)	TEM, HRTEM, XRD, XPS	Cr(VI), nitrate (co-contaminant)	Z-Fe/Pd could effectively reduce Cr(VI) (121 mg/g) and nitrate (95.5 mg/g)	(He et al., 2020)
30	IONPs (iron oxide nanoparticles)/alginate beads	UV-vis spectroscopy, FTIR	Cr (VI)	90% of Cr(VI) has been degraded (beads size—2 mm, pH—2.5)	(Sathya et al., 2020)

nanoparticles with the removal efficiencies of 68.9 mg/g (45°C) and 34.0 mg/g (25°C) for Pb^{2+} and Cu^{2+}, respectively. The researchers also reported that the removal mechanism was highly pH-dependent (Rajput et al., 2017).

18.3.2 Nanocomposite

A nonbiodegradable pollutant Pb(II) was alleviated by the GO-Fe_3O_4-CS-EDTA (graphene oxide derivative and chitosan-based) nanocomposite. The adsorption efficiency of the nanocomposite was found to be 93.98% of Pb(II). The Langmuir model showed the maximum adsorption efficiency of 666.66 mg/g which is higher than the recently developed Pb(II) adsorbents (Foroughi and Azqhandi, 2020). A group of researchers proclaimed that Cr(VI) was completely removed in 80 min by the BiOBr-Ti_3C_2 (bismuth oxyhalides/titanium carbide) nanocomposite (Huang et al., 2019a). Montmorillonite (MMT) had been incorporated with ZnO (zinc oxide) nanoparticles to make a novel ZnO/MMT nanocomposite. The nanocomposite was used in the adsorption of heavy metals such as Pb and Cu. Owing to its durability and stability, the ZnO/MMT nanocomposite can be used at least thrice in the adsorption of heavy metals. Adsorption percentages of ZnO/MMT were 89.5% (Cu^{2+}) and 97.2% (Pb^{2+}) which are higher than the adsorption percentages of MMT with adsorption percentages 40.7% (Cu^{2+}) and 51.2% (Pb^{2+}) (Abubakar et al., 2017). The MWCNTs-Fe_3O_4 (multiwalled carbon nanotubes-iron oxide) nanocomposite was used for the removal of Hg(II). According to the Langmuir isotherm model, the removal capacity was found to be 238.78 mg/g of Hg(II) (Sadegh et al., 2018).

18.3.3 Nanoscale zero-valent iron

Nano zero-valent iron (nZVI) has been coated with rhamnolipid (RL) to immobilize river contaminants (Cd and Pb). RL was added to control the contents of organic matter and to prevent the aggregation of nZVI. Inferences were made that RnZVI effectively immobilized Cd ions higher than Pb ions. Therefore, the mobility of the Cd ions had been greatly reduced due to the immobilization capacity of RnZVI (Xue et al., 2018a). Since nZVI agglomerates easily and has poor stability, the nZVI-Cu bimetal has been synthesized to increase stability and reduce the formation of iron oxides which may lead to agglomeration. The nZVI-Cu bimetal has been assessed to check the reduction of Cr(VI) and the results exhibited that 99% of Cr(VI) has been adsorbed (Zhu et al., 2016). A study shows high adsorption of As(III), Cd(II), and Pb(II) by Z-NZVI (zeolite supported) which was synthesized from liquid-phase iron (III) salts reduction (Li et al., 2018).

18.3.4 Carbon nanotube

Carbon nanotubes have strong and high adsorption capacity which can be used to reduce the toxicity and bioavailability of heavy metals. The uptake efficiency of Cd(II) by CNTs was found to increase to 1.482 mg/g from 1.205 mg/g, with reduced k_2 (pseudo-second-order) values to 0.0070 g/(mg min) (Sun et al., 2015). The PET/CNT (polyethylene terephthalate/carbon nanotube) electrode has been utilized in the removal of multi-metal contaminants (Cd, Cu, Ni, Pb, and Zn) in kaolin. The experimental adsorption efficiencies of PET/CNTs were brought about to be 89.7% of Cd, 63.6% of Cu, 90.7% of Ni, 19.2% of Pb, and 88.7% of Zn (Yuan et al., 2016). The toxicity of Hg(III) is found to be higher than other heavy metals. Multiwalled carbon nanotubes (MWCNTs) have been assessed to eradicate toxic heavy metals. Nearly 99.1% of Hg(II) has been removed using MWCNTs, where the results confirmed the adsorption efficiency (Alijani et al., 2015).

18.3.5 Other nanomaterials

Magnetite nanospheres have been synthesized with hollow interiors using ferric chloride as a precursor. The adsorption capacity has been evaluated using the Langmuir isotherm model. The adsorption efficiencies of magnetite nanospheres were found to be ~ 9 mg/g of Cr^{6+} and ~ 19 mg/g of Pb^{2+} (Kumari et al., 2015). Uranium (U) (VI) ions are potentially harmful and toxic metals that reside in wastewater which destroys the well-being of the living system. A novel h-Fe_3O_4@Au/polydopamine (magnetic/polydopamine hybrid) nanosphere was used to remediate U(VI) ions. The nanosphere exhibited a good adsorption efficiency of 82.9 mg/g of U(VI) ions (Xu et al., 2020). Nitrogen-doped carbon dots (N-CDs) exhibit their excellent catalytic property in the bioaccumulation of Cd^{2+} by *Arabidopsis thaliana*. The studies on the interaction between *A. thaliana*, N-CDs and Cd^{2+} disclose that the bioaccumulation of Cd was enhanced (58.3%) by the catalytic activity of N-CDs in *A. thaliana* (Youssef et al., 2019).

Easily recyclable nano adsorbent TO(2,2,6,6-tetremethylypiperidine-1-oxyl)-CNF (cellulose nanofibril)/TMPTAP (trimethylyolpropane-tris-(2-methyl-1-aziridine)/PEI (polyethyleneimine) aerogel has been synthesized to remove Cu(II) ions. Through the Langmuir isotherm model, the removal capacity of TO-CNF/TMPTAP/PEI aerogel was found to be 484.44 mg/g of Cu(II) (Mo et al., 2019).

18.4 Mechanisms of heavy metals degradation by nanomaterials

The degradation mechanism will be different for different nanomaterials as the properties of nanomaterials vary. Adsorption of heavy metals by nanomaterials has been evaluated using adsorption isotherms and kinetics. Double exponential model, intraparticle diffusion model, pseudo first-order kinetics, and pseudo second-order kinetics are frequently used to evaluate the adsorption efficiency. Similarly, adsorption isotherms are being used to analyze the interaction between the adsorbent and the adsorbate. The commonly used adsorption isotherms are Dubinin-Radushkevich (D-R), Freundlich, Langmuir, and Temkin isotherm models (Emmanuel and Rao, 2008; Wadhawan et al., 2020; Zu et al., 2020). Adsorption of heavy metals (adsorbate) by nanomaterials (adsorbent) is of two forms: (1) physisorption (van der Waals interaction) and (2) chemisorption (chemical bond formation) (Aigbe and Osibote, 2020). In a study, CA (calcium alginate)-Ni/Fe (bi-metallic nanoparticles) beads were used to remove Cu(II). The mechanism of Cu(II) removal involves the adsorption of Cu(II) on CA-Ni/Fe beads (by pseudo second-order) and reduction of Cu(II) by Ni/Fe bimetallic nanoparticles (pseudo first order) (Kuang et al., 2015b). The magnetic nanomaterial (Fe_3O_4@SiO_2) was used to remediate Cu(II), Pb(II), and Cd(II). The nanomaterials exhibited high affinity to adsorb Cu(II) more than Pb(II) and Cd(II), which was studied using the Langmuir isotherm. The adsorption capacity (qm) of Cu(II), Pb(II), and Cd(II) was found to be 5273, 148.1, and 93.1 L/mmol, respectively (Wang et al., 2010). The general heavy metal degradation mechanism is depicted in Fig. 18.2.

18.5 Toxicity and limitations of nanomaterials

Nanomaterials have been grabbing attention among researchers due to their beneficial features in different fields. Though nanomaterials are being used in remediation, drug delivery, wastewater treatment, etc., several studies have focused on their downside effects (Karakoti et al., 2006). Nanoparticles deftly disperse in the environment due to their nanosize and higher mobility, which may turn toxic to the environment. Accumulation of nanoparticles that existed in the environment due to living organisms is another concern (Diao and Yao, 2009). Oxidation of nZVI by

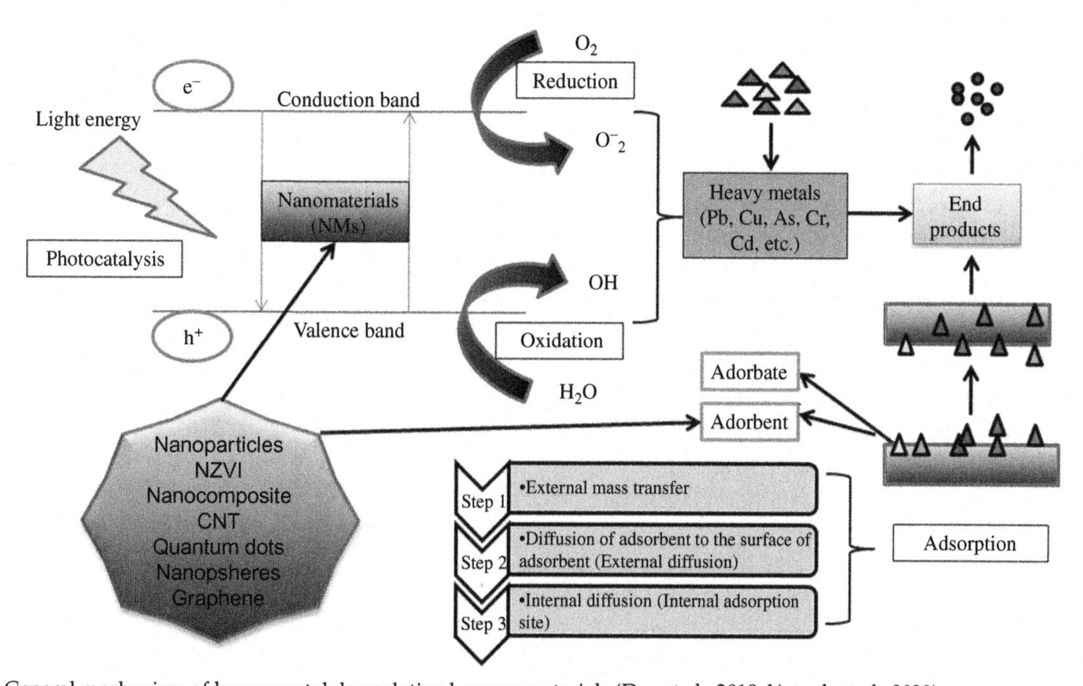

FIG. 18.2 General mechanism of heavy metal degradation by nanomaterials (Das et al., 2018; Yaqoob et al., 2020).

certain bacteria (sulphate reducing bacteria) leads to the generation of ROS (reactive oxygen species) which may cause oxidative stress (Ma et al., 2013). Reduction in the transcription rate in plants had been noticed when the concentration of nZVI was high (Goornavar et al., 2015). MWCNTs induce apoptosis and cause damage to DNA in the roots of *Allium cepa*, a probable cause of nanotubes penetration into the cells (Ghosh et al., 2011).

18.6 Biosafety assessment

Poly(methacrylate citric acid) as a nano adsorbent has been used in the study to remove heavy metals in aqueous solution. Higher adsorption of Cu^{2+}, Cd^{2+} and Pb^{2+} has been achieved by PCA nano adsorbent. The removal efficiency was found to be greater than 90%. Considering biosafety, cell toxicity studies were performed in mice to check the toxicity of nano adsorbent. The results showed that PCA can be used to remove heavy metals and it is not toxic to the living system (Zhang et al., 2020). When it comes to remediation, the component which is clearing up the contaminants or pollutants should not have negative effects. Several studies have confirmed about the biosafety of mesoporous silica nanoparticles, yet there is no sufficient information about the biosafety due to the dependency on several characteristics such as shape, size of pores, dimensions of the particle, interactions between biological sources-nanosized materials, and crystallinity (Farjadian et al., 2019). Considering the biosafety of nanoparticles is as important as checking their efficiency in using them. Various biological models have been explored to evaluate the biosafety of nanoparticles. *Caenorhabilitis elegans* is a nematode that is used to assess the risks of nanoparticles. The study has been done to know the effects of nanoparticles on *C. elegans*, including defense mechanism, reproduction, functions of neuron and intestine, and the mechanism behind those effects. The inference made on the nematode was that it could be used to assess the biosafety of nanoparticles as it is highly sensitive to pollutants. Reference samples are important to study the reliability and getting standard results of biosafety of nanoparticles *C. elegans* could be used as an in vivo assay model to check the toxicity of nanoparticles (Wu et al., 2019a). Researchers are effectively working to develop eco-safe nanomaterials. SSA (serum amyloid A), CRP (C-reactive protein), and haptoglobin are acute-phase proteins, which have been developed to predict the toxicity of nanomaterials in living organisms (Higashisaka et al., 2011).

18.7 Conclusion

Having lost their quality by pollutants, natural resources need to be protected from hazardous heavy metals, which have become a global concern. The rise in human activities without proper awareness amongst them has led to an increase in the level of pollutants in the environment. The presence of heavy metal tailings in the soil above permissible limits creates greater chaos as they might move to agricultural lands. Therefore, these pollutants cause severe threat to the living system. Since numerous methods are available to eradicate heavy metals, the limitations of those methods such as usage of high-cost materials, time-consuming, and emergence of secondary pollutants after treating primary pollutants are major issues in using them in clearing up the heavy metals.

The need for newer, sustainable, and cost-effective technology to eradicate heavy metal tailings has made nanoremediation a perfect choice to treat heavy metals. Nanomaterials are unique, smaller in size, have a higher surface to volume ratio, have reduced clean up time, cost-effective, have greater efficiency, and are highly reactive with the targeted contaminants. Nanoparticles, nanocomposites, CNTs, quantum dots, nanospheres, nanoribbons, and nanosheets are widely used for the remediation of heavy metals. Nanocomposites could be used in the removal of heavy metals since the adsorption capacity of nanocomposites is high when compared to other nanomaterials. Nanocomposites are reinforced and multifunctional materials that can overcome several external environmental factors protecting their reactivity with the contaminants. Nanomaterials are extremely important to make soil and water resources free of pollutants. Novel nanomaterials should be explored and employed with higher stability, effectual and reducing possible risks by eliminating negative effects. It is important to generate eco-safety nanomaterials to reduce the risks of new pollutants. Furthermore, meticulous research is needed to develop novel nanomaterials to eradicate hazardous heavy metal tailings.

References

Abubakar, H., Ahmad, M.B., Zobir, M., Azowa, N., Musa, A., Saleh, T.A., 2017. Nanocomposite of ZnO with montmorillonite for removal of lead and copper ions from aqueous solutions. Process Saf. Environ. Prot. 109, 97–105.

Ahmed, M., Elektorowicz, M., Hasan, S.W., 2019. GO, SiO_2, and SnO_2 nanomaterials as highly efficient adsorbents for Zn^{2+} from industrial wastewater—a second stage treatment to electrically enhanced membrane bioreactor. J. Water Process Eng. 31, 1–11.
Aigbe, U.O., Osibote, O.A., 2020. A review of hexavalent chromium removal from aqueous solutions by sorption technique using nanomaterials. J. Environ. Chem. Eng. 104503.
Alijani, H., Shariatinia, Z., Aroujalian, A., 2015. Water assisted synthesis of MWCNTs over natural magnetic rock : an effective magnetic adsorbent with enhanced mercury (II) adsorption property. Chem. Eng. J. 281, 468–481.
Alonso, J., Gallego, J.R., Lobo, M.C., Gil-díaz, M., Baraga, D., 2020. Zero valent iron and goethite nanoparticles as new promising remediation techniques for As-polluted soils. Chemosphere 238, 12624. https://doi.org/10.1016/j.chemosphere.2019.124624.
Alothman, Z.A., Habila, M.A., Sheikh, M., Al-Qahtani, K.M., AlMasoud, N., Al-Senani, G.M., Al-kadhi, N.S., 2020. Fabrication of renewable palm-pruning leaves based nanocomposite for remediation of heavy metals pollution. Arab. J. Chem. 13, 4936–4944.
Amerh, A., Adjei, R., Anokye, J., Banunle, A., 2020. Municipal waste dumpsite : Impact on soil properties and heavy metal concentrations, Sunyani. Ghana. Sci. African 8, e00390.
An, B., Zhao, D., 2012. Immobilization of As (III) in soil and groundwater using a new class of polysaccharide stabilized Fe-Mn oxide nanoparticles. J. Hazard. Mater. 211–212, 332–341. https://doi.org/10.1016/j.jhazmat.2011.10.062.
Baruah, J., Chaliha, C., Kalita, E., Nath, B.K., Field, R.A., Deb, P., 2020. Modelling and optimization of factors influencing adsorptive performance of agrowaste-derived nanocellulose iron oxide nanobiocomposites during remediation of arsenic contaminated groundwater. Int. J. Biol. Macromol. 164, 53–65.
Cai, C., Zhao, M., Yu, Z., Rong, H., Zhang, C., 2019. Utilization of nanomaterials for in-situ remediation of heavy metal (loid) contaminated sediments: a review. Sci. Total Environ. 662, 205–217.
Cai, X., Gao, Y., Sun, Q., Chen, Z., Megharaj, M., Naidu, R., 2014. Removal of co-contaminants Cu(II) and nitrate from aqueous solution using kaolin-Fe/Ni nanoparticles. Chem. Eng. J. 244, 19–26. https://doi.org/10.1016/j.cej.2014.01.040.
Chi, Z., Hao, L., Dong, H., Yu, H., Liu, H., Wang, Z., Yu, H., 2020. The innovative application of organosolv lignin for nanomaterial modification to boost its heavy metal detoxification performance in the aquatic environment. Chem. Eng. J. 382, 1–11.
Chu, W., Dang, N., Kok, Y., Yap, K.I., Phang, S., Convey, P., 2019. Heavy metal pollution in Antarctica and its potential impacts on algae. Polar Sci. 20, 75–83.
Das, S., Chakraborty, J., Chatterjee, S., Kumar, H., 2018. Prospects of biosynthesized nanomaterials for the remediation of organic and inorganic environmental contaminants. Environ. Sci. Nano 5, 2784–2808. https://doi.org/10.1039/C8EN00799C.
Deng, S., Li, Z., Huang, J., Yu, G., 2010. Preparation, characterization and application of a Ce–Ti oxide adsorbent for enhanced removal of arsenate from water. J. Hazard. Mater. 179, 1014–1021. https://doi.org/10.1016/j.jhazmat.2010.03.106.
Diao, M., Yao, M., 2009. Use of zero-valent iron nanoparticles in inactivating microbes. Water Res. 43, 5243–5251. https://doi.org/10.1016/j.watres.2009.08.051.
Emmanuel, K.A., Rao, A.V., 2008. Adsorption of Mn (II) from aqueous solutions using *Pithacelobium dulce* carbon. Rasayan J. Chem. 1, 840–852.
Esmat, M., Farghali, A.A., Khedr, M.H., El-sherbiny, I.M., 2017. Alginate-based nanocomposites for efficient removal of heavy metal ions. Int. J. Biol. Macromol. 102, 272–283. https://doi.org/10.1016/j.ijbiomac.2017.04.021.
Fang, Y., Lv, X., Xu, X., Zhu, J., Liu, P., Guo, L., Yuan, C., Cui, B., 2020. Three-dimensional nanoporous starch-based material for fast and highly efficient removal of heavy metal ions from wastewater. Int. J. Biol. Macromol. 164, 415–426.
Farjadian, F., Roointan, A., Mohammadi-samani, S., Hosseini, M., 2019. Mesoporous silica nanoparticles: synthesis, pharmaceutical applications, biodistribution, and biosafety assessment. Chem. Eng. J. 359, 684–705.
Foroughi, M., Azqhandi, M.H.A., 2020. A biological-based adsorbent for a non-biodegradable pollutant: modeling and optimization of Pb(II) remediation using GO-CS-Fe_3O_4-EDTA nanocomposite. J. Mol. Liq. 318, 114077.
Ghosh, M., Chakraborty, A., Bandyopadhyay, M., Mukherjee, A., 2011. Multi-walled carbon nanotubes (MWCNT): induction of DNA damage in plant and mammalian cells. J. Hazard. Mater. 197, 327–336. https://doi.org/10.1016/j.jhazmat.2011.09.090.
Gil-díaz, M., Gonzalez, A., Alonso, J., Lobo, M.C., 2016. Evaluation of the stability of a nanoremediation strategy using barley plants. J. Environ. Manage. 165, 150–158. https://doi.org/10.1016/j.jenvman.2015.09.032.
Goornavar, V., Biradar, S., Ezeagwu, C., Ezeagwu, D., Hall, J.C., Ramesh, G.T., 2015. Toxicity of raw and purified single-walled carbon nanotubes in rat's lung epithelial and cervical cancer cells. J. Nanosci. Nanotechnol. 15, 2105–2114. https://doi.org/10.1166/jnn.2015.9524.
Grieger, K., Hjorth, R., Rice, J., Kumar, N., Bang, J., 2015. Nano-remediation: tiny particles cleaning up big environmental problems. IUCN. http://cmsdata.iucn.org/downloads/nanoremediation.pdf.
Harraz, F.A., Abdel-salam, O.E., Mostafa, A.A., Mohamed, R.M., Hanafy, M., 2013. Rapid synthesis of titania-silica nanoparticles photocatalyst by a modified sol-gel method for cyanide degradation and heavy metals removal. J. Alloys Compd. 551, 1–7.
Hayati, B., Maleki, A., Naja, F., Daraei, H., Gharibi, F., Mckay, G., 2016. Synthesis and characterization of PAMAM/CNT nanocomposite as a super-capacity adsorbent for heavy metal (Ni^{2+}, Zn^{2+}, As^{3+}, Co^{2+}) removal from wastewater. J. Mol. Liq. 224, 1032–1040. https://doi.org/10.1016/j.molliq.2016.10.053.
He, Y., Lin, H., Luo, M., Liu, J., Dong, Y., Li, B., 2020. Highly efficient remediation of groundwater co-contaminated with Cr(VI) and nitrate by using nano-Fe/Pd bimetal-loaded zeolite: process product and interaction mechanism *. Environ. Pollut. 263, 1–11.
Higashisaka, K., Yoshioka, Y., Yamashita, K., Morishita, Y., Fujimura, M., Nabeshi, H., Nagano, K., Abe, Y., Kamada, H., Tsunoda, S., Yoshikawa, T., Itoh, N., Tsutsumi, Y., 2011. Acute phase proteins as biomarkers for predicting the exposure and toxicity of nanomaterials. Biomaterials 32, 3–9. https://doi.org/10.1016/j.biomaterials.2010.08.110.
Huang, D., Wu, J., Wang, L., Liu, X., Meng, J., Tang, X., Tang, C., Xu, J., 2019c. Novel insight into adsorption and co-adsorption of heavy metal ions and an organic pollutant by magnetic graphene nanomaterials in water. Chem. Eng. J. 358, 1399–1409.
Huang, Q., Liu, Y., Cai, T., Xia, X., 2019a. Simultaneous removal of heavy metal ions and organic pollutant by BiOBr/Ti_3C_2 nanocomposite. J. Photochem. Photobiol. A Chem. 375, 201–208.
Huang, Q., Liu, Y., Cai, T., Xia, X., 2019b. Simultaneous removal of heavy metal ions and organic pollutant by BiOBr/Ti_3C_2 nanocomposite. J. Photochem. Photobiol. A Chem. 375, 201–208. https://doi.org/10.1016/j.jphotochem.2019.02.026.
Joneidi, Z., Mortazavi, Y., Memari, F., Roointan, A., Chahardouli, B., 2019. Biomedicine & pharmacotherapy the impact of genetic variation on metabolism of heavy metals: genetic predisposition? Biomed. Pharmacother. 113, 108642.

Karakoti, A.S., Hench, L.L., Seal, S., 2006. The potential toxicity of nanomaterials—the role of surfaces. J. Miner. Met. Mater. Soc. 58, 77–82.
Karn, B., Kuiken, T., Otto, M., 2009. Nanotechnology and in situ remediation: a review of the benefits and potential risks. Environ. Health Perspect. 117, 1823–1831. https://doi.org/10.1289/ehp.0900793.
Kuang, Y., Du, J., Zhou, R., Chen, Z., Megharaj, M., Naidu, R., 2015a. Calcium alginate encapsulated Ni/Fe nanoparticles beads for simultaneous removal of Cu(II) and monochlorobenzene. J. Colloid Interface Sci. 447, 85–91. https://doi.org/10.1016/j.jcis.2015.01.080.
Kuang, Y., Du, J., Zhou, R., Chen, Z., Megharaj, M., Naidu, R., 2015b. Calcium alginate encapsulated Ni/Fe nanoparticles beads for simultaneous removal of Cu(II) and monochlorobenzene. J. Colloid Interface Sci. 447, 85–91.
Kumar, S., Panwar, J., Gupta, S., 2017. Enhanced heavy metal removal using silver-yttrium oxide nanocomposites as novel adsorbent system. J. Environ. Chem. Eng. 5, 5801–5814. https://doi.org/10.1016/j.jece.2017.11.007.
Kumari, M., Pittman, C.U., Mohan, D., 2015. Heavy metals [chromium (VI) and lead (II)] removal from water using mesoporous magnetite (Fe_3O_4) nanospheres. J. Colloid Interface Sci. 442, 120–132.
Li, Z., Wang, L., Meng, J., Liu, X., Xu, J., Wang, F., Brookes, P., 2018. Zeolite-supported nanoscale zero-valent iron: new findings on simultaneous adsorption of Cd (II), Pb (II), and As (III) in aqueous solution and soil. J. Hazard. Mater. 344, 1–11.
Lindqvist, O., 2000. Environmental impact of mercury and other heavy metals. J. Power Sources 57, 3–7.
Ma, J., Xia, M., Zhu, S., Wang, F., 2020. A new alendronate doped HAP nanomaterial for Pb^{2+}, Cu^{2+} and Cd^{2+} effect absorption. J. Hazard. Mater. 400, 1–11.
Ma, X., Gurung, A., Deng, Y., 2013. Phytotoxicity and uptake of nanoscale zero-valent iron (nZVI) by two plant species. Sci. Total Environ. 443, 844–849.
Mahmoud, M.E., Allam, E.A., El-sharkawy, R.M., Soliman, M.A., Saad, E.A., El-khatib, A.M., 2020. Nano-manganese oxide-functionalized-oleyl amine as a simple and low cost nanosorbent for remediation of ZnII/CoII and their radioactive nuclides 65 Zn and 60 Co from water. Appl. Radiat. Isot. 159, 1–8.
Mary, J., Karthik, C., Ganesh, R., Kumar, S.S., Prabakar, D., Kadirvelu, K., Pugazhendhi, A., 2018. Biological approaches to tackle heavy metal pollution: a survey of literature. J. Environ. Manage. 217, 56–70. https://doi.org/10.1016/j.jenvman.2018.03.077.
Mishra, S., Singh, A.K., Singh, J.K., 2020. Ferrous sulfide and carboxyl-functionalized ferroferric oxide incorporated PVDF-based nanocomposite membranes for simultaneous removal of highly toxic heavy-metal ions from industrial ground water. J. Memb. Sci. 593, 1–14. https://doi.org/10.1016/j.memsci.2019.117422.
Mittal, T., 2013. Significant manipulations of nanotechnology in water purification. Int. J. Enhanc. Res. Sci. Technol. Eng. 2, 1–7.
Mo, L., Pang, H., Tan, Y., Zhang, S., Li, J., 2019. 3D multi-wall perforated nanocellulose-based polyethylenimine aerogels for ultrahigh efficient and reversible removal of Cu (II) ions from water. Chem. Eng. J. 378, 1–14.
Mohammad, N., Taghizadeh, M., Taghizadeh, A., Abdi, J., Hayati, B., Akbar Shekarchi, A., 2019. Bio-based magnetic metal-organic framework nanocomposite: ultrasound- assisted synthesis and pollutant (heavy metal and dye) removal from aqueous media. Appl. Surf. Sci. 480, 288–299. https://doi.org/10.1016/j.apsusc.2019.02.211.
Ou, J., Sheu, Y., Tsang, D.C.W., Sun, Y., Kao, C., 2020. Application of iron/aluminum bimetallic nanoparticle system for chromium-contaminated groundwater remediation. Chemosphere 256, 127158.
Pinilla, P., Alonso, J., Lobo, M.C., 2017. Viability of a nanoremediation process in single or multi-metal (loid) contaminated soils. J. Hazard. Mater. 321, 812–819. https://doi.org/10.1016/j.jhazmat.2016.09.071.
Rai, P.K., Lee, S.S., Zhang, M., Tsang, Y.F., Kim, K.H., 2019. Heavy metals in food crops: health risks, fate, mechanisms, and management. Environ. Int. 125, 365–385. https://doi.org/10.1016/j.envint.2019.01.067.
Rajput, S., Singh, L.P., Pittman, C.U., Mohan, D., 2017. Lead (Pb^{2+}) and copper (Cu^{2+}) remediation from water using superparamagnetic maghemite (c-Fe_2O_3) nanoparticles synthesized by Flame Spray Pyrolysis (FSP). J. Colloid Interface Sci. 492, 176–190.
Sadegh, H., Ali, G.A.M., Salam, A., Makhlouf, H., Feng, K., Alharbi, N.S., Agarwal, S., Kumar, V., 2018. MWCNTs-Fe_3O_4 nanocomposite for Hg (II) high adsorption efficiency. J. Mol. Liq. 258, 345–353.
Sadeghi, M.H., Ahmadzadeh, M., Mohammadi, T., 2020. One-dimensional graphene for efficient aqueous heavy metal adsorption: rapid removal of arsenic and mercury ions by graphene oxide nanoribbons (GONRs). Chemosphere 253, 1–10.
Sathya, S., Ragul, V., Priya, V., Singh, L., Ahamed, M.I.N., 2020. An in vitro study on hexavalent chromium [Cr (VI)] remediation using iron oxide nanoparticles based beads. Environ. Nanotechnol. Monit. Manag. 14, 1–5.
Sharma, J., 2019. Nanoremediation. Int. J. Life Sci. Technol. 12, 1–6.
Sharma, S., Kaur, A., Kaur, I., 2018. Heavy metal contamination in soil, food crops and associated health risks for residents of Ropar wetland, Punjab, India and its environs. Food Chem. 255, 15–22.
Sun, W., Jiang, B., Wang, F., Xu, N., 2015. Effect of carbon nanotubes on Cd (II) adsorption by sediments. Chem. Eng. J. 264, 645–653.
Sun, Y., Li, H., Guo, G., Semple, K.T., Jones, K.C., 2019. Soil contamination in China: current priorities, defining background levels and standards for heavy metals. J. Environ. Manage. 251, 109512.
Tepanosyan, G., Sahakyan, L., Belyaeva, O., Asmaryan, S., Saghatelyan, A., 2018. Continuous impact of mining activities on soil heavy metals levels and human health. Sci. Total Environ. 639, 900–909.
Tian, S., Gong, Y., Ji, H., Duan, J., Zhao, D., 2020. Efficient removal and long-term sequestration of cadmium from aqueous solution using ferrous sulfide nanoparticles: performance, mechanisms, and long-term stability. Sci. Total Environ. 704, 135402.
Vitkova, M., Komakare, M., Tejnecky, V., Sillerova, H., 2015. Interactions of nano-oxides with low-molecular-weight organic acids in a contaminated soil. J. Hazard. Mater. 293, 7–14.
Wadhawan, S., Jain, A., Nayyar, J., Kumar, S., 2020. Role of nanomaterials as adsorbents in heavy metal ion removal from waste water: a review. J. Water Process Eng. 33, 1–17. https://doi.org/10.1016/j.jwpe.2019.101038.
Wang, J., Zheng, S., Shao, Y., Liu, J., Xu, Z., Zhu, D., 2010. Amino-functionalized Fe_3O_4@SiO_2 core-shell magnetic nanomaterial as a novel adsorbent for aqueous heavy metals removal. J. Colloid Interface Sci. 349, 293–299. https://doi.org/10.1016/j.jcis.2010.05.010.
Wang, P., Sun, Z., Hu, Y., Cheng, H., 2019a. Leaching of heavy metals from abandoned mine tailings brought by precipitation and the associated environmental impact. Sci. Total Environ. 695, 133893. https://doi.org/10.1016/j.scitotenv.2019.133893.
Wang, T., Liu, Y., Wang, J., Wang, X., Liu, B., Wang, Y., 2019b. In-situ remediation of hexavalent chromium contaminated groundwater and saturated soil using stabilized iron sulfide nanoparticles. J. Environ. Manage. 231, 679–686.

Wu, T., Xu, H., Liang, X., Tang, M., 2019a. Caenorhabditis elegans as a complete model organism for biosafety assessments of nanoparticles. Chemosphere 221, 708–726.

Wu, Y., Pang, H., Liu, Y., Wang, X., Yu, S., Fu, D., Chen, J., Wang, X., 2019b. Environmental remediation of heavy metal ions by novel-nanomaterials: a review. Environ. Pollut. 246, 608–620.

Xu, K., Wu, J., Fang, Q., Bai, L., Duan, J., Li, J., Xu, H., Hui, A., Hao, L., Xuan, S., 2020. Magnetically separable h-Fe_3O_4@Au/polydopamine nanosphere with a hollow interior: a versatile candidate for nanocatalysis and metal ion adsorption. Chem. Eng. J. 398, 1–14.

Xue, W., Huang, D., Zeng, G., Wan, J., Zhang, C., Xu, R., Cheng, M., Deng, R., 2018a. Nanoscale zero-valent iron coated with rhamnolipid as an effective stabilizer for immobilization of Cd and Pb in river sediments. J. Hazard. Mater. 341, 381–389. https://doi.org/10.1016/j.jhazmat.2017.06.028.

Xue, W., Peng, Z., Huang, D., Zeng, G., Wan, J., Xu, R., 2018b. Nanoremediation of cadmium contaminated river sediments: microbial response and organic carbon changes. J. Hazard. Mater. 359, 290–299. https://doi.org/10.1016/j.jhazmat.2018.07.062.

Yang, Y., Zhang, W., Wang, S., Zhang, H., Zhang, Y., Zhang, W., Wang, S., 2020. Response of male reproductive function to environmental heavy metal pollution in a free-living passerine bird, *Passer montanus*. Sci. Total Environ. 747, 141402.

Yaqoob, A.A., Parveen, T., Umar, K., Ibrahim, M.N.M., 2020. Role of nanomaterials in the treatment of wastewater: a review. Water (Switzerland) 12, 1–30. https://doi.org/10.3390/w12020495.

Youssef, A.M., El-naggar, M.E., Malhat, F.M., El, H.M., 2019. Efficient removal of pesticides and heavy metals from wastewater and the antimicrobial activity of fMWCNTs/PVA nanocomposite film. J. Clean. Prod. 206, 315–325. https://doi.org/10.1016/j.jclepro.2018.09.163.

Yuan, L., Li, H., Xu, X., Zhang, J., Wang, N., Yu, H., 2016. Electrokinetic remediation of heavy metals contaminated kaolin by a CNT-covered polyethylene terephthalate yarn cathode. Electrochim. Acta 213, 140–147.

Zhang, F., Zhu, Z., Dong, Z., Cui, Z., Wang, H., Hu, W., Zhao, P., Wang, P., Wei, S., Li, R., Ma, J., 2011. Magnetically recoverable facile nanomaterials: synthesis, characterization and application in remediation of heavy metals. Microchem. J. 98, 328–333. https://doi.org/10.1016/j.microc.2011.03.005.

Zhang, X., Wang, X., Qiu, H., Kong, D., Han, M., Guo, Y., 2020. Poly(methacrylate citric acid) with good biosafety as nanoadsorbents of heavy metal ions. Colloids Surf. B Biointerf. 187, 1–7.

Zhang, Y., Ni, S., Wang, X., Zhang, W., Lagerquist, L., Qin, M., Willfor, S., Xu, C., Fatehi, P., 2019. Ultrafast adsorption of heavy metal ions onto functionalized lignin-based hybrid magnetic nanoparticles. Chem. Eng. J. 372, 82–91.

Zhu, F., Li, L., Ma, S., Shang, Z., 2016. Effect factors, kinetics and thermodynamics of remediation in the chromium contaminated soils by nanoscale zero valent Fe/Cu bimetallic particles. Chem. Eng. J. 302, 663–669. https://doi.org/10.1016/j.cej.2016.05.072.

Zu, S., Ahmad, N., Norharyati, W., Salleh, W., Fauzi, A., Yusof, N., Zamri, M., Yusop, M., Aziz, F., 2020. Adsorptive removal of heavy metal ions using graphene-based nanomaterials: toxicity, roles of functional groups and mechanisms. Chemosphere 248, 1–16.

CHAPTER

19

Excess fluoride issues and mitigation using low-cost techniques from groundwater: A review

Adane Woldemedhin Kalsido[a,b], *Beteley Tekola*[a,c], *Beshah Mogessie*[a,d], *and Esayas Alemayehu*[a,e]

[a]African Centre of Excellence for Water Management, Addis Ababa University, Addis Ababa, Ethiopia [b]Department of Hydraulic & Water Resources Engineering, College of Engineering and Technology, Wachemo University, Hossana, Ethiopia [c]School of Chemical and Bio Engineering, Addis Ababa University, Addis Ababa, Ethiopia [d]Water Development Commission, Ministry of Water, Irrigation, and Energy, Addis Ababa, Ethiopia [e]Faculty of Civil & Environmental Engineering, Jimma Institute of Technology, Jimma University, Jimma, Ethiopia

OUTLINE

Cost Effective Technologies for Solid Waste and Wastewater Treatment
https://doi.org/10.1016/B978-0-12-822933-0.00004-8

19.1 Introduction

Fluoride ion concentration in drinking water has been recognized as the foremost issue worldwide, imposing a significant threat to living being's health (Reddy and Fichtner, 2015). Fluorine has an extraordinary tendency to induce attraction by charged ions like metallic elements (Hiroyuki Fyjimoto, 1993). Consequently, the impact of this halide on mineralized tissues like bone and teeth resulting in organic process alternations is of clinical significance. This can attract the utmost quantity of halide that gets deposited as calcium-fluorapatite crystals. This consists chiefly of crystalline hydroxyl apatite. The particle gets substituted by halide ion since apatite is more stable than hydroxyl apatite. Thus, an outsized quantity of halide gets certain in these tissues, and solely, a tiny low quantity is excreted through sweat, piss, and stool.

As stated by Janardhana Raju et al. (2009), the problem of excess fluoride in drinking water gets more attention. For instance, the dissolution of natural minerals in the rocks and soils creates the formation of fluoride when it interacts with water. High fluoride concentrations can be built up in groundwaters, which have long residence times in the host aquifers. In addition to the natural means of getting fluoride, there are also different anthropogenic sources like drinking water fluoridation, fluoridated toothpaste, untested bottled water, fluoride supplements, and some foods. Comprehensive fluoride concentration distribution on foods is reported by Fyjimoto et al. (1993) for further clarification. It is a well-known fact that ingestion of permissible amount of fluoride helps to avoid tooth decay through mineralization and remineralization. Furthermore, it helps to keep strength of backbone and avoid early deformation. Ingestion of excessive amount of fluoride through water and food can have a risk of health on humans and animals. Dental fluorosis, skeletal fluorosis, thyroid difficulties, neurological disorders are major health effects and others like high blood pressure, decreased fertility and puberty in females, bone cancer, and cardiovascular problems may be caused by ingesting of high fluoride level through food and water.

There are numerous fluoride removal technologies that were reported by different scholars: these are coagulation-precipitation, adsorption technology, membrane technology, the Nalgonda technique, electrocoagulation, dialysis, electrodialysis, and nanofiltration methods. Thus, the aim of this book chapter is to give insight on fluoride toxicity and eco-friendly mitigation technique for readers and experts working on this subject. Furthermore, it identifies the gaps and challenges concerning eco-friendly sorption techniques.

19.2 Materials and methods

19.2.1 General strategy

Data sources like journals, articles, books, and other webpages were referred to formulate the book chapter. Relevant documents in fluoride toxicity and mitigation methods are used. The consistency of the document highlights the recent progress of the fluoride toxicity, its extent, and effect on both local and global perspectives.

Accordingly, the published articles during the past decades till today in the area of fluoride are collected, studied, and assessed. Furthermore, the comprehensiveness of local context is considered to balance on technological, economic, social, and operational perspectives of the eco-friendly sorption technology. The adsorptive capacity of sorbents ($mgF\,g^{-1}$), the concentration range of F, the sorbent regenerability, its locally availability, and its cost with respect to the permissible limit reduction of fluoride were set as baseline for the feasibility of the eco-friendly sorption technologies.

19.2.2 Searching in literature database

Publications of nearly 725 were searched, out of which about 120 are selected after discarding the rest of papers due to insufficient scientific standard, deepness, and focus. The selection of the papers was performed based on relevance to the topic, timeliness, comparability, and novelty. The literature search for this exhaustive review was performed using web of scientific databases, ScienceDirect, Scopus database, ResearchGate, and Google Scholar. Over 100 years old publications were referred with emphasis on the years after 1919 is followed during reviewing about fluoride health effects. To do so, fluoride, toxicity, mitigation measures, low-cost adsorbents, hydroxyapatite (HAP), bentonite as fluoride adsorbent, diatomite as fluoride adsorbent, and batch and fixed adsorption were used as keywords.

19.3 Chemistry of fluorine

As reported by Peckham and Awofeso (2014), the fluoride ion is basic. Hence, hydrofluoric acid is a weak acid in water solution. Still, water is not an inert solvent in this case: when less basic solvents such as anhydrous acetic acid are used, HF_2 is the strongest of the hydrohalogenic acids. In addition, due to the basicity of the fluoride ion, soluble fluorides give basic water solutions. Since fluoride ion is a Lewis base, and has a high affinity to certain elements like aluminum and other trivalent and divalent cations, the reaction in the aqueous solution is rapid.

Fluorine and hydrogen can replace each other wherever it is found. This substitution of hydrogen with fluorine and vice versa for inorganic compounds offers a very large number of various compounds. Literatures reveal that an estimate of 1/5 of pharmaceutical compounds and 30% of agrochemical compounds contain fluorine as the main constituents. Bharti (2017) stated: "direct cellular poisons derived from Fluorine and fluorides by interfering with calcium metabolism and enzyme mechanism."

Furthermore, fluorides form an insoluble precipitate with calcium and lower plasma calcium. Hydrofluoric acid (HF_2) is straight corrosive to tissues. Inflammation and necrosis of mucous membranes may be occurred by neutral fluorides in 1%–2% concentrations.

The fluorine's reactivity makes it excellent for countering with and breaking down other compounds. Many industries like the microelectronics use fluorine to etch circuit patterns onto silicon and tungsten (Vithanage and Bhattacharya, 2015).

19.4 Production of fluorine

The very first study conducted by Argo et al. (1919) revealed that industrial production of fluorine involves the electrolysis of hydrogen fluoride in the presence of potassium fluoride (Eqs. 19.1 and 19.2). F_2 (g) formed at the anode and H_2 (g) gas at the cathode. At the same environment, the KF converts to KHF_2, which is the actual electrolyte. This KHF_2 aids electrolysis by greatly increasing the electrical conductivity of the solution.

$$HF + KF \rightarrow KHF_2 \tag{19.1}$$

$$2KHF_2 \rightarrow 2KF + H_2 + F_2 \tag{19.2}$$

The essential amount of HF_2 for the electrolysis is obtained as a byproduct of the production of phosphoric acid. Phosphate-containing minerals contain significant amounts of calcium fluorides, such as fluorite. Upon treatment with sulfuric acid, these minerals release hydrogen fluoride (Eq. 19.3):

$$CaF_2 + H_2SO_4 \rightarrow 2HF + CaSO_4 \tag{19.3}$$

19.5 Sources of fluoride

There are many different sources of either fluorine (F) or fluoride (F^-), but the main sources are the weathering of rocks, industrial emissions, and atmospheric deposition (Liu et al., 2014). Commonly, fluoride is found from sources like minerals, geochemical stores and on the other hand, a large proportion of the discharge of fluoride into subsoil recharge is through the degradation of rocks containing fluorine (Jacks et al., 2005). F^- is among the more abundant elements in earth crust and is present in various rocks with a range of approximately 100–1000 $\mu g\,mg^{-1}$, with $\mu g\,mg^{-1}$ being a typical value. High concentrations of F are present in granites, felsic, quartz monzonites, syenites, biotite, and granodiorites. F-containing rocks such as muscovite, pegmatites, amphibolites, and biotite micas supply F to groundwater and soil by different processes such as soil forming and weathering (Bhat et al., 2015).

There are also anthropogenic sources of fluoride such as the release of hydrogen fluoride (HF_2) to the environment or the discharge of fluoride to water bodies with various human activities, e.g., motorization, fluoridation of drinking water supplies, industrialization, and utilization of F-containing pesticides (Ghosh et al., 2012). Industries which produce phosphate fertilizers are the major source of fluoride as anthropogenic source. A considerable amount of F, e.g., 3.5%, is present in fertilizer made from phosphate extracted from rock, but this percentage is reduced in the manufacturing process.

Other man-made sources for fluoride are the current use of chemicals, such as hydrogen fluoride, calcium fluoride (CaF_2), phosphate manures, sodium fluoride (NaF), and sulfur hexafluoride (SF_6). When fluoride is emitted into the air, there is a direct exposure to wind and carried to the soil surface and vegetation. The contamination of soil with fluoride is basically due to use of fluoride-containing fertilizers such as phosphorous fertilizers. The fluoride content of soil usually varies from 150 to 400 $\mu g\,mg^{-1}$, and the value may rise in heavy clay soil up to 1000 $mg\,kg^{-1}$. The polluted soil eventually affects human beings through the inhalation of vaporized soil and with direct drinking of water contaminated with fluoride by its passage through the soil (Ziarati et al., 2019). Moreover, the use and production of the ceramic, zinc, phosphate fertilizers, and steel industries and energy plants are noteworthy sources of fluoride pollution in the environment (Paul et al., 2011).

19.6 Fluoride health effects on humans

The consumption of fluoride by human beings is commonly through the oral cavity and absorbed through the gastrointestinal tract and rarely through inhalation and dermal absorption.

Ingestion of excessive amount of fluoride may cause toxic and harmful effects. It is a vital to consider that the major source of fluoride toxicity remains oral hygiene products. According to fluoride poisoning data collected by the American Association of Poison Control (AAPC), toothpaste ingestion remains the main source of toxicity followed by fluoride-containing mouthwashes and supplements as reported by Cavalli and Flório (2018) in Table 19.1. The highest proportion (more than 80%) of the cases of fluoride toxicity was reported in children below the age of 6.

These reported toxicities are because the swallowing reflex in children is not completely developed and fluoride toothpastes are flavored, which results in voluntary toothpaste swallowing. The possible mechanisms of fluoride toxicity as per the explanation and detailed research work conducted by Barbier et al. (2010) are the following:

- When fluoride interacts with moisture, hydrofluoric acid is formed, and it results in burning of tissues due to low pH.
- The formation of chemical complexes due to calcium with fluoride leads to hypocalcemia, and physiological nerve functioning is resulted.
- Due to the fact that inhibition of enzymes required physiological functioning of cells, cellular poisoning may happen.
- The formation of electrolyte imbalance due to hypocalcemia and hyperkalemia, the cardiac rhythm disturbances may occur.
- Due to the reaction between fluoride and oxygen, hydrogen peroxide may be formed and this may disrupt the metabolism leading to toxic amount of fluoride in the body. Furthermore, the excessive production of free radicles may be the result that disturbs the antioxidant formation.
- Acute toxic effects and chronic effects are categorized as result of excessive amount of fluoride in the body.

One recent study by Zuo et al. (2018) on the potential adverse effects of overdose fluoride on various organisms, accumulation of excess fluoride in the environment poses serious health risks to plants, animals, and humans. This endangers human health, affects organism growth and development, and negatively impacts the food chain, thereby affecting ecological balance. Currently, the number of researchers focused on the molecular mechanisms linked with fluoride toxicity. The outputs of the research works reveal that fluoride can regulate intracellular redox homeostasis, induce oxidative stress, lead to mitochondrial damage and endoplasmic reticulum stress, and alter gene expression.

Though very serious fluoride poisoning is rarely reported, however, it may be fatal. It usually occurs due to the accidental consumption of fluoride solution or fluoride salts wrongly perceived as sugar solution or powdered eggs. The symptoms of acute fluoride toxicity depend upon the type and chemical nature of the ingested compound, the age, and the elapsed time between exposure and the beginning of management (Khandare et al., 2017).

TABLE 19.1 Percentage of reported cases of fluoride toxicity.

Cause of fluoride toxicity	Percentage of cases (%)
Toothpaste	68
Mouth rinses	17
Fluoride supplements	15

Adapted from Cavalli, A.M., Flório, F., 2018. Children's Menu Diversity: Influence on Fluoride Absorption and Excretion.

Long-term excessive fluoride (F) intake is known to be toxic and can damage reproductive tissues. Chronic toxicity of fluoride is more common than acute toxicity (Wei et al., 2016). The harmful effects of ingestion of fluoride at chronic level depend not only on the duration and dose but also on several other factors such as renal function, nutritional status, and interactions with other trace elements (Chai et al., 2017).

19.6.1 Dental fluorosis

The association between excessive ingestion of fluoride and dental mottling (fluorosis) was initially discovered over a century ago by Frederick Sumner McKay a practicing dentist in Colorado Springs area and G. V. Black. Dental fluorosis is the most sensitive and the earliest indicator of chronic fluoride toxicity. Although fluoride is an important element for caries prevention, the chronic intake of fluoride greater than $1\,mg\,L^{-1}$ or $0.1\,mg\,kg^{-1}$ daily during the period of tooth development interferes with the process of enamel and dentin formation and leads to dental fluorosis. The mechanism of dental fluorosis is very complex and not fully understood (Aimee et al., 2017).

The severity of dental fluorosis depends not only on excessive consumption of fluoride but also on the timing and duration of excessive fluoride consumption, plasma concentration of fluoride, type of fluoride consumed, renal function, and genetic factors.

Therefore, the following measures should be incorporated to be safer in order to prevent fluorosis:

- The fluoride level in the drinking water should be regulated between 0.5 and $1\,mg\,L^{-1}$ as suggested by the World Health Organization.
- Low-fluoride dentifrices ($500\,mg\,L^{-1}$) are indicated for children living in fluoridated areas.
- Supervised brushing and a smear layer of low-fluoride toothpaste should be applied on the brush.
- Following these precautionary measurements, the chances of fluorosis and related lesions will be reduced.

19.6.2 Skeletal fluorosis

Chronic fluoride exposure at more than the recommended levels either by ingestion, inhalation, or a combination of both results in skeletal fluorosis. This condition is characterized by an increase in bone mass and density because of deposition of excess fluoride within the bone matrix (Peicher and Maalouf, 2017; Shukla, 2016).

Radiographically, skeletal fluorosis may appear as osteosclerosis and calcification of ligaments. The neurological symptoms that occur because of fluoride toxicity are due to abnormal bone outgrowths. Primary symptoms of skeletal fluorosis usually occur in fluoride doses greater than $4\,mg\,L^{-1}$. While the crippling skeletal fluorosis is rare and is associated with intake of water with fluoride level greater than $10\,mg\,L^{-1}$, it results in a remarkable limitation of joint movements and deformities of major joints and spine leading to neurological problems (Mohammadi et al., 2017).

The severity of skeletal fluorosis depends on the amount of water intake, quality of water, renal disease, and dietary factors for instance calcium-rich diet, which has a protective effect and prevents toxic effects of fluoride on bones (Ramesh et al., 2017).

19.6.3 Miscellaneous effects

In addition to the negative effects stated earlier, excessive fluoride ingestion affects multiple body systems with distractions in the gastrointestinal system, abnormal sensations in toes and fingers, causes hematological manifestations including red blood cell deformation, liver, respiratory functions, and excretory system, neurological manifestations such as depression, excessive thirst, reduction in immune response, and headache (Ullah et al., 2017).

Key harmful effects are summarized below:

- Fluoride level ($> 3\,mg\,L^{-1}$) affects the reproductive system. Animals are more susceptible to chronic fluoride toxicity especially male reproductive system. This can be due to the production of free radicals that result in structural and histological changes in the reproductive system that disturb sexual functions.
- High fluoride level affects thyroid function due to rise in calcitonin activity.
- Chronic fluoride toxicity adversely affects both cell-mediated and humoral immunity; for instance, it destroys the white cell energy reservoirs that are required for phagocytosis of foreign agents and by inhibition of antibody formation.

The overall effects of ingesting high fluoride levels in humans and animals are shown in the flowchart (Fig. 19.1).

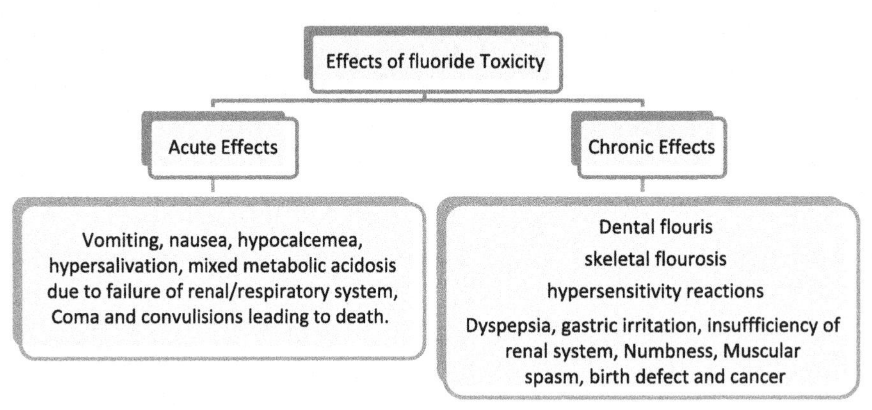

FIG. 19.1 Classification of toxic effects due to excessive ingestion of fluoride.

19.7 Defluoridation methods

Defluoridation is a process of removal of fluoride ion from drinking water. All the defluoridating method may broadly be classified and discussed here below:

To overcome the hazardous health impact of fluorosis, different approaches for defluoridations are available. These are membrane separation processes, coagulation-precipitation, ion exchange, adsorption techniques, electrodialysis, and electrochemical process (El Din Mahmoud and Fawzy, 2016). Each technology has their benefits and limitations and worked productively under ideal condition to remove fluoride to more noteworthy range. All the above approaches are examined briefly with their advantages and shortcomings.

19.8 Coagulation-precipitation method

Liming was the old-style method of removing fluoride from drinking water, and accompanying precipitation was fluorites. The precipitation and coagulation process with iron (III), activated alumina, alum sludge, and calcium has been widely investigated (Dargahi et al., 2016). The Nalgonda technique which uses lime and alum are the most usually utilized coagulants for defluoridation of water. Lime stimulation causes fluoride to precipitate as insoluble calcium fluoride and elevates the pH to 11–12. A leftover of 8.0 $mgF L^{-1}$ leaves from lime, and it is constantly connected with alum treatment to guarantee the best possible fluoride removal. As a first step, precipitation happens by lime dosing which is trailed by a second step in which alum is added to bring about coagulation. This technique is one of the widespread techniques broadly used for defluoridation of water in developing countries. The entire operation of a commonly used "fill and draw type" defluoridation unit for small community (around 200 people) can be completed within 2–3 h, with a number of batch performances in a day (Ayoob et al., 2008). Fig. 19.2 illustrates the scheme of mixing alum and lime based on Nalgonda technique. The mixture is stirred, and the precipitates containing fluoride settle as sludge at the bottom of the solution (Yami et al., 2018). When alum is added to water, two reactions happen. In the first reaction, alum reacts with an alkalinity's portion to deliver insoluble aluminum hydroxide

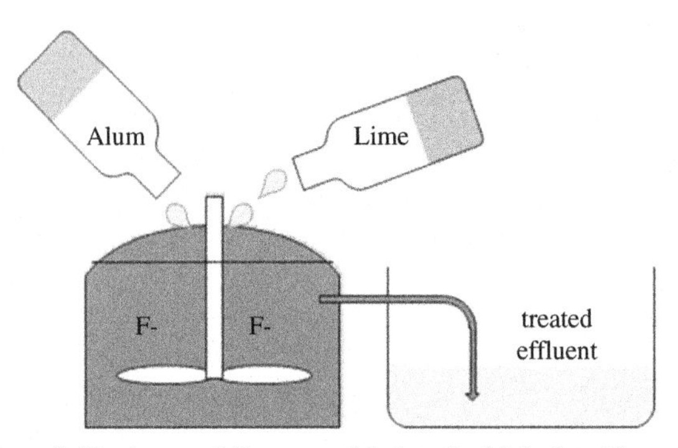

FIG. 19.2 Principle of Nalgonda technique (left): alum and lime are added to the high fluoride water, the mixture is stirred and precipitates containing fluoride settle as sludge to the bottom of the solution.

TABLE 19.2 Basic operational parameters of the Nalgonda technique.

Operational parameters	Values
Effective PH range	5.5–7.5
Bleaching powder requirement	$3\,mg\,L^{-1}$
Suitability	$20\,L\,day^{-1}$ utilization

TABLE 19.3 Advantages and shortcomings of coagulation-precipitation methods.

Advantages	Shortcomings
Most widely used technique	High consumption of alum
The technology is more practical	Concentrate issue
Easy to use and operate	Health-related issue with consumption of aluminum with finished water
The system is easy to understand	The utilization of aluminum sulfate as coagulant expands the sulfate ion concentration greatly which prompting cathartic impacts in human

$[Al(OH)_3]$. In the second reaction, alum reacts with fluoride ions in the water. Best fluoride removal is proficient at pH range of 5.5–7.5 as indicated in Table 19.2 (Gan et al., 2019).

In general, the advantages and shortcomings of coagulation-precipitation methods are summarized in Table 19.3.

19.9 Membrane process

The membrane separation process is more well-known for defluoridation of groundwater, wastewater treatment, and seawater desalination. The most normally utilized membrane separation processes for removal of fluoride are reverse osmosis, nanofiltration, Donnan dialysis and electrodialysis. Each process is discussed below.

19.9.1 Reverse osmosis technology

Reverse osmosis technology is a physical process in which the anions are removed by applying pressure on the feed water to direct it through the semipermeable membrane. Reverse osmosis operates at higher pressure with more protruding rejection of dissolved solids.

The membrane rejects the ions taking into account the size and electrical charge. Reverse osmosis technology is the reverse of natural osmosis as a result of applied hydraulic pressure to the high concentration side of the solution, and it forces solvent filter through the foils or tubes of the membrane, in contradiction of a pressure drop into the lower-concentration solution. In reverse osmosis, utilizing a mechanical pump, pressure is applied to a solution via one side of the semipermeable membrane to overcome inalienable osmotic pressure. The process likewise removes soluble and particulate matter, incorporating salt from seawater in desalination.

Generally speaking, the RO process has proven to be a very efficient process for defluoridation of drinking water supplies because it works at very low pressure and, in addition to fluoride, other inorganic pollutants are also removed without making any additional effort (Ingole and Patil, 2012). The advantages of RO system are summarized as follows:

- Proven technology.
- It removes chemical and biological contamination to the permissible limit of WHO standards.
- Recognized and desired technology communities—willingness to pay.
- Widespread implementation—supply chain and operational benefits.

Drawbacks of the RO system:

- The RO system is more costly to install when compared with other defluoridation technologies.
- It may remove nutritional minerals (like magnesium, potassium, and calcium) that our body needs.
- Removing minerals drops the pH of drinking water, which increase free radicals, with the protentional of the cancer risk.
- Heavily contaminated brine.

19.9.2 Nanofiltration membrane process

As stated by Hu and Dickson (2006), nanofiltration (NF) is the later innovation among all the membrane processes utilized for defluoridation of water. The essential contrast in the NF and RO membrane separation is that NF has somewhat bigger pores than those utilized for reverse osmosis and offer less resistance to entry of both solvent and solute. As an outcome, pressures needed are much lower, energy prerequisites are less, removal of solute is substantially less thorough, and flow is faster. Singh et al. (2016) also confirmed nanofiltration makes use of a similar overall development as reverse diffusion. For nanofiltration, the membranes have slightly larger pores than those used for reverse diffusion and provide less resistance to passage each of solvent and of solutes. As a result, the pressures required are much lower, the energy required is reduced, and the NF may run at lower pressures while still producing the same permeate flow.

The property of nanofiltration relative to reverse diffusion may be a specific advantage, and experimental and theoretical analyses are being dedicated to getting a mechanism of substance retention to facilitate production and choice of targeted membranes in addition to optimization of conditions. The NF performance is usually described in terms of the pure water flux (Jw), the solution flux (Jv), and the rejection (R). In comparison to other types of fluoride removal systems, investigations on NF's performance on fluoride removal from water are relatively scarce.

19.9.3 Dialysis

Dialysis separates solutes by transport of the solutes through a membrane instead of utilizing a membrane to hold the solutes while water goes through it as in reverse osmosis and nanofiltration. The membrane pores are a great deal less prohibitive than those for nanofiltration, and the solute can be driven through by either the Donnan effect or a connected electric field.

Donnan dialysis, also known as diffusion dialysis, is similar to ion-exchange membrane but differs from electro-membrane processes in that the driving force is not an electric current, but rather a difference in chemical potential. Concentration difference is the most obvious driving force for ion transport in Donnan dialysis. A negative ion can be driven out of a feed solution through Donnan dialysis is equipped with anion exchange membrane by utilizing a second alkaline stream (Ktari et al., 1993).

Donnan dialysis (DD) is being applied to such diverse problems as enrichment of trace levels of ions, metal separations, water softening, and recovery of metals from waste streams (Cox and DiNunzio, 1977; Wilson and DiNunzio, 1981) and the common schematic diagram is shown in Fig. 19.3.

19.9.4 Electrodialysis

Electrodialysis (ED) is a process that removes ionic components under the driving force of an electric current from aqueous solutions through ion-exchange membranes. ED separates ionic contaminants from water like reverse osmosis, except current, rather than pressur. Generally, ED is not suitable for rural water treatment because of use of energy.

The ED processes presented in this literature are mainly explained in two different ways. These are with a pretreatment and a chemical pretreatment. When we see the ED without pretreatment, the precipitation risk should be minimized. In this case, it has been evaluated by Majewska-Nowak et al. (2015) considering two steps. The first one

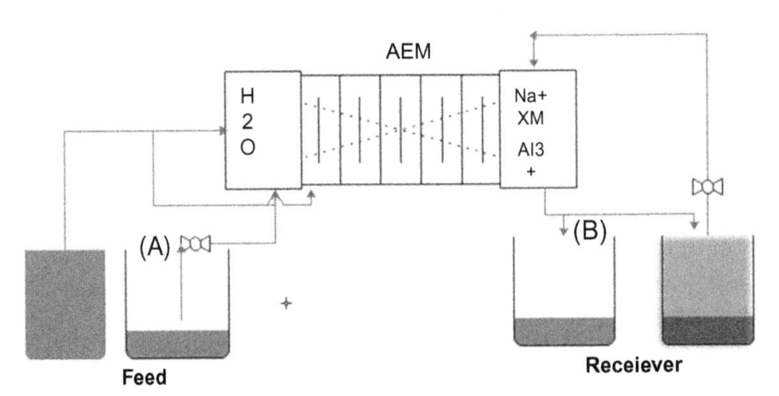

FIG. 19.3 Schematic flow diagram of Donnan dialysis system: circuit (A) open receiver, circiut (B), closed receiver.

is with 10 V and ACS-CMX membrane is used to stop the bivalent ions. And the second one is a 10 V with a conventional membrane AFN-CMX in order to remove the ionic contaminants. Since the second method requires chemical additives in the pretreatment step and, consequently, has more impact on the environment than the first one, the first method which is without pretreatment is technically easy and preferable for ED process. However, the high-risk precipitation of the bivalent ions in the concentrate compartment is the major drawback of ED process.

19.10 Ion-exchange process

As explained by Zarrabi (2014), fluoride can be removed from water supplies with a strongly fundamental anion-exchange resin containing quaternary ammonium functional groups. The removal takes place according to the following reaction (Eq. 19.4):

$$\text{Matrix}-\text{NR}_3+\text{Cl}^-+\text{F}^-\rightarrow\text{Matrix}-\text{NR}_3+\text{F}^-+\text{Cl}^- \tag{19.4}$$

The fluoride ions substitute the chloride ions of the resin. This process proceeds until every one of the sites on the resin is possessed. Water that is supersaturated with dissolved sodium chloride salt can be used which backwashed the resin. New chloride ions then substitute the fluoride ions prompting recharge of the resin and beginning the process once more. The driving force for the substitution of chloride ions from the resin is the stronger electronegativity of the fluoride ions.

19.11 Electrocoagulation process

Electrocoagulation (EC) is a technique for applying direct current to sacrificial electrodes that are submerged in an aqueous solution. EC is a straightforward and efficient technique to remove the flocculating agent produced by electro-oxidation of a sacrificial anode and generally made of iron or aluminum. In this process, the treatment is performed without including any chemical coagulant or flocculants.

The mechanism of fluoride removal by EC process is carried out through an electrochemical coprecipitation of fluoro-aluminum complexes (Eqs. 19.5 and 19.6; Fig. 19.4) and by a chemical substitution reaction as stated in Sandoval et al. (2014).

$$\text{nAl}^{3+}+(3n-m)\text{OH}^-+\text{mF}^-\rightarrow\left[\text{Al}_n\text{F}_m(\text{OH})_{3n-m}\right]_{(s)} \tag{19.5}$$

$$\text{Al}_n(\text{OH})_{3n(s)}+\text{mF}^-+\text{mOH}^-\rightarrow\text{Al}_n\text{F}_m(\text{OH})_{3n-m(s)} \tag{19.6}$$

The EC process is governed by the following process at the anode (Eq. 19.7) and the cathode (Eq. 19.8), respectively. In the bulk solution, at the pH 7, the aluminum flocs in the form of aluminum hydroxide and aluminum oxides will be formed as illustrated in Eqs. (19.9) and (19.10). These aluminum aggregates interact with the pollutants to produce flocs and motivate the removal of pollutants.

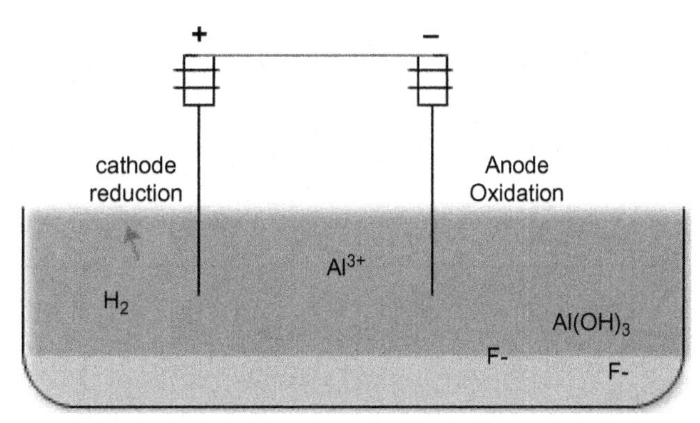

FIG. 19.4 Schematic principle of electrocoagulation.

$$Al_{(s)} + Al^{3+} + 3e^{-} \quad (19.7)$$

$$2H_2O + 2e^{-} + H_{2(g)} + 2OH^{-} \quad (19.8)$$

$$Al^{3+} + 3H_2O \rightarrow Al(OH)_{3(s)} + 3H^{+} \quad (19.9)$$

$$2Al^{3+} + 3H_2O \rightarrow Al_2O_{3(s)} + 6H^{+} \quad (19.10)$$

In literature by Grich et al. (2019), it is also showed that the removal of fluoride is not attractive at $J > 7\,mA\,cm^{-2}$ (Fig. 19.4), because the massive H_2 and O_2 bubble production at both the anode and the cathode, in the filter press cells, causes the rupture of aluminum aggregates.

Due to surface activity, the fluoride ion gets preferentially adsorbed on the bed surface thereby causing a reduction of fluoride ion in the exit stream. In this particular review, an adsorptive method is discussed focusing on two adsorbents.

19.12 Adsorption techniques

The applicability of adsorption process depends on the adsorbents and their physical and chemical properties. The used sorbent should be expected to have high selectivity, high adsorption capacity, and long life. Furthermore, the used sorbent should be available in abundance at economical costs. A wide variety of adsorbents are commercially available and successfully used for the removal of organic and inorganic pollutants (Mahmoud et al., 2016, 2018c, 2020). A large number of adsorbent materials have been developed and tested, mostly in the laboratory, for the treatment of fluoride-contaminated water including manganese-oxide-coated alumina, bone charcoal, fired clay chips, fly ash, calcite, sodium-exchanged montmorillonite-Na^+, ceramic adsorbent, laterite, granular ferric hydroxides, cerium-based adsorbent (CTA), CeO_2-TiO_2/SiO_2 surface composites, schwertmannite (Xiuru et al., 1998; Kumar et al., 2009; Eskandarpour et al., 2008), unmodified pumice, bauxite, zeolites, fluorspar, iron-oxide-coated sand, calcite, activated quartz, and activated carbon (Biswas et al., 2009; Ma et al., 2009). Also, a large number of low-cost adsorbents such as bentonite, kaolinite, acid activated red mud, chitosan, biomass, fly ash, diatomite, and industrial wastes are used for fluoride removal (Farrah et al., 1987; Ghorai and Pant, 2005; Centeno et al., 2016; Apambire et al., 1997). The advantages and limitation of the adsorption method is indicated in Table 19.4.

Many natural and synthesized sorbent materials have been tried in the past to find out an efficient and economical treatment techniques (Mahmoud et al., 2018b; Mahmoud and Fawzy, 2015; Nasr et al., 2017). In this review, a low-cost modified and unmodified bentonite clay material, HAP and diatomite minerals, has been reviewed in consideration to address the problem of developing countries water quality and scarcity.

19.13 Bentonite clay

Bentonite is a common group of clay minerals, which is a hydrous aluminum silicate, and it has been reported as an economical material for adsorption of fluoride from water. Bentonite is an aluminum phyllosilicate clay material consisting mostly of montmorillonite. It is reported that bentonite is a material derived from the alteration, over geological periods, of glassy material emitted from volcanoes (tuff or ash), or from the alteration of silica-bearing rocks such as granite and basalt; this requires hydration (taking up or combination with water) and a loss of alkalis and bases (Hosterman and Patterson, 1992). In addition to its constituent mineral montmorillonite, bentonite can have a variety of accessory minerals depending on the nature of formation. These are attapulgite, kaolin, mica, illite, quartz, feldspar, calcite, and gypsum.

TABLE 19.4 The main points of interest and restrictions of adsorption process.

Advantages	Limitations
Locally available adsorbent materials to use	Process is dependent on pH
High efficiency	Presence of sulfate, phosphate, bicarbonate, etc. results in ionic competition
Cost-effective	Regeneration and disposal are required for fluoride-laden material

However, its low adsorption capacity is the major problem for its possible application as adsorbent for fluoride-rich groundwater. Therefore, it is a vital activity to modify the surface of bentonite clay to increase its adsorption capacity and overcome its drawbacks for fluoride removal.

19.13.1 Modification of clay materials using acids and bases

Literature survey suggests the fact that the chemical nature and pore structure of clay materials generally influence their adsorptive capability and their regenerability. In order to increase their adsorptive capacity, modification of the surface of clay materials can be conducted chemically as well as physically.

Basically, the modification can alter the clay structure to enlarge its surface area, therefore increasing the adsorption capacities (Toor, 2011). The chemical composition of clay minerals varies depending upon the origin influencing the layer charge, cation exchange capacity, adsorption capacity, and morphology. These factors play a significant role in the modification of natural clays (Steudel et al., 2009). Inorganic acids, bases, salts, and surfactants are the major chemicals used for modification of clay minerals. Furthermore, mixture of chemical and physical treatment processes was also suggested to modify the surface and structure of clay minerals (Ismadji et al., 2015).

Various researchers conducted a study on modification of the surface of clay minerals through impregnation using acids, bases, and salts. Usually, inorganic acids such as hydrochloric acid (HCl), sulfuric acid (H_2SO_4), and nitric acid (HNO_3) were used for that purpose. While for the modification using base, the solution, the alkali, and the alkali earth metal hydroxide such as NaOH, KOH, $Ca(OH)_2$, or $Al(OH)_3$ were employed as impregnating agents. The anion salts such as potassium dihydrogen phosphate (KH_2PO_4), orthophosphate, tri-polysulphate, and sodium tetraoxosulphate(VI) were among the chemicals used for surface modification of several clay minerals (Ajemba and Onukwuli, 2014; Bergaya and Lagaly, 2001; Ismadji et al., 2015).

Temuujin et al. (2006) studied the influence of acid concentration on the pore structure, crystalline phases, and chemical composition of montmorillonite. The chemical composition of the clay used in their study was 40% montmorillonite, 40% amorphous compounds, 4% cristobalite, 5% feldspar, 2% calcite, and traces of quartz. Since the clay contains large amount of impurities, the purification of clay increased the montmorillonite content to 92.8%. The acid concentration and the duration of the acid modification strongly affected the pore structure and the acidity of the montmorillonite. H^+ in the surface exchange leads to the acidity of the montmorillonite clay for the sites (Brønsted acid sites). These acid sites were generated by the exchange of interlamellar cations (Mg^{2+} and Al^{3+}) of the octahedral sheet of montmorillonite with protons (H^+) from the sulfuric acid. With increasing acid concentration and impregnation time, considerably greater amounts of octahedral Al^{3+} and Mg^{2+} were removed, leading to a decrease in the Lewis acidity.

The major drawbacks of the clay minerals for the use of adsorption system are their low adsorptive capacity. Toor and Jin (2012) studied the modification of sodium bentonite used thermal activation, acid activation, and combined acid and thermal activation. On that study, surface modification was done on Congo red bentonite clay and the effects of process parameters of the activation such as contact time, dye concentration, bentonite dosage, pH, and temperature on the adsorption performance were assessed. As a result, the bentonite modified by combining an acid and thermal activation gave a better adsorption performance compared to standalone thermal activation and acid activation.

19.13.2 Modification of clay materials using surfactants

The intercalation of surfactant ions onto the clay mineral structure is a vital conversion of the initially hydrophilic clay mineral particles to hydrophobic organoclays. Surfactants are organic compounds which possess polar and nonpolar characteristics. Surfactants are a big group of surface-active compounds with many applications. Surfactants are divided based on their dissociation in water: cationic surfactants, anionic surfactants, and nonionic surfactants.

Cationic surfactants mostly are linear alkyl amines and alkyl ammonium substances (fatty amine salts and quaternary ammoniums). Anionic surfactants include alkylbenzene sulfonates, soaps, dialkyl sulfosuccinate, lauryl sulfate, and lignosulfonates. The nonionic surfactants are not dissociated in water, and these surfactants include ethoxylated linear alcohols, ethoxylated alkyl phenols, fatty acid esters, amine, and amide derivatives (Ismadji et al., 2015).

Quaternary ammonium salt in the alkane solvent environment was studied for the surface modification of sodium montmorillonite as cation surfactant by Yu et al. (2014). The intercalation and ion-exchange reaction in alkane solvents have shown the change in the surface structure and properties of the montmorillonite (Yunfei Xi et al., 2005).

Shariati-Rad et al. (2014) studied the modification of bentonite using glycol bis-*N*-cetylnicotinate dibromide, a cationic gemini surfactant. The XRD and FTIR results revealed that the cationic gemini surfactant successfully intercalated into the bentonite layer. For comparison, the authors also modified the bentonite using conventional cationic surfactant (cetyltrimethyl ammonium bromide). It is observed that the gemini surfactant-modified bentonite has a better adsorption capacity than CTA-bentonite. However, the rapid decrease in the adsorption efficiency of gemini surfactant-modified bentonite at pH above 6 was observed. This is due to the fact that the ester groups in the gemini surfactant molecules are possibly hydrolyzed in alkaline environment.

The strong hydrophobic interaction between the adsorbed surfactants and organic pollutants facilitated was found due to the gemini surfactants containing longer spacer group arranged relatively loose in the interlayer being though, and hence, it derived to the higher adsorption capacity of organoclays (Wang, 2018). The adsorption performance of modified bentonite was evaluated through adsorption of 2-naphthol from aqueous phase solution. The experimental results of adsorption indicate that the optimum condition for 2-naphthol removal was achieved at pH 6 (Wang, 2018).

Kurniawan et al. (2011) have utilized the extract from berry soap for the preparation of organo-bentonite. The bentonite was used as the raw material for rarasaponin-bentonite preparation. The bentonite was obtained from bentonite mining located in Pacitan, East Java province, Indonesia. Prior to the modification, the impurities of bentonite were removed by impregnation using hydrogen peroxide solution. A combined microwave radiation method was used for the impregnation to prepare the rarasaponin-bentonite/organo-bentonite. After intercalation using rarasaponin, the increment of basal spacing (d_{001}) of the bentonite was observed (1.42 for bentonite and 2.67 nm for rarasaponin-bentonite). Methylene blue was used as a adsorbate for examining the adsorption capacity of rarasaponin-bentonite. The highest adsorption capacity of rarasaponin-bentonite obtained in their study was 256 $mg\,g^{-1}$ while for raw bentonite was 194 $mg\,g^{-1}$ at 60°C. The modification of clay minerals as adsorbent has received much attention over the past decades, especially in adsorption and separation processes (Zhou et al., 2012).

19.14 Review of literature on both modified and unmodified bentonite clay for fluoride removal

Ma et al. (2005) synthesized granular Zr-loaded bentonite (GZLB) prepared with polyvinyl alcohol (PVA) for the removal of fluoride from aqueous solutions. Also, the regenerated spent bentonite by alum revealed the increase of the defluoridation capacity of prepared sorbent. Considering the pH 6.97 and inlet fluoride concentration of 6.34 $mg\,L^{-1}$, the adsorption capacity of 755 $mg\,kg^{-1}$ was obtained.

Modification of bentonite clay with an electropositive atom (using lanthanum, magnesium, and manganese), in order to enhance its adsorption capacity for fluoride ions from drinking water, was conducted by Kamble et al. (2009). High fluoride uptake was observed at 10% La-bentonite for defluoridation of drinking water as compared to Mg-bentonite, Mn-bentonite, and bare bentonite clay. At an adsorbent dose of 1.0 $g\,L^{-1}$, the maximum loading capacity of 10% La-bentonite was found to be 1.4 $mg\,g^{-1}$. Furthermore, the uptake of fluoride in acidic pH was higher as compared to alkaline pH and maximum fluoride removal was found at pH 5. The fluoride removal was decreased, due to the competition for adsorbent's active sites by excessive amount of hydroxyl ions at higher pH. The increase in the amount of positive charge in the presence of cations on the oxide surfaces and/or forming of a positively charged surface may helped the uptake of fluoride. The Langmuir maximum sorption capacity of 10% La-bentonite for fluoride was reported as 4.24 $mg\,g^{-1}$. The fluoride uptake mechanism was explained through exchange of OH^- ions from 10% La-bentonite surface, which was shown from the results of pH. Fluoride removal by montmorillonite clay has been evaluated by Karthikeyan et al. (2005).

Effects of its particle size, temperature, and pH were investigated, and fluoride uptake was found maximum at pH 2, and it decreased with increase in pH. Particle size of 75 μm showed maximum uptake of fluoride when compared to the other particle sizes. At different temperatures, the Langmuir maximum sorption capacity was varied between 1.48 and 1.91 $mg\,g^{-1}$. The presence of trivalent ion (HCO_3^-) has adversely affected the adsorption of fluoride, and the study also reveals that adsorption occurred on the surface as well as through intraparticle diffusion pattern of the adsorbent material. Using X-ray diffraction, it is confirmed that the deposition of fluoride on the surface of the clay material was occurred. FTIR analysis also exhibited the involvement of hydroxyl group present on the surface as mechanism for the adsorption.

Another study was conducted by Tor (2006) for the defluoridation of aqueous solution using montmorillonite. At pH below 5, adsorption was not favorable due to the formation of weakly ionized hydrofluoric acid in acidic conditions, but at pH 6 the maximum removal of fluoride was obtained. The fluoride saturation capacity of montmorillonite was reported 0.263 $mg\,g^{-1}$ at room temperature. The uptake of fluoride by the adsorbent was due to the interaction between the metal oxides at the surface of montmorillonite and fluoride ions. Also, desorption was made by washing the adsorbent with a solution pH of 12.

Vhahangwele et al. (2014) introduced Al^{3+} bentonite clay (Alum-bent) by ion exchange of base cations on the matrices of bentonite clay for the removal of fluoride. Modification of bentonite clay with Al^{3+} was performed in batch experiments. Parameters optimized include time, dosage, and Al^{3+} concentration. X-ray fluorescence, X-ray diffraction, energy-dispersive X-ray spectrometry attached to scanning electron microscopy, Brunauer-Emmett-Teller analysis, cation exchange capacity (CEC) by ammonium acetate method, and pH_{pzc} by solid addition method were performed for the physicochemical characterization of raw and modified bentonite clay.

The effect of contact time, adsorbent dosage, adsorbate concentration, and pH were evaluated for fluoride removal, in batch modes at (26 ± 2)°C room temperature. The adsorption capacity of fluoride by modified bentonite clay was observed to be 5.7 mg g^{-1}. The highest adsorption of fluoride was reached at 30 min, 1 g of dose, 60 mg L^{-1} of adsorbate concentration, pH of 2–12, and a solid/liquid (S/L) ratio of 1:100. On the kinetic studies, pseudo-second-order model was well fitted than pseudo-first-order fluoride adsorption. Monolayer and multilayer adsorption was confirmed due to the Langmuir, and Freundlich adsorption isotherms were well fitted. Alum-bent showed good stability in removing fluoride from groundwater to below the prescribed limit as stipulated by the World Health Organization. We can say that Alum-bent is a potential defluoridation adsorbent for the application of point-of-use devices for defluoridation of fluoride-rich water in rural areas especially for the low-income developing countries. And hence, Alum-bent is a promising adsorbent with a high adsorption capacity for fluoride that may replace the expensive water defluoridation technologies.

Thakre et al. (2010) have studied the removal of fluoride by modification using magnesium chloride to enhance its adsorptive capacity. The Mg^{2+}-modified bentonite (MB) was characterized by using XRD and SEM techniques. Various operational parameters such as adsorbent dose, contact time, pH, effect of co-ions, and initial fluoride concentration were experimented by batch mode. The maximum fluoride removal capacity of 2.26 mg g^{-1} was observed over wide range of pH and initial fluoride concentration of 5 mg L^{-1}, which is much better than the raw bentonite. Langmuir adsorption isotherm and pseudo-first-order kinetics models were well fitted into the experimental data. Thermodynamic study reveals that fluoride adsorption on MB is reasonably spontaneous and endothermic process. MB exhibited meaningfully high fluoride removal in synthetic water as compared to real groundwater. The study also considers desorption of MB in which that almost all the loaded fluoride was desorbed (~ 97%) using 1 M NaOH solution. The maximum fluoride removal of 73% after regeneration was applied. Therefore, chemical modification enhances the fluoride removal efficiency of bentonite and it works as an effective adsorbent for defluoridation of water.

Mn^{2+}-modified bentonite clay was also used for groundwater defluoridation (Mudzielwana et al., 2018). The effect of contact time at various adsorbent dosages, adsorption isotherms, and the effect of pH on fluoride removal were evaluated by batch experiments. At first 30 min of contact time and increased adsorbent dosage, the percentage removal was achieved. The data fitted better to pseudo-second-order reaction indicating that F^- adsorption occurred via chemisorption. Both surface and intraparticle diffusion processes were involved during the F^- adsorption process as evaluation made by Weber-Morris model of intraparticle diffusion. Furthermore, at pH 4 the optimum of 75.2% fluoride removal was achieved. The electrostatic attraction and ion-exchange process are the primary processes responsible for fluoride adsorption at acidic pH and at moderate to alkaline pH levels. Therefore, the study verified that Mn^{2+}-modified bentonite clay has potential for application in defluoridation of groundwater.

On other study conducted by Gitari et al. (2013), the use of raw unprocessed bentonite clay and its Fe^{3+}-modified form for fluoride adsorption through a series of batch adsorption experiments was evaluated. Incarnation of Fe^{3+} on bentonite was attained by mixing the crushed clay with 80 ppm Fe^{3+} solution for 15 min at ratio of 2 g 100 mL^{-1}. It was observed that the raw bentonite clay increases the pH of the F^- solution as opposed to the Fe^{3+}-modified bentonite. Fe^{3+} bentonite exhibited 100% F^- removal as different to raw bentonite, while the raw bentonite clay reveals release of F^- over the same pH range. This is critical to consider for the application of the adsorbent for defluoridation at the normal pH of the groundwater with no need for adjustment. Iron (III)-modified bentonite was effective in fluoride removal in high fluoride groundwater samples. The experimental data fitted well to Langmuir adsorption model which indicates a monolayer coverage of the adsorbent. Room temperature was also found to be favorable for the adsorption process. The results of the work indicated that iron (III)-modified bentonite has applicable for the use of onsite water defluoridation systems for household use in rural areas in South Africa.

Assaoui et al. (2018) investigated using natural bentonite clay (NBC) through batch adsorption experiments for wastewater defluoridation. X-ray diffraction (XRD), X-ray fluorescence (XRF), energy-dispersive X-ray attached to scanning electron microscopy (SEM-EDX), BET specific surface area (SSAN2BET) analysis, and Fourier-transform infrared spectrometry (FTIR) were applied for the determination of the mineralogical and physicochemical characterization of the adsorbent. Batch adsorption experiments were conducted to evaluate the effect of various operational parameters such as contact time, initial fluoride concentration, adsorbent dose, and initial pH solution at room temperature $(25 \pm 2$°C).

The experimentation results reveal that 30 min of contact time was sufficient for attaining equilibrium. For 5 mg L^{-1} and 2 g L^{-1} of initial fluoride concentration and adsorbent dose, the maximum wastewater defluoridation (52.2%) was obtained in the acidic conditions (pH = 2). The pseudo-second-order kinetic adsorption model was fitted well to the experimental data. The maximum adsorption capacity was achieved by applying NBC to be 4.4 mg g^{-1}. The Freundlich isotherm model was fitted well with the adsorption data confirming multilayer adsorption with heterogeneous energetic distribution of active sites and with interaction between adsorbed molecules. The adsorption mechanism might be through ion-exchange adsorption, accompanied by interactions of fluoride with metal ions forming metal-F complexes.

In the study conducted by Zhang et al. (2013), new adsorbent namely bentonite/chitosan beads has been synthesized and studied for its defluoridation efficiency. Following the preparation of beads using inverse suspension polymerization method, the bentonite/chitosan beads (bentonite dosage of 3.0 g) were used for defluoridation and showed an adsorption capacity of 0.895 mg g^{-1} whereas chitosan beads had only 0.359 mg g^{-1}. The maximum defluoridation capacity of 1.164 mg g^{-1} was observed at optimal pH value of 5. The adsorption capacity was evaluated by the process parameters such as effect of temperature, contact time, and initial fluoride concentration of the adsorbent. The mineralogical and physicochemical characterizations were also made by scanning electron microscopy (SEM), energy-dispersive X-ray analysis (EDAX), and Fourier-transform infrared spectrometry (FTIR). Freundlich isotherm model and pseudo-second-order kinetic model were fitted well with the adsorption data.

In another study done by Zhang et al. (2014), a novel adsorbent, La(III)-loaded bentonite/chitosan beads (La-BCB), was prepared for defluoridation from aqueous solution. To examine the defluoridation behavior of the adsorbent, the effects of various parameters such as dosage of La (III), pH, temperature, contact time, initial fluoride concentration, and presence of coexisting anions were investigated. At pH 5 and temperature of 30°C, the maximum defluoridation capacity of La-BCB of 2.87 mg g^{-1} was achieved. The characteristics of La-BCB were analyzed by scanning electron microscopy (SEM), energy-dispersive spectroscopy (EDX), and Fourier-transform infrared spectroscopy (FTIR). Both Langmuir and Freundlich isotherm models were fitted well with equilibrium fluoride adsorption data, and pseudo-second-order kinetic and particle and intraparticle diffusion models were also fitted well. The presence of coexisting ions (carbonates and bicarbonates) reduced the defluoridation capacity of La-BCB while sulfate, nitrate, and chloride showed slight effect. Only a 17% loss was observed from the used adsorbent, but the rest was successfully regenerated using sodium hydroxide as a regenerant.

As shown in Table 19.5, bentonite clay was chemically modified using Al^{3+} by Vhahangwele et al. (2014). The aluminum-modified bentonite (Al-MB) exhibited significantly high fluoride removal efficiency over a wide pH range of 2–12. Maximum fluoride removal capacity of 5.6 mg g^{-1} was achieved at maximum fluoride concentration of 60 mg L^{-1} which is much better than the raw bentonite.

For low concentrations of fluoride, bentonite clay modified by Thakre et al. (2010), is effective. In a wide pH range of 3–10, the magnesium-incorporated bentonite (MB) showed significantly high fluoride removal efficiency. Coexisting anions had no major effect on fluoride removal efficiency of MB except bicarbonate which has showed slight effect. A maximum fluoride removal capacity of 2.26 mg g^{-1} was achieved at an initial fluoride concentration of 5 mg L^{-1}, which was reported better than the unmodified bentonite.

The order of fluoride removal is presented as follows: Bentonite > Charfines > Kaolinite > Lignites > Nirmali Seeds.

TABLE 19.5 Comparison of different types of adsorbents used as fluoride adsorbent in batch and column study.

Adsorbent	Experiment type	pH	Concentration range (mg L^{-1})	Adsorption capacity (mg g^{-1})	References
Al^{3+}-modified bentonite clay	Batch	2–12	60	5.7	Vhahangwele et al. (2014)
Mg^{2+}-modified bentonite clay	Batch	3–11	5	2.26	Thakre et al. (2010)
Mn^{2+}-modified bentonite	Batch	4	5	2.4	Mudzielwana et al. (2018)
Fe^{3+}-modified bentonite clay	Batch	2	10	2.91	Gitari et al. (2013)
Natural bentonite clay	Batch	2	5	4.4	Assaoui et al. (2018)
Bentonite/chitosan beads	Batch	5	100	1.164	Zhang et al. (2013)
La^{3+} bentonite/chitosan beads	Batch	5	15	2.87	Zhang et al. (2014)
Zr-loaded bentonite	Batch	6.97	6.34	0.755	Ma et al. (2005)
Granular acid-treated bentonite	Batch/column	4.9	6	0.009	Ma et al. (2011)

19.15 Apatite materials: HAP

The HAP is a highly crystallized material. Its specific surface area was $0.052\,m^2 g^{-1}$ and very close to that of calcite, quartz, and fluorspar. The gradual uptake of the fluoride in HAP could be due to ion exchange rather than intraparticle diffusion. The fluoride exchanged with some elements inside HAP, giving a highest adsorption capacity. The most abundant of phosphatic minerals is HAP ($Ca_{10}(PO_4)_6(OH)_2$) and is usually used as a raw material in industrial chemistry. In recent extensive studies, HAP was used as bone and tooth implants since its chemical composition and crystallographic structure are similar to those of human hard tissue (Suchanek and Yoshimura, 1998).

Spinelli et al. (1971), stated that the sorption capacity of HAP was found to increase with decrease in pH (below pH 7) and increase in surface area and more suitable for low initial concentration of fluoride. The crystal structure of the apatite formed at high concentrations of fluoride was found to be hexagonal plate-like, while the apatite formed at low F^- concentrations was found to be dendritic or needle-like.

In another study by Fan et al. (2003), they found the defluoridation efficiency of the materials to be in the order: tricalcium phosphate, TCP (87.0%) > HAP (68.0%) > bone char, BC (66.4%). Combination of free phosphate with BC or HAP removed 95% F^- from high initial concentrations. High fluoride removal ability of HAP makes it a good choice for application in defluoridation.

Fan et al. (2003) established the order of F^- adsorption capacities of various materials as: HAP > Fluorspar > Quartz (activated using ferric ions) > Calcite > Quartz.

The "fill, mix, and filter" technique, which involves mixing F-containing water with Ca^{2+} and PO_4^3 ions and then putting it into contact with BC saturated with F, exhibited 95%–98%removal of F (Dahi, 1997).

The investigation on F^- removal from contaminated water also started simultaneously with the study of the causes of dental fluorosis. A related work on the reaction between powdered enamel and F^- solutions was studied with X-ray diffraction (XRD) technique (Nath and Dutta, 2015).

Waghmare and Arfin (2015) indicated that the main constituent of teeth and bone is HAP, which has been proved to be a good adsorbent for F^-. It is a general perception that for every naturally occurring problem, people try to find out the remediation of the problem by applying the tricks hidden in the nature. Thus, F^- removal was also tried in a natural way in 1937 by using degreased bone as teeth and bone absorb fluoride a lot. The hydroxide group of the HAP of the degreased bones is replaced by anion exchange with F^-, forming FAP. The FAP was reconverted to HAP by treatment with NaOH. A controlled reaction between phosphoric acid and lime was found to give two products, namely calcium phosphate and HAP both capable of removing fluoride. A similar removal process was reported where phosphate and calcium were added to F^- containing water to form FAP. The removal improved with increase in the pH above 10 by addition of $Ca(OH)_2$ (Guan et al., 2015).

Rawat and Patel (2018) and Nath and Dutta (2015) have explained the chemical reactions occurring in these defluoridation processes are suggested to be as follows by Eqs. (19.11) and (19.12):

$$NaF + 3H_3PO_4 + 0.5CaCl_2 + 4.5Ca(OH)_2 \rightarrow Ca_5(PO_4)_3F_{(s)} + NaCl + 9H_20 \tag{19.11}$$

$$Ca_5(PO_4)_3OH_{(s)} + NaF \rightarrow Ca_5(PO_4)_3F_{(s)} + NaOH \tag{19.12}$$

Nath and Dutta (2015) contributed significantly to understand the chemistry of the formation of FAP from HAP and the stability of the different composition of the compounds formed during the course of the reactions under different conditions. The works concluded that the conversion of HAP to FAP took place through the formations of different apatite compounds depending upon the system's pH, temperature, and the chemicals used as the source of F^-, e.g., NaF, and KF.

Nanomaterials are well known because of their enhanced properties. The decrease in the particle size causes increase in surface area which improves the adsorption properties of nanomaterials (Mahmoud, 2020). HAP nanocomposite with chitin showed greater F^- removal capacity ($2840\,mg\,kg^{-1}$ of F^-) than the single nano-HAP ($1296\,mg\,kg^{-1}$ of F^-) (Sundaram et al., 2008). If the treatment cost of F^- removal by HAP or nano-HAP can be made affordable for common people, then defluoridation by using HAP will have a potentiality to be used worldwide. However, due to their nano-size, nanoparticles may create new hazards for environment and human health if it is present in the water after treatment. Therefore, it should be separated from the water after treatment (Mahmoud et al., 2018a).

19.15.1 Mechanism of fluoride removal by HAP

The elimination of fluoride from using HAP has been proven to be by the precipitation of FAP and CaF_2. Furthermore, through surface adsorption or exchange of F^- with surface hydroxyl groups. The way of fluoride removal by Ca^{2+} and PO_4^{3-} ions and successive contact with bone char was suggested to be the saturated bones char catalyzed formation of FAP and formation of CaF_2 to some extent (Dahi, 1997).

Fluoroapatite (FAP) was formed at the surface of HAP through adsorption followed by an F^-/OH^- exchange process at low concentrations of F^-. Additionally, calcium fluorite (CaF_2) was formed on HAP by a surface precipitation process at higher concentrations of F^-.

Fluoride ion became adsorbed on the surface of HAP when it came into contact with fluoride aqueous solution, and then the adsorbed fluoride exchanged with the OH group present on the apatite surface. The fluoride adsorption into HAP occurs because of F^-/OH^- exchange reactions.

The Langmuir adsorption isotherm analysis indicated that adsorption of F^- ion on HAP also hinders dissolution of the HAP. The adsorption increased with the duration of exposure to HAP and increased with decrease in pH. An increase in the temperature increased the defluoridation by HAP. However, crystal growth of FAP over HAP was found to be a complicated process at low super saturation level and the surface hydroxyl groups are exchanged by F^- ions in this process. Fan et al. (2003) described the uptake of F^- on HAP with pseudo-first-order and second-order ion-exchange mechanisms, whereas the uptake on the other materials was through a pseudo-second-order surface adsorption (Nath and Dutta, 2015).

Ca-deficient HAP which is a by-product of phosphate wastewater treatment had been used to remove F^- ions in the presence of coexisting Ca^{2+}, Mg^{2+}, and Cl^- ions, and it was found that the removal followed pseudo-second-order kinetics, and the adsorption of F^- occurs mainly through ion exchange. The method was prescribed as applicable for treatment of high F^- contamination with a removal capacity of 85%. The adsorption of F^- was also reported to follow Freundlich isotherm with chemisorption taking place on heterogeneous surface. Table 19.6 summarizes the sorption parameters stated in literature to remove F-from aqueous solutions.

Generally, the mechanisms of fluoride uptake by HAP formation have been studied by many researchers. The following are some of the proposed mechanisms as shown using Eqs. (19.13) and (19.14):

(a) OH-/F-exchange on HAP:

$$Ca_{10}(PO_4)_6(OH)_2 + 2F^- = Ca_{10}(PO_4)6F_2 + 2OH^- \quad (19.13)$$

The reaction was considered to occur mainly at the crystal surfaces. At higher concentration (100 ppm) of fluoride, the exchange was accompanied by the partial disintegration of the crystal lattice and the formation of CaF_2 (fluorite).

$$Ca_{10}(PO_4)_6(OH)_2 + 20F^- = 10CaF_2 + 6PO_4^{3-} + 2OH^- \quad (19.14)$$

TABLE 19.6 Adsorption capacities and other parameters for the removal of fluoride by apatite and hydroxyapatite.

Type of hydroxyapatite	Concentration range ($mg\,L^{-1}$)	Dose ($mg\,L^{-1}$)	Adsorption capacity ($mg\,g^{-1}$)	pH	References
Nano-hydroxyapatite	3–80	4	4.575	5–6	Sundaram et al. (2008)
Biogenic apatite			4.99		Suchanek and Yoshimura (1998)
Treated biogenic apatite			6.849		
Geogenic apatite			0.014		
Hydroxyapatite	3–80	20	1.432	5	Sundaram et al. (2008)
Nano-sized hydroxyapatite	50	0.2	3.44	5	Wallace et al. (2019)
Cellulose-hydroxyapatite nanocomposites	10	4	4.2	6.5	Yu et al. (2013)
Al-modified hydroxyapatite (Al-HAP)	5–50	0.5	32.57	7	Nie et al. (2012)
Nano-hydroxyapatite	10–50	1	19.74–40.81	2–11	Chen et al. (2016)

(b) Model of Spinelli where the mechanism proposed by Spinelli et al. (1971), for fluoride uptake by HAP involves three processes:
- Precipitation of fluorapatite
- Dissolution of the solid HAP followed by precipitation of fluoroapatite (recrystallization)
- Removal of fluoride ions by adsorption and ion exchange.

(c) Ramsey et al. (1973) studied the uptake of fluoride by HAP as a function of solution pH and concluded that the effect of changing the pH of the fluoride solution is to increase the F content in the HAP by partial dissolution of HAP at pH 4.0 leading to the formation of CaF_2 and transformation of CaF_2 into fluoroapatite as the pH increases.

(d) Theoretical model of Nelson and Higuchi derived a mathematical model based on diffusion and chemical reactions at moving boundaries. They suggested that CaF_2 forms initially at the HAP solution interface and that the solid boundary progresses.

It was also concluded that the defluoridation efficiency increased with increase in the dose of HAPs and contact time but decreased with increase in initial fluoride concentration and pH. The nHAP as per examined by ultrasonic and microwave technique and observed its role in defluoridation. The role of nHAP in defluoridation was investigated using ultrasonic and microwave techniques. Nano-HAP composites synthesized by cellulose were prepared in NaOH/thiourea/urea/H_2O solution via in situ hybridization. HAP synthesized by aluminum (Al-HAP) was also used for defluoridation. In addition, to produce with high purity, HAP manufactured using (PG) nanoparticles was used. Nano-HAPs derived from phosphogypsum (PG) reveals good adsorption capacity for fluoride.

Gao et al. (2009) investigated the size-dependent defluoridation characteristics of synthesized HAP from aqueous solution. In their study, it was observed that the HAP with smaller size particles possesses the better fluoride removal capacity than bigger size particles. The Langmuir adsorption capacity of different size HAP ranged between 0.295 and 0.489 mg g^{-1}. The experimental data from nano-sized HAP were quite suitable with pseudo-second-order kinetics and Freundlich isotherm models, and it also found that adsorption process was basically physisorption and endothermic in nature according to the thermodynamic study. The fluoride removal efficiency increases with an increase in the adsorbent dose and contact time, but it decreases with an increase of initial fluoride concentration and pH of solution. Crushed limestone was used for defluoridation of water and found it very efficient for removal of fluoride (Nath and Dutta, 2010).

Jiménez-Reyes and Solache-Ríos (2010) used HAP for fluoride retention from aqueous solutions. Maximum removal was achieved in the pH range 5.0–7.3 with 16 h of contact time using 0.1 g of HAP for 25 mL of solution. The adsorption data followed the Freundlich isotherm and pseudo-second-order kinetics. The fluoride uptake by HAP was due to chemical adsorption.

Garg and Chaudhari (2012) developed a magnesium-substituted hydroxyapatite (Mg-HAp) adsorbent for defluoridation of potable water. The characteristics of the Mg-HAp were analyzed using an X-ray diffractometer (XRD). Batch adsorption experiments were conducted to study the effects of various parameters, such as pH, contact time, initial fluoride concentration, and coexisting ions, on the fluoride adsorption capacity of Mg-HAp. The authors observed that an increase in pH reduced the adsorption capacity. The presence of coexisting anions did not significantly affect the fluoride adsorption. The experimental data followed the Freundlich isotherm and pseudo-second-order kinetics.

Mourabet et al. (2015) have investigated the defluoridation capacity of HAP by using response surface methodology. The 88.86% of fluoride was removed from solution of 20 mg L^{-1} of initial fluoride concentration under optimal conditions of pH as 4.16, temperature as 39°C, and adsorbent dose as 0.28 g. The maximum Langmuir adsorption capacity was noted to be as 3.12 mg g^{-1}. The experimental data fitted well with Langmuir and Freundlich isotherms models as well as pseudo-second-order kinetic model, and the adsorption process by HAP was found to be endothermic in nature as per dynamic study.

Jayarathne et al. (2015) used natural crystalline appetite for fluoride adsorption from aqueous solution. The adsorption process was heavily dependent on pH, and the optimal pH for maximum removal was observed to be 6. It is also illustrative that the adsorption increased with adsorbent dose because of the increase in adsorbent sites. The highest adsorption was achieved at pH 6 for a solution containing 15 mg L^{-1} fluoride within 10 min of contact time. The adsorption data showed a monolayer adsorption capacity of 0.212 mg g^{-1} and were best fit by the Langmuir isotherm. The experimental data obeyed pseudo-second-order kinetics.

Mondal et al. (2016) assessed the defluoridation potential of magnesium-incorporated hydroxyapatite (M-i-HAP) having a surface area of 46.62 m^2 g^{-1}. The highest removal of fluoride (94.5%) by M-i-HAP was achieved under the

TABLE 19.7 Advantages and limitations of using hydroxyapatite in F^- removal.

Advantages	Limitations
- HAP can remove F^- up to 95% by precipitation as FAP and CaF_2, and through adsorption - It possesses the highest capacity for adsorption of F^- compared to other common adsorbents, the order of decreasing capacity being: HAP > Fluorspar > Quartz (activated using ferric ions) > Calcite > Quartz - It works within acceptable pH range for drinking water and does not add any other contaminant to the water - The F^- removal capacity increases with increase in surface area and temperature and requires a small dose of the adsorbent - Nano-HAP composite with chitin has capacity of holding as high as 2840 mg of F^- per kg of the adsorbent	- The removal capacity is poor at high pH, i.e., above 7.5 - The removal capacity decreases after regeneration - The formation of FAP is slow at high (above 5.5 mg L^{-1}) and low F^- concentrations (below 3.4 mg L^{-1})

optimum conditions of 303 K, pH 7, 3 h of contact time, and 10 g L^{-1} adsorbent dose for a 10 mg L^{-1} fluoride solution. The adsorption process followed the Langmuir isotherm, with an adsorption capacity of 1.16 mg g^{-1}, and a pseudo-second-order kinetic model. The desorption study showed that 91% of the adsorbent could be regenerated using 0.1 M NaOH solution.

Among all the tested materials by Fan et al. (2003), HAP presented a unique adsorption pattern and the highest fluoride adsorption capacity. For example, at an initial fluoride concentration of 0.064 mg L^{-1}, the adsorption limits were 0.022 mg g^{-1} on HAP, 0.0071 mg g^{-1} on fluorspar, 0.005 mg g^{-1} on activated quartz, 0.0035 mg g^{-1} on calcite, and 0.0013 mg g^{-1} on quartz. The fluoride adsorption equilibrium reached in about 2 min on quartz, calcite, and fluorspar, but took much longer time (150 min) on HAP.

Very old and recent studies indicated that HAP is an electrically conductive material. The OH group is considered as the charge carrier and can move in an electrical field, even it is a structural element of HAP and forms an hydrogen bond with the nearest of the $[PO_4]_3$ ions (Wallace et al., 2019; Chen et al., 2016). Therefore, their adsorption capacities follow the order: Hydroxyapatite > Fluorspar > Quartz activated using ferric ions > Calcite > Quartz.

19.15.2 Advantages and limitations of the use of HAP

The salient points regarding the advantages and the limitations of HAP as a material for F^- removal can be summarized in Table 19.7.

19.16 Diatomaceous earth materials

Diatomaceous earth rock is of sedimentary origin. It is a loose, earthy or loosely cemented porous, and lightweight rock, mainly formed by skeletons of diatom algae: diatomea and radiolaria. It is a microscopic diatom alga whose size ranges from 0.75 to 1500 m. This rock sometimes is called infusorial earth, kieselguhr, or mountain meal. $SiO_2 \cdot nH_2O$ is the main components of the siliceous armor. It belongs to the group of silica-bearing materials (Paschen, 1986). Moreover, it is explained as natural material formed from deposits of seas or lakes derived from the remains of diatoms, which grew and die. Diatomaceous earth has been used for strengthening and stiffening of materials. Recently, it is used as an absorbent, catalysts, and cloud seeding (Zhaolun et al., 2005).

The unique physical characteristics of diatomite make it abundant in many areas of the world, and it has high permeability (0.1–10 mD) and porosity (35%–65%) (Murer et al., 1997), small particle size, low thermal conductivity and density (Hassan et al., 1999), and high surface area (Gao et al., 2009).

Diatomite's surface characteristics of solubility, hydrophobia, charge, acidity, ion exchange, and adsorption enable it to generate active hydroxyl groups in the presence of water, which is partially structurally related to the crystal mesh of the diatomite (Yuan et al., 2001).

The chemical composition of diatomite (usually 70%–90%) is silica. The remaining compositions include alumina (0.6%–8%), iron oxide (0.2%–3.5%), alkali metal oxides, Na_2O and MgO (less than 1%), CaO (0.3%–3%), and a minor amount of other impurities, such as P_2O_5, and TiO_2. Typical contaminants like the deposits consisting of shell, sediments such as clay and fine sand, organic and inorganic materials, carbonates, and metals oxides were also observed.

Many researchers in recent years have widely examined for the development of low-cost and effective adsorbents for defluoridation of groundwater. Among these, diatomaceous earth is the most widely used adsorbents.

Diatomaceous earth (DE) is used to filter drinking water, sugar, honey, and syrups without altering their natural properties. The natural or the modified form of DE has been already used for the removal of fluoride from aqueous solution.

Under the research conducted by Datsko et al. (2011), the adsorption capacity of natural (D1) and chemically structure-modified diatomite (DMA) in the removal of fluorine ions from highly concentrated fluorine solutions (up to 0.3 mol L^{-1}) under static conditions at room temperature is investigated. The effect of different parameters—solution pH, initial fluorine concentration, sorbent weight, and particle surface charge density—is examined to determine the adsorption properties of DMA under different process conditions. It is shown that the solution pH plays a crucial role in the removal of fluorine from solutions. At pH of 4.5–5.5, it is observed that fluoride removal was efficient. Under equilibrium conditions, upon the saturation of the DMA surface with fluorine ions, the adsorption capacity of DMA achieves 58 mmol g^{-1} of sorbent; this value is 5.5 times higher than that of unmodified D1. Fluorine adsorption isotherms for DMA samples are derived; equilibrium adsorption data are modeled using a two-stage Langmuir model; and it is shown that the experimental and calculated data on fluorine adsorption are in good agreement: correlation coefficient R^2 for the unmodified and chemically modified diatomite samples is 0.9952 and 0.9687, respectively. The fluorine adsorption mechanism is also investigated by the same author. The mineralogical and physicochemical study reveals that the diatomite-NaF-H_2O system is characterized by the occurrence not only of physical adsorption and ion exchange but also of the chemical bonding of the fluorine ions with the active sites of the sorbent surface, i.e., the formation of weakly soluble fluorine compounds with Al on chemically modified diatomite (DMA) and with Ca on unmodified (AlF_3, Na_3AlF_6, CaF_2). They employed X-ray diffraction (XRD), Fourier transform-infrared (FT-IR), (BET) surface analyzer, and chemical studies to characterize diatomite of local origin loaded with aluminum compounds. A solution of aluminum salt and ammonia plus NaOH solution have been used for surface modification of diatomite through subsequent heating treatment. During the treatment with NaOH, the amorphous surface silica partially dissolves and forms an aluminosilicate compound at the addition of an aluminum salt. On the surface of the diatomite and on the inner surface of the macro- and larger mesopores, the obtained material is deposited, which yielded specific surface area of 81.8 m^2 g^{-1}, which is 2.5 times larger than the initial values of diatomite (37.5 m^2 g^{-1}). Aluminosilicate compound was precipitated, and its concentration equal to 0.34 g of Al/g of aluminosilicate contributes to the development of a porous structure in the treated diatomite. Following this, the volume of the mesopores increases from 0.029 to 0.079 cm^3 g^{-1} and that of the micropores from 0.012 to 0.027 cm^3 g^{-1}. The emergence of new lines in the X-ray diffraction patterns reveals that the qualitative changes in the composition of the obtained sorbent which are characteristic for aluminosilicates and more peak values in the IR spectra corresponding to the stretching vibrations of Si-O-Al were observed. At the initial fluorine concentration equal to 0.15 mol L^{-1}, the adsorption capacity increases from 8.9 to 57.6 mgF g^{-1}. Therefore, the selectivity of the obtained adsorbent with respect to fluoride ions increases significantly.

Izuagie et al. (2017) is exhibited that fluoride removal with aluminum-iron oxide-modified DE the maximum in 50 min using a dosage of 0.3 g 100 mL^{-1} in 10 mg L^{-1} initial fluoride concentration 82.3% removal attained. At 5 min of contact time, 81.3% of fluoride removal was attained. At a sorbent dosage of 0.6 g 100 mL^{-1} contacted with the same concentration for 50 min, a substantial amount of fluoride (93.1%) was removed from the aqueous solution. A values 7.633 mg g^{-1} was optimum adsorption capacity of the adsorbent using a solution containing initially 100 mg L^{-1} fluoride. The reduced values of fluoride content to 0.928 mg L^{-1} were obtained by contacting the sorbent at a dosage of 0.6 g 100 mL^{-1} with field water containing 5.53 mg L^{-1} at 200 rpm for 50 min which is below the upper limit of WHO guideline of 1.5 mg L^{-1} fluoride in drinking water. Both Langmuir and Freundlich isotherms were sorption data fitted but better with the former. Pseudo-second-order kinetic agreed with the sorption data indicating the adsorption mechanism is chemisorption.

Wambu et al. (2011) have entirely studied on adsorption of fluoride (F^-) from aqueous solutions using ATDE obtained from Kenya. The process parameters like initial fluoride concentration, adsorbent dosage, contact time, temperature, pH, and coexisting anions were assessed. Due to the increase of temperature from 293 to 303 K, adsorption onto ATDE increased strongly from about 40% to over 92%. The increase in concentration of OH^- ions does affect F adsorption onto acid treated diatomaceous earth strongly. The involvement of chloride during fluoride adsorption did not have any competing effect. At 400 mg L^{-1} initial fluoride concentrations using 0.5 g mL^{-1} ATDE batch loading, complete fluoride removal (100% adsorption) could be achieved with temperature of 303–313 K and pH = 3.4 ± 0.2. Both Freundlich and Langmuir models were fitted well with adsorption data correlated. A 51.1 mg g^{-1} maximum

F^- adsorption capacity of ATDE was obtained indicating that the mineral could be used as an inexpensive adsorbent for the removal of F ions from groundwater.

Gitari et al. (2017) have studied on the modification of tri-metal Mg/Ce/Mn oxide-modified diatomaceous earth (DE) at optimal conditions for adsorption of fluoride. The effects of various operational parameters like sorbent dosage of 0.6 g 100 mL^{-1}, contact time of 60 min, mixing speed of 200 rpm, and temperature 297 K were assessed.

For solutions containing initial fluoride concentration of 10–60 mg L^{-1}, the fluoride removal was > 93%. At an initial fluoride concentration of 100 m mg L^{-1}, a fluoride absorption capacity of 12.63 mg g^{-1} was obtained. The fluoride removal was 93% for solutions with an initial pH range of 4–11, an initial fluoride content of 9 mg L^{-1}, and a sorbent dose of 0.6 g.

Evaluation of the effect of competing ions on fluoride removal revealed that carbonates would reduce the amount of fluoride removed from solution, while other anions such as phosphate, nitrate, and sulfate had no observable effect. Potassium sulfate solution was the most suitable regenerant for the spent Mg/Ce/Mn oxide-modified diatomaceous earth compared to sodium carbonate and sodium hydroxide. An exchange of hydroxyl groups on surface of the DE sorbent with fluoride ions from solution was the mechanism of fluoride removal at pH > 5.45 (pH_{pzc} = 5.45). Both Langmuir isotherm and pseudo-second-order models were fitted well to the adsorption data.

Oladoja and Helmreich (2016) used aluminate (A) prepared through a self-propagation combustion method and got incorporated, in situ, into the framework of diatomaceous earth (DE) to produce a composite material (A-DE) that is capable of serving as a reactive filter material for groundwater defluoridation. Aqua defluoridation potential of the A-DE was evaluated and optimized in a batch process. The XRD analysis of the A-DE showed that amorphous A was incorporated into the DE. The results of the kinetic analysis of the defluoridation process showed that the time-concentration profiles, at different initial F^- concentrations, were best described by the pseudo-second-order kinetic equation. The appraisal of the performance efficiency of the A-DE, via equilibrium isotherm analysis, showed that the monolayer sorption capacity of the A-DE was 18.22 mg g^{-1}, and Freundlich isotherm equation gave better description of the defluoridation process. Variation in ionic strength, pH, anionic interference, and organic load had minimal effects on the extent of aqua defluoridation by the A-DE.

The influence of pH value, diatomite dose, poly aluminum chloride dose, contact time, and initial concentration on defluoridation was investigated by static experiment. The results showed that the optimum defluoride turned up when pH value of the raw water was 6, the diatomite dose was 3.0 g L^{-1}, and the contact time was 15 min. The largest removal rate can reach 97.18%. It was also found that the adsorptive capacity of diatomite could be improved by reducing its dose. The largest adsorptive capacity can reach 1.78 mg g^{-1}, and the equilibrium fluoride concentration is 0.66 mg g^{-1}. These experimental results show that diatomite has great potential applications in defluoridation of drinking water.

19.16.1 Mechanism of fluoride removal by diatomite

The F^- adsorption isotherm could be classified as a type three high affinity (H_3) isotherm, according to Giles classification of adsorption isotherms, based on the initial part of the curve being perpendicular to the concentration axis. Type H_3 isotherm describes a system of high affinity for the solute by substrate. The vertical initial portion of the curve indicates that the adsorbent has got such high affinity for the adsorbate that has < 200 mg L^{-1} F concentration, the ion is completely adsorbed, or at least there is no measurable amount remaining in solution. The H_3 isotherm has been associated with adsorption of large units, such as ionic micelles or polymeric molecules, but single-ion systems' H-type isotherms are consistent with ion-exchange mechanisms where the adsorbate exchanges with others of much lower affinity for the surface. This indicates that anionic exchange protocol could be the main adsorption mechanism in F^- adsorption on M1. The higher charge-to-volume ratio in F ions could also make them more energetically favored in the interaction with the substrate materials. This would be important for adsorption of F from aqueous streams in the presence of other anions with larger spatial radius, selective removal of F being desirable.

Kır et al. (2015) have investigated for fluoride removal by an acidic activated diatomite in order to increase the percentage of fluoride removal. It is generally known that adsorption capacity of the adsorbent can be improved with increased acidity. Thus, acid activated diatomite material was used as an adsorbent on this specific research here discussed in detail, and examples of modified and unmodified diatomites used in F^- removal are illustrated in Table 19.8.

TABLE 19.8 Comparison of modified and nonmodified diatomites for fluoride removal.

Adsorbent	Experiment	pH	Concentration range ($mg\,L^{-1}$)	Adsorption capacity ($mg\,g^{-1}$)	References
Hydrous ferric oxide-modified diatomite	Batch	4–4.5	180–78.56	NA (56.36%)	Jang et al. (2006)
Diatomite modified aluminum chloride	Batch	6	4.217	1.78 $mg\,g^{-1}$	
Calcium aluminate (CA)-diatomaceous earth (DE)	Batch	4–6.5	4.338.4	18.22 $mg\,g^{-1}$	Oladoja and Helmreich (2016)
Tri-metal Mg/Ce/Mn oxide-modified DE	Batch	4–11	10–60	12.69	Gitari et al. (2017)
Acid treated diatomaceous earth (ATDE)	Batch	3.4	400	51.1 (100%)	Wambu et al. (2011)
Al/Fe oxide-modified DE	Batch	8.6	100	7.633	Izuagie et al. (2017)
Aluminum salt and ammonia-modified diatomite	Batch	–	0.15 $mol\,L^{-1}$	57.6	Datsko et al. (2011)

19.17 Conclusions and future research areas

The following concluding remarks are derived from the review:

I. Modification of minerals with metal oxide is a crucial matter for the actual application of diatomaceous earth (DE) or bentonite clay minerals for the removal of fluoride from groundwater.
II. Tri-metal and di-metal compounds are found to be more effective on filling the pores of diatomite and bentonite minerals that enhance the surface area to adsorb anions.
III. HAP is an efficient defluoridating agent. In situ production of HAP from low-cost materials may be an even better defluoridation technique. With proper sludge management, defluoridation by HAP has a high potential for application.
IV. To maximize the adsorption capacity (low capacity of adsorption) of both diatomite and bentonite, surface modification has shown significant changes in the physicochemical and adsorption structural characteristics of the samples; i.e., the ratio of the micro- and mesopores and also the value of the specific surface increase, the apparent density decreases, and the porosity of the sorbent grows, and hence, modification of the low-cost adsorbents is a paramount job for researchers on the field to have a good defluoridating agent.
V. Depending on the literatures' result on defluoridating water, among the reviewed adsorbents in batch experimentation, HAP takes the lead: HAP > Bentonite > Diatomite, but the selectivity of the adsorbent depends on the availability and feasibility at the local level.
VI. In most of the literatures that have been reviewed, the mode of experimentation is based on batch experimental setups. The adsorption capacity of adsorbents gained from the batch experiment is valuable in giving basic information about the effectiveness of the adsorbents. Nevertheless, the data obtained from batch studies may not be appropriate for continuous processes where contact time is insufficient for the achievement of the equilibrium. Hence, there is huge interest to conduct defluoridation study in a column mode.

Acknowledgments

Africa Center of Excellence for Water Management (Addis Ababa University, Ethiopia) and Indian Institute of Technology (New Delhi, India) are acknowledged for their support.

References

Aimee, N.R., van Wijk, A.J., Maltz, M., Varjao, M.M., Mestrinho, H.D., Carvalho, J.C., 2017. Clin. Oral Invest. 21, 1811–1820.
Ajemba, R., Onukwuli, O., 2014. Fizykochemiczne Problemy Mineralurgii—Physicochem. Problems Mineral Process. 50, 349–358.
Apambire, W.B., Boyle, D.R., Michel, F.A., 1997. Environ. Geol. 33, 13–24.
Argo, W.L., Mathers, F.C., Humiston, B., Anderson, C.O., 1919. Trans. Am. Electrochem. Soc. 35, 335–349.
Assaoui, J., Kheribech, A., Hatim, Z., 2018. Inorg. Chem. Indian J. 13, 1–16.
Ayoob, S., Gupta, A., Bhat, V.T., 2008. Crit. Rev. Environ. Sci. Technol. 38, 401–470.
Barbier, O., Arreola-Mendoza, L., Del Razo, L.M., 2010. Chem.-Biol. Interact. 188, 319–333.
Bergaya, F., Lagaly, G., 2001. Appl. Clay Sci. 19, 1–3.
Bharti, V.K., 2017. Peertechz J. Environ. Sci. Toxicol. 2 (1), 021–032.
Bhat, N., Jain, S., Asawa, K., Tak, M., Shinde, K., Singh, A., Gandhi, N., Gupta, V.V., 2015. J. Clin. Diagn. Res. 9, ZC63–ZC66.

Biswas, K., Gupta, K., Ghosh, U.C., 2009. Chem. Eng. J. 149, 196–206.
Cavalli, A.M., Flório, F., 2018. J. Contemp. Dent. Pract. 19 (1), 30–36.
Centeno, J., Finkelman, R., Selinus, O., 2016. Geosciences 6, 8.
Chai, L., Wang, H., Zhao, H., Dong, S., 2017. Bull. Environ. Contamin. Toxicol. 98, 496–501.
Chen, L., Zhang, K.-S., He, J.-Y., Xu, W.-H., Huang, X.-J., Liu, J.-H., 2016. Chem. Eng. J. 285, 616–624.
Cox, J.A., DiNunzio, J.E., 1977. Anal. Chem. 49, 1272–1275.
Dahi, E., 1997. Development of the contact precipitation method for appropriate defluoridation of water. Proceedings of the 2nd International Workshop on Fluorosis Prevention and Defluoridation of Water, pp. 128–137.
Dargahi, A., Atafar, Z., Mohammadi, M., Azizi, A., Almasi, A., Ahagh, M.M.H., 2016. Int. J. Pharm. Technol. 8, 16772–16778.
Datsko, T.Y., Zelentsov, V., Dvornikova, E., 2011. Surf. Eng. Appl. Electrochem. 47, 530–539.
El Din Mahmoud, A., Fawzy, M., 2016. In: Ansari, A.A., Gill, S.S., Gill, R., Lanza, G.R., Newman, L. (Eds.), Phytoremediation: Management of Environmental Contaminants, Volume 3. Springer International Publishing, Cham, pp. 209–238.
Eskandarpour, A., Onyango, M.S., Ochieng, A., Asai, S., 2008. J. Hazard. Mater. 152, 571–579.
Fan, X., Parker, D.J., Smith, M.D., 2003. Water Res. 37, 4929–4937.
Fan, X., Parker, D.J., Smith, M.D., 2003. Water Res. 37, 4929–4937.
Farrah, H., Slavek, J., Pickering, W., 1987. Soil Res. 25, 55–69.
Fyjimoto, H., Maeda, T., Yoshikawa, M., 1993. J. Florine Chem. 60, 69–77.
Gan, Y., Wang, X., Zhang, L., Wu, B., Zhang, G., Zhang, S., 2019. Chemosphere 218, 860–868.
Gao, S., Cui, J., Wei, Z., 2009. J. Fluor. Chem. 130, 1035–1041.
Garg, P., Chaudhari, S., 2012. Proceedings of the 2012 International Conference on Future Environment and Energy (ICFEE 2012), Singapore. pp. 26–28.
Ghorai, S., Pant, K.K., 2005. Sep. Purif. Technol. 42, 265–271.
Ghosh, A., Mukherjee, K., Ghosh, S.K., Saha, B., 2012. Res. Chem. Intermed. 39, 2881–2915.
Gitari, W.M., Ngulube, T., Masindi, V., Gumbo, J.R., 2013. Desalin. Water Treat. 53, 1578–1590.
Gitari, W.M., Izuagie, A.A., Gumbo, J.R., 2017. Arab. J. Chem. 13, 1–16. https://doi.org/10.1016/j.arabjc.2017.01.002.
Grich, N.B., Attour, A., Mostefa, M.L.P., Guesmi, S., Tlili, M., Lapicque, F., 2019. Electrochim. Acta 316, 257–265. https://doi.org/10.1016/j.electacta.2019.05.130.
Guan, X., Zhou, J., Ma, N., Chen, X., Gao, J., Zhang, R., 2015. Desalin. Water Treat. 55, 440–447.
Hassan, M., Ibrahim, I., Ismael, I., 1999. Chin. J. Geochem. 18, 233–241.
Hosterman, J.W., Patterson, S.H., 1992. Report 1522. pubs.er.usgs.gov.
Hu, K., Dickson, J.M., 2006. J. Membr. Sci. 279, 529–538.
Ingole, N.W., Patil, S.S., 2012. J. Eng. Res. Stud. 3 (1), 111–119.
Ismadji, S., Soetaredjo, F.E., Ayucitra, A., 2015. Clay Materials for Environmental Remediation. Springer, pp. 39–56.
Izuagie, A.A., Gitari, W.M., Gumbo, J.R., 2017. Water Sci. Technol. Water Supply 17, 688–697.
Jacks, G., Bhattacharya, P., Chaudhary, V., Singh, K.P., 2005. Appl. Geochem. 20, 221–228.
Janardhana Raju, N., Dey, S., Das, K., 2009. Curr. Sci. 96, 979–985.
Jang, M., Min, S.-H., Kim, T.-H., Park, J.K., 2006. Environ. Sci. Technol. 40, 1636–1643.
Jayarathne, A., Weerasooriya, R., Chandrajith, R., 2015. Environ. Earth Sci. 73, 8369–8377.
Jiménez-Reyes, M., Solache-Ríos, M., 2010. J. Hazard. Mater. 180, 297–302.
Kamble, S.P., Dixit, P., Rayalu, S.S., Labhsetwar, N.K., 2009. Desalination 249, 687–693.
Karthikeyan, G., Pius, A., Alagumuthu, G., 2005. Indian J. Chem. Technol. 12 (3), 263–272.
Khandare, A., Rao Gourineni, S., Validandi, V., 2017. Environ. Monit. Assess. 189 (11), 579.
Kır, E., et al., 2015. Removal of fluoride from aqueous solution by natural and acid-activated diatomite and ignimbrite materials. Desalin. Water Treat. 57 (46).
Ktari, T., Larchet, C., Auclair, B., 1993. J. Membr. Sci. 84, 53–60.
Kumar, E., Bhatnagar, A., Ji, M., Jung, W., Lee, S.H., Kim, S.J., Lee, G., Song, H., Choi, J.Y., Yang, J.S., Jeon, B.H., 2009. Water Res. 43, 490–498.
Kurniawan, A., Sutiono, H., Ju, Y.-H., Soetaredjo, F.E., Ayucitra, A., Yudha, A., Ismadji, S., 2011. Microporous Mesoporous Mater. 142, 184–193.
Liu, X., Wang, B., Zheng, B., 2014. Chin. J. Geochem. 33, 277–279.
Ma, Y., Shi, F., Zheng, X., Ma, J., Yuan, J., 2005. Harbin Inst. Technol. (New Series) 12, 236–240.
Ma, Y., Wang, S.G., Fan, M., Gong, W.X., Gao, B.Y., 2009. J. Hazard. Mater. 168, 1140–1146.
Ma, Y., Shi, F., Zheng, X., Ma, J., Gao, C., 2011. J. Hazard. Mater. 185, 1073–1080.
Mahmoud, A.E.D., 2020. Handbook of Nanomaterials and Nanocomposites for Energy and Environmental Applications. Springer.
Mahmoud, A.E.D., Fawzy, M., 2015. J. Bioremediat. Biodegrad. 6, 1–7.
Mahmoud, A.E.D., Fawzy, M., Radwan, A., 2016. Int. J. Phytoremed. 18, 619–625.
Mahmoud, A.E.D., Franke, M., Stelter, M., Braeutigam, P., 2018a. Evaluation of bisphenol A removal from water bodies using Eco-friendly technique. In: 4th International Congress on Water, Waste and Energy Management. Sciknowledge, Spain.
Mahmoud, A.E.D., Stolle, A., Stelter, M., 2018b. ACS Sustain. Chem. Eng. 6, 6358–6369.
Mahmoud, A.E.D., Stolle, A., Stelter, M., Braeutigam, P., 2018c. Abstracts of Papers of the American Chemical Society. American Chemical Society, Washington, DC.
Mahmoud, A.E.D., Franke, M., Stelter, M., Braeutigam, P., 2020. Powder Technol. 366, 629–640.
Majewska-Nowak, K., Grzegorzek, M., Kabsch-Korbutowicz, M., 2015. Environ. Prot. Eng. 41, 67–81.
Mohammadi, A.A., Yousefi, M., Yaseri, M., Jalilzadeh, M., Mahvi, A.H., 2017. Sci. Rep. 7, 17300.
Mondal, P., Mehta, D., George, S., 2016. Desalin. Water Treat. 57, 27294–27313.
Mourabet, M., El Rhilassi, A., El Boujaady, H., Bennani-Ziatni, M., El Hamri, R., Taitai, A., 2015. J. Saudi Chem. Soc. 19, 603–615.
Mudzielwana, R., Gitari, M.W., Akinyemi, S.A., Msagati, T.A.M., 2018. S. Afr. J. Chem. 71, 15–23.

Murer, A., McClennen, K., Ellison, T., Timmer, R., Larson, D., Wolcott, K., Walker, T., Thomsen, M., 1997. SPE Western Regional Meeting. Society of Petroleum Engineers.

Nasr, M., Mahmoud, A.E.D., Fawzy, M., Radwan, A., 2017. Appl. Water Sci. 7, 823–831.

Nath, S.K., Dutta, R.K., 2010. *Enhancement of limestone defluoridation of water by acetic and citric acids in fixed bed reactor*. CLEAN Soil Air Water 38 (7), 614–622. https://doi.org/10.1002/clen.200900209.

Nath, S.K., Dutta, R.K., 2015. Desalin. Water Treat. 53, 2070–2085.

Nie, Y., Hu, C., Kong, C., 2012. J. Hazard. Mater. 233, 194–199.

Oladoja, N., Helmreich, B., 2016. J. Water Process Eng. 9, 58–66.

Paschen, S., 1986. Erzmetall 39, 158–161.

Paul, E.D., Gimba, C.E., Kagbu, J.A., Ndukwe, G.I., Okibe, F.G., 2011. Basic Appl. Chem. 6, 33–38.

Peckham, S., Awofeso, N., 2014. TheScientificWorldJournal 2014, 293019.

Peicher, K., Maalouf, N.M., 2017. Calcif. Tissue Int. 101 (5), 545–548. https://doi.org/10.1007/s00223-017-0305-0.

Ramesh, M., Malathi, N., Ramesh, K., Aruna, R., Kuruvilla, S.J., 2017. Pharm. Biol. Sci. 9, 88–91.

Ramsey, A.C., Duff, E.J., Paterson, L., Stuart, J.L., 1973. Caries Res. 7, 231–244.

Rawat, N., Patel, V.K., 2018. Water Remediation. Springer, pp. 211–224.

Reddy, M.A., Fichtner, M., 2015. Fluoride Cathodes for Secondary Batteries. Elsevier, pp. 51–76.

Sandoval, M., Fuentes-Ramirez, R., Nava, J., Rodríguez, I., 2014. Sep. Purif. Technol. 134, 163–170.

Shariati-Rad, M., Irandoust, M., Amri, S., Feyzi, M., Ja'fari, F., 2014. Int. Nano Lett. 4, 91–101.

Shukla, A., 2016. Indian J. Rheumatol. 11, 171–173.

Singh, J., Singh, P., Singh, A., 2016. Arab. J. Chem. 9, 815–824.

Spinelli, M., Brudevold, F., Moreno, E., 1971. Arch. Oral Biol. 16, 187–203.

Steudel, A., Batenburg, L.F., Fischer, H., Weidler, P.G., Emmerich, K., 2009. Appl. Clay Sci. 44, 105–115.

Suchanek, W., Yoshimura, M., 1998. J. Mater. Res. 13, 94–117.

Sundaram, C.S., Viswanathan, N., Meenakshi, S., 2008. J. Hazard. Mater. 155, 206–215.

Temuujin, J., Senna, M., Jadambaa, T., Burmaa, D., Erdenechimeg, S., MacKenzie, K.J.D., 2006. J. Chem. Technol. Biotechnol. 81, 688–693.

Thakre, D., Rayalu, S., Kawade, R., Meshram, S., Subrt, J., Labhsetwar, N., 2010. J. Hazard. Mater. 180, 122–130.

Toor, M.K., 2011. https://digital.library.adelaide.edu.au/dspace/bitstream/2440/66283/9/01front.pdf.

Toor, M., Jin, B., 2012. Chem. Eng. J. 187, 79–88.

Tor, A., 2006. Desalination 201, 267–276.

Ullah, R., Zafar, M.S., Shahani, N. Iran., 2017. J. Basic Med. Sci. 20, 841–848.

Vhahangwele, M., Mugera, G.W., Tholiso, N., 2014. Toxicol. Environ. Chem. 96, 1294–1309.

Vithanage, M., Bhattacharya, P., 2015. Fluoride in the environment: sources, distribution and defluoridation. 13 Spriger, pp. 131–147.

Waghmare, S.S., Arfin, T., 2015. Int. J. Innov. Res. Sci. Eng. Technol. 4, 8090–8102.

Wallace, A.R., Su, C., Sun, W., 2019. Environ. Eng. Sci. 36, 634–642.

Wambu, E.W., Onindo, C.O., Ambusso, W.J., Muthakia, G.K., 2011. Mater. Sci. Appl. 2, 1654.

Wang, Z., 2018. Mod. Approach. Oceanogr. Petrochem. Sci. 1, 9–11. https://doi.org/10.32474/MAOPS.2018.01.000106.

Wei, R., Luo, G., Sun, Z., Wang, S., Wang, J., 2016. Chemosphere 153, 419–425.

Wilson, R.L., DiNunzio, J.E., 1981. Anal. Chem. 53, 692–695.

Xiuru, Y., Kuanxiu, S., Jianping, W., Liuchang, H., Zhaohui, Y., 1998. J. Rare Earths 16, 279–280.

Yami, T., Chamberlain, J., Zewge, F., Sabatini, D., 2018. Performance enhancement of Nalgonda technique and pilot testing electrolytic defluoridation system for removing fluoride from drinking water in East Africa. 12 AJOL, pp. 357–369.

Yu, X., Tong, S., Ge, M., Zuo, J., 2013. Carbohydr. Polym. 92, 269–275.

Yu, W.H., Ren, Q.Q., Tong, D.S., Zhou, C.H., Wang, H., 2014. Appl. Clay Sci. 97–98, 222–234.

Yuan, P., Wu, D.-Q., Lin, Z.-Y., Diao, G.-Y., Peng, J., Wei, J.-F., 2001. Guang pu xue yu guang pu fen xi= Guang pu 21, 783–786.

Yunfei Xi, R.L.F., He, H., Kloprogge, T., Bostrom, T., 2005. Am. Chem. Soc. 21, 8675–8680.

Zarrabi, M., 2014. Environ. Eng. Manage. J. 13, 205–214.

Zhang, Y., Wang, D., Liu, B., Gao, X., Xu, W., Liang, P., Xu, Y., 2013. Am. J. Anal. Chem. 04, 48–53.

Zhang, Y., Xu, Y., Cui, H., Liu, B., Gao, X., Wang, D., Liang, P., 2014. J. Rare Earths 32, 458–466.

Zhaolun, W., Yuxiang, Y., Xuping, Q., Jianbo, Z., Yaru, C., Linxi, N., 2005. Environ. Chem. Lett. 3, 33–37.

Zhou, C.-H., Zhang, D., Tong, D.-S., Wu, L.-M., Yu, W.-H., Ismadji, S., 2012. Chem. Eng. J. 209, 223–234.

Ziarati, P., El-Esawi, M., Sawicka, B., Umachandran, K., Mahmoud, A.E.D., Hochwimmer, B., Vambol, S., Vambol, V., 2019. J. Med. Discov. 4, 1–16.

Zuo, H., Chen, L., Kong, M., Qiu, L., Lu, P., Wu, P., Yang, Y., Chen, K., 2018. Life Sci. 198, 18–24.

CHAPTER

20

Cost-effective biogenic-production of inorganic nanoparticles, characterizations, and their antimicrobial properties

Kishore Kumar Kadimpati[a] *and Narasimha Golla*[b]

[a]Department of Pharmaceutical Biotechnology, Narayana Pharmacy College, Nellore, Andhra Pradesh, India [b]Applied Microbiology Laboratory, Department of Virology, Sri Venkateswara University, Tirupati, Andhra Pradesh, India

OUTLINE

Abbreviations

AgNPs silver nanoparticles
ZnONPs zinc nanoparticles
AuNPs gold nanoparticles
IC_{50} inhibitory concentration
DMF *N,N*-dimethylformamide
DNA deoxyribo nucleic acids

Cost Effective Technologies for Solid Waste and Wastewater Treatment
https://doi.org/10.1016/B978-0-12-822933-0.00017-6

ZOI zone of inhibition
PEG polyethylene glycol
TEM transmission electron microscopy
SEM scanning electron microscopy
DLS dynamic light scattering
XRD X-ray diffraction
DPPH 1-diphenyl-1-2-picrylhydrazyl
SERS surface enhanced Raman spectroscopy
MIC minimum inhibitory concentration
FTIR Fourier transform infrared spectroscopy

20.1 Introduction

Nanobiotechnology is a new science of nanoparticle synthesis using biotechnological applications. It is a multifaceted field that covers various facets of biology, physics, engineering, and chemistry. Nanoinorganic antibacterial agents have gained interest for bacterial control, owing to their outstanding safety, thermal stability, sustainability, and huge outside dimensions. Nanobiotechnology is a novel branch of current science utilizing metal oxides, that is, copper, gold, silver, platinum, and zinc chiefly. It is endowed with new paraphernalia for the rapid detection of diseases. Molecular treatment show enormous potential to transform biology, pharmacy, and medicine. Many investigations have proved that various inorganic nanoparticles exhibited antimicrobial activities against Gram-positive, Gram-negative bacteria, and fungus. The nanosize of nanomaterials make them easy to into a cell and give a huge surface area to deliver chemical species; these are important physical properties for effectively inhibiting/killing microorganisms. A number of nanoparticles (NPs) have clearly shown themselves to be successful antimicrobial agents against numerous pathogenic microorganisms. This chapter runs through recent studies worldwide by researchers into green nanobiotechnological approaches, characterizations, and the catalytic and antibacterial behavior of biogenic nanoparticles. Further, this chapter also summarizes the potential applications of biogenic nanomaterials with a focus on nanomedicine.

20.2 Various methods for nanomaterial synthesis

Physical and chemical methods: NPs cannot be produced on a large scale by these techniques owing to many obstacles such as huge amounts of organic toxic liquids, released hazardous wastes, highly laborious processes, and huge energy consumption. Various techniques and approaches for nanoparticle synthesis are enlisted in Table 20.1.

Unfortunately, several NP synthesis procedures involve hazardous materials and have little material conversion, high energy inputs, and difficulty in further purification. Chemical and physical NP synthesis cannot be expanded easily to large-scale production because of many drawbacks such as the (i) the presence of toxic organic solvents, (ii) the production of hazardous byproducts and intermediary compounds, (iii) high energy utilization, (iv) particles may produce aggregates, and (v) capping agents must be used. Moreover, other issues are also initiated such as

TABLE 20.1 List of various techniques for the synthesis of nanoparticles.

Sl. No	Physical methods	Chemical methods	Biological methods
1	High energy ball mill	Sol-gel synthesis	Polymeric assisted synthesis
2	Inert gas condensation	Microemulsion	Microorganism assisted
3	Pulse vapor deposition	Hydrothermal synthesis	Fungus assisted
4	Laser pyrolysis	Polysol synthesis	Natural gum assisted
5	Flash spray pyrolysis	Chemical vapor synthesis	Plant extract assisted
6	Electro-spraying	Plasma enhanced chemical vapor deposition	Biotemplate assisted
7	Melt mixing	Photochemical	

(i) NP toxicity, (ii) decreased particle yield, (iii) deformation in particle structure, and (iv) particle growth inhibition. In chemical synthesis of inorganic nanoparticles major problems are (i) nanocomposites are formed, (ii) increase toxicity and reactivity of particles, (iii) reaction conditions and reaction kinetics and (iv) harmful to human beings and ecology.

20.2.1 Physical methods

Metal nanoparticles are usually prepared through evaporation condensation, in which a tube furnace is used at atmospheric conditions. At the center of the furnace, a boat containing the source material is placed, which is vaporized into carrier gas. By using evaporation/condensation techniques, several nanoparticles/nanomaterials are produced with Au, Ag, Pb, etc. With the use of tube furnace in AgNP production, huge energy is required to attain the thermal stability of the material, desired temperature and also occupies large area in the facility, altogether it is not economically feasible. Another physical process to produce AgNPs is the laser ablation of silver bulk material in solution (solution may be water or ethanol). In laser ablation technique, the beam is equipped to colloid for the formation of nanomaterials and doesn't require chemical reagents. The main parameters influencing NP genesis are beam power, the duration of ablation, and the pore spot (Mafune et al., 2000). Sonochemical synthesis is another method applied for AgNPs; long chain polymers bind with nanoparticles. In this method, thermal stability and nanoparticle stability are well established without the formation of chemical pollutants. But sonochemical methods require reducing/capping agents and take much more time to form small nanoparticles. Also, buffering the chemical synthesis (pH 9–11) under sonication conditions leads to the formation of clusters of Ag nanoparticles.

Thermal decomposition also can produce AgNPs by decomposing the silver-organic complexes at higher temperatures, leading to silver nanoparticles. Capping agents are compulsory, and mixing is required at rigid conditions with nitrogen gas. In the thermal decomposition process also energy required to heat, and reflux for 3 h time (Navaladian et al., 2007) leads to cost enhancement. Another physical method is the electrochemical technique used for AgNPs production with high purity. Here, two types stabilizers are utilized: (i) electrostatic stabilizer (usually organic monomers), and (ii) steric stabilizers (polymeric compounds) (Yu et al., 1997; Yin et al., 2003). Few researchers are used polyethylene glycol (PEG) stabilizer for AgNPs which are easy permeable through diffusion into small molecules and are also very stable (Shkilnyy et al., 2009). Microwave irradiation (MW) is another technique in which the frequency range of 300 MHz to 300 GHz was used to heat the sample with H_2O (reducing agent) to orient the electrical field. The dipolar atoms try to reorient in the electric field, and lose energy by molecular friction as heat. Gao et al. (2006) synthesized AgNPs using DMF as the reducing factor with PVP irradiated at 140°C for 3 h to achieve a decahedron shape and an 80 nm size. Pal et al. (2009) synthesized AgNPs by the microwave irradiation of $AgNO_3$ as a precursor with ethanol as the reducing agent and polyvinylpyrrolidone (PVP) as the stabilizing agent. This achieved spherical and monodispersible particles of 10 nm. Overall, the physical methods use huge amounts of energy while being laborious with a high equipment cost and release of toxicants.

20.2.2 Photochemical method

This technique is a physicochemical method in which light energy accelerates the reduction for AgNP production. Huang and Yang (2008) used an inorganic clay suspension stabilizing factor to prevent aggregation in the photoreduction of $AgNO_3$. Photo-irradiation disintegrated into smaller size further similar distribution of size was achieved. In photo-irradiation method also the equipment cost experimental conditions are also costly.

20.2.3 Chemical methods

AgNPs can be synthesized by a chemical reaction with higher yields at an even lower cost. For the chemical reaction process, metal precursors (silver salt), stabilizing agents (PVP, sodium oleate, PVA etc.), and reducing factors (ethylene glycol, $NaBH_4$, glucose etc.) are required along with other parameters such as pH and temperature (Chen and Zhang, 2012; Dang et al., 2012; Patil et al., 2012). The conversion of silver salt into a colloidal solution occurred in two stages, that is, nucleation and consequent growth. In the formation of nuclei, the main consideration is all nuclei are formed at a similar time with monodospersed AgNPs with similar size distribution, later there may be subsequent growth likely to be occurred. By observing all the steps including the agents one can considered as there is releasing of toxic chemicals and sludge residue formed after the process (Song and Kim, 2009).

20.2.4 Biological/green synthesis of nanoparticles

Nanomaterial synthesis by green methods has attracted researchers and scholars worldwide, as this technology is ecofriendly and could produce the required dimensions with greater stability. In a conventional process, three ingredients are required to formulate nanoparticles; metal salts, reducing/stabilizing agents, and capping factors. In the biogenic approach, reducing or stabilizing agents and capping factors are replaced by biomolecules, which come from the extracts of plant parts, bacteria, microalgae, yeast, fungi, etc. This process eliminates expensive chemicals, instead using naturally abundant molecules; it finished in a single step at ambient heat and normal atmospheric pressure, leading to low energy consumption. In biogenic/green synthesis, using natural molecules as the capping and stabilizing factors consumes low energy while generating environmentally benign products and byproducts. This process minimizes the production of toxic effluents and intermediates and creates no/less negative environment, so that it is green and ecofriendly. Further, this process creates good stability and NP dimensions while enhancing the porosity of NPs, which are used in drug delivery and diagnostics. The process was nonlaborious, used low amounts of energy, was nontoxic, and can finish the process in limited time. Due to the regulation of pH and temperature, the nanoparticle shape and size can be managed. The above points show that the biogenic syntheses of nanoparticles is a better technology than conventional processes.

20.3 Classification of NPs

Based on various dimensions and sizes, nanomaterials are classified as (i) 0D (zero dimensional), (ii) 1D (one dimensional), (iii) 2D (two dimensional), and (iv) 3D (three dimensional) (Yu et al. (2015). Nanomaterials with a 1–100nm diameter and a spherical nature called 0D nanomaterials (quantum dots, fullerenes, nanoparticles etc.). The nanomaterials with two dimensions in the nanosize are 1D (nanofibers, nanowires, nanoribbons, nanobelts, and nanotubes). The 2D nanomaterials have one dimension restricted to the nanosize while the other two dimensions are larger than 100nm (graphene, nanosheets, nanolayers). Lastly, the nanomaterials with all three dimensions larger than 100nm but in the nanosize are called 3D nanomaterials (mesoporous carbon, MOF, metal foams, and grapheme aerogels) (Huang et al., 2013; He and Chen, 2015; Yu et al., 2015). The European Commission defined nanomaterials as "natural, incidental/artificial matter comprising particles in an aggregate or unbound or agglomerate state where the particles 50% or more in size distribution, number and external dimensions are in size of 1–100 nm" (Commission, 2011).

20.4 Green synthesis approach for inorganic nanoparticles

In the biological synthesis of NPs, several microorganisms such as fungi, yeast, bacteria, viruses, and plant extracts are used. NP biosynthesis is classified into various types according to the biological material used; (i) biosynthesis by microorganisms, (ii) biosynthesis by biomolecules as templates, and (iii) biosynthesis by plant extracts. Biosynthesis by microorganisms is of two types, intracellular or extracellular, depending upon the NP location. Biosynthesis of NPs makes use of bacteria, fungi, plant extracts were appeared as unadorned, replaceable to chemical and physical approach. This process is cost effective, environmentally acceptable, and can be produced in a large scale. Furthermore, chemicals, energy, temperature, toxic chemicals, and high pressures are not needed. Several biomaterials/biomolecules, plant extracts, microorganisms such as fungus, bacteria, microalgae, and secretions of microorganisms have been used for the synthesis of biogenic nanoparticles. Jaidev and Narasimha (2010) have synthesized silver NPs (3–30nm) by using the fungus *Aspergillus niger*, examined for antimicrobial behavior. They were characterized by EDAX, TEM, and FTIR, and were investigated for antibacterial behavior against Gram-positive bacteria *Staphylococcus* sp. and *Bacillus* sp. and Gram-negative bacteria *E. coli*. The results revealed that AgNPs-*A. niger* showed efficient activity.

20.4.1 Silver nanoparticles

Silver is the best option because of its low lethality profile and outstanding tissue tolerance as well as fewer adverse events of chemical agents and antibiotics. It shows two different antimicrobial mechanisms. The first one explains that denaturation of proteins especially in disulfide bonds, turn into disrupt the protein chains (structural proteins in bacteria) by AgNPs catalytic activity. The second one demonstrates oxidization where Ag ions produces reactive oxygen, attach to the surface of the bacteria, inhibits multiplication, and lead to cell deterioration (Spies, 1999). Synthesized AgNPs could maintain the above characteristics and are formed through surfactant-stabilization in several colloidal processes (Lee et al., 2006). The strong disinfectant and antibacterial behavior of AgNPs can lead to cytotoxic activity in cells. In the literature, cytotoxicity can be found with 200 mg L^{-1} of AgNPs.

The application of NPs in the micropropogation of tissue culture, an important aspect is the Nps should not phytotoxic meanwhile works against microorganisms. To checkout phytotoxicity the trail testing on plant cell culture would be important to fix the successful treatment and this may be varying with plant type and Nps concentration. AgNPs are poorly penetrated into the tissue culture, thereby these are nontoxic, but antibiotics are penetrated easily (Reed et al., 1995). AgNPs readily releases Ag^{+} ions complexed with sulfhydryl (–SH) groups of proteins of microorganisms can cause DNA unwinding and inhibit the respiratory processes and interacted with hydrogen bonding processes leads to inhibition of cell multiplication (Davod et al., 2011). Further, environmental factors influence AgNP antibacterial properties, including pH, concentration, and natural organic matter (Fabrega et al., 2009). The interaction of AgNP surfaces with bacteria is prevented by organic matter through the sorption process (Cho et al., 2005). AgNPs are promising materials for micropropogation due to its low concentration, nontoxic for plant and human tissues, and low cost (Mohammad et al., 2014). Several reports clearly stated that AgNPs are strongly recommended for disinfestations of tissue cultures in micropropogation techniques for substitution or replacement of other toxic chemicals and antibiotics.

Nanoparticles made by synthetic polymers suffer from a lack of uniformity in size, size dispersion, biodegradability, and biocompatibility. To answer these issues, researchers investigated NP biosynthesis with biopolymers such as gum Acacia (Mohan et al., 2007), gum gellan (Dhar et al., 2008), carboxymethylated-curdlan, fucoidan (Leung et al., 2010), and alginate (Pal et al., 2005) as reducing/stabilizing factors. The nanoparticles stabilized by those polymeric compounds conveyed several advantages such as biodegradation, controlled size, morphology, and dispersion (Pal et al., 2005). *Cochlospermum gossypium* is Indian tree gum used as a stabilizing factor for the biosynthesis of silver NPs (Kondaiah et al., 2018) used for antibacterials.

Few plants have medicinal activities that have been utilized as reducing, capping, and stabilizing agents for NP biosynthesis. In this view, several workers used plant parts/whole plants or even weed plants or wastes from the agro, oil, and fruit industries for the biogenic synthesis of nanoparticles that have multiple applications from biomedicine to wastewater treatment technologies. *Allium ampeloprasum* leaf extract was used for the biogenic synthesis of AgNPs as a reducing agent. This was analyzed using TEM, EDAX, FTIR, UV-visible (UV-vis) spectrophotometry, and a zetasizer. The UV absorption bands in AgNPs were noted as 420–440 nm, 2–43 nm size, and quasispherical, spherical, ellipsoidal, and hexagonal structures. These AgNPs exhibited catalytic activity from 4-nitrophenol to 4-aminophenol, and antioxidant activities were proved through DPPH and (2, 2′-azino-bis (3-ethylbenzothiazoline-6-sulphonic acid) ABTS+· radicals. Inductively coupled plasma-mass spectrometry (ICP-MS) was used for the examination and it was noted as 99.5% (Khoshnamvand et al., 2019). In a similar way, *Abutilon indicum* leaf extract was used for the biogenic synthesis of AgNPs as the capping agent. It contains flavonoids, terpenes, and alkaloids and was used for ulcers as well as antiinflammatory, hypoglycemic, and antileprotic treatment. These chemical constituents are accountable for the silver nanoparticle formation (Mata et al., 2015). The AgNPs obtained were characterized through TEM, XRD, EDAX, FTIR, and UV-vis. The UV-vis spectral peaks were noted as λ_{max} at 455 nm while the particle size was noted as 30 nm with a globular shape and crystallinity. These AgNPs were tested for antioxidant, anticancer, and antibacterial activities. A total of 50% of DPPH radical scavenging was achieved at 60 μg/mL of AgNPs when compared to the original rutin. The antibacterial behavior was examined on the AgNP concentration range between 100 and 300 μg/mL. The zone of inhibition was noted on *E. coli*, *P. fluorescence*, *B. cereus*, *S. typhi*, *S. aureus*, and *S. flexneri* as 26, 24, 14, 2, 18, and 16 mm, respectively. The cytotoxic studies on dose and time dependence were confirmed on MDCK and COLO 205 cell lines. This showed a very low IC_{50} of 4 and 3 μg/mL after 28–48 h; on MDCK cells it did not show any cytotoxicity, but after 75 μg/mL it was cytotoxic.

Rafique et al. (2019) used *Albizia procera* leaf extract as the reducing factor for the biogenic synthesis of AgNPs; they were examined for catalytic and antibacterial behavior. Formed AgNPs were characterized by XRD, TEM, AFM, and UV-vis, and the morphology was noted as 6.18 nm with a globular shape and crystallinity. These AgNPs efficiently degraded methylene blue successfully and showed antimicrobial behavior on *E. coli* and *S. aureus* bacterial pathogens. Stable AgNPs were formed by the reaction of *Albizia adianthifolia* leaf extract as reducing factor with $AgNO_3$then characterized through TEM, DLS, XRD, FTIR (Gengan et al., 2013). The UV-vis spectroscopy spectral peak was 448 nm and resins and glycosides are present in the plant extracts responsible for obtaining AgNPs. The AgNPs showed an average of 4–35 nm while the hydrodynamics were 80.27 nm and the zeta potential was − 24.7 mV. FTIR spectra revealed that the capping/stabilizing agents from plant extracts were protein molecules, saponins, and glycosides. The toxicity was examined on A549 lung cells and peripheral lymphocytes (PLs) by three types of assays,

MTT, ATP, and lactate dehydrogenase, at 10 and 50 μg/mL. After 6 h exposure of 10 and 50 μg/mL of AgNPs, the cell viability was recorded as 21 and 73 on A549 lung cell lines while on PLs, 10 and 50 μg/mL of AgNPs showed 117% and 109%. The lactate dehydrogenage was significantly altered at 50 μg/mL AgNPs from 2.43 ± 0.04 to 0.77 ± 0.04. Gomathi et al. (2017) used the leaf extract of *Datura stramonium* for the biogenic process of AgNPs characterized by TEM, SEM, EDAX, FTIR, and UV-vis; the spectral resonance peak was noted at 444 nm, the size of the AgNPs was 15–20 nm, and they were spherical with a crystalline nature. These AgNPs exhibited antibacterial behavior on *Escherichia coli* and *S. aureus*.

Khan et al. (2016) utilized agro waste fruit peel for AgNP biosynthesis using $AgNO_3$ and aqueous Lychee (*Litchi chinensis*) extract containing tannins, epicatechin, and procyanidin A2. Prepared AgNPs were analyzed by UV-vis, HRTEM, EDX, XRD, and FTIR; they had a globular shape of 4–8 nm size. AgNPs exhibited antibacterial behavior against *Escherichia coli*, *S. aureus*, and *B. subtilus*; the MICs are 62.5 and 125 μg/mL for *Bacillus subtilus* and *Staphalococcus aureus*, respectively. They suggested the mechanism for antibacterial behavior was cytoplasmic membrane deterioration and also proved the DPPH free radical scavenger. The AgNPs were examined for methylene blue, and the photocatalysis results were revealed as 99.24% due to the spherical shape, small size, and high dispersion. Banana peel extract as agro waste was checked for AgNP biosynthesis and analyzed by TEM, XRD, EDAX, FTIR, and UV-vis (Ibrahim, 2015). The optimum conditions were noted as a precursor ($AgNO_3$) of 1.75 mM and a banana (*Musa acuminata*) peel aqueous extract of 20.4 mg dry mass incubated at 72 h with 4.5 pH. Also, the reaction time was 5 min between 40 and 100°C. The formed AgNPs were globular with a crystalline structure and a size of 23.7 nm. The FTIR resonance peaks were shown at 433 nm. The chemical functional groups available on the banana peel extract are responsible for the formation of AgNPs and were acts as capping and reducing agent. These AgNPs showed potential antibacterial behavior on *Bacillus subtilis*, *S. aureus*, *Escherichia coli*, and *Pseudomonas aeruginosa* and the MICs were 6.8, 5.1, 3.4, and 1.70 μg/mL respectively. The MBCs were 10.2, 10.2, 5.1, and 5.1 μg/mL, respectively, and levofloxacin enhanced the antibacterial activity about 1.32-fold. Coconut (*Cocos nucifera*) oil cake waste was utilized as a reducing agent for AgNP biosynthesis and investigated for antibacterial activity (Govarthanan et al., 2016) on multiantibiotic-resistant bacteria isolated from livestock wastewater. The formed AgNPs were analyzed through SEM, XRD, EDAX, and FTIR; the particle was 10–70 nm, spherical, and crystalline. The proteins present in the cake accounted for AgNP formation and excellent antibacterial behavior was exhibited on *Aeromonas* sp., *Acinetobacter* sp., and *Citrobacter* sp. isolated from real wastewater.

Citrus sinensis peel extract was used for AgNP biogenic synthesis as a reducing agent and was analyzed by FESEM, TEM, EDAX, XRD, FTIR, DLS, and UV-vis (Kaviya et al., 2011). They attempted to investigate the temperature effect, that is, at 25°C, and enhanced temperatures of 60°C were tried; the AgNPs formed were 35 and 10 nm respectively. These AgNPs showed antibacterial behavior against *P. aeruginosa*, *S. aureus*, and *Escherichia coli*. Annu et al. (2018) investigated AgNP biogenic synthesis using reducing agents, that is, waste peels of fruits of *Citrus limon*, *Citrus limetta*, and *Citrus sinensis*. The formed AgNPs were analyzed through SEM, XRD, TEM, DLS, and UV-vis; the spectral resonance peak was noted as 400 nm. The AgNP shape was globular and crystalline with a size range from 4 to 46 nm. It was also examined for antibacterial and antioxidant behavior. The ability for apoptosis was checked using AgNPs and was successful in G0/G1 phases of the cancer cell cycle. It also acts against *E. coli* and *S. aureus* effectively. Dwivedi and Gopal (2010) used herb weed (a troublesome weed for crops) *Chenopodium album* leaf aqueous extract (reducing agent) for AuNP and AgNP biogenic synthesis. The formed AuNPs and AgNPs were characterized through TEM, XRD, EDAX, FTIR, and UV-vis; the spectral resonance peak was observed at 540 and 460 nm for AuNPs and AgNPs, respectively. The NP size was 10–30 nm while quasispherical shapes and crystallinity were observed.

Naraginti and Li (2017) prepared multifunctional AuNPs and AgNPs through biogenic synthesis using *Actinidia deliciosa* fruit extract; they were investigated for catalytic, antibacterial, anticancer, and antioxidant behavior. The formed nanoparticles were analyzed through TEM, XRD, EDAX, XPS, and FTIR; the sizes were 25–40 nm and 7–20 nm for AuNPs and AgNPs with a globular morphology. Cytotoxic studies showed that AgNPs at 350 μg/mL treated HCT116 cell lines with 78% viability and for AuNPs it was 71%, confirmed by assay MTT. These NPs showed antibacterial behavior against *P. aeruginisa*, and the mechanism predicted was cell membrane deterioration. Biodegradation experiments showed 4-nitophenol and methlyne blue reduction. The FTIR peaks proved that as proteins, aminoacids (cysteine residues) are accountable for capping the AuNPs.

Reddi et al. (2014) attempted to produce AgNPs through a biogenic process using *Acoruscalamus* rhizome extract; it was evaluated for antibacterial, anticancer and antioxidant behavior. The formed AgNPs were analyzed by UV-vis, SEM, EDAX, DLS, and FTIR; they were found to be 31.83 nm with a spherical shape and crystallinity. The availability of alcohol/phenol and aromatic amine groups in the extract of *Acorus calamus* is responsible for the capping process. The AgNP antibacterial behavior showed that the log phase of *Escherichia coli* inhibited growth. An MTT cytotoxic assay was done through on HeLa and A549 cells while apoptotic death was examined by acridine orange/ethidium bromide and annexin V-Cy3 staining techniques. The IC_{50} noted after the first and second day showed 92.48 and 69.44 μg/mL in HeLa cells and 53.2 and 32.1 μ g/mL for A549 cell lines. A rapid physical method was reported in the microwave irradiation technique for AgNP green synthesis by a reducing agent such as the leaf extract of *Biophytum sensitivum* (Joseph and Mathew, 2015). The formed AgNPs were analyzed through FTIR, HR-TEM, XRD, UV-vis, and the UV-vis spectra of AgNPs was noted as 406 nm while the diameter was 19.06 nm with a globular morphology and a crystalline nature. These AgNPs are excellent catalysts for methylene blue and methyl orange degradation in $NaBH_4$ at the reaction mixture.

Phull et al. (2016) used methanolic extract of rhizomes of *Bergenia ciliate* as the reducing agent in the biosynthesis of AgNPs and analyzed that through UV-vis, TEM, XRD, and FTIR. The formed AgNPs were globular, 35 nm in size, and with a crystalline structure; they showed better antioxidant behavior than a crude extract. These AgNPs potentially worked on bacterial and fungus strain types when compared to a crude extract. The cytotoxicity was reported as LD_{50} 33.92 μg/mL on *Artemia salina* (brine shrimp). Anbu et al. (2019) investigated the temperature effect at 37°C and 50°C on the green synthesis of AgNPs by *Platycodon grandiflorum* leaf extract. The formed AgNPs were analyzed through AFTEM, AFSEM, XRD, XPS, FTIR, AFM, and UV-vis spectra. The results revealed that the plasmon resonance peaks at 442 and 457 nm, the size of Ag nanoparticles were noted at 37°C and 50°C as 19 and 21 nm, respectively. The optimum temperature for the formation of AgNPs was reported at 50°C. They showed excellent stability, morphology, and structure while efficiently working against *E. coli* and *Bacillus subtilis*.

Allium sativum L, *Camellia sinensis* L., and *Curcuma longa* L. plants were used in the green synthesis of AgNPs as reducing and stabilizing agents (Selvan et al., 2018). The AgNPs were analyzed through FTIR, SEM, XRD, TEM, and UV-vis. The UV-vis spectral resonance peaks were observed as 450 nm and antioxidant activities were established by ABTS, DPPH, p-NDA, H_2O_2, and DMSO scavenging assays. The vitro cytotoxicity behavior of AgNPs was examined for four human breast adenocarcinoma (MCF-7), cervical (HeLa), epithelioma (Hep-2), and lung (A549) cancers, and one normal human dermal fibroblast cell lines by MTT assay. The results revealed that the *Curcuma longa* extract showed good antioxidant potential and cytotoxic behavior. Dhand et al. (2016) prepared AgNPs using seeds of *Coffea arabica* extract, then analyzed them by SEM, XRD, EDAX, FTIR, DLS, and UV-vis and investigated their antibacterial behavior. UV-vis spectral resonance peaks was observed at 459 nm and the AgNP size was measured as 20–30 nm; they had a spherical shape with a clear crystalline powder. AgNPs exhibited excellent antibacterial behavior on *E. coli* and *S. aureus* with good ZOI. *Ziziphora tenuior* L. is a medicinal plant that shows antimicrobial, wound healing, antiseptic, and expectorant activities. Sadeghi and Gholamhoseinpoor (2015) studied the stability of biogenic synthesized AgNPs using *Ziziphora tenuior*. The AgNPs formed were analyzed through SEM, TEM, XRD, EDAX, FTIR, and UV-vis; the UV-vis spectral peak was observed at 460 nm; the particle size was between 8 and 40 nm and the mean size was 20 nm; it had a globular morphology and crystalline nature. The FTIR results revealed that corresponding amines, aldehydes, and carboxylic acids in plant extracts were responsible for NP formation. The *Camellia sinensis* plant is called white tea (Wt), and it contains polyphenols such as epigallocatechingallate, a powerful antioxidant and reducing agent. *Camellia sinensis* was used for AgNP green synthesis (Haghparasti and Shahri, 2018) from an $AgNO_3$ precursor into polymer montmorillonite as a protective agent. The formed Wt/Ag@Mt nanocomposite was analyzed through FTIR, SEM, TEM, XRD, EDAX, and UV-vis. The antioxidant and cytotoxic behavior was compared with and without lamellar inorganic polymer montmorillonite. The nanocomposites were crystalline, face-centered cubic structures and the Wt/AgNPs and Wt/AgNPs@Mt were 19.77 and 15.87 nm, respectively. In vitro cytotoxicity studies revealed that on MOLT-4 cells, the nontoxic concentration was below 40 μg/mL and it is dose dependent. The IC_{50} of Wt/AgNPs@Mt was 0.0039 μM compared to doxorubicin and cisplatin, which were 2.13329 and 0.013 μM, respectively. Further, this bio Wt/Ag@Mt nanocomposite exhibited excellent antioxidant behavior through DPPH scavenging. Various plants/plant extracts have been utilized for the green synthesis of AgNPs (Table 20.2).

Alfuraydi et al. (2019) investigated the *Sesamum indicum* oil cake used for AgNP biogenic synthesis, and it was analyzed through TEM, XRD, EDAX, and UV-vis. It was examined for antimicrobial antitumor behavior by using breast cancer cell lines (MCF-7). The formed AgNPs were globular in morphology with a diameter between 6.6 and 14.8 nm and a crystal nature. The AgNPs MIC was noted as 500 ng/mL on *P. aeruginosa*, *K. pneumonia*, and *Escherichia coli*. The AgNPs of 2.5 and 7.5 μg/mL concentration were treated with MCF-7 cells. It was found that viable cells region 72.02 and 56.97% at 2.5 and 7.5 μg/mL. Apoptosis was noted as 11.81 and 7.42% and late apoptosis was 15.18 and 31.19%; necrosis was 1.20 and 4.85% at 2.5 and 7.5 μg/mL. In case of control the viable cells region 73.72%, apoptosis 10.82%, late apoptosis 14.54% and necrosis was 1.58%. These AgNPs prepared with *Sesamum indicum* showed excellent antibacterial and antitumor activities. *Origanum vulgare* is a temperate herb containing carvacrol, terpinen, quercetin, and apigenin that is traditionally used as a hypoglycemic, antioxidant, and potential antimicrobial agent. The reducing agent is aqueous leaf extract of *Origanum vulgare*, $AgNO_3$ is as precursor for the green synthesis of silver nanoparticles. The AgNPs formed were analyzed through FTIR, FE-SEM, XRD, and UV-vis; the SPR band was

TABLE 20.2 Capping and reducing agents with their sources and the morphology of the AgNPs formed.

Various biological resources	Shape	AgNPs (nm)	References
Nervalia zeylanica	Spherical	34.2	Vijayan et al. (2019)
Zingiber officinale (ginger)	Spherical	10	Yang et al. (2017)
Indigofera tinctoria leaf	Spherical	16.4	Vijayan et al. (2018)
Salmalia malabarica	Spherical	7	Murali et al. (2016)
Eichhornia crassipes plant	Spherical	00002.14	Oluwafemi et al. (2016)
Euphorbiaceae latex	Spherical	5–10	Rajkuberan et al. (2017)
Eucalyptus globulus Leaf	Spherical	1.9–4.2	Ali et al. (2015)
E. coli culture supernatant	Spherical	8–9	Koilparambil et al. (2016)
sodium alginate	Spherical	10	Shao et al. (2018)
alkali hydrolyzed pectin	Spherical	2.9	Su et al. (2019)
Saraca Indica leaf extract	Spherical	5–50	Perugu et al. (2016)
Aerva lanata	Spherical	18.62	Kanniah et al. (2020)
Azadirachta indica leaf	Spherical	40–130	Roy et al. (2017)
Aloe vera Plant Extract	Spherical	5–50	Logaranjan et al. (2016)
Pleurotus florida	Irregular	5–50	Kaur et al. (2018)
Fraxinus excelsior leaf	Irregular	25–40	Parveen et al. (2016)
L-Cysteine	Spherical	10	Panhwar et al. (2018)
Carboxymethylated cashew	Spherical	92	Araruna et al. (2020)
Biophytum sensitivum	Spherical	19.6	Augustine et al. (2016)
L-Lysine	Spherical	26.3	Han et al. (2017)
Carboxymethyl cellulose-gelatin	Spherical	12	Pedroza-Toscano et al. (2017)
Rosa canina	Spherical	9.75	Gulbagca et al. (2019)
Averrhoa bilimbi leaf	Spherical	20–50	Sagadevan et al. (2019)
Parkia speciosa Husk pods	Spherical	20–50	Fatimah (2016)

observed at 430 nm, the average particle size was 136 ± 10.09 nm, and the shape was globular. The cytotoxic behavior was investigated on A549 lung cell lines and it was shown to be dose dependent. The LD_{50} was noted as 0.1 g/mL. The AgNPs exhibited ZOI of more than 10 mm against *Salmonella* sp., *S. dysenteriae*, *S. paratyphi*, *A. hydrophila*, *E. coli*, and *E. coli* (EP).

Azadirachta indica consists of the phytoconstituents terpenoids and flavanones that are utilized as reducing agents for AgNP green synthesis. Verma and Mehata (2016) used *Azadirachta indica* (Neem) leaf extract for the green synthesis of silver nanoparticles; it was characterized by SEM, XRD, EDAX, FTIR, and UV-vis. A UV-vis spectrum (SPR) was noted at 400 nm along with a spherical shape of 4 nm and a crystalline nature. The formed AgNPs were examined for antibacterial behavior against *E. coli and S. aureus* using 0, 2, 8, 10, and 12 μg/mL. The ZOI was observed at 12 μg/mL and showed 6 mm. They clearly disclosed the influence of temperature and pH on the green synthesis of silver nanoparticles. The yield and morphology between pH 9–13 was feasible and the size was decreased with increased temperature. Vasanth et al. (2014) examined the green synthesis of AgNPs using *Moringa oleifera* stem bark aqueous extract as the reducing agent; it contains phenols, caffeoylquinic acid, β-sitosterol, kaempferol, and quercetin. These AgNPs were analyzed by AF-SEM, TEM, FTIR, and UV-vis. The UV-vis spectrum peak was at 420 nm with a size around 40 nm. The mean particle dimensions were noted as 38 nm with a spherical and crystalline nature. The AgNPs were examined through treated HeLa cells and evaluated utilizing DAPI (4, 6-diamino-2-phenylindole) staining; the apoptosis by ROS genesis was observed.

Kim et al. (2018) explained the green synthesis of silver nanoparticles using an $AgNO_3$ precursor and *Laminaria japonica* algal extract as the reducing factor. They utilized the steam autoclave (hydrothermal process) at elevated

121°C for 20 min. This system was upgraded through various considerations, that is, temperature, pH, extract concentration, and $AgNO_3$ ratios. The AgNPs were characterized by SEM, XRD, FTIR, EDAX, and UV-vis; the SPR was noted as 405 nm, the size was approximately 31 nm, and the average particle size was 23 nm. A germination assay was conducted to estimate the toxicity on *Triticum aestivum* and *P. mungo* seeds at concentrations up to 80 ppm. Significant observations were not noted on the germination of seeds. Philip (2010a) studied the biogenic synthesis of gold nanoparticles and AgNPs using *Hibiscus rosa* as the reducing agent and $HAuCl_4$ and $AgNO_3$ as the precursors. The formed AgNPs and AuNPs were analyzed through FTIR, SEM, TEM, XRD, and UV-vis. The UV-vis SPR peak was noted as 535 nm for the complex, the particle size was ~ 14 nm, and the shape was spherical and crystalline. FTIR analysis reported formed stable AgNPs and AuNPs due to the functional groups, that is, $-NH_2$ and $-COOH$, respectively.

Biosynthesized AgNPs from kondagogu gum (Aruna and Sashidhar, 2018) were used as the reducing agent, and the antibacterial intensity on Gram-positive and Gram-negative bacteria was examined. This study comprehensively included various susceptibility tests such as microbroth dilution, cytoplasmic content leakage, cell wall permeability, growth kinetics, antibiofilm activity, etc.; biocompatibility studies were also attempted with HeLa cells. The MICs are very low for Gram-negative (*E. coli*) and Gram-positive (*S. aureus*). The biosynthesized AgNPs were more potent bactericidal candidates. The mechanism of antibacterial behavior concluded as due to leakage of cellular content, membrane permeability and also dose dependent manner.

20.4.2 Zinc nanoparticles

Zinc nanoparticles (ZnO) are already using as fungicides, adsorbents, and in medical sunscreen because they exhibit good catalytic capacity and efficient adsorption capability while preventing UV rays from sunlight. These are used in ceramics, rubber processing, sensors, electronics, wastewater treatment, biology, environmental protection, and the pharmaceutical industry. ZnO nanoparticles exhibit unique chemical and optical activities that can be simply altered by changing the morphology. ZnO nanoparticles exhibit antibacterial, fungicidal, larvicidal, and antidiabetic activities. In the synthesis of ZnONPs, several procedures were involved, that is, physical, chemical, and green processes. Among them, the green process is ecofriendly and economical.

Divyapriya et al. (2014) biosynthesized ZnO nanoparticles with the ethanolic extract of *Murraya koenigii*. Parthasarathy et al. (2017) performed the ecofriendly synthesis of nZnO using *Curcuma neilgherrensis* methanolic leaf extract. Methanol leaf extract was found to possess alkaloids, flavonoids, steroids, phenols, tannin, and carbohydrates. These nanoparticles were examined for antibacterial behavior and determined to be successful. ZnONPs were biosynthesized through *Trifolium pratense* flower extract and investigated for antibacterial behavior by Dobrucka and Dugaszewska (2016). Gunalan et al. (2012) biosynthesized nZnO particles, investigated zone inhibitory activity, and concluded that they have a stronger inhibitory effect than chemically synthesized nanoparticles. This study used the following microorganisms; *Staphylococcus aureus*, *Serratia marcescens*, *Proteus mirabilis*, *Citrobacter freundii*, *Aspergillus flavus*, *Aspergillus nidulans*, *Trichoderma harzianum*, and *Rhizopus stolonifer*. ZnONPs are synthesized by the sol-gel technique with *Citrus aurantifolia*. The formulated nanoparticles were 50–200 nm with hexagonal and spherical shapes (Samat and Nor, 2013). Similarly, ZnONPs were prepared using *Aloe barbadensis* Miller leaf extract as the capping factor, and the average particle size was 15 nm. The antibacterial activity was noted against clinical isolates Gram-positive (*Staphylococcus aureus*), Gram-negativeve (*E. coli*), *P. aeruginosa*, and *Staphylococcus aureus* (methicillin resistant). The MBC and MIC values were noted as 2300 and 2700 and 2200 and 2400 μg/mL (Ali et al., 2016). In another work, ZnONPs were biosynthesized using *Aloe vera* extract as the capping agent and zinc nitrate (Gunalan et al., 2012) as the precursor; they were 40 and 25 nm. They were tested for antimicrobial behavior and exhibited enhanced biocidal activity against various microorganisms when compared to conventional ZnONPs. Ambika and Sundrarajan (2015) explained the synthesis of ZnONPs using zinc nitrate hexahydrate with *Vitex negundo* plant extract as the precursor. The plant contains water-soluble chemical constituents such as alkaloids, terpenoids, flavonoid, flavones, and phenolic chemical constituents. Among them, flavones are responsible for the formation of 75–80 nm particles with a spherical shape. ZnONPs showed antibacterial activity on Gram-positive and Gram-negative bacteria (*S. aureus* and *E. coli*).

The *Anisochilus carnosus* wallich leaves showed antiulcer activity and contain alkaloids, phenols, tannins, and flavonoids. This extract was used as a reducing agent for the biosynthesis of ZnONPs by Anbuvannan et al. (2015a). The DLS results revealed that the mean diameter of the zinc nanoparticles was between 20 and 40 nm. The EDAX results confirmed the available components as only oxygen and zinc. ZOIs of 9, 7, 6, and 10 mm were recorded against *S. aureus*, *E. coli*, *S. paratyphi*, and *V. cholerae*, respectively. They also showed photocatalytic activity by electron donor/carrier from (ZnO) and degraded methylene blue as the blue color changed to colorless. These authors reported a similar work on a plant used as folk medicine (Anbuvannan et al., 2015b), *Phyllanthus niruri,* containing terpenoids, tannins, saponins, polyphenols, lignans, flavonoids, coumarins, and alkaloids. *Corymbia citriodora* is a tall tree with the leaf extract containing the chief chemical constituents. Citronellal was used as reducing agent for ZnONP green synthesis; they were also examined for photocalytic activity. The formed ZnONPs were characterized through Raman spectroscopy, SEM, UV-vis, EDAX, and TGA. They were 64 nm and degraded methylene blue under visible light irradiation (Zheng et al., 2015). The ZnONPs were prepared from the extract of rambutan (*Nephelium lappaceum* L.) fruit peel; the formed ZnONPs degraded methyl orange by 83.99% within 2 h exposure through UV irradiation.

Azizi et al. (2016) selected the flower extract of *Anchusa italic* (reducing agent) for ZnONP production and tested the annealing temperature effect on antimicrobial and structural characteristics. They explained that when the annealing temperature was between 100 and 200°C, the size of the zinc nanoparticles was ~ 8 nm and reached ~ 14 nm with a hexagonal shape and crystalline nature. *Anchusa italic* containing tannins, triterpenes, saponins, flavonoids, and alkaloids showed antioxidant, antiinflammatory, anticancer, antiviral, and antimicrobial properties. The fatty acids present in the *Anchusa italic* flower extract are considered for ZnONP formation. The antimicrobial intensity was reduced with increased annealing temperature against Gram-positive (*Bacillus megaterium, S. aureus*) and Gram-negative (*Salmonella typhimurium, E. coli*). ZnONPs (60 μg/mL) prepared at 100°C and 200°C were compared against four microorganisms; the ZOIs were reduced from 15.9, 16.8, 15.9, and 15.1 mm to 13.6, 14.6, 12.9, and 14.4 mm, respectively. Cytotoxic results on Verocells indicated that the activity is dose dependent; the IC_{50} was 142 μg/mL, and it is safe and has low toxicity.

Azadirachta indica leaf extract was used as reducing agent for the formation of ZnONPs. This was analyzed by SEM, FESEM, AFM, X-RD, EDAX, and FTIR, and the results were an 18 nm size and crystal structure (Elumalai and Velmurugan, 2015). The antimicrobial intensity was dose dependent, the concentration increased (50, 100, 200 μg/mL) leads increase in activity owing to enhanced formation of H_2O_2 concentration from the action of ZnONPs and alone ZnO or *Azadirachta indica* extract were not shown this much of antimicrobial intensity. In another work, Elumalai et al. (2015) used *Vitex trifolia* L. leaf extract for ZnONP formation. The FTIR peak at 476 cm^{-1} formed because of the Zn–O (metal-oxygen). They performed experiments on methylene blue degradation and antimicrobial activities. That optimum decoloration efficiency was observed with 30 mg of ZnONPs by 90 min. The ZOI (mm) at 200 μg/mL was noted as more for all microorganisms (*S. aureus, B. subtilis, P. aeruginosa, P. mirabilis, E. coli, C. tropicalis,* and *Candida albicans*) compared to a standard drug. The MIC values of ZnONPs were between 6.25 and 50 μg/mL while the MBC/MFC was 12.5–50 μg/mL. The reason beyond the increased antimicrobial activity is expected that Vitrifolin A (major chemical constituent (32.9%) binds with ZnONPs surfaces enhanced the antimicrobial behavior. Madan et al. (2016) prepared ZnONPs using *Azadirachta indica* extract with $ZnNO_3$; they showed multifunctional and different morphologies while the reaction was complete within 4 min. These ZnONPs were characterized by SEM, X-PS, TEM, FTIR, EDAX, and X-RD and the size was between 9 and 40 nm with various shapes. ZnONPs were tested for antibacterial, antioxidant, and photocatalytic behavior. The MIC values are noted as 150 and 200 μg/mL on *K. aerogenes* and *S. aureus,* respectively while the IC_{50} of DPPH activity was 8355 μg/mL. The percentage of scavenging capacity was up to 92% and bullet-shaped ZnONPs showed that methylene blue degraded by photocatalytic activity under 2 h exposure of UV light.

Rosa canina fruit extract was used as a reducing agent and zinc nitrate as the precursor for the formation of ZnONPs. They structural and colloidal properties were analyzed by SEM, FTIR, EDAX, X-RD, and DLS (Jafarirad et al., 2016). The ZnONPs were examined for antibacterial behavior on *E. coli, Listeria monocytogenses,* and *S. typhimurium.* The efficacy and anticancer intensity were also studied by an MTT assay with a concentration of 50 and 100 μg/mL ZnONPs with low toxicity. A natural sweetener, *Stevia leaf* is a native American plant. The leaf extract was used as a stabilizing factor with zinc acetate at 80–90°C for the formation of blackish to green gel. This was then calcinized at 600°C for 2 h to form white nanoparticle ZnO. These particles were characterized by SEM, FTIR, EDAX, X-RD, TGA, and zeta potential; they were 10–90 nm with a rectangular to spherical shape. These ZnONPs were tested for antiprotozoal activity against *Leishmaniasis major* and antimicrobial behavior against *Staphylococcus aureus* and *Escherichia coli* (Khatami et al., 2018). *Polygala tenuifolia* is a plant that contains xanthones, saponins, and oligosaccharides, and it is traditionally used to treat insomnia, asthma, palpitations, and rhinitis. Nagajyothi et al. (2015) formulated ZnONPs with *P. tenuifolia* as the reducing agent and they were characterized by SEM, TEM, EDAX, FTIR, and TGA; the ZnONPs were 33–73 nm and spherical. ZnONPs were tested for antioxidant (DPPH) and antiinflammatory activities. Antioxidant and antiinflammatory activities were showed as dose dependent as increased the concentration increases the percent activity, suppressing both mRNA and protein expression, and also suppress the COX-2, iNOS, IL-ib, IL-6 (Interleukins) and TNF-α.

ZnONPs were biosynthesized using *Solanum nigrum* as the reducing agent and zinc nitrate as the precursor. The particles formed were in the range of 20–30 nm with hexagonal structures (Ramesh et al., 2015). ZnONPs were characterized by SEM, TEM, UV-vis, FTIR, TGA, and XPS studies. Antibacterial properties were also investigated, and the ZOIs were 11, 7, 17, and 18 mm on *Vibrio cholera*, *E. coli*, *Samonelal paratyphi*, and *S. aureus*, respectively. Sharma (2016) reported on the biosynthesis of ZnONPs with *Carica papaya* milk (CPM) latex as the stabilizing agent; the formed particles were various shapes and sizes. ZnONPs were characterized by SEM, TEM, FTIR, and UV absorption spectrophotometry. The latex concentration directly influenced the formation of various shaped morphologies of particles. At 10 mL latex, ZnONPs were hexagonal pyramids with sharp end boundaries. At 15 mL of latex, a hexagonal shape was formed along with open tips, and 20 mL of latex gave aggregated prism shapes. At 25 mL latex, nanobuds formed and at 30 mL latex concentration, nanoflowers were formed with petal-like structures. This might be the involvement of cysteine proteinases, carbohydrates, and lipids present as organic compounds in the latex. ZnO nanoflowers showed good antibacterial intensity on *P.aeruginosa* and *Staphylococcus aureus*, *Klebsiella aerogenes*, and *Pseudomonas desmolyticum*. At the 0.4 mg/mL concentration, the ZOIs were noted as 6.62, 9.0, 7.25, and 6.12 mm for *Psuedomonas aeruginosa* and *S. aureus*, *Klebsiella aerogenes*, and *Pseudomonas desmolyticum*, respectively. The ZnO nanoflowers also exhibited photocatalytic activity on Alizarin Red-S. The factors influenced are (i) surface imperfections are higher, (ii) crystallite size is low, (iii) wide band space, (iv) electron-hole pair combination reduction capabilities all together they acts as efficient photo-catalytic material and also ZnO nano-flower can be recycled for several cycles.

The ZnONPs also exhibited other properties antioxidant, photodegradative, photoluminescence and dyes degradation. Suresh et al. (2015) synthesized ZnONPs using *Artocarpus gomezianus* (AG) extract as the stabilizing agent by combustion synthesis. The ZnONPs were tested through TEM, SEM, XRD, and UV-vis; they effectively degraded methylene blue and had DPPH antioxidant efficiency. Priyabrata et al. (2016) biosynthesized AgNPs and ZnONPs using the mangrove plants *Heritiera fomes* and *Sonneratia apetala* as stabilizing agents and evaluated HF-AgNPs, HF-ZnONPs, SA-AgNPs, and SA-ZnONPs for biomedical uses. All nanoparticles that is, HF-AgNPs (400 nm), HF-ZnONPs (40–50 nm), SA-AgNPs (20–30 nm), and SA-ZnONPs (400–500 nm) were examined by SEM, XRD, TEM, and UV-vis. The chief chemical constituents present in the mangroves were oximes and heterocyclic compounds that were accountable for NP formation. These NPs exhibited moderate antioxidant activities while the antiinflammatory potential was compared; ZnONPs and AgNPs have 79% and 69%, respectively. The ZOI of ZnONPs and AgNPs produced from two different mangrove plants were between 9 and 16 mm against *S. aureus*, *S. flexneri*, *V. cholera*, *Escherichia coli*, *S. epidermidis*, and *B. subtilis*. The DPPH activity of AgNPs and ZnONPs was considerable with the IC_{50} noted as between 53.64 and 169.71 μg/mL. This was compared with catechol IC_{50} at 52.89 μg/mL. Dobrucka and Dugaszewska (2016) used the extract of the *Trifolium pratense* flower for ZnONP biosynthesis as the stabilizing agent. ZnONPs were analyzed by using FTIR, EDAX, TEM, XRD, and UV-vis, and the antimicrobial behavior was investigated against *E. coli*, *S. aureus*, and *P. aeruginosa*. Vijayakumar et al. (2015) used the leaf extract of *Plectranthus amboinicus* for ZnONPs (20–50 nm) synthesis, and this was characterized by TEM, XRD, FTIR, and UV-vis absorption. They evaluated ZnONP efficacy on the control of *Staphylococcus aureus* biofilm (methicillin-resistant) (MRSA) and this was achieved at 8–10 μg/mL concentration. These nanoparticles at 8 and 10 μg/mL showed 100% mortality of *Anopheles stephensi*, *Culex quinquefasciatus*, and *C. tritaeniorhynchus*. They explained that the mortality may be owing to the decomposition, epithelial rupture, and cellular vacuolization in the gut.

20.4.3 Copper nanoparticles

CuO is already well established and used as an antimicrobial, antifouling, antibiotic, and antifungal agent when incorporated in coatings, plastics, and textiles (Apostolov et al., 2014). CuONPs were biosynthesized by Nasrollahzadeh et al. (2015) using *Anthemis nobilis*, *Calotropis gigantea* was used by Sharma et al. (2015), *Cinnamomum camphora* was used by Huang et al. (2007), *Carica papaya* was utilized by Sankar et al. (2014), *Aloe vera* was utilized by (Vijay Kumar et al., 2015), and *Emblica officinalis* was used by Ankamwar et al. (2005).

20.4.4 Gold nanoparticles

Gold nanoparticles (AuNPs) are compatible to human tissues and show high stability even in biological fluids. AuNPs have gained superior consideration due to their new optical and biocompatible (Anwar and Sherif, 2013) properties. AuNPs exhibit notable claims in catalysis, sensor development, bioimaging, pharmaceutical formulations, etc. (Sang-Woo, 2006; Shon and Cutler, 2004). AuNPs have been actively studied for antimicrobial activity and compared to various antibiotics such as cefaclor, ampicillin, vancomycin, etc., on Gram-negative and Gram-positive bacteria. The mechanism of AuNPs is the inhibition of peptidoglycan layer synthesis, a rigorous increase of ATP and membrane damage (Beyth et al., 2015). AuNP fusion with antibiotics shows the development of pores on the cell wall, consequently damaging the integrity of the cell walls (Ravishankar and Jamuna, 2011) and eventually leaking cellular organelles, causing cell death. AuNPs with NIR (near-infrared radiation) at characteristic absorption of radiation, the photothermal heating occur and this can kill the cancer cells and also bacteria. According to the literature, the green synthesis of AuNPs was also done by *Salmalia malabarica* (Bhagavanth et al., 2015), *Ocimum sanctum* (Philip, 2010a,b; Philip and Unni, 2011), the leaf of *Diopyros kaki* and *Magnolia kobus* gum (Jae et al., 2009), and the *Acacia nilotica* leaf (Rakhi et al., 2013). AuNPs are promising alternatives to antibiotics/toxic chemical agents that are non-toxic and tolerable with plant and human tissues, according to the literature. AuNPs are preferred in tissue culture, especially in micropropagation. A few microbial species used as capping/reducing agents are shown in Table 20.3.

As an interested for clinicians and scientists for suppression of multi drug resistance (MDR) strains gold nanoparticles are promising material for MDR therapy. Killing MDR strains by the hydrophobic functionalization of gold nanoparticles was achieved by Li et al. (2013). It worked on 11 clinical MDR stains and succeeded in suppressing their growth. Recently, several workers took advanced steps, that is, capping with biological materials or antibiotics to enhance the antibacterial intensity. Kanamycin-capped AuNPs (Kan-AuNPs) synthesized by Payne et al. (2016) showed efficient antibacterial activity on microorganisms (*S. bovis*, *S. epidermidis*, *E. aerogenes*, *P. aeruginosa*, *Y. pestis*). These works confirmed that AuNP fusion noticeably enhanced the efficacy of the antibiotic against Kan-resistant bacteria. Kondaiah et al. (2019) biosynthesized AuNPs using kondagagu gum as the capping agent and proved their antibacterial activity.

TABLE 20.3 Capping and reducing agents with their sources and the morphology of AuNPs formed.

Various biological resources	Shape	AuNPs (nm)	References
Pseudomonas fluo-rescens 417	Spherical	5–50	Husseiny et al. (2007)
Escherichia coli	Spherical	11.8–130	Honary et al. (2012)
Escherichia coli DH5a	Spherical	20	Suganya et al. (2015)
Klebsiella pneumonia	Spherical	5–35	Kumar et al. (2014)
Bacillus stearothermophilus	Triangular	5–30	Luo et al. (2014)
Yarrowia lipolytica (NCIM3589)	Triangular	15	Tripathia et al. (2014)
Spirulina platensis	Spherical	~ 5	Issazade et al. (2013)
Shewanellao neidensis	Spherical	2–50	Suresh et al. (2011)
Staphylococcus epidermidis	Spherical	20–25	Ogi et al. (2010)
Thermomonospora	Spherical	8	Fayaz et al. (2011)
Penicillium crustosum	Spherical	100	Roy and Das (2016)
Streptomyces hygroscopicus	Spherical	20	Husseiny et al. (2007)
Rhizopus oryzae	Spherical	5–65	Sanghi et al. (2011)
Candida guilliermondii	Spherical	50–70	Waghmare et al. (2014)
Trichoderma harzianum	Spherical	26–34	Roy et al. (2016)
Streptomyces fulvissimus	Spherical	20–50	Balagurunathan et al. (2011)
Pycnoporus sanguineus	Triangular	26–60	Du et al. (2015)
Botrytis cinerea	Triangular, Hexagonal	1–100	Vijayaraghavana et al. (2011)
Stoechospermum marginatum	Hexagonal	18.7–93.7	Singh et al. (2013)
Spirulina platensis	Spherical	20–30	Rajathi et al. (2012)
Magnusiomycesingens LH-F1	Sphere, triangle	80	Venkatesan et al. (2014)
Pithophoraoedogonia	Spherical	32.06	Nangia et al. (2009)
Padina tetrastromatica	Thin planner	8.3–25	Li and Zhang (2016)
Gordoniaamarae	Spherical	15–40	Elavazhagan and Arunachalam (2011)
Galaxaura elongata	Spherical Triangular	3.85–77.13	Mohsen et al. (2012)

Instead of ecofriendly capping agents, a few scientists tried with antibiotics for elevated antibacterial activity. Roshmi et al. (2015) established that extremely stable AuNPs designed by a Bacillus sp. and functionalized with ciprofloxacin, rifampicin, gentamycin, and vancomycin were highly efficient against coagulase-negative staphylococci (CoNS), that is, *S. epidermidis* and *Staphylococcus haemolyticus*. Lima et al. (2013) experimented with AuNPs distributed on zeolites, and accomplished an outstanding biocide to kill food pathogens such as *E. coli* and *S. typhi* in a short time. Thirumurugan et al. (2012) tried to observe AuNPs linked bacteriocin-like peptides for enhanced effect against food bacteria and confirmed as enhanced antibacterial intensities on both combination of bacteriocin with AuNPs and bacteriocin-nisin linked AuNPs. Kanchi et al. (2018) utilized the biodiesel industry waste cake *Jatropha* after oil removal as a reducing agent and the $HAuCl_4$ precursor for the green synthesis of gold nanoparticles; this was characterized by SEM, FTIR, TEM, X-RD, and UV-vis. Various parameters were studied to optimize the AuNP size and morphology, that is, reaction temperature, pH, and precursor/reducer ratio. They observed that the triangle, hexagonal, and globular shapes of AuNPs, the ~ 14nm size, and the proteins inside the cake of *Jatropha* waste were the exact causes for the stabilized AuNP formation. Barboza-Filho et al. (2012) reported on a natural rubber membrane with AuNPs and found their activity against *Leishmania brasiliensis*, a protozoan parasite. AuNPs in rubber membranes decrease the population growth and attached to membrane surfaces, they can lead to the preparation of flexible dressings for skin lesions while stimulating angiogenesis in damaged tissues.

Freshwater algae *Prasiola crispa* was used as the capping agent for AuNP biosynthesis; they were crystallite and spherical with a size of 9.8nm. AuNPs were analyzed through XRD, TEM, FTIR, UV-vis, and DLS (Sharma et al., 2014). Special material, bee glue is combination of proposils collected from several plant buds with honeybee secretions which is utilized for construction of hives in honeycomb and were used as reducing agent in green synthesis of silver and gold nanoparticles (Roy et al., 2010). This material is actually a resinous material used by honeybees to avoid microbial attacks on the honeycomb. Two solvents were used to extract the bee glue (water and ethanol) for the gold and silver nanoparticles formation and also tested the total polyphenol content and efficiency of pinocembrin and galangin. The nanoparticles were analyzed through SEM, TEM, UV-vis, and FTIR cyclic voltammetry. The diameter of the NPs was ~ 20nm, various shapes of spheroids, nano-rods, nano-wedges and nano-prisms and various geometrical shapes: triangles, hexagons, squares and trapezoids were formed. The FTIR cyclic voltammetry reports established the pinocembrin and galangin.

A few workers also tried to bioreduce/accumulate the noble metals with intra- or extracellular accumulation/reduction. *Shewanella algae* is a mesophilic bacteria used as a reducing agent for the biosynthesis of gold nanoparticles at a pH range of 2.0–7.0, 25°C, and H_2 as the electron donor (Konishi et al., 2007). The reduction reaction of 1mM $AuCl_4$ was completed in 30min and AuNPs of size 10–20nm were formed at pH7 in the periplasmic space of bacteria. The AuNP size (15–200nm) depends on the pH of the reaction solution, based on cells at 2.8 pH and 2.0 pH, size was 20nm were formed interacellularly, 100–200nm crystalline nanotriangles and 350nm were formed extracellularly. Baker and Sathish (2015) biosynthesized AuNPs by using *P. veronii* and *Annona squamosa* L as reducing agents; they were characterized by TEM, XRD, FTIR, and UV-vis. The AuNPs formed were crystalline and 5–25nm. The AuNP antimicrobial behavior was evaluated on *Escherichia coli* and *S.aureus* with 10^6 cell density and DNA binding studies were also conducted. AuNPs were treated with 10ng of DNA incubated at 37°C. The mixture was then examined using electrophoresis. The bands were deformed and damaged DNA with light color band indicated that the DNA binding.

Priyanka et al. (2016) selected the bacteria *Sporosarcina koreensis* for the microbial synthesis of AgNPs and AuNPs. The biosynthesized NPs were analyzed using UV-vis; the maximum absorbance was 531 and 424nm for AuNPs and AgNPs, respectively. The AgNPs were globular (102nm) and the AuNPs were 92.4nm, as analyzed through XRD, EDAX, and TEM. The antimicrobial behavior of AgNPs has been examined based on the MBC and MIC of *Vibrio parahaemolyticus*, *E. coli*, *Salmonella enterica*, *S. aureus*, *B. cereus*, and *B. anthracis*. Further, AgNPs with a combination of antibiotics, such as vancomycin, lincomycin, rifampicin, oleandomycin, penicillin G, and novobiocin, have shown that 3µg/mL sufficiently enhanced the antimicrobial efficacy; biofilms were also inhibited on *S. aureus*, *P. aeruginosa*, and *E. coli* with a concentration of 6µg/mL. AuNPs exhibited catalytic activity with sodium borohyride on 4-nitrophenol to 4-aminophenol.

Zhang et al. (2016) used the yeast *Magnusiomyces ingens* LH-F1 strain (2.2mg/mL) as the reducing agent for the biosynthesis of AuNPs by 1.0mM $HAuCl_4$ as the precursor. AuNPs are formed with spherical (10–80nm), triangle, and hexagonal shapes. AuNPs showed tremendous catalytic behavior along with $NaBH_4$ on 4-nitrophenol, 3-nitrophenol, and 2-nitrophenol to aminophenols. Begum et al. (2009) chose black tea leaf extract as the reducing factor for AgNP and AuNP biosynthesis; three solvents were selected, water, ethyl acetate, and dichloromethane. Among them, water and ethyl acetate efficiently reduced the $AuCl_4$ into various shapes such as spheres, trapezoids, prisms, and rods, but the dichloromethane tea extract did not produced nanoparticles. They explained the presence of polyphenols, quercetin, etc., in the first two solvents and dichloromethane solvent was not able to extract these chemical constituents.

The formed nanoparticles were analyzed through FTIR, TEM, and UV absorption spectra; a ~ 20 nm size was formed. The phytocostituents, that is, polyphenols and flavonoids, are accountable for NP formation. The plant extracts of *Angelica archangelica* roots, *Hypericum perforatum* herb, and *Hamamelis virginiana* bark were used for AuNP green synthesis (Pasca et al., 2014). The NPs were analyzed through TEM, XRD, FTIR, and UV-vis. The AuNPs formed at 8 pH were 4–8 nm; the shapes were spherical, oval, and polyhedral.

20.5 Characterization of nanoparticles

20.5.1 Ultraviolet-visible spectrophotometry

The UV-vis spectrophotometer was used to obtain spectra at the resolution of 300–800 nm. 3 mL of the NP samples were withdrawn in various time periods and the absorbance was examined. The reduction of metal ions was monitored by measuring the reduction medium diluted with the required distilled water to the sample and UV-vis spectral analysis was carried out using a UV-vis spectrophotometer between 300 nm and 600 nm.

The above figure shows the optimization studies of AgNPs. In this work, AgNPs were synthesized using CMGK (*Cochlospermum gossypium*) functionalization with carboxy methylation. The concentration of CMGK was changed (0.1%–0.5%), with a stable concentration of $AgNO_3$, to determine the CMGK intensity in the formation of silver nanoparticles. Fig. 20.1B shows the corresponding changes in surface plasmon resonance bands by increasing the CMGK concentration. As the CMGK concentration increased from 0.1 to 0.5%, the SPR peak increased as the particle size of the AgNPs decreased. It is affirmed that by an increase in the concentration of CMGK, the formation of AgNPs also increased. The effect of the $AgNO_3$ quantity was also examined on the formation of AgNPs, and it was found that increased concentration of Ag^+ ions was more available and the capping efficiency of CMGK was decreased. The spectral band (SPR) is an indicator for formation and NP reduction. The researcher can use as an indicator for the reduction process while optimizing the various parameters on the NPs biogenic synthesis, i.e., temperature, pH, reaction time, reducing agent concentration and metal precursor concentration.

20.5.2 FT-IR analysis

To identify the functional groups present on the biomolecules and metal-organic interfaces, FTIR is the best tool, as the sample is mixed with KBr and form a pellet in the ratio of 1:100. Then record the spectrum in the between 400 and 4000 cm^{-1} at a resolution of 0.2.

The formation of CMGK capped AuNPs and AgNPs was analyzed by FTIR spectrophotometry. Fig. 20.2 shows the CMGK and CMGK stabilized gold and silver nanoparticle spectroscopic results. The major bands (Fig. 20.2) showed that the IR spectra of CMGK (*Cochlospermum gossypium*) bands were present at 1035, 1225, 1415, 1615, 1733, 2960, and 3430 cm^{-1}. The peaks noted as 2960 and 1035 cm^{-1} were due to the stretching vibrations of the C–H and O–H groups. The strong peaks noticed at 1733, 1416, and 1615 cm^{-1} were attributed to COO^- (carboxylation) group asymmetrical and symmetrical stretching vibrations related to the CMGK gum. The peaks at 1035 and 1225 cm^{-1} are designated as C–O–C stretching vibrations of CMGK's ether and alcohol groups. Fig. 20.2 (curve b) revealed the FTIR of the CMGK spectrum capped AgNPs, which shows the characteristic peaks at 1026, 1220, 1406, 1591, 1727, and 3375 cm^{-1}. After the reduction process, the shifts in the peaks were observed due to CMGK stabilized AgNP formation, and

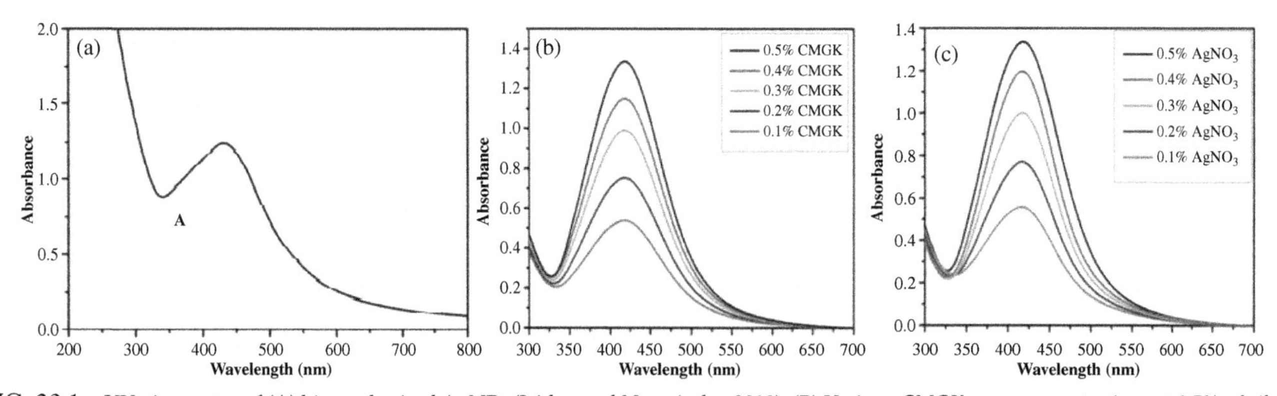

FIG. 20.1 UV-vis spectra of (A) biosynthesized AgNPs (Jaidev and Narasimha, 2010); (B) Various *CMGK* gum concentrations at 0.5% of silver nitrate, (Kondaiah et al., 2018); (C) Various concentrations of $AgNO_3$ at 0.5% of *CMGK* gum (Kondaiah et al., 2018).

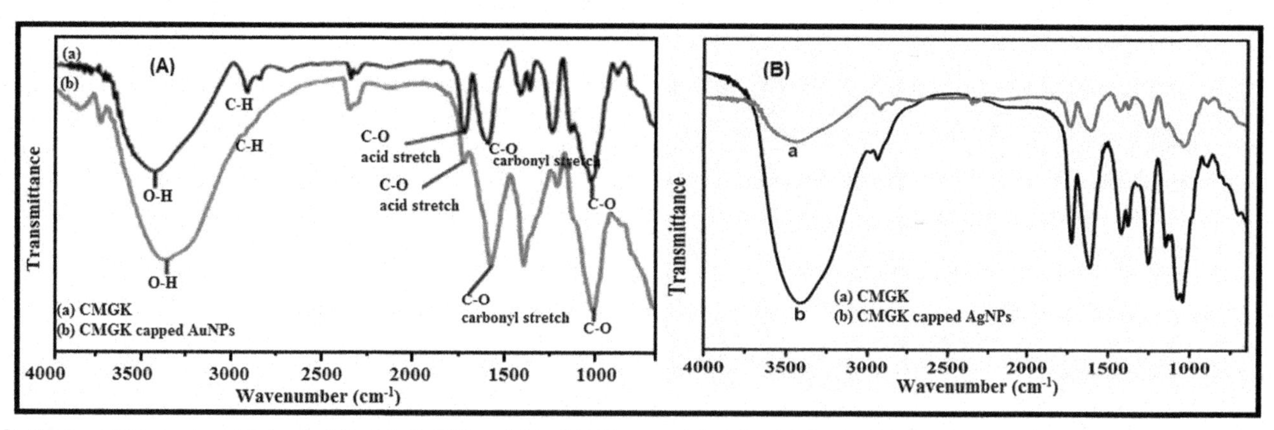

FIG. 20.2 FTIR spectra: (A) for AuNPs (a) CMGK and (b) CMGK-capped AuNPs (B) for AgNPs (a) CMGK and (b) CMGK-capped AgNPs (Kondaiah et al., 2018, 2019).

they shifted from 3430 to 3375, 1733 to 1727, 1615 to 1591, and 1415 to 1410 cm^{-1}; the other bands found remained the same. These shifted spectral peaks suggested the binding of AgNPs in the capping process with –OH and –COOH groups of *Cochlospermum gossypium*. Based on these peak shifts, that is, the –OH and –COOH groups, one can confirm that both these groups were involved in the green synthesis and stabilization of AgNPs (Kondaiah et al., 2018). The FTIR spectra of plant/microbial extracts gives information about the available functional groups before and after the reduction/capping process. The peaks shifting/deleting will guide us as in the process which functional groups are responsible for caping/reduction process.

20.5.3 Zeta potential studies

The surface zeta potentials of AgNPs were measured using the laser zeta meter (Malvern zeta seizer 2000, Malvern). For this, liquid samples of AgNPs (5 mL) were diluted with water using NaCl as the suspending electrolyte solution (2×10^{-2} M NaCl). The sample was stirred about 30 min and the equilibrium pH was recorded. The zeta potential of the particles was measured. Determination of the hydrodynamic particle size of AgNPs using DLS depends on the laser diffraction method with multiple scattering techniques. In the suspension of small AgNPs, the number of phytochemicals per particle is high; the hydrodynamic size acquired from DLS is more prominent than that from TEM. The zeta potential (ζ) and hydrodynamic particle size of AgNPs were quantified by DLS technology, and the zeta potential (ζ) was utilized for the confirmation of charge on AgNPs. The confirmed zeta potential indicates the presence of repulsive forces and can be utilized to assess nanoparticles stability (Fig. 20.3). Nanoparticles with severe negative/positive surface charges are stable. A flat-out zeta possible estimation of $\pm$ 30 mV is an overall sign that the colloidal solutions are highly stable. AgNPs have a mean size of nanoparticles of 29 nm and a zeta potential of – 18.5 mV value; the CMGK stabilized AgNPs had a zeta potential of – 18.7 mV (Kondaiah et al., 2018). These findings suggest that NPs stabilized with phytochemicals carried negative charges and the NPs were stable due to the powerful repulsion between the NPs.

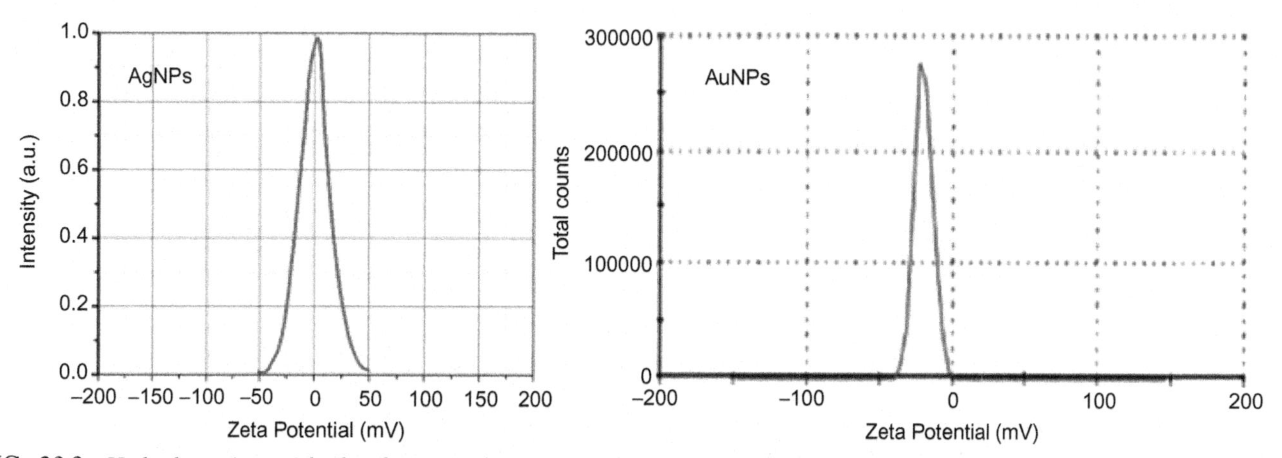

FIG. 20.3 Hydrodynamic particle distribution and zeta potential by dynamic light scattering technique (A) for AgNPs; (B) for AuNPs (Kondaiah et al., 2018, 2019).

20.5.4 XRD analysis

The elemental structure and crystallinity morphology was determined by using drop-coated films of silver nitrate above a glass plate and employed with X-RD (diffractometer). The structure of the formed AgNPs was studied using the XRD technique, and the XRD pattern of AgNPs is shown in Fig. 20.4. The diffraction bands with 2θ, theta values 38.32, 44.50, 64.75, and 77.05 were indexed as (111), (201), (222), and (312), correspondingly. In the XRD pattern, the peaks are indexed with a face-centered cubic assembly. The peak (111) plane is the most intense compared to other peaks in the X-RD spectrum, which reveals the permissible growth of AgNPs along (111) directions. The determined lattice constant has been established as 0.4085 nm, that was compatible with the master worth a = 0.4086 nm, synthesized AgNPs were assessed through Scherer's equation ($D = 0.9\lambda/\beta cos\theta$) (Rodríguez-León et al., 2013; Chung et al., 2016). Where D is the crystalline size, β is the full width at half maximum, λ is the wavelength of the x-ray (1.5406 A°), and θ is the inclination (Bragg's). The crystallite size was calculated to be 8.6 nm, which about effectively correlates with the reports of TEM.

20.5.5 TEM analysis

The AgNP size and shape were detected with TEM operating at 15 kV. For TEM measurements, grids were prepared by putting one drop of bioreduced AgNP solution onto a copper grid coated with carbon; this was later dried at room temperature, then micrographs were recorded. The AgNPs were added to a (2.5% w/v) aqueous/glutaraldehyde mixture, centrifuged, and suspended in 1500 μL of 0.1 M of 7.2 pH phosphate buffer at 4°C, then post fixed in 1% of osmium tetra oxide at 4°C in 0.1 M phosphate buffer (60 min) for TEM (HITACHI 7500). Samples were dehydrated using acetone (analytical graded series). Micrographs were taken at 100 kV with TEM. Fig. 20.5 shows the TEM images of the synthesized AgNPs. Synthesized silver nanoparticles have globular shapes with smooth surfaces and are

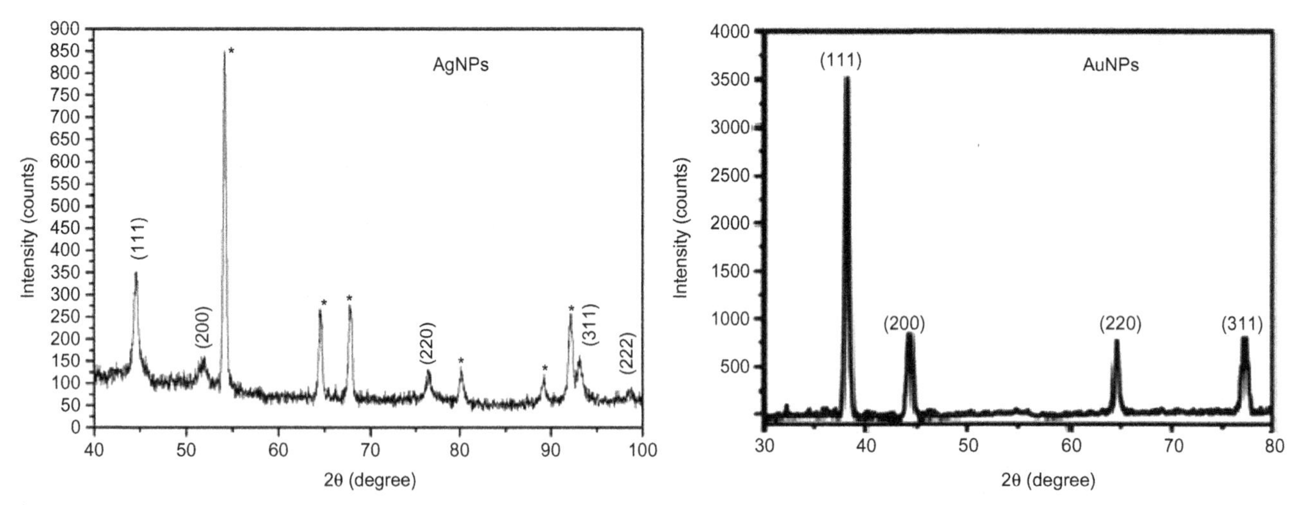

FIG. 20.4 XRD pattern of AgNPs and AuNPs by microwave irradiation (Kondaiah et al., 2018).

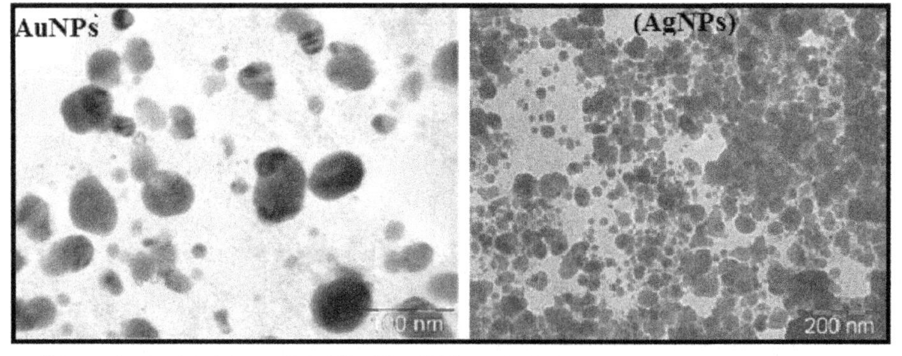

FIG. 20.5 TEM image of biosynthesized AuNps and AgNPs (Kondaiah et al., 2018, 2019).

well distributed. This study clearly demonstrates that the average size dispersion was 9±2nm. The lattice space of 0.232nm compares to planes of Ag (111), which shows that the favored confront of AgNPs is with planes (111). TEM confirmed that the microwave-aided synthesized AuNps and AgNPs were single crystals (Kondaiah et al., 2019).

20.6 Antimicrobial activity of biosynthesized nanometal oxide particles

In the micropropagation of tissue cultures for plants, the major problem associated with this is microbial contamination. Bacterial contamination is difficult to detect, as they may not show noticeable symptoms as they remain in plant tissue (Viss et al., 1991). In later stages, tissues reduce multiplication (Leifert and Casselles, 2001), have a poor rooting rate, and may die. Prevention is only method to avoid the microbial contamination by using antimicrobial agents. In the micropropogation technique, one of the essential steps is the eradication of microorganisms. In this stage, various chemical agents have been utilized for controlling in vitro contamination. The efficacy and duration of activity is low by chemical agents; moreover, they are toxic in nature. In practice, several antibiotics have been utilized such as tetracycline, vancomycin, streptomycin, nalidixic acid, gentamicin, etc., or a blend of these antibiotics to prevent/kill the microorganisms (Cornu and Michel, 1987). But these agents are high cost, heat liable, low efficacy, and have a specific or narrow activity against bacteria or fungi. Moreover, these may alter the growth of plant tissue. Further, the long-time exposure of plant tissues for antibiotics causes mutation in genes (Bhojwani and Razdan, 1996).

Nanoinorganic antibacterial agents have gained interest for bacterial control, owing to their outstanding safety, thermal stability, sustainability, and high surface area. Nanobiotechnology is a novel branch of the current science and applications of metal oxide NPs, that is, copper, gold, platinum, silver, and zinc, which have come to be chiefly used. It is endowed with new paraphernalia for the rapid detection of diseases and molecular treatment shows enormous potential to transform biology, pharmacy, and medicine. Many investigations have concluded that various metal oxide NPs established antimicrobial behavior on Gram-positive and Gram-negative bacteria as well as fungus. The nanosize to insert into a cell and the high surface area to release their chemical species are the key physical properties of NPs for the effective inhibition/killing of microorganisms. It is clear that a number of metal NPs were successful antimicrobial agents in opposition to numerous pathogenic microorganisms.

The mechanisms of inorganic oxide nanoparticle for antibacterial activity are inadequately explained however the scientist proposed mechanisms are (i) induction of oxidative stress, (ii) nonoxidative and (iii) metal ion release mechanisms. Several studies proposed that the electric charge on the bacterial wall surface is rapidly neutralized by nanoparticles, and changes in integrity and penetrability eventually lead to bacterial death (Jung et al., 2008). Further, ROS were generated, inhibiting the antioxidative defense mechanism of the cell membrane and leading to mechanical damage. Existing research reports revealed that the basic antibacterial effects of NPs are divided into: (i) cell membrane disruption; (ii) ROS genesis; (iii) penetration into cell membrane; and (iv) interaction with cellular organelles, that is, DNA, ribosomes, proteins, and plasmids. Biofilm (consortia of bacteria) prevention is attained by a nanosize and a greater surface area to mass ratio. Also, the nanoparticle shape has a noteworthy effect on biofilm destruction (Lellouche et al., 2012).

20.6.1 ROS generation of nanoparticles

The imperative antibacterial mechanism of nanoparticles is oxidative stress (ROS) have tough redox potential by molecules and active intermediates, several types of nano-particles produce dissimilar varieties of reducing oxygen molecules (ROS). The reducing oxygen superoxide radicals are four types; singlet oxygen (O_2), hydrogen peroxide (H_2O_2), and (· OH) hydroxyl radical and (O^{2-}) superoxide radicals that exhibit various intensities of activities and dynamics. For a paradigm, magnesium oxide and calcium oxide nanoparticles able to produce O_2 while zinc oxide NPs can produce H_2O_2 and · OH. Studies have explained that H_2O_2 and O^{2-} can produce low acute oxidative stress reactions that can be deactivated by endogenous antioxidants; those are catalase and superoxide enzymes, whereas –OH and O_2 can produce greater acute oxidative stress reactions that lead to microbial death. ROS production by nanoparticles is intensified by defect sites, restructuring, and oxygen vacancies in the crystal (Malka et al., 2013). In general, ROS production and clearance in microorganisms are balanced; an unbalanced state produces greater oxidative stress, which damages the cellular organelles (Li et al., 2012; Peng et al., 2013). Moreover, ROS is advantageous to escalating the gene expression intensity of oxidative proteins, which is a key instrument in bacterial cell apoptosis (Wu et al., 2011). Further ROS can damage proteins and destroy the periplasmic enzymes, which are vital to maintain normal physiology and morphology in microorganisms (Padmavathy and Vijayaraghavan, 2011).

Nanoparticles can release ROS by dissimilar mechanisms and the photocatalytic hypothesis is most acceptable. In the photocatalytic process the nano-metal oxides releasing highly electrons (e^-) and holes (valency band, H) on and inside of the catalytic nanoparticles due to light irradiation energy superior than or equivalent to the band gap leads to stimulation and transition to the conduction band. In case ZnO nanoparticles hydroxyl radical (–OH) is formed due to oxidation when interacted with H_2O or OH^{+-} on the facets remains with H^+. After that, the superoxide (O_2^-) radical is formed due to electronic interaction by means of O_2 on the surface of ZnO (Fig. 20.6). The mechanism is the vital active components of cells responsible for sustaining the standard physiological and morphological functions in microorganisms were destroyed by (Yu et al., 2014) generated ROS. TiO_2 nanoparticles produce pairs of electron-holes (Depan and Misra, 2014) after light absorption, which immediately react with air and water on the surface, then release extremely reactive ROS that leads to the damage of intracellular organic components of bacterium.

Matai et al. (2014) reported on *Campylobacter jejuni* for mechanism of ROS that under UV-visible light ZnO nanoparticles released extremely reactive ROS. Further hydroxyl and superoxide radicals (negatively charged) on the surface of the cell well, no way to permeated into intracellular areas of bacteria, H_2O_2 can pass right through the cell membrane. They reported that nanoparticles altered the cell shape from spiral to spherical and caused cellular leakage. Seil and Webster (2012) applied ultrasonic treatment with ZnONPs were reported as H_2O split into H^+ and then reacted with dissolved oxygen with water, forms to genesis of H_2O_2 thereby induce the reactive oxygen species formation. They explained that formed ROS, upon the ultrasonic treatment dissociated the colony forming structures and promote the NPs permeation, then easily permeate the cell membrane, quickly release of NPs leads to antimicrobial intensity. Xu et al. (2013) demonstrated that NPs additionally have antibacterial action in obscurity. ZnO delivered a negligible quantity of · OH, which is the primary antimicrobial species created by zinc oxide when invigorated by light. Oxygen vacancies situated on the exterior of ZnO assume a noteworthy function in delivering H_2O_2. In heterogeneous catalysis, a nano-metal oxide relies upon the thickness of the dynamic site. Strong deformities are regularly considered as the dynamic locales of heterogeneous catalysis. Nano ZnO antibacterial activity effectively influenced by crystal size, lattice constant, direction, and exterior concentration was reported.

20.6.2 Lipid and protein interactions on cell membrane

The main function of the cell membranes of bacteria and fungi is to permit nutrients into cells for their growth. Cellular components floated in the cytoplasm facilitate the normal physiology and protection from outer environmental challenges. Silver nanoparticles require crossing the cell membrane by interacting with the membrane and altering the integrity and fluidity of membrane, then entering into the cell cytoplasm. The bacterial surface and silver nanoparticles or silver ions interact and interfere in the respiratory chain and obstruct the energy. Generally, the assembly and maintenance of a cell envelope or plasma membrane precisely depends on the energy. The bacteria is of two types depending on the cell membrane. If layers of lipopolysaccharides are present, those are called Gram-negative. If layers of peptidoglycans are present, they are called Gram-positive.

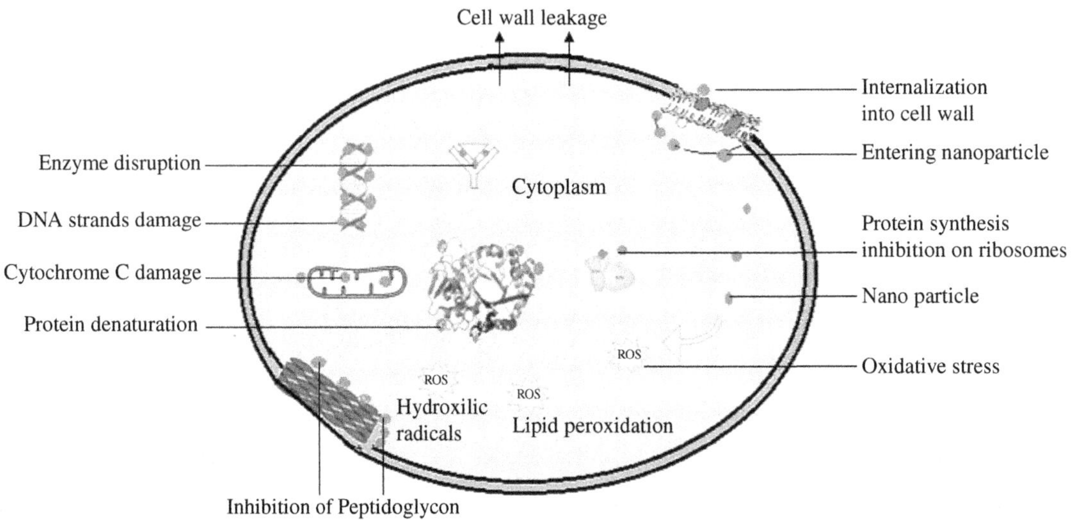

FIG. 20.6 Schematic representation of mechanisms proposed on antibacterial activity by NPs.

Collectively, the outer membranes are composed of proteins and lipids. Ag^+ ions can interact with protein to form a complex having electron donors that contain phosphorous, nitrogen, sulfur, and oxygen atoms. An interaction of Ag^+ with proteins leads to the inactivation of membrane-bound proteins and enzymes. Numerous proteins present in the bacterial membrane contain sulfur. An interaction between silver ions/sulfur-containing proteins causes damage to membrane-bound proteins and enzymes, leading to the inactivation of the integration of the cell membrane (Pal et al., 2007). The lipids present in the bacterial wall may also interact with Ag^+ ions and increase the trans/cis ration of unsaturated fatty acids (Hachicho et al., 2014). This isomerization of unsaturated membrane fatty acids and modification in membrane lipids leads to changes in fluidity and integrity. Therefore, penetrating into the cell membrane and leaked cellular organelles leads to cell death (Rajesh et al., 2015). In this route, lipids and polysaccharides interact, damage the cell wall integrity, damage fluidity, and release the Ag^+ ions with the O_2, phosphorous, and nitrogen.

20.6.3 DNA interaction: Damage and repair

Deoxyribonucleic acids (DNA) are composed of nucleobases (adenine, thymine, guanine, cytosine, etc.). Those containing nitrogen at the cyclic rings may interact by nanoparticles, which cause DNA changes. DNA polymerase and RNA polymerases are the DNA binding proteins occupied by nanoparticles with sites and hamper the progress of protein by side by side of the bases of DNA. This can affect the competitive inhibition of genetic functions by electrostatic interaction and sometimes may interfere in signaling. Ouda (2014) worked on the AgNp mechanism of DNA binding using *Proteus* and *Klebsiella* sp., pathogenic microorganisms evaluated on the agarose gel electrophoresis. Both organisms showed destructive effects on the genome. Proteus was splits as three bands with the lengths of 4.9, 5.,5 and 6.3 cm quoted *m.w.* as 950, 850 and 740 base pairs respectively. *Klebsiella* normal and AgNps treated DNA was also tested in normal the band shown 1.5 cm designates as molecular weight (*m.w.*) 2500 bp, whereas in second case 3 bands with the length of 4.8, 5.4 and 6.2 cm quoted *m.w.* as 950, 850 and 740 base pairs respectively. The work discovered that the genome was splits and bands are came on gel electrophoresis indicated that the underneath mechanism is AgNPs condensation of DNA, that causes to irreversible loss of replication and denaturation of DNA lead to bacterial growth inhibition. The binding sites present on the nucleobases of DNA (Watson-Crick model) are depicted in Fig. 20.7. AgNPs can bind to the N7 position of the nucleobases, that is, adenine and guanine, whereas in thymine and cytosine it is at the N3 position. Also, there is a possible bind at the minor groove of the double helix DNA. Further, AgNPs acting as interposers irreversibly and cause serious helical deformation. Tainted and twisted inclinations of DNA lead to changes in DNA base loading (Rahban et al., 2010). Vishnupriya et al. (2013)

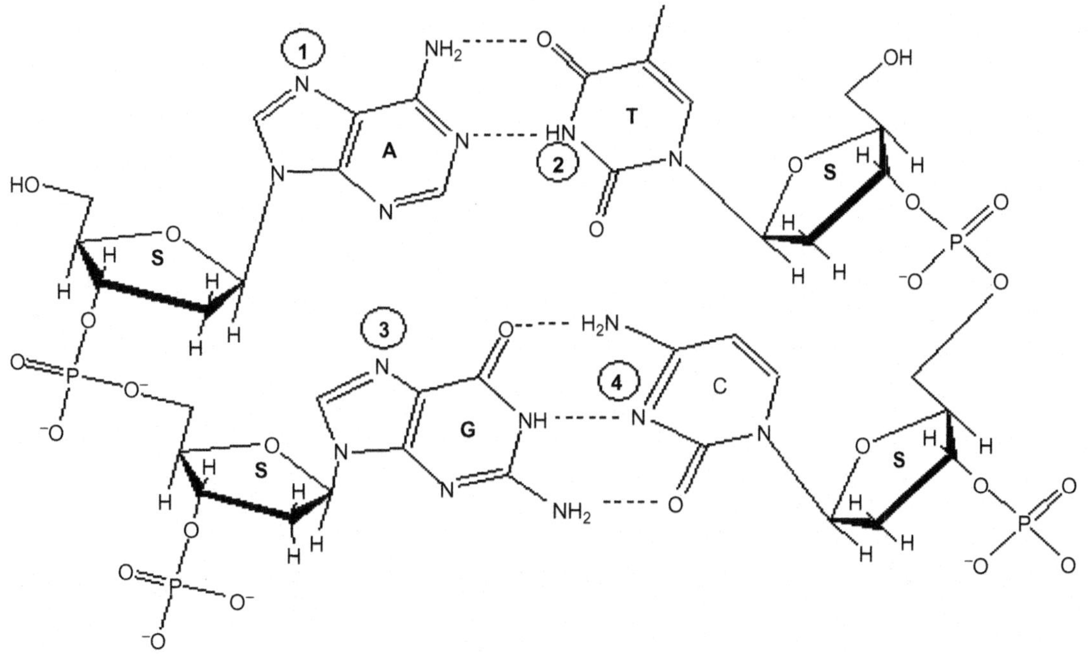

FIG. 20.7 Binding sites for silver nanoparticles on nucleobases in double stranded DNA (as per Watson-Crick model). (A=adenine; T=thymine; G=Guanine; C=Cytosine; S=ribose sugars; 1, 2, 3, and 4 binding sites on N7 for adenine and guanine, N3 for thymine and cytosine).

conducted experiments on the interaction studies of AgNPs (25±8.5nm) on *Escherichia coli* (ATCC 25922) with the aid of surface enhanced Raman (SERS) spectroscopy juxtaposition of plasma resonance imaging. They observed the AgNP dispersion inside the cell by hyperspectral imaging against incubation time. The reported results showed that the incubation of AgNPs with *E. coli* was attachment and evidenced by hyperspectral imaging. The degradation of DNA and molecular changes were observed by Raman spectroscopy. Further the report also explained that the site of interaction particularly at exocyclic nitrogen available on the nucleoside bases led to changes occurred in DNA.

Chen et al. (2011) prepared thermosensitive gel (S-T gel) using a glycerol and ethylparaben mixture along with surfactant pluronic F127 and F68 for embedding AgNPs to study NP effects on the DNA of *E. coli*, *P. aeruginosa*, and *S. aureus*. They explained that AgNPs cause the occurrence of "pits" on the cell wall through them AgNPs were enters into periplasm then devastate the membrane led to coalesce and coagulate the cytoplasm resulted outflow of cellular contents ultimately damage the microbes. Further, they confirmed that DNA condensation leads to no replication and denaturation of DNA occurred on *E. coli*, *S. aureus*, and *P. aeruginosa* cells with AgNPs embedded on S-T gel. Radzig et al. (2013) developed AgNPs with 8.3nm diameter and studied them for antimicrobial intensity on biofilms of *E. coli* AB1157, *P. aeruginosa* PAO1, and *S. proteamaculans* 94 at 4–5μg/mL concentration, 10μg/mL, and 10–20μg/mL, respectively. The results revealed that porins were participated in the movement of AgNPs into the bacteria. Li et al. (2013) examined the *Sac* I– linearized plasmid DNA interaction studies with 12 various nanoparticles by determining the energy by the Derjaguin-Landau-Verwey-Overbeek (DLVO) model to envisage the affinity potential. They correlated the theoretical expectations and experimental results for the affinity potential of nanoparticles with DNA. The AFM (atomic force microscopy) and in vitro PCR (polymerase chain reactor) results revealed that AgNPs have high affinity and powerfully suppressed DNA replication. Pramanik et al. (2016) studied AgNP interaction on *E. coli* and *Micrococcus lysodeikticus* DNA and they reported that double-stranded DNA has more potential to interact specifically. They further studied DNA melting and binding with Hoechst (minor groove binder with DNA) confirmed that unwinding of DNA bases and interacted at the minor groove of DNA. They reported that AgNPs bind robustly with *E. coli* and *Micrococcus lysodeikticus* DNAs at GC nucleoside base pairs. AgNPs as well as Ag^+ ions can be inhibiting respiratory enzymes leading to ROS genesis. Oxidized DNA precursor reactions with DNA may cause lesions to the DNA. A DNA repair procedure may recover the integrity of the damaged DNA, considering oxidative stress as a significant mechanism for binding DNA.

20.7 Future directions and conclusions

Biogenic techniques were contemplated as safer alternatives to conventional physical and chemical methods of nanosynthesis due to their ecofriendly nature and low cost. The phytosynthesis of metal NPs is considered as a good, cheap, and green approach. The wastes from agro, fruit, and pulp industries are also to be utilized as capping/reducing agents for the biogenic synthesis of NPs. In the biomedical sector, there is a great deal of interest in NPs. They are used in aesthetic dental applications, the development of antibacterial coatings, chronic nonhealing wound treatment, hydro fiber dressing, antioxidant formulations, antiinflammatory applications, and the suppression of autacoid release in inflammatory reactions. Further, these nanoparticles are proposed for food packaging and preservation applications. Nanobiotechnology is in its infancy in nanomedicine to develop nanoimaging, nanodrug delivery systems, and nanodiagnosis.

There is a need for research to develop the nanobiotechnological rapid diagnosis of viral infections and cancers in early stages. The advances of nanobiotechnology in the biomedical field were for the development of novel medical products and in tissue generation, which have attracted the interest of researchers and scientists. Recently, AgNP coatings were applied as thin films on personal protective equipment (PPE) kits of health professionals. With this, silver nanoparticle thin film coatings are to be applied to hospital infrastructures and protective kits and in general, this process takes commercially high budgets. Researchers and scientists should focus on nanoparticle thin film coating technologies on the large scale to reach the common man and health systems with low cost. The food-water nexus is a sustainable developmental goal. In order to fulfill this, food must be stored for a longer tenure. AgNPs helps to preserve food in packing materials. Research is intensively required for further development to commercialize large-scale production at an affordable cost. Through this, the global food demand can be met by storing food in commercially developed packing equipment and achieving a sustainable development goal (SDG 1). The funding agencies/governments must financially support young researchers and scientists working on societal problems. Researchers and scientists are also focusing on fulfilling risk assessment, developing standard operating procedures, and doing life cycle assessments to follow the regulatory authorities, regulations, and recommendations.

Acknowledgment

The author (KKK) is thankful to the Department of Science and Technology (SERB) for a research grant (Letter No. SERB/F/4631/2013-2014).

References

Alfuraydi, A.A., Ranjitsingh, A.J., Devanesan, S., Al-Ansari, M., AlSalhi, M.S., 2019. Eco-friendly green synthesis of silver nanoparticles from the sesame oil cake and its potential anticancer and antimicrobial activities. J. Photochem. Photobiol. B Biol. 192, 83–89.

Ali, K., Ahmed, B., Dwivedi, S., Saquib, Q., Al-Khedhairy, A.A., Musarrat, J., 2015. Microwave accelerated green synthesis of stable silver nanoparticles with *Eucalyptus globulus* leaf extract and their antibacterial and antibiofilm activity on clinical isolates. PLoS One 10 (7), e0131178.

Ali, K., Dwivedi, S., Azam, A., Saquib, Q., Al-Said, M.S., Alkhedhairy, A.A., Musarrat, J., 2016. *Aloe vera* extract functionalized zinc oxide nanoparticles as nanoantibiotics against multi-drug resistant clinical bacterial isolates. J. Colloid Interface Sci. 472, 145–156.

Ambika, S., Sundrarajan, M., 2015. Antibacterial behaviour of *Vitex negundo* extract assisted ZnO nanoparticles against pathogenic bacteria. J. Photochem. Photobiol. B Biol. 146, 52–57.

Anbu, P., Gopinath, S.C.B., Yun, H.-S., Lee, C.-G., 2019. Temperature-dependent green biosynthesis and characterization of silver nanoparticles using balloon flower plants and their antibacterial potential. J. Mol. Struct. 1177, 302–309.

Anbuvannan, M., Ramesh, M., Viruthagiri, G., Shanmugam, N., Kannadasan, N., 2015a. *Anisochilus carnosus* leaf extract mediated synthesis of zinc oxide nanoparticles for antibacterial and photocatalytic activities. Mater. Sci. Semicond. Process. 39, 621–628.

Anbuvannan, M., Ramesh, M., Viruthagiri, G., 2015b. Synthesis, characterization and photocatalytic activity of ZnO nanoparticles prepared by biological method. Spectrochim. Acta A Mol. Biomol. Spectrosc. 143, 304–308.

Ankamwar, B., Chinmay, D., Absar, A., Murali, S., 2005. Biosynthesis of gold and silver nanoparticles using *Emblica officinalis* fruit extract, their phase transfer and transmetallation in an organic solution. J. Nanosci. Nanotechnol. 10, 1665–1671.

Annu, I.S., Ahmed, S., Kaur, G., Sharma, P., Singh, S., 2018. Fruit waste (peel) as bio-reductant to synthesize silver nanoparticles with antimicrobial, antioxidant and cytotoxic activities. J. Appl. Biomed. 16, 221–231.

Anwar, M.K.A., Sherif, A.M., 2013. The gold nanoparticle size and exposure duration effect on the liver and kidney function of rats: In vivo. Saudi J. Biol. Sci. 20, 177–218.

Apostolov, A.T., Apostolova, I.N., Wesselinowa, J.M., 2014. Dielectric constant of multiferroic pure and doped CuO nanoparticles. Solid State Commun. 192, 71–74.

Araruna, F.B., de Oliveira, T.M., Quelemes, P.V., de Araújo, A.R.N., Plácido, A., Vasconcelos, A.G., de Paula, R.C.M., Mafud, A.C., de Almeida, M.P., Delerue-Matos, C., Mascarenhas, Y.P., Eaton, P., de Souza, A.L.J.R., da Silva, D.A., 2020. Antibacterial application of natural and carboxymethylated cashew gum-based silver nanoparticles produced by microwave-assisted synthesis. Carbohydr. Polym. 241, 115260.

Aruna, J.K., Sashidhar, R.B., 2018. Biogenic silver nanoparticles synthesized with rhamnogalacturonan gum: antibacterial activity, cytotoxicity and its mode of action. Arab. J. Chem. 11, 313–323.

Augustine, R., Augustine, A., Kalarikkal, N., Thomas, S., 2016. Fabrication and characterization of biosilver nanoparticles loaded calcium pectinate nano-micro dual-porous antibacterial wound dressings. Prog. Biomater. 5, 223–235.

Azizi, S., Mohamad, R., Bahadoran, A., Bayat, S., Rahim, R.A., Ariff, A., Saad, W.Z., 2016. Effect of annealing temperature on antimicrobial and structural properties of bio-synthesized zinc oxide nanoparticles using flower extract of *Anchusa italica*. J. Photochem. Photobiol. B 161, 441–449.

Baker, S., Sathish, S., 2015. Biosynthesis of gold nanoparticles by *Pseudomonas veronii* AS41G inhabiting *Annona squamosa* L. Spectrochim. Acta A Mol. Biomol. Spectrosc. 150, 691–695.

Balagurunathan, M.R., Ramaswamy, B.R., Velmurugan, D., 2011. Biosynthesis of gold nanoparticles from actinomycete *Streptomyces viridogens* strain (HM10). Indian J. Biochem. Biophys. 48, 331–335.

Barboza-Filho, G.C., Cabrera, F.C., Dos Santos, R.J., Saja Saez, J.A.D., Job, A.E., 2012. The influence of natural rubber/Au nanoparticle membranes on the physiology of *Leishmania brasiliensis*. Exp. Parasitol. 130, 152–158.

Begum, N.A., Mondal, S., Basu, S., Laskar, R.A., Mandal, D., 2009. Biogenic synthesis of Au and Ag nanoparticles using aqueous solutions of black tea leaf extracts. Colloids Surf. B: Biointerfaces 71, 113–118.

Beyth, N., Houri-Haddad, Y., Domb, A., Khan, W., Hazan, R., 2015. Alternative antimicrobial approach: nano-antimicrobial materials. Evid. Based Complement. Alternat. Med., 1–16. 246012.

Bhagavanth, R.G., Madhusudhan, A., Ramakrishna, D., Ayodhya, D., Venkatesham, M., Veerabhadram, G., 2015. Catalytic reduction of methylene blue and Congo red dyes using green synthesized gold nanoparticles capped by *Salmalia malabarica* gum. Int. Nano Lett. 5, 215–222.

Bhojwani, S.S., Razdan, M.K., 1996. Plant Tissue Culture: Theory and Practice, Revised ed. Elsevier Science, Netherlands.

Chen, M., Yang, Z., Wu, H., Pan, X., Xie, X., Wu, C., 2011. Antimicrobial activity and the mechanism of silver nanoparticle thermosensitive gel. Int. J. Nanomedicine 6, 2873–2977.

Chen, S.F., Zhang, H., 2012. Aggregation kinetics of nanosilver in different water condition. Adv. Nat. Sci. Nanosci. Nanotechnol. 3. 035006-1–7.

Cho, K.H., Park, J.E., Osaka, T., Park, S.G., 2005. The study of antibacterial activity and preservative effects of nano silver ingredient. Electrochim. Acta 51, 956–960.

Chung, I.M., Park, I., Seung-Hyun, K., Thiruvengadam, M., Rajakumar, G., 2016. Plant-mediated synthesis of silver nanoparticles: their characteristic properties and therapeutic applications. Nanoscale Res. Lett. 11, 40.

Commission E, 2011. Commission recommendation of 18 October 2011 on the definition of nanomaterial. Off. J., 38–40.

Cornu, D., Michel, M.F., 1987. Bacteria contaminants in shoot cultures of *Prunus avium* L. choice and phytotoxicity of antibiotics. Acta Hortic. 212, 83–86.

Dang, T.M.D., Le, T.T.T., Blance, E.F., Dang, M.C., 2012. Influence of surfactant on the preparation of silver nanoparticles by polyol method. Adv. Nat. Sci. Nanosci. Nanotechnol. 3. 035004-1–4.

Davod, T., Reza, Z., Ali, V.A., Mehrdad, C., 2011. Effects of nanosilver and nitroxin bio-fertilizer on yield and yield components of potato minitubers. Int. J. Agric. Biol. 13, 986–990.

Depan, D., Misra, R.D., 2014. On the determining role of network structure titania in silicone against bacterial colonization: mechanism and disruption of biofilm. Mater. Sci. Eng. C Mater. Biol. Appl. 34, 221–228.

Dhand, V., Soumya, L., Bharadwaj, S., Chakra, S., Bhatt, D., Sreedhar, B., 2016. Green synthesis of silver nanoparticles using *Coffea arabica* seed extract and its antibacterial activity. Mater. Sci. Eng. C 58, 36–43.

Dhar, S., Reddi, E.M., Shiras, A., Pokharkar, V., Prasad, B.L.V., 2008. Natural gum reduced/stabilized gold nanoparticles for drug delivery formulations. Chem. Eur. J. 14 (33), 10244–10250.

Divyapriya, S., Sowmia, C., Sasikala, S., 2014. Synthesis of zinc oxide nanoparticles and antimicrobial activity of *Murraya Koenigii*. World J. Pharm. Pharm. Sci. 3 (12), 1635–1645.

Dobrucka, R., Dugaszewska, J., 2016. Biosynthesis and antibacterial activity of ZnO nanoparticles using *Trifolium pratense* flower extract. Saudi J. Biol. Sci. 23, 517–523.

Du, L., Xu, Q., Huang, M., Feng, L.X., 2015. Synthesis of small silver nanoparticles under light radiation by fungus *Penicillium oxalicum* and its application for the catalytic reduction of methylene blue. Mater. Chem. Phys. 47 (14), 40–47.

Dwivedi, A.D., Gopal, K., 2010. Biosynthesis of silver and gold nanoparticles using *Chenopodium album* leaf extract. Colloids Surf. A: Physicochem. Eng. Aspects 369, 27–33.

Elavazhagan, T., Arunachalam, K.D., 2011. *Memecylon edule* leaf extract mediatedgreen synthesis of silver and gold nanoparticle. Int. J. Nanomed. 6, 1265–1278.

Elumalai, K., Velmurugan, S., 2015. Green synthesis, characterization and antimicrobial activities of zinc oxide nanoparticles from the leaf extract of *Azadirachta indica* (L.). Appl. Surface Sci. 345, 329–336.

Elumalai, K., Velmurugan, S., Ravi, S., Kathiravan, V., Adaikala Raj, G., 2015. Bio-approach: plant mediated synthesis of ZnO nanoparticles and their catalytic reduction of methylene blue and antimicrobial activity. Adv. Powder Technol. 26 (6), 1639–1651.

Fabrega, J., Fawcett, S.R., Rensha, J.C., Lead, J.R., 2009. Silver nanoparticle impact on bacterial growth: effect of pH, concentration, and organic matter. Environ. Sci. Technol. 43, 7285–7290.

Fatimah, I., 2016. Green synthesis of silver nanoparticles using extract of *Parkia speciosa Hassk* pods assisted by microwave irradiation. J. Adv. Res. 7 (6), 961–969.

Fayaz, M., Mashihur, G.M., Kalaichelvan Rahman, P.T., 2011. Biosynthesis of silver and gold nanoparticles using thermophilic bacterium *Geobacillusstearo thermophilus*. Process Biochem. 46, 1958–1962.

Gao, Y., Jian, P., Song, L., Wang, J.X., Liu, L.F., Xiang, Y.J., Zhang, Z.X., Zhao, X.W., Dou, X.Y., Luo, S.D., Zhou, W.Y., Xie, S.S., 2006. Studies on silver nanodecahedrons synthesized by PVP-assisted N, N-dimethylformamide (DMF) reduction. J. Cryst. Growth 289, 376–380.

Gengan, R.M., Anand, K., Phulukdaree, A., Chuturgoon, A., 2013. A549 lung cell line activity of biosynthesized silver nanoparticles using *Albizia adianthifolia* leaf. Colloids Surf. B: Biointerfaces 105, 87–91.

Gomathi, M., Rajkumar, P.V., Prakasam, A., Ravichandran, K., 2017. Green synthesis of silver nanoparticles using *Datura stramonium* leaf extract and assessment of their antibacterial activity. Resour. Efficient Technol. 3, 280–284.

Govarthanan, M., Seo, Y.-S., Lee, K.-J., Jung, I.-B., Ju, H.-J., Kim, J.S., Cho, M., Kannan, S.K., Oh, B.-T., 2016. Low-cost and eco-friendly synthesis of silver nanoparticles using coconut (*Cocos nucifera*) oil cake extract and its antibacterial activity. Artif. Cells Nanomed. Biotechnol. 44 (8), 1878–1882.

Gulbagca, F., Ozdemir, S., Gulcan, M., Sen, F., 2019. Synthesis and characterization of *Rosa canina*-mediated biogenic silver nanoparticles for anti-oxidant, antibacterial, antifungal, and DNA cleavage activities. Heliyon 5 (12), e02980.

Gunalan, S., Sivaraj, R., Rajendran, V., 2012. Green synthesized ZnO nanoparticles against bacterial and fungal pathogens. Prog. Nat. Sci. Mater. Int. 22 (6), 693–700.

Hachicho, N., Hoffmann, P., Ahlert, K., Heipieper, H.J., 2014. Effect of silver nanoparticles and silver ions on growth and adaptive response mechanisms of *Pseudomonas putida* mt-2. FEMS Microbiol. Lett. 355, 71–77.

Haghparasti, Z., Shahri, M.M., 2018. Green synthesis of water-soluble nontoxic inorganic polymer nanocomposites containing silver nanoparticles using white tea extract and assessment of their in vitro antioxidant and cytotoxicity activities. Mater. Sci. Eng. C 87, 139–148.

Han, G.Z., Gao, K.L., Wu, S.R., Zhang, Y., 2017. A facile green synthesis of silver nanoparticles based on poly-L-lysine. J. Nanosci. Nanotechnol. 17 (2), 1534–1537.

He, S., Chen, W., 2015. 3D graphene nanomaterials for binder-free supercapacitors: scientific design for enhanced performance. Nano 7, 6957–6990.

Honary, S., Fathabad, E.G., Paji, Z.K., Eslamifar, M., 2012. A novel biological synthesis of gold nanoparticle by enterobacteriaceae family. Trop. J. Pharm. Res. 11, 887–891.

Huang, H., Yang, Y., 2008. Preparation of silver nanoparticles in inorganic clay suspensions. Compos. Sci. Technol. 68, 2948–2953.

Huang, J., Li, Q., Sun, D., Lu, Y., Su, Y., Yang, X., Wang, H., Wang, Y., Shao, W., He, N., Hong, J., Chen, C., 2007. Biosynthesis of silver and gold nanoparticles by novel sundried *Cinnamomum camphora* leaves. Nanotechnology 18, 105104–105115.

Huang, Y.-L., Gong, Y.-N., Jiang, L., Lu, T.-B., 2013. A unique magnesium-based 3D MOF with nanoscale cages and temperature dependent selective gas sorption properties. Chem. Commun. 49, 1753–1755.

Husseiny, M.I., Aziz, M.A.E., Badr, Y., Mahmoud, M.A., 2007. Biosynthesis of gold nanoparticles using *Pseudomonas aeruginosa*. Spectrochim. Acta Part A 67, 1003–1106.

Ibrahim, H.M.M., 2015. Green synthesis and characterization of silver nanoparticles using banana peel extract and their antimicrobial activity against representative microorganisms. J. Radiat. Res. Appl. Sci. 8, 265–275.

Issazade, K., Jahanpour, N., Pourghorbanali, F., Raeisi, G., Faekhondeh, J., 2013. Heavy metals resistance by bacterial strains. Ann. Biol. Res. 4, 60–63.

Jae, Y.S., Hyeon-Kyeong, J., Beom, S.K., 2009. Biological synthesis of gold nanoparticles using *Magnolia kobus* and *Diopyros kaki* leaf extracts. Process Biochem. 44 (10), 1133–1138.

Jafarirad, S., Mehrabi, M., Divband, B., Kosari-Nasab, M., 2016. Biofabrication of zinc oxide nanoparticles using fruit extract of *Rosa canina* and their toxic potential against bacteria: a mechanistic approach. Mater. Sci. Eng. C 59, 296–302.

Jaidev, L.R., Narasimha, G., 2010. Fungal mediated biosynthesis of silver nanoparticles, characterization and antimicrobial activity. Colloids Surf. B: Biointerfaces 81, 430–433.

Joseph, S., Mathew, B., 2015. Microwave-assisted green synthesis of silver nanoparticles and the study on catalytic activity in the degradation of dyes. J. Mol. Liq. 204, 184–191.

Jung, W.K., Koo, H.C., Kim, K.W., Shin, S., Kim, S.H., Park, Y.H., 2008. Antibacterial activity and mechanism of action of the silver ion in *Staphylococcus aureus* and *Escherichia coli*. Appl. Environ. Microbiol. 74 (7), 2171–2178.

Kanchi, S., Kumar, G., Lo, A.-Y., Tseng, C.-M., Chen, S.-K., Lin, C.-Y., Chin, T.-S., 2018. Exploitation of deoiled *Jatropha* waste for gold nanoparticles synthesis: a green approach. Arab. J. Chem. 11 (2), 247–255.

Kanniah, P., Radhamani, J., Chelliah, P., Muthusamy, N., Balasingh, E.J.J.S., Thangapandi, R., Balakrishnan, J.S., Shanmugam, R., 2020. Green synthesis of multifaceted silver nanoparticles using the flower extract of *Aerva lanata* and evaluation of its biological and environmental applications. ChemistrySelect 5 (7), 2322–2331.

Kaur, T., Kapoor, S., Kalia, A., 2018. Synthesis of silver nanoparticles from *Pleurotus Florida*, characterization and analysis of their antimicrobial activity. Int. J. Curr. Microbiol. Appl. Sci. 7 (7), 4085–4095.

Kaviya, S., Santhanalakshmi, J., Viswanathan, B., Muthumary, J., Srinivasan, K., 2011. Biosynthesis of silver nanoparticles using citrus sinensis peel extract and its antibacterial activity. Spectrochim. Acta A 79, 594–598.

Khan, A.U., Yuan, Q., Wei, Y., Khan, Z.H., Tahir, K., Khan, S.U., Ahmad, A., Khan, S., Nazir, S., Khan, F.U., 2016. Ultra-efficient photocatalytic deprivation of methylene blue and biological activities of biogenic silver nanoparticles. J. Photochem. Photobiol. B: Biol. 159, 49–58.

Khatami, M., Alijani, H.Q., Heli, H., Sharifi, I., 2018. Rectangular shaped zinc oxide nanoparticles: green synthesis by stevia and its biomedical efficiency. Ceram. Int. 44 (13), 15596–15602.

Khoshnamvand, M., Huo, C., Liu, J., 2019. Silver nanoparticles synthesized using *Allium ampeloprasum* L. leaf extract: characterization and performance in catalytic reduction of 4- nitrophenol and antioxidant activity. J. Mol. Struct. 1175, 90–96.

Kim, D.-Y., Saratale, R.G., Shinde, S., Syed, A., Ameen, F., Ghodake, G., 2018. Green synthesis of silver nanoparticles using *Laminaria japonica* extract: characterization and seedling growth assessment. J. Clean. Prod. 172, 2910–2918.

Koilparambil, D., Kurian, L.C., Vijayan, S., Manakulam, S.J., 2016. Green synthesis of silver nanoparticles by *Escherichia coli*: analysis of antibacterial activity. J. Water Environ. Nanotechnol. 1 (1), 63–74.

Kondaiah, S., Bhagavanth, R.G., Babu, P., Kishore, K.K., Narasimha, G., 2018. Microwave-assisted synthesis of silver nanoparticles and their application in catalytic, antibacterial and antioxidant activities. J. Nanostruct. Chem. 8 (2), 179–188.

Kondaiah, S., Narasimha, G., Bhagavanth, R.G., Babu, P., Mushtaq, H., Syed, S.H., Kishore, K.K., 2019. Eco-friendly synthesis of gold nanoparticles using carboxymethylated gum *Cochlospermum gossypium* (CMGK) and their catalytic and antibacterial applications. Chem. Pap. 73 (7), 1695–1704.

Konishi, Y., Tsukiyama, T., Tachimi, T., Saitoh, N., Nomura, T., Nagamine, S., 2007. Microbial deposition of gold nanoparticles by the metal-reducing bacterium *Shewanella algae*. Electrochim. Acta 53, 186–192.

Kumar, D.A., Palanichamy, V., Roopan, S.M., 2014. Green synthesis of silver nanoparticles using *Alternanthera dentata* leaf extract at room temperature and their antimicrobial activity. Spectrochim. Acta Part A 127, 168–171.

Lee, K.J., Lee, Y.I., Shim, I.K., Joung, J., Oh, Y.S., 2006. Direct synthesis and bonding origins of monolayer-protected silver nanocrystals from silver nitrate through in situ ligand exchange. J. Colloids Interface Sci. 304, 92–97.

Leifert, C., Casselles, A.C., 2001. Microbial hazards in plant tissue and cell cultures in vitro cell. Dev. Biol. Plant 37, 133–139.

Lellouche, J., Friedman, A., Gedanken, A., Banin, E., 2012. Antibacterial and antibiofilm properties of yttrium fluoride nanoparticles. Int. J. Nanomed. 7, 5611–5624.

Leung, T.C.Y., Wong, C.K., Xie, Y., 2010. Green synthesis of silver nanoparticles using biopolymers, carboxymethylated-curdlan and fucoidan. Mater. Chem. Phys. 121 (3), 402–405.

Li, K., Zhao, X., Hammer, B.K., Du, S., Chen, Y., 2013. Nanoparticles inhibit DNA replication by binding to DNA: modeling and experimental validation. ACS Nano 7 (11), 9664–9674.

Li, L., Zhang, Z., 2016. Biosynthesis of gold nanoparticles using green alga *Pithophoraoedogonia* with their electrochemical performance for determining carbendazim in soil. Int. J. Electrochem. Sci. 11, 4550–4559.

Li, Y., Zhang, W., Niu, J., Chen, Y., 2012. Mechanism of photogenerated reactive oxygen species and correlation with the antibacterial properties of engineered metal-oxide nanoparticles. ACS Nano 6 (6), 5164–5173.

Lima, E., Guerra, R., Lara, V., Guzmán, A., 2013. Gold nanoparticles as efficient antimicrobial agents for *Escherichia coli* and *Salmonella typhi*. Chem. Cent. J. 7 (11), 1–7.

Logaranjan, K., Raiza, A.J., Gopinath, S.C.B., Chen, Y., Pandian, K., 2016. Shape and size-controlled synthesis of silver nanoparticles using *Aloe vera* plant extract and their antimicrobial activity. Nanoscale Res. Lett. 11 (1), 520.

Luo, P., Liu, Y., Xia, Y., Xu, H., Xie, G., 2014. Aptamer biosensor for sensitive detection of toxin A of *Clostridium difficile* using gold nanoparticles synthesized by *Bacillus stearothermophilus*. Biosens. Bioelectron. 54, 217–221.

Madan, H.R., Sharma, S.C., Udayabhanu, Suresh, D., Vidya, Y.S., Nagabhushana, H., Rajanaik, H., Anantharaju, K.S., Prashantha, S.C., Maiya, P.S., 2016. Facile green fabrication of nanostructure ZnO Plates, bullets, flower, prismatic tip, closed pine cone: their antibacterial, antioxidant, photoluminescent and photocatalytic properties. Spectrochim. Acta Part A: Mol. Biomol.r Spectrosc. 152, 404–416.

Mafune, F., Kohno, J., Takeda, Y., Kondow, T., Sawabe, H., 2000. Formation and size control of silver nanoparticles by laser ablation in aqueous solution. J. Phys. Chem. B 104, 9111–9117.

Malka, E., Perelshtein, I., Lipovsky, A., et al., 2013. Eradication of multi-drug resistant bacteria by a novel Zn-doped CuO nanocomposite. Small 9 (23), 4069–4076.

Mata, R., Nakkala, J.R., Sadras, S.R., 2015. Biogenic silver nanoparticles from *Abutilon indicum*: their antioxidant, antibacterial and cytotoxic effects in vitro. Colloids Surf. B: Biointerfaces 128, 276–286.

Matai, I., Sachdev, A., Dubey, P., Kumar, S.U., Bhushan, B., Gopinath, P., 2014. Antibacterial activity and mechanism of Ag-ZnO nanocomposite on *S. aureus* and GFP-expressing antibiotic resistant *E. coli*. Colloids Surf. B Biointerfaces 115, 359–367.

Mohammad, M.A., Abbas, Y., Mehdi, H.M., Somayeh, B., 2014. Effects of antimicrobial activity of silver nanoparticles on in vitro establishment of G×N15 (hybrid of almond–peach) rootstock. J. Genet. Eng. Biotechnol. 12, 103–110.

Mohan, Y.M., Raju, K.M., Sambasivudu, K., Singh, S., Sreedhar, B., 2007. Preparation of acacia-stabilized silver nanoparticles: a green approach. J. Appl. Polym. Sci. 106 (5), 3375–3381.

Mohsen, A.M.A., Hrdina, R., Burgert, L., Benes, L., 2012. Green synthesis of hyaluronan fibers with silver nanoparticles. Carbohydr. Polym. 89, 411–422.

Murali, K.I., Bhagavanth, R.G., Veerabhadram, G., Madhusudhan, A., 2016. Eco-friendly green synthesis of silver nanoparticles using *Salmalia malabarica*: synthesis, characterization, antimicrobial, and catalytic activity studies. Appl. Nanosci. 6, 681–689.

Nagajyothi, P.C., Cha, S.J., Yang, I.J., Sreekanth, T.V.M., Kim, K.J., Shin, H.M., 2015. Antioxidant and anti-inflammatory activities of zinc oxide nanoparticles synthesized using *Polygala tenuifolia* root extract. J. Photochem. Photobiol. B Biol. 146, 10–17.

Nangia, Y., Wangoo, N., Goyal, N., Shekhawat, G., Suri, C.R., 2009. A novel bacterial isolate *Stenotrophomonas maltophiliaas* living factory for the synthesis of gold nanoparticles. Microb. Cell Fact 8 (52). https://doi.org/10.1186/1475-2859-8-52.

Naraginti, S., Li, Y., 2017. Preliminary investigation of catalytic, antioxidant, anticancer and bactericidal activity of green synthesized silver and gold nanoparticles using *Actinidia deliciosa*. J. Photochem. Photobiol. B 170, 225–234.

Nasrollahzadeh, M., Maham, M., Sajadi, S.M., 2015. Green synthesis of CuO nanoparticles by aqueous extract of *Gundelia tournefortii* and evaluation of their catalytic activity for the synthesis of N-monosubstituted ureas and reduction of 4-nitrophenol. J. Colloid Interface Sci. 455, 245–253.

Navaladian, S., Viswanathan, B., Viswanath, R.P., Varadarajan, T.K., 2007. Thermal decomposition as route for silver nanoparticles. Nanoscale Res. Lett. 2, 44–48.

Ogi, T., Saitoh, N., Nomura, T., Konishi, Y., 2010. Room-temperature synthesis of gold nanoparticles and nanoplates using *Shewanella algae* cell extract. J. Nanopart. Res. 12, 2531–2539.

Oluwafemi, O.S., Mochochoko, T., Leo, A.J., Mohan, S., Jumbam, D.N., Songca, S.P., 2016. Microwave irradiation synthesis of silver nanoparticles using cellulose from *Eichhornia crassipes* plant shoot. Mater. Lett. 185, 576–579.

Ouda, S.H., 2014. Some nanoparticles effects on *Proteus* sp. and *Klebsiella* sp. isolated from water. Am. J. Infect. Dis. Microbiol. 2, 4–10.

Padmavathy, N., Vijayaraghavan, R., 2011. Interaction of ZnO nanoparticles with microbes–a physio and biochemical assay. J. Biomed. Nanotechnol. 7 (6), 813–822.

Pal, A., Esumi, K., Pal, T., 2005. Preparation of nanosized gold particles in a biopolymer using UV photoactivation. J. Colloid Interface Sci. 288 (2), 396–401.

Pal, A., Shah, S., Devi, S., 2009. Microwave-assisted synthesis of silver nanoparticles using ethanol as a reducing agent. Mater. Chem. Phys. 114, 530–532.

Pal, S., Tak, Y.K., Song, J.M., 2007. Does the antibacterial activity of silver nanoparticles depend on the shape of the nanoparticle? A study of the gram-negative bacterium *Escherichia coli*. Appl. Environ. Microbiol. 73, 1712–1720.

Panhwar, S., Hassan, S.S., Mahar, R.B., Canlier, A., Sirajuddin, A.M., 2018. Synthesis of l-cysteine capped silver nanoparticles in acidic media at room temperature and detailed characterization. J. Inorg. Organomet. Polym. Mater. 28 (3), 863–870.

Parthasarathy, G., Saroja, M., Venkatachalam, M., Evanjelene, V.K., 2017. Biological synthesis of zinc oxide nanoparticles from leaf extract of *Curcuma neilgherrensis* wight. Int. J. Mater. Sci. 12 (1), 74–86.

Parveen, M., Ahmad, F., Malla, A.M., Azaz, S., 2016. Microwave-assisted green synthesis of silver nanoparticles from *Fraxinus excelsior* leaf extract and its antioxidant assay. Appl. Nanosci. 6, 267–276.

Pasca, R.-D., Horovitz, O., Mocanu, A., Cobzac, S.-C., Petean, I., Maria, T.-C., 2014. Biogenic syntheses of gold nanoparticles using plant extracts. Partic. Sci. Technol. Int. J. 32, 131–137.

Patil, R.S., Kokate, M.R., Jambhale, C., Pawar, S.M., Han, S.H., Kolekar, S.S., 2012. One-pot synthesis of PVA-capped silver nanoparticles their characterization and biomedical application. Adv. Nat. Sci. Nanosci. Nanotechnol. 3. 015013-1–7.

Pedroza-Toscano, M.A., López-Cuenca, S., Rabelero-Velasco, M., Moreno-Medrano, E.D., Mendizabal-Ruiz, A.P., Salazar-Peña, R., 2017. Silver nanoparticles obtained by semicontinuous chemical reduction using carboxymethyl cellulose as a stabilizing agent and its antibacterial capacity. J. Nanomater. 2017. https://doi.org/10.1155/2017/1390180, 1390180.

Peng, Z., Ni, J., Zheng, K., Shen, Y., Wang, X., He, G., Jin, S., Tang, T., 2013. Dual effects and mechanism of TiO2 nanotube arrays in reducing bacterial colonization and enhancing C_3H_{10} T1/2 cell adhesion. Int. J. Nanomedicine 8, 3093–3105.

Perugu, S., Nagati, V., Bhanoori, M., 2016. Green synthesis of silver nanoparticles using leaf extract of medicinally potent plant *Saraca indica*: a novel study. Appl. Nanosci. 6, 747–753.

Philip, D., 2010a. Green synthesis of gold and silver nanoparticles using *Hibiscus rosa sinensis*. Phys. E. 42, 1417–1424.

Philip, D., 2010b. Rapid green synthesis of spherical gold nanoparticles using *Mangifera indica* leaf. Spectrochim. Acta A Mol. Biomol. Spectrosc. 77, 807–810.

Philip, D., Unni, C., 2011. Extracellular biosynthesis of gold and silver nanoparticles using Krishna tulsi (*Ocimum sanctum*) leaf. Phys. E Low-Dimen. Syst. Nanostruct. 43, 1318–1322.

Phull, A.-R., Abbas, Q., Ali, A., Raza, H., Kim, S.J., MD, Z., Haq, I.-U., 2016. Antioxidant, cytotoxic and antimicrobial activities of green synthesized silver nanoparticles from crude extract of *Bergenia ciliate*. Future J. Pharm. Sci. 2, 31–36.

Pramanik, S., Chatterjee, S., Saha, A., Devi, P.S., Kumar, G.S., 2016. Unraveling the interaction of silver nanoparticles with mammalian and bacterial DNA. J. Phys. Chem. B 120 (24), 5313–5324.

Priyabrata, T., Rout, G.K., Sushanto, G., Gitishree, D., Krishna, P., Hrudayanath, T., Jayanta, K.P., 2016. Photomediated green synthesis of silver and zinc oxide nanoparticles using aqueous extracts of two mangrove plant species, *Heritiera fomes* and *Sonneratia apetala* and investigation of their biomedical applications. J. Photochem. Photobiol. B 163, 311–318.

Priyanka, S., Hina, S., Ju, K.Y., Ramya, M., Chao, W., Chun, Y.D., 2016. Extracellular synthesis of silver and gold nanoparticles by *Sporosarcina koreensis* DC4 and their biological applications. Enzyme Microb. Technol. 86, 75–83.

Radzig, M.A., Nadtochenko, V.A., Koksharova, O.A., Kiwi, J., Lipasova, V.A., Khmela, I.A., 2013. Antibacterial effects of silver nanoparticles on gram-negative bacteria: influence on the growth and biofilms formation, mechanisms of action. Colloids Surf. B Biointerfaces 102, 300–306.

Rafique, M., Sadaf, I., Tahir, M.B., Rafique, M.S., Nabi, G., Iqbal, T., Sughra, K., 2019. Novel and facile synthesis of silver nanoparticles using *Albizia procera* leaf extract for dye degradation and antibacterial applications. Mater. Sci. Eng. C 99, 1313–1324.

Rahban, M., Divsalar, A., Saboury, A.A., Golestani, A., 2010. Nanotoxicity and spectroscopy studies of silver nanoparticle: calf thymus DNA and K562 as targets. J. Phys. Chem. C 114 (13), 5798–5803.

Rajathi, A.F., Parthiban, C., Kumar, G.V., Anantharaman, P., 2012. Biosynthesis of antibacterial gold nanoparticles using brown alga, *Stoechospermum marginatum* (kutzing). Spectrochim. Acta A Mol. Biomol. Spectrosc. 99, 166–173.

Rajesh, S., Dharanishanthi, V., Kanna, A.V., 2015. Antibacterial mechanism of biogenic silver nanoparticles of *Lactobacillus acidophilus*. J. Exp. Nanosci. 10, 1143–1152.

Rajkuberan, C., Prabukumar, S., Sathishkumar, G., Wilson, A., Ravindran, K., 2017. Sivaramakrishnan S, Facile synthesis of silver nanoparticles using *Euphorbia antiquorum* L. latex extract and evaluation of their biomedical perspectives as anticancer agents. J. Saudi Chem. Soc. 21 (8), 911–919.

Rakhi, M., Braja Gopal, B., Nabasmita, M., 2013. *Acacia nilotica* (Babool) leaf extract mediated size-controlled rapid synthesis of gold nanoparticles and study of its catalytic activity. Int. Nano Lett. 3 (53), 1–6.

Ramesh, M., Anbuvannan, M., Viruthagiri, G., 2015. Green synthesis of ZnO nanoparticles using *Solanum nigrum* leaf extract and their antibacterial activity. Spectrochim. Acta A Mol. Biomol. Spectrosc. 136, 864–870.

Ravishankar, R.V., Jamuna, B.A., 2011. Nanoparticles and their potential application as antimicrobials. In: Méndez, V. (Ed.), Science against microbial pathogens, communicating current research and technological advances. Badajoz: Formatex, pp. 197–209.

Reddi, N.J., Rani, M., Kumar, G.A., Rani, S.S., 2014. Biological activities of green silver nanoparticles synthesized with *Acorous calamus* rhizome extract. Eur. J. Med. Chem. 85, 784–794.

Reed, B.M., Buckley, P.M., DeWilde, T.N., 1995. Detection and eradication of endophytic bacteria from micropropagated mint plants in vitro cell. Dev. Biol. Plant. 31, 53–57.

Rodríguez-León, E., Iñiguez-Palomares, R., Navarro, R.E., Herrera-Urbina, R., Tánori, J., Iñiguez-Palomares, C., Maldonado, A., 2013. Synthesis of silver nanoparticles using reducing agents obtained from natural sources (*Rumex hymenosepalus* extracts). Nanoscale Res. Lett. 8, 318.

Roy, N., Mandal, D., Mondal, S., Laskar, R.A., Basu, S., Begum, N.A., 2010. Biogenic synthesis of Au and Ag nanoparticles by Indian propolis and its constituents. Colloids Surf. B: Biointerfaces 76, 317–325.

Roy, P., Das, B., Mohanty, A., Mohapatra, S., 2017. Green synthesis of silver nanoparticles using *azadirachta indica* leaf extract and its antimicrobial study. Appl. Nanosci. 7, 843–850.

Roshmi, T., Soumya, K.R., Jyothis, M., Radhakrishnan, E.K., 2015. Effect of biofabricated gold nanoparticle-based antibiotic conjugates on minimum inhibitory concentration of bacterial isolates of clinical origin. Gold Bull. 48, 63–71.

Roy, S., Das, T.K., 2016. Effect of biosynthesised silver nanoparticles on the growth and some biochemical parameters of *Aspergillus foetidus*. J. Env. Chem. Eng. 4, 1574–1583.

Roy, S., Das, T.K., Prasad, G., Maiti, U., 2016. Basu, microbial biosynthesis of nontoxic gold nanoparticles. Mater. Sci. Eng. B 203, 41–51.

Sadeghi, B., Gholamhoseinpoor, F., 2015. A study on the stability and green synthesis of silver nanoparticles using *Ziziphora tenuior* (Zt) extract at room temperature. Spectrochim. Acta A Mol. Biomol. Spectrosc. 134, 310–315.

Sagadevan, S., Vennila, S., Singh, P., Lett, J.A., Johan, M.R., Marlinda, A.R., Muthiah, B., Lakshmipathy, M., 2019. Facile synthesis of silver nanoparticles using *Averrhoa bilimbi* L and Plum extracts and investigation on the synergistic bioactivity using in vitro models. Green Process Synth. 8, 873–884.

Samat, N.A., Nor, M.R., 2013. Sol–gel synthesis of zinc oxide nanoparticles using *Citrus aurantifolia* extracts. Ceram. Int. 39 (S1), S545–S548.

Sanghi, R., Verma, P., Puri, S., 2011. Enzymatic formation of gold nanoparticles using *Phanerochaete chrysosporium*. Adv. Chem. Eng. Sci. 1, 154–162.

Sang-Woo, J., 2006. Adsorption of bipyridine compounds on gold nanoparticle surfaces investigated by UV-Vis absorbance spectroscopyand surface enhanced raman scattering. Spectrosc. Lett. 39 (1), 85–96.

Sankar, R., Manikandan, P., Malarvizhi Fathima, V.T., Shivashangari, K.S., Ravikumar, V., 2014. Green synthesis of colloidal copper oxide nanoparticles using *Carica papaya* and its application in photocatalytic dye degradation. Spectrochim. Acta Part A Mol. Biomol. Spectrosc. 121, 746–750.

Seil, J.T., Webster, T.J., 2012. Antibacterial effect of zinc oxide nanoparticles combined with ultrasound. Nanotechnology 23 (49), 495101.

Selvan, D.A., Mahendiran, D., Kumar, R.S., Rahiman, A.K., 2018. Garlic, green tea and turmeric extracts-mediated green synthesis of silver nanoparticles: phytochemical, antioxidant and in vitro cytotoxicity studies. J. Photochem. Photobiol. B Biol. 180, 243–252.

Shao, Y., Wu, C., Wu, T., Yuan, C., Chen, S., Ding, T., Ye, X., Hu, Y., 2018. Green synthesis of sodium alginate-silver nanoparticles and their antibacterial activity. Int. J. Biol. Macromol. 111, 1281–1292.

Sharma, B., Purkayastha, D.D., Hazra, S., Gogoi, L., Bhattacharjee, C.R., Ghosh, N.N., Rout, J., 2014. Biosynthesis of gold nanoparticles using a freshwater green alga, *Prasiola crispa*. Mater. Lett. 116, 94–97.

Sharma, S.C., 2016. ZnO nano-flowers from Carica papaya milk: degradation of alizarin RedS dye and antibacterial activity against *Pseudomonas aeruginosa* and *Staphylococcus aureus*. Optics-Int. J. Light Electron Opt. 127 (16), 6498–6512.

Sharma, J.K., Akhtar, M.S., Ameen, S., Srivastava, P., Singh, G., 2015. Green synthesis of CuO nanoparticles with leaf extract of Calotropis gigantea and its dye-sensitized solar cells applications. J. Alloy Compd. 632, 321–325.

Shkilnyy, A., Souće, M., Dubois, P., Warmont, F., Saboungi, M.L., Chourpa, I., 2009. Poly(ethylene glycol)-stabilized silver nanoparticles for bioanalytical applications of SERS spectroscopy. Analyst 134 (9), 1868–1872.

Shon, Y.S., Cutler, E., 2004. Aqueous synthesis of alkane thiolated protected Ag nanoparticles using Bunte salts. Langmuir 20, 6626–6630.

Singh, M., Kalaivani, R., Manikandan, S., Sangeetha, N., Kumaraguru, A.K., 2013. Facile green synthesis of the variable metallic gold nanoparticle using *Padina gymnospora*, a brown marine macroalga. Appl. Nanosci. 3, 141–151.

Song, J.Y., Kim, B.S., 2009. Rapid biological synthesis of silver nanoparticles using plant leaf extracts. Bioprocess Biosyst. Eng. 32, 79–84.

Spies, T., 1999. Analysis of the acute cytotoxicity of the Erlanger silver catheter. Infection 27, S34–S37.

Su, D.-L., Li, P.-J., Ning, M., Li, G.-Y., Shan, Y., 2019. Microwave assisted green synthesis of pectin based silver nanoparticles and their antibacterial and antifungal activities. Mater. Lett. 244, 35–38.

Suganya, K.S.U., Govindaraju, K., Kumar, G., Manickan, E., 2015. Blue green alga mediated synthesis of gold nanoparticles and its antibacterial efficacy against Gram positive organisms. Mater. Sci. Eng. C Mater. Biol. Appl. C 47, 351–356.

Suresh, A.K., Pelletier, D.A., Wang, W., Broich, M.L., Moon, J.W., Gu, B., Allison, D.P., Joy, D.C., Phelps, T.J., Doktycz, M.J., 2011. Biofabrication of discrete spherical gold nanoparticles using the metal-reducing bacterium *Shewanella oneidensis*. Acta Biomater. 7, 2148–2152.

Suresh, D., Shobharani, R.M., Nethravathi, P.C., Pavan Kumar, M.A., Nagabhushana, H., Sharma, S.C., 2015. *Artocarpus gomezianus* aided green synthesis of ZnO nanoparticles: luminescence, photocatalytic and antioxidant properties. Spectrochim. Acta Part A: Mol. Biomol. Spectrosc. 141, 128–134.

Thirumurugan, A., Ramachandran, S., Tomy, N.A., Jiflin, G.J., Rajagomathi, G., 2012. Biological synthesis of gold nanoparticles by *Bacillus subtilis* and evaluation of increased antimicrobial activity against clinical isolates. Korean J. Chem. Eng. 29 (12), 1–5.

Tripathia, R.M., Gupta, R.K., Singh, P., Bhadwal, A.S., Shrivastava, A., Kumar, N., Shrivastav, B.R., 2014. Ultra-sensitive detection of mercury(II) ions in water sample using gold nanoparticles synthesised by *Trichoderma harzianum* and their mechanistic approach. Sens. Actuators B Chem. 204, 637–646.

Vasanth, K., Ilango, K., MohanKumar, R., Agrawal, A., Dubey, G.P., 2014. Anticancer activity of *Moringa oleifera* on mediated silver nanoparticles human cervical carcinoma cells by apoptosis induction. Colloids Surf. B: Biointerfaces 117, 354–359.

Venkatesan, J., Manivasagan, P., Kim, S.K., Kirthi, A.V., Marimuthu, S., Rahuman, A.A., 2014. Marine algae-mediated synthesis of gold nanoparticles using a novel *Ecklonia cava*. Bioprocess Biosyst. Eng. 37 (8), 1591–1597.

Verma, A., Mehata, M.S., 2016. Controllable synthesis of silver nanoparticles using neem leaves and their antimicrobial activity. J. Radiat. Res. Appl. Sci. 9, 109–115.

Vijay Kumar, P.P.N., Shameem, U., Kollu, P., Kalyani, R.L., Pammi, S.V.N., 2015. Green synthesis of copper oxide nanoparticles using *Aloe vera* leaf extract and its antibacterial activity against fish bacterial pathogens. Bio Nano Sci 5, 135–139.
Vijayakumar, S., Vinoj, G., Malaikozhundan, B., Shanthi, S., Vaseeharan, B., 2015. *Plectranthus amboinicus* leaf extract mediated synthesis of zinc oxide nanoparticles and its control of methicillin resistant *Staphylococcus aureus* biofilm and blood sucking mosquito larvae. Spectrochim. Acta A Mol. Biomol. Spectrosc. 137, 886–891.
Vijayan, R., Joseph, S., Mathew, B., 2019. Green synthesis of silver nanoparticles using *Nervalia zeylanica* leaf extract and evaluation of their antioxidant, catalytic, and antimicrobial potentials. Part. Sci. Technol. 37, 809–819.
Vijayan, R., Joseph, S., Mathew, B., 2018. *Indigofera tinctoria* leaf extract mediated green synthesis of silver and gold nanoparticles and assessment of their anticancer, antimicrobial, antioxidant and catalytic properties. Artif. Cells Nanomed. Biotechnol. 46, 861–871.
Vijayaraghavana, K., Mahadevana, A., Sathishkumar, M., Pavagadhi, S., Balasubramanian, R., 2011. Biosynthesis of Au (0) from Au(III) via biosorption and bioreduction using brown marine alga *Turbinaria conoides*. Chem. Eng. J. 167, 223–227.
Vishnupriya, S., Chaudhari, K., Jagannathan, R., Pradeep, T., 2013. Single-cell investigations of silver nanoparticle–bacteria interactions. Part. Part. Syst. Charact. 30, 1056–1062.
Viss, P.R., Brooks, E.M., Driver, J.A., 1991. A simplified method for the control of bacterial contamination in woody plant tissue culture in vitro. Cell Dev Biol. Plant 27, 42–49.
Waghmare, S.S., Arvind, M.D., Sadowski, Z., 2014. Biosynthesis, optimization, purification and characterization of gold nanoparticles. Afr. J. Microbiol. Res. 8, 138–146.
Wu, B., Zhuang, W.Q., Sahu, M., Biswas, P., Tang, Y.J., 2011. Cu-doped TiO_2 nanoparticles enhance survival of *Shewanella oneidensis* MR-1 under ultraviolet (UV) light exposure. Sci. Total Environ. 409 (21), 4635–4639.
Xu, X., Chen, D., Yi, Z., et al., 2013. Antimicrobial mechanism based on H_2O_2 generation at oxygen vacancies in ZnO crystals. Langmuir 29 (18), 5573–5580.
Yang, N., Li, F., Jian, T., Liu, C., Sun, H., Wang, L., Xu, H., 2017. Biogenic synthesis of silver nanoparticles using ginger (*Zingiber officinale*) extract and their antibacterial properties against aquatic pathogens. Acta Oceanol. Sin. 36, 95–100.
Yin, B., Ma, H., Wang, S., Chen, S., 2003. Electrochemical synthesis of silver nanoparticles under protection of poly(Nvinylpyrrolidone). J. Phys. Chem. B 107 (34), 8898–8904.
Yu, Y., Chang, S., Lee, C., Wang, C., 1997. Gold nanorods: electrochemical synthesis and optical properties. J. Phys. Chem. B 101 (34), 6661–6664.
Yu, Z., Tetard, L., Zhai, L., Thomas, J., 2015. Supercapacitor electrode materials: nanostructures from 0 to 3 dimensions. Energy Environ. Sci. 8, 702–730.
Yu, J., Zhang, W., Li, Y., Wang, G., Yang, L., Jin, J., Chen, Q., Huang, M., 2014. Synthesis, characterization, antimicrobial activity and mechanism of a novel hydroxyapatite whisker/nano zinc oxide biomateriali. Biomed. Mater. 10 (1), 015001.
Zhang, X., Qu, Y., Shen, W., Wang, J., Li, H., Zhang, Z., Li, S., Zhou, J., 2016. Biogenic synthesis of gold nanoparticles by yeast *Magnusiomyces ingens* LH-F1 for catalytic reduction of nitrophenols. Colloids Surf. A: Physicochem. Eng. Aspects 497, 280–285.
Zheng, Y., Fu, L., Han, H., Wang, A., Cai, W., Yu, J., Yang, J., Peng, F., 2015. Green biosynthesis and characterization of zinc oxide nanoparticles using *Corymbia citriodora* leaf extract and their photocatalytic activity. Green Chem. Lett. Rev. 8 (2), 59–63.

Further reading

Ashajyothi, C., Prabhurajeshwar, C., Handral, H.K., Chandrakanth, K.R., 2016. Investigation of antifungal and anti-mycelium activities using biogenic nanoparticles: an eco-friendly approach. Environ. Nanotechnol. Monit. Manage. 5, 81–87.
Karnan, T., Selvakumar, S.A.S., 2016. Biosynthesis of ZnO nanoparticles using rambutan (*Nephelium lappaceum* L.) peel extract and their photocatalytic activity on methyl orange dye. J. Mol. Struct. 1125, 358–365.
Lakshmi, J.V., Sharath, R., Chandraprabha, M.N., Neelufar, E., Abhishikta, H., Malyasree, P., 2012. Synthesis, characterization and evaluation of antimicrobial activity of zinc oxide nanoparticles. J. Biochem. Technol. 3 (5), S151–S154.
Ngoepe, N.M., Mbita, Z., Mathipa, M., Mketo, N., Ntsendwana, B., Hintsho-Mbita, N.C., 2018. Biogenic synthesis of ZnO nanoparticles using *Monsonia burkeana* for use in photocatalytic, antibacterial and anticancer applications. Ceram. Int. 44 (14), 16999–17006.
Rajeshkumar, S., Kumar, S.V., Ramaiah, A., Agarwal, H., Lakshmi, T., Roopan, S.M., 2018. Biosynthesis of zinc oxide nanoparticles using *Mangifera indica* leaves and evaluation of their antioxidant and cytotoxic properties in lung cancer (A549) cells. Enzym. Microb. Technol. 117, 91–95.
Sankar, R., Karthik, A., Prabu, A., Karthik, S., Shivashangari, K.S., Ravikumar, V., 2013. *Origanum vulgare* mediated biosynthesis antibacterial of silver nanoparticles for its and anticancer activity. Colloids Surf. B: Biointerfaces 108, 80–84.
Syed Baker, S., Satish, S., 2015. Biosynthesis of gold nanoparticles by *Pseudomonas veronii* AS41G inhabiting *Annona squamosa* L. Spectrochim. Acta A Mol. Biomol. Spectrosc. 150, 691–695.
Vinod, V.T.P., Sashidhar, R.B., 2011. Bioremediation of industrial toxic metals with gum kondagogu (Cochlospermum gossypium): a natural carbohydrate biopolymer. Ind. J. Biotechnol. 10 (1), 113–120.

Index

Note: Page numbers followed by *f* indicate figures and *t* indicate tables.

N

O

P

Printed in the United States
by Baker & Taylor Publisher Services